Probability and
Random Processes

Probability and Random Processes

With Applications to Signal Processing and Communications

Scott L. Miller

Professor
Department of Electrical Engineering
Texas A&M University

Donald Childers

Professor Emeritus
Department of Electrical and Computer Engineering
University of Florida

ELSEVIER
ACADEMIC
PRESS

Amsterdam • Boston • Heidelberg • London • New York • Oxford
Paris • San Diego • San Francisco • Singapore • Sydney • Tokyo

Senior Acquisition Editor:	*Barbara Holland*
Project Manager:	*Troy Lilly*
Associate Editor:	*Tom Singer*
Marketing Manager:	*Linda Beattie*
Cover Design:	*Eric DeCicco*
Composition:	*Cepha Imaging Private Limited*
Cover Printer:	*Phoenix Color Corporation*
Interior Printer:	*The Maple-Vail Book Manufacturing Group*

Elsevier Academic Press
200 Wheeler Road, Burlington, MA 01803, USA
525 B Street, Suite 1900, San Diego, California 92101-4495, USA
84 Theobald's Road, London WC1X 8RR, UK

This book is printed on acid-free paper. ∞

Library of Congress Cataloging-in-Publication Data
Miller, Scott L.
  Probability and random processes : with applications to signal
  processing and communications / Scott L. Miller and Donald Childers.
      p. cm.
  Includes index.
  ISBN-13: 978-0-12-172651-5   ISBN-10: 0-12-172651-7 (hardcover : alk. paper)
  1. Signal processing-Mathematics. 2. Probabilities. 3. Stochastic
processes. I. Childers, Donald G. II. Title.
  TK5102.9.M556 2004
  621.382′2′0151–dc22

                                                              2004010367

British Library Cataloguing in Publication Data
A catalogue record for this book is available from the British Library

ISBN-13: 978-0-12-172651-5
ISBN-10: 0-12-172651-7 (hardcover : alk. paper)

For all information on all Elsevier Academic Press Publications
visit our Web site at www.books.elsevier.com

Transferred to Digital Printing, 2011

Printed and bound in the United Kingdom

Contents

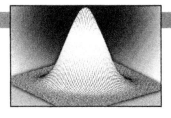

Preface

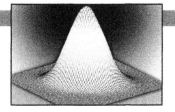

This book is intended to be used as a text for either undergraduate level (junior/senior) courses in probability or introductory graduate level courses in random processes that are commonly found in Electrical Engineering curricula. While the subject matter is primarily mathematical, it is presented for engineers. Mathematics is much like a well-crafted hammer. We can hang the tool on our wall and step back and admire the fine craftmanship used to construct the hammer, or we can pick it up and use it to pound a nail into the wall. Likewise, mathematics can be viewed as an art form or a tool. We can marvel at the elegance and rigor, or we can use it to solve problems. It is for this latter purpose that the mathematics is presented in this book. Instructors will note that there is no discussion of algebras, Borel fields, or measure theory in this text. It is our belief that the vast majority of engineering problems regarding probability and random processes do not require this level of rigor. Rather, we focus on providing the student with the tools and skills needed to solve problems. Throughout the text we have gone to great effort to strike a balance between readability and sophistication. While the book provides enough depth to equip students with the necessary tools to study modern communication systems, control systems, signal processing techniques, and many other applications, concepts are explained in a clear and simple manner that makes the text accessible as well.

It has been our experience that most engineering students need to see how the mathematics they are learning relates to engineering practice. Toward that end, we have included numerous engineering application sections throughout the text to help the instructor tie the probability theory to engineering practice. Many of these application sections focus on various aspects of telecommunications since this community is one of the major users of probability theory, but there are applications to other fields as well. We feel that this aspect of the text can be very useful for accreditation purposes for many institutions. The Accreditation Board for Engineering and Technology (ABET) has stated that all electrical engineering programs should provide their graduates with a knowledge of probability and statistics including applications to electrical engineering. This text provides not only the probability theory, but also the applications to electrical engineering and a modest amount of statistics as applied to engineering.

A key feature of this text, not found in most texts on probability and random processes, is an entire chapter devoted to simulation techniques. With the advent of powerful, low-cost, computational facilities, simulations have become an integral part of both academic and industrial research and development. Yet, many students have major misconceptions about how to run simulations. Armed with the material presented in our chapter on simulation, we believe students can perform simulations with confidence.

It is assumed that the readers of this text have a background consistent with typical junior level electrical engineering curricula. In particular, the reader should have a knowledge of differential and integral calculus, differential equations, linear algebra, complex variables, discrete math (set theory), linear time-invariant systems, and Fourier transform theory. In addition, there are a few sections in the text that require the reader to have a background in analytic function theory (e.g., parts of Section 4.10), but these sections can be skipped without loss of continuity. While some appendices have been provided with a review of some of these topics, these presentations are intended to provide a refresher for those who need to "brush up" and are not meant to be a substitute for a good course.

For undergraduate courses in probability and random variables, we recommend instructors cover the following sections:

Chapters 1–3: all sections,

Chapter 4: sections 1–6,

Chapter 5: sections 1–7 and 9,

Chapter 6: sections 1–3,

Chapter 7: sections 1–5.

These sections, along with selected application sections, could easily be covered in a one semester course with a comfortable pace. For those using this text in graduate courses in random processes, we recommend that instructors briefly review Chapters 1–7 focussing on those concepts not typically taught in an undergraduate course (e.g., Sections 4.7–4.10, 5.8, 5.10, 6.4, and 7.6) and then cover selected topics of interest from Chapters 8–12.

We consider the contents of this text to be appropriate background material for such follow-on courses as Digital Communications, Information Theory, Coding Theory, Image Processing, Speech Analysis, Synthesis and Recognition, and similar courses that are commonly found in many undergraduate and graduate programs in Electrical Engineering. Where possible, we have included engineering application examples from some of these topics.

Introduction

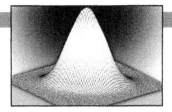

The study of probability, random variables, and random processes is fundamental to a wide range of disciplines. For example, many concepts of basic probability can be motivated through the study of games of chance. Indeed, the foundations of probability theory were originally built by a mathematical study of games of chance. Today, a huge gambling industry is built on a foundation of probability. Casinos have carefully designed games that allow players to win just enough to keep them hooked, while keeping the odds balanced slightly in favor of the "house." By nature, the outcomes of these games are random, but the casino owners fully understand that as long as the players keep playing, the theory of probability guarantees—with very *high* probability—that the casino will always come out ahead. Likewise, those playing the games may be able to increase their chances of winning by understanding and using probability.

In another application of probability theory, stock investors spend a great deal of time and effort trying to predict the random fluctuations in the market. Day traders try to take advantage of the random fluctuations that occur on a daily basis, whereas long-term investors try to benefit from the gradual trends that unfold over a much longer time period. These trends and fluctuations are random in nature and hence can be described only in a probabilistic fashion. Another business built on managing random occurrences is the insurance industry. Insurance premiums are calculated based on a careful study of the probabilities of various events happening. For example, the car insurance salesman has carefully evaluated the inherent risk of various classes of drivers and will adjust the premiums of each class according to the probabilities that those drivers will have an accident. In yet another application, a meteorologist tries to predict future weather events based on current and past meteorological conditions. Since these events are random, the weather forecast will often be presented in terms of probabilities (e.g., there is a 40 percent chance, or probability, of rain on Tuesday).

Since the theory of probability and random processes finds such a wide range of applications, students require various levels of understanding depending on the particular field they are preparing to enter. For those who wish to improve their proficiency at card games, a firm understanding of discrete probability may be sufficient. Those going into operations management need to understand queueing theory and therefore Markov and related random processes. A telecommunications engineer needs to have a firm understanding of models of noise and the design of systems to minimize the effects of noise.

This book is not intended to serve the needs of all disciplines, but rather is focused on preparing students entering the fields of electrical and computer engineering. One of the main goals of the text is to prepare the student to study random signals and systems. This material is fundamental to the study of digital signal processing (voice, image, video, etc.), communications systems and networks, radar systems, power systems, and many other applications within the engineering community. With this readership in mind, a background consistent with most electrical and computer engineering curricula is assumed. That is, in addition to fundamental mathematics including calculus, differential equations, linear algebra, and complex variables, the student is assumed to be familiar with the study of deterministic signals and systems. We understand that some readers may be very strong in these areas, while others may need to "brush up." Accordingly, we have included a few appendices that may help those who need a refresher and also provide a quick reference for significant results.

Throughout the text, the reader will find many examples and exercises that utilize MATLAB. MATLAB is a registered trademark of the MathWorks, Inc.; it is a technical software computing environment. Our purpose for introducing computer-based examples and problems is to expand our capabilities so that we may solve problems too tedious or complex to do via hand calculations. Furthermore, MATLAB has nice plotting capabilities that can greatly assist the visualization of data. MATLAB is used extensively in practice throughout the engineering community; therefore, we feel it is useful for engineering students to gain exposure to this important software package. Examples in the text that use MATLAB are clearly marked with a small computer logo.

Before diving into the theory of discrete probability in the next chapter, we first provide a few illustrations of how the theory of probability and random processes is used in several engineering applications. At the end of each subsequent chapter, the reader will find engineering application sections that illustrate how the material presented in that chapter is used in the real world. These sections can be skipped without losing any continuity, but we recommend that the reader at least skim through the material.

1.1 A Speech Recognition System

Many researchers are working on methods for computer recognition of speech. One application is to recognize commands spoken to a computer. Such systems are presently available from several vendors. A simple speech recognition system might use a procedure called template matching, which may be described as follows. We define a vocabulary, or a set of possible words for a computerized dictionary. This restricts the number of possible alternatives that must be recognized. Then a template for each word is obtained by digitizing the word as it is spoken. A simple dictionary of such templates is shown in Figure 1.1. The template may be the time waveform, the spectrum of the word, or a vector of selected features of the word. Common features might include the envelope of the time waveform, the energy, the number of zero crossings within a specified interval, and the like.

Speech recognition is a complicated task. Factors that make this task so difficult include interference from the surroundings, variability in the amplitude and duration of the spoken word, changes in other characteristics of the spoken word such as the speaker's pitch, and the size of the dictionary to name a few. In Figure 1.2, we have illustrated some of the variability that may occur when various speakers say the same word. Here we see that the waveform templates may vary considerably from speaker to speaker. This variability may be described by the theory of probability and random processes, which in turn may be used to develop models for speech production and recognition. Such models may then be used to design systems for speech recognition.

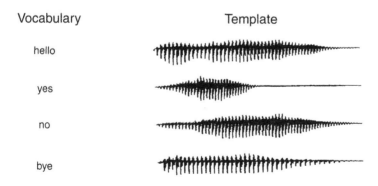

Figure 1.1 A simple dictionary of speech templates for speech recognition.

Templates

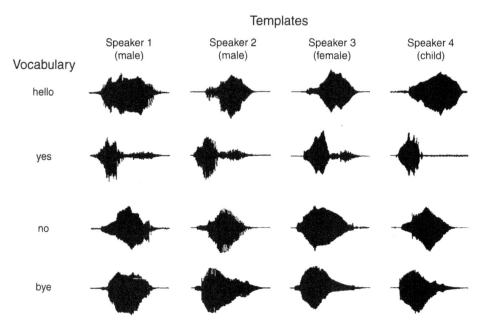

Figure 1.2   Variations in speech templates for different speakers.

1.2 A Radar System

A classical problem drawing heavily on the theory of probability and random processes is that of signal detection and estimation. One example of such a problem is a simple radar system, such as might be used at an airport to track local air traffic. A known signal is converted to an electromagnetic wave and propagated via an antenna. This wave will reflect off an aircraft and return back to the antenna, where the signal is processed to gather information about the aircraft. In addition to being corrupted by a random noise and interference process, the returning signal itself may exhibit randomness. First, we must determine if there is a reflected signal present. Usually, we attempt to maximize the probability of correctly detecting an aircraft subject to a certain level of false alarms. Once we decide that the aircraft is there, we attempt to estimate various random parameters of the reflected signal to obtain information about the aircraft. From the time of arrival of the reflected signal, we can estimate the distance of the aircraft from the radar site. The frequency of the returned signal will indicate the speed of the aircraft. Since the desired signal is corrupted by noise and interference, we can never estimate these various parameters exactly. Given sufficiently accurate models for these random

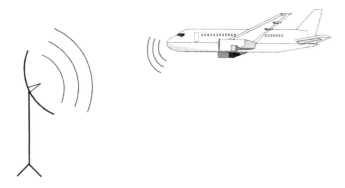

Figure 1.3 A radar system.

disturbances, however, we can devise procedures for providing the most accurate estimates possible. We can also use the theory of probability and random processes to analyze the performance of our system.

1.3 A Communication Network

Consider a node in a computer communication network, such as that depicted in Figure 1.4, that receives packets of information from various sources and must forward them along toward their ultimate destinations. Typically, the node has a

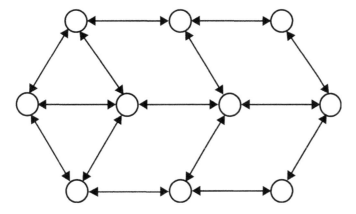

Figure 1.4 Nodes and links in a communications network.

fixed, or at least a maximum, rate at which it can transmit data. Since the arrival of packets to a node will be random, the node will usually have some buffering capability, allowing the node to temporarily store packets that it cannot forward immediately. Given a random model of the arrival process of packets at a node, the theory of probability and random processes developed in this text will allow the network designer to determine how large a buffer is needed to insure a minimal probability of buffer overflow (and a resulting loss of information). Or, conversely, given a set buffer size, a limit on the amount of traffic (i.e., throughput) that the node can handle can be determined. Other random quantities such as the delay a packet encounters at the node can also be statistically characterized.

On a less local basis, when information is generated at one of the nodes with a specified destination, a route must be determined to get the packet from the source to the destination. Some nodes in the network may be more congested than others. Congestion throughout the network tends to be very dynamic, so the routing decision must be made using probability. Which route should the packet follow so that it is least likely to be dropped along the way? Or, maybe we want to find the path that will lead to the smallest average delay. Protocols for routing, flow control, and the like are all based in the foundations of probability theory.

These few examples illustrate the diversity of problems that probability and random processes may model and thereby assist in the development of effective design solutions. By firmly understanding the concepts in this text, the reader will open up a vast world of engineering applications.

Introduction to Probability Theory

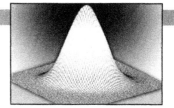

Many electrical engineering students have studied, analyzed, and designed systems from the point of view of steady-state and transient signals using time domain or frequency domain techniques. However, these techniques do not provide a method for accounting for variability in the signal nor for unwanted disturbances such as interference and noise. We will see that the theory of probability and random processes is useful for modeling the uncertainty of various events (e.g., the arrival of telephone calls and the failure of electronic components). We also know that the performance of many systems is adversely affected by noise, which may often be present in the form of an undesired signal that degrades the performance of the system. Thus, it becomes necessary to design systems that can discriminate against noise and enhance a desired signal.

How do we distinguish between a deterministic signal or function and a stochastic or random phenomenon such as noise? Usually, noise is defined to be any undesired signal, which often occurs in the presence of a desired signal. This definition includes deterministic as well as nondeterministic signals. A deterministic signal is one that may be represented by parameter values, such as a sinusoid, which may be perfectly reconstructed given an amplitude, frequency, and phase. Stochastic signals, such as noise, do not have this property. While they may be approximately represented by several parameters, stochastic signals have an element of randomness that prevents them from being perfectly reconstructed from a past history. As we saw in Chapter 1 (Figure 1.2), even the same word spoken by different speakers is not deterministic; there is variability, which can be modeled as a random fluctuation. Likewise, the amplitude and/or phase of a stochastic signal cannot be calculated for any specified future time instant, even though the entire past history of the signal may be known. However, the amplitude and/or phase of a random signal can be predicted to occur with a specified probability, provided certain factors are known. The theory of probability provides a tool to model and analyze phenomena that occur in many diverse fields, such as communications, signal processing, control, and computers. Perhaps the major reason for studying probability and random processes is to be able to model complex systems and phenomena.

2.1 Experiments, Sample Spaces, and Events

The relationship between probability and gambling has been known for some time. Over the years, some famous scientists and mathematicians have devoted time to probability: Galileo wrote on dice games; Laplace worked out the probabilities of some gambling games; and Pascal and Bernoulli, while studying games of chance, contributed to the basic theory of probability. Since the time of this early work, the theory of probability has become a highly developed branch of mathematics. Throughout these beginning sections on basic probability theory, we will often use games of chance to illustrate basic ideas that will form the foundation for more advanced concepts. To start with, we will consider a few simple definitions.

DEFINITION 2.1: An *experiment* is a procedure we perform (quite often hypothetical) that produces some result. Often the letter E is used to designate an experiment (e.g., the experiment E_5 might consist of tossing a coin five times).

DEFINITION 2.2: An *outcome* is a possible result of an experiment. The Greek letter xi (ξ) is often used to represent outcomes (e.g., the outcome ξ_1 of experiment E_5 might represent the sequence of tosses heads-heads-tails-heads-tails; however, the more concise HHTHT might also be used).

DEFINITION 2.3: An *event* is a certain set of outcomes of an experiment (e.g., the event C associated with experiment E_5 might be $C = \{$all outcomes consisting of an even number of heads$\}$).

DEFINITION 2.4: The *sample space* is the collection or set of "all possible" distinct (collectively exhaustive and mutually exclusive) outcomes of an experiment. The letter S is used to designate the sample space, which is the universal set of outcomes of an experiment. A sample space is called discrete if it is a finite or a countably infinite set. It is called continuous or a continuum otherwise.

The reason we have placed quotes around the words *all possible* in Definition 2.4 is explained by the following imaginary situation. Suppose we conduct the experiment of tossing a coin. It is conceivable that the coin may land on edge. But experience has shown us that such a result is highly unlikely to occur. Therefore, our sample space for such experiments typically excludes such unlikely outcomes. We also require, for the present, that all outcomes be distinct. Consequently, we are considering only the set of simple outcomes that are collectively exhaustive and mutually exclusive.

EXAMPLE 2.1: Consider the example of flipping a fair coin once, where fair means that the coin is not biased in weight to a particular side. There are two possible outcomes, namely, a head or a tail. Thus, the sample space, S, consists of two outcomes, $\xi_1 = H$ to indicate that the outcome of the coin toss was heads and $\xi_2 = T$ to indicate that the outcome of the coin toss was tails.

EXAMPLE 2.2: A cubical die with numbered faces is rolled and the result observed. The sample space consists of six possible outcomes, $\xi_1 = 1, \xi_2 = 2, \ldots, \xi_6 = 6$, indicating the possible faces of the cubical die that may be observed.

EXAMPLE 2.3: As a third example, consider the experiment of rolling two dice and observing the results. The sample space consists of 36 outcomes, which may be labelled by the ordered pairs $\xi_1 = (1,1), \xi_2 = (1,2), \xi_3 = (1,3), \ldots, \xi_6 = (1,6), \xi_7 = (2,1), \xi_8 = (2,2), \ldots, \xi_{36} = (6,6)$; the first component in the ordered pair indicates the result of the toss of the first die, and the second component indicates the result of the toss of the second die. Several interesting events can be defined from this experiment, such as

$A = \{$the sum of the outcomes of the two rolls $= 4\}$,
$B = \{$the outcomes of the two rolls are identical$\}$,
$C = \{$the first roll was bigger than the second$\}$.

An alternative way to consider this experiment is to imagine that we conduct two distinct experiments, with each consisting of rolling a single die. The sample spaces (S_1 and S_2) for each of the two experiments are identical, namely, the same as Example 2.2. We may now consider the sample space, S, of the original experiment to be the combination of the sample spaces, S_1 and S_2, which consists of all possible combinations of the elements of both S_1 and S_2. This is an example of a combined sample space.

EXAMPLE 2.4: For our fourth experiment, let us flip a coin until a tails occurs. The experiment is then terminated. The sample space consists of a collection of sequences of coin tosses. Label these outcomes as $\xi_n, n = 1, 2, 3, \ldots$. The final toss in any particular sequence is a tail and terminates the sequence. The preceding tosses prior to the occurrence of the tail must be heads. The possible outcomes that may occur are

$$\xi_1 = (T), \; \xi_2 = (H, T), \; \xi_3 = (H, H, T), \; \ldots .$$

Note that in this case, n can extend to infinity. This is another example of a combined sample space resulting from conducting independent but identical experiments. In this example, the sample space is countably infinite, while the previous sample spaces were finite.

EXAMPLE 2.5: As a last example, consider a random number generator that selects a number in an arbitrary manner from the semi-closed interval $[0, 1)$. The sample space consists of all real numbers, x, for which $0 \leq x < 1$. This is an example of an experiment with a continuous sample space. We can define events on a continuous space as well, such as

$A = \{x < 1/2\}$,
$B = \{|x - 1/2| < 1/4\}$,
$C = \{x = 1/2\}$.

Other examples of experiments with continuous sample spaces include the measurement of the voltage of thermal noise in a resistor or the measurement of the (x, y, z) position of an oxygen molecule in the atmosphere. Examples 2.1 to 2.4 illustrate discrete sample spaces.

There are also infinite sets that are uncountable and that are not continuous, but these sets are beyond the scope of this book. So for our purposes, we will consider only the preceding two types of sample spaces. It is also possible to have a sample space that is a mixture of discrete and continuous sample spaces. For the remainder of this chapter, we shall restrict ourselves to the study of discrete sample spaces.

A particular experiment can often be represented by more than one sample space. The choice of a particular sample space depends upon the questions that are to be answered concerning the experiment. This is perhaps best explained by recalling Example 2.3 in which a pair of dice was rolled. Suppose we were asked to record after each roll the sum of the numbers shown on the two faces. Then, the sample space could be represented by only eleven outcomes, $\xi_1 = 2$, $\xi_2 = 3$, $\xi_3 = 4$, ..., $\xi_{11} = 12$. However, the original sample space is in some way more fundamental, since the sum of the die faces can be determined from the numbers on the die faces. If the second representation is used, it is not sufficient to specify the sequence of numbers that occurred from the sum of the numbers.

2.2 Axioms of Probability

Now that the concepts of experiments, outcomes, and events have been introduced, the next step is to assign probabilities to various outcomes and events. This requires a careful definition of probability. The words *probability* and *probable* are commonly used in everyday language. The meteorologist on the evening news may say that rain is probable for tomorrow or he may be more specific and state that the chance (or probability) of rain is 70 percent. Although this sounds like a precise statement, we could interpret it in several ways. Perhaps it means that about 70 percent of the listening audience will experience rain. Or, maybe if tomorrow could be repeated many times, 70 percent of the tomorrows would have rain while the other 30 percent would not. Of course, tomorrow cannot be repeated and this experiment can be run only once. The outcome will be either rain or no rain. The meteorologist may like this interpretation since there is no way to repeat the experiment enough times to test the accuracy of the prediction. However, there is a similar interpretation that can be tested. We might say that any time a day with similar meteorological conditions presents itself, the following day will have rain 70 percent of the time. In fact, it may be his or her past experience with the given meteorological conditions that led the meteorologist to the prediction of a 70 percent chance of rain.

It should be clear from our everyday usage of the word *probability* that it is a measure of the likelihood of various events. So, in general terms, probability is a function of an event that produces a numerical quantity that measures the likelihood of that event. There are many ways to define such a function, which we could then call probability. In fact, we will find several ways to assign probabilities to various events, depending on the situation. Before we do that, however, we start with three axioms that any method for assigning probabilities must satisfy:

AXIOM 2.1: For any event A, $\Pr(A) \geq 0$ (a negative probability does not make sense).

AXIOM 2.2: If S is the sample space for a given experiment, $\Pr(S) = 1$ (probabilities are normalized so that the maximum value is unity).

AXIOM 2.3a: If $A \cap B = \varnothing$, then $\Pr(A \cup B) = \Pr(A) + \Pr(B)$.

As the word *axiom* implies, these statements are taken to be self-evident and require no proof. In fact, the first two axioms are really more of a self-imposed convention. We could have allowed for probabilities to be negative, or we could have normalized the maximum probability to be something other than one. However, this would have greatly confused the subject and we do not consider these possibilities. From these axioms (plus one more to be presented shortly), the entire theory of probability can be developed. Before moving on to that task, a corollary to Axiom 2.3a is given.

COROLLARY 2.1: Consider M sets $A_1, A_2, \ldots, A_M$ that are mutually exclusive, $A_i \cap A_j = \varnothing$ for all $i \neq j$,

$$\Pr\left(\bigcup_{i=1}^{M} A_i\right) = \sum_{i=1}^{M} \Pr(A_i). \tag{2.1}$$

PROOF: This statement can be proved using mathematical induction. For those students who are unfamiliar with this concept, the idea behind induction is to show that if the statement is true for $M = m$, then it must also hold for $M = m + 1$. Once this is established, it is noted that by Axiom 2.3a, the statement applies for $M = 2$, and hence it must be true for $M = 3$. Since it is true for $M = 3$, it must also be true for $M = 4$, and so on. In this way we can prove that Corollary 2.1 is true for any finite M. The details of this proof are left as an exercise for the reader (see Exercise 2.1). ∎

Unfortunately, the proof just outlined is not sufficient to show that Corollary 2.1 is true for the case of an infinite number of sets. That has to be accepted on faith and is listed here as the second part of Axiom 2.3.

AXIOM 2.3b: For an infinite number of mutually exclusive sets, A_i, $i = 1, 2, 3, \dots, A_i \cap A_j = \varnothing$ for all $i \neq j$,

$$\Pr\left(\bigcup_{i=1}^{\infty} A_i\right) = \sum_{i=1}^{\infty} \Pr(A_i). \tag{2.2}$$

It should be noted that Axiom 2.3a and Corollary 2.1 could be viewed as special cases of Axiom 2.3b. So, a more concise development could be obtained by starting with Axioms 2.1, 2.2, and 2.3b. This may be more pleasing to some, but we believe the approach given here is a little easier to follow for the student learning the material for the first time.

The preceding axioms do not tell us directly how to deal with the probability of the union of two sets that are not mutually exclusive. This can be determined from these axioms as follows.

THEOREM 2.1: For any sets A and B (not necessarily mutually exclusive),

$$\Pr(A \cup B) = \Pr(A) + \Pr(B) - \Pr(A \cap B). \tag{2.3}$$

PROOF: We give a visual proof of this important result using the Venn diagram shown in Figure 2.1. To aid the student in the type of reasoning needed to complete proofs of this type, it is helpful to think of a pile of sand lying in the sample space shown in Figure 2.1. The probability of the event A would then be analogous to the mass of that subset of the sand pile that is above the region A and likewise for the probability of the event B. For the union of the two events, if we simply added the mass of the sand above A to the mass of the sand above B, we would double count that region that is common to both sets. Hence, it is necessary to subtract the

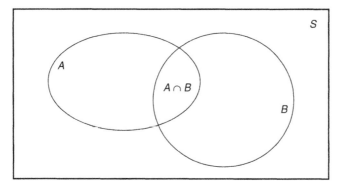

Figure 2.1 Venn diagram for proof of Theorem 2.1.

probability of $A \cap B$. We freely admit that this proof is not rigorous. It is possible to prove Theorem 2.1 without having to call on our sand analogy or even the use of Venn diagrams. The logic of the proof will closely follow what we have done here. The reader is led through that proof in Exercise 2.2.   ■

Many other fundamental results can also be obtained from the basic axioms of probability. A few simple ones are presented here. More will be developed later in this chapter and in subsequent chapters. As with Theorem 2.1, it might help the student to visualize these proofs by drawing a Venn diagram.

THEOREM 2.2: $\Pr(\overline{A}) = 1 - \Pr(A)$.

PROOF:
$$1 = \Pr(S) = \Pr(A \cup \overline{A}) \qquad \text{(by Axiom 2.2)}$$
$$= \Pr(A) + \Pr(\overline{A}) \qquad \text{(by Axiom 2.3a)}$$
$$\therefore \Pr(\overline{A}) = 1 - \Pr(A).$$
                                                                                    ■

THEOREM 2.3: If $A \subset B$, then $\Pr(A) \leq \Pr(B)$.

PROOF: See Exercise 2.4.   ■

2.3 Assigning Probabilities

In the previous section, probability was defined as a measure of the likelihood of an event or events that satisfy the three Axioms 2.1–2.3. How probabilities are assigned to particular events was not specified. Mathematically, any assignment that satisfies the given axioms is acceptable. Practically speaking, we would like to assign probabilities to events in such a way that the probability assignment actually represents the likelihood of occurrence of that event. Two techniques are typically used for this purpose and are described in the following paragraphs.

In many experiments, it is possible to specify all of the outcomes of the experiment in terms of some fundamental outcomes, which we refer to as *atomic outcomes*. These are the most basic events that cannot be decomposed into simpler events. From these atomic outcomes, we can build more complicated and more interesting events. Quite often we can justify assigning equal probabilities to all atomic outcomes in an experiment. In that case, if there are M mutually exclusive exhaustive atomic events, then each one should be assigned a probability of $1/M$. Not only does this make perfect common sense, it also satisfies the mathematical requirements of the three axioms that define probability. To see this, we label the M atomic outcomes of an experiment E as $\xi_1, \xi_2, \cdots, \xi_M$. These atomic events are taken to be mutually exclusive and exhaustive. That is, $\xi_i \cap \xi_j = \varnothing$ for all $i \neq j$, and $\xi_1 \cup \xi_2 \cup \cdots \cup \xi_M = S$. Then by Corollary 2.1 and Axiom 2.2,

$$\Pr(\xi_1 \cup \xi_2 \cup \cdots \cup \xi_M) = \Pr(\xi_1) + \Pr(\xi_2) + \cdots + \Pr(\xi_M) = \Pr(S) = 1 \qquad (2.4)$$

If each atomic outcome is to be equally probable, then we must assign each a probability of $\Pr(\xi_i) = 1/M$ for there to be equality in the preceding equation. Once the probabilities of these outcomes are assigned, the probabilities of some more complicated events can be determined according to the rules set forth in Section 2.2. This approach to assigning probabilities is referred to as the *classical approach*.

EXAMPLE 2.6: The simplest example of this procedure is the coin flipping experiment of Example 2.1. In this case, there are only two atomic events, $\xi_1 = H$ and $\xi_2 = T$. Provided the coin is fair (again, not biased towards one side or the other), we have every reason to believe that these two events should be equally probable. These outcomes are mutually exclusive and collectively exhaustive (provided we rule out the possibility of the coin landing on its edge). According to our theory of probability, these events should be assigned probabilities of $\Pr(H) = \Pr(T) = 1/2$.

EXAMPLE 2.7: Next consider the dice rolling experiment of Example 2.2. If the die is not loaded, the six possible faces of the cubicle die are reasonably taken to be equally likely to appear, in which case, the probability assignment is $\Pr(1) = \Pr(2) = \cdots = \Pr(6) = 1/6$. From this assignment we can determine the probability of more complicated events, such as

$\Pr(\text{even number is rolled}) = \Pr(2 \cup 4 \cup 6)$

$$= \Pr(2) + \Pr(4) + \Pr(6) \quad \text{(by Corollary 2.3)}$$
$$= 1/6 + 1/6 + 1/6 \quad \text{(by probability assignment)}$$
$$= 1/2.$$

EXAMPLE 2.8: In Example 2.3, a pair of dice were rolled. In this experiment, the most basic outcomes are the 36 different combinations of the six atomic outcomes of the previous example. Again, each of these atomic outcomes is assigned a probability of 1/36. Next, suppose we want to find the probability of the event $A = \{\text{sum of two dice} = 5\}$. Then,

$$\Pr(A) = \Pr((1,4) \cup (2,3) \cup (3,2) \cup (4,1))$$

$$= \Pr(1,4) + \Pr(2,3) + \Pr(3,2) + \Pr(4,1) \quad \text{(by Corollary 2.1)}$$

$$= 1/36 + 1/36 + 1/36 + 1/36 \quad \text{(by probability assignment)}$$

$$= 1/9.$$

EXAMPLE 2.9: In this example we will use the MATLAB command rand to simulate the flipping of coins and the rolling of dice. The command rand(m, n) creates a matrix of m rows and n columns, where each element of the matrix is a randomly selected number equally likely to fall anywhere in the interval (0,1). By rounding this number to the nearest integer, we can create a randomly selected number equally likely to be 0 or 1. This can be used to simulate the flipping of a coin if we interpret 0 as "tails" and 1 as "heads" or vice versa. Similarly, if we multiply rand(1) by 6 and round up to the nearest integer, we will get one of the numbers $1, 2, \ldots, 6$ with equal probability. This can be used to simulate the rolling of a die. Try running the following script in MATLAB.

```
%   Simulation of coin flipping and die tossing.
coin_flip=round(rand(1))      % Simulate flip of a coin.
die_toss=ceil(6*rand(1))      % Simulate toss of one die.
dice_toss=ceil(6*rand(1,2))   % Simulate toss of two dice.
```

You should find that each time you run this script, you get different (random) looking results. With any MATLAB command, if you want more information on what the command does, type help followed by the command name at the MATLAB prompt for detailed information on that command. For example, to get help on the rand command, type help rand.

Care must be taken when using the classical approach to assigning probabilities. If we define the set of atomic outcomes incorrectly, unsatisfactory results may occur. In Example 2.8, we may be tempted to define the set of atomic outcomes as the different sums that can occur on the two die faces. If we assign equally likely probability to each of these outcomes, then we arrive at the assignment

$$\Pr(\text{sum} = 2) = \Pr(\text{sum} = 3) = \cdots = \Pr(\text{sum} = 12) = 1/11. \qquad (2.5)$$

Anyone with experience in games involving dice knows that the likelihood of rolling a 2 is much lower than the likelihood of rolling a 7. The problem here is that the atomic events we have assigned are not the most basic outcomes and can be decomposed into simpler outcomes, as demonstrated in Example 2.8.

This is not the only problem encountered in the classical approach. Suppose we consider an experiment that consists of measuring the height of an arbitrarily chosen student in your class and rounding that measurement to the nearest inch. The atomic outcomes of this experiment would consist of all the heights of the students in your class. However, it would not be reasonable to assign an equal probability to each height. Those heights corresponding to very tall or very short students would be expected to be less probable than those heights corresponding to a medium height. So, how then do we assign probabilities to these events? The problems associated with the classical approach to assigning probabilities can be overcome by using the *relative frequency approach*.

The relative frequency approach requires that the experiment we are concerned with be repeatable, in which case, the probability of an event, A, can be assigned by repeating the experiment a large number of times and observing how many times the event A actually occurs. If we let n be the number of times the experiment is repeated and n_A be the number of times the event A is observed, then the probability of the event A can be assigned according to

$$\Pr(A) = \lim_{n \to \infty} \frac{n_A}{n}. \qquad (2.6)$$

This approach to assigning probability is based on experimental results and thus has a more practical flavor to it. It is left as an exercise for the reader (see Exercise 2.6) to confirm that this method does indeed satisfy the axioms of probability and is thereby mathematically correct as well.

EXAMPLE 2.10: Consider the dice rolling experiment of Examples 2.3 and 2.8. We will use the relative frequency approach to assign the probability of the event, $A = \{\text{sum of two dice} = 5\}$. We simulated the tossing of two dice using the following MATLAB code. The results of this dice tossing simulation are shown in Table 2.1.

Table 2.1   Simulation of Dice Tossing Experiment.

n	1,000	2,000	3,000	4,000	5,000	6,000	7,000	8,000	9,000	10,000
n_A	96	200	314	408	521	630	751	859	970	1,095
n_A/n	0.096	0.100	0.105	0.102	0.104	0.105	0.107	0.107	0.108	0.110

```
% Simulation code for dice tossing experiment.
n=1000;                    % Number of times to toss the dice.
die1=ceil(6*rand(1,n));    % Toss first die n times.
die2=ceil(6*rand(1,n));    % Toss second die n times.
dice_sum=die1+die2;        % Compute sum of two tosses.
nA=sum(dice_sum==5);       % Count number of times sum = 5.
pA=nA/n                    % Display relative frequency.
```

The next to last line of MATLAB code may need some explanation. The double equal sign asks MATLAB to compare the two quantities to see if they are equal. MATLAB responds with 1 for "yes" and 0 for "no." Hence the expression dice_sum==5 results in an n element vector where each element of the vector is either 0 or 1 depending on whether the corresponding element of dice_sum is equal to 5 or not. By summing all elements of this vector, we obtain the number of times the sum 5 occurs in n tosses of the dice.

   To get an exact measure of the probability of an event using the relative frequency approach, we must be able to repeat the event an infinite number of times—a serious drawback to this approach. In the dice rolling experiment of Example 2.8, even after rolling the dice 10,000 times, the probability of observing a 5 was measured to only two significant digits. Furthermore, many random phenomena in which we might be interested are not repeatable. The situation may occur only once, and hence we cannot assign the probability according to the relative frequency approach.

2.4 Joint and Conditional Probabilities

Suppose that we have two sets, A and B. We saw a few results in the previous section that dealt with how to calculate the probability of the union of two sets, $A \cup B$. At least as frequently, we are interested in calculating the probability of the intersection of two sets, $A \cap B$. This probability is referred to as the joint probability of the sets A and B, $\Pr(A \cap B)$. Usually, we will use the notation $\Pr(A, B)$.

This definition and notation extends to an arbitrary number of sets. The joint probability of the sets $A_1, A_2, \ldots, A_M$, is $\Pr(A_1 \cap A_2 \cap \cdots \cap A_M)$ and we use the simpler notation $\Pr(A_1, A_2, \ldots, A_M)$ to represent the same quantity.

Now that we have established what a joint probability is, how does one compute it? To start with, by comparing Axiom 2.3a and Theorem 2.1, it is clear that if A and B are mutually exclusive, then their joint probability is zero. This is intuitively pleasing, since if A and B are mutually exclusive, then $\Pr(A, B) = \Pr(\emptyset)$, which we would expect to be zero. That is, an impossible event should never happen. Of course, this case is of rather limited interest, and we would be much more interested in calculating the joint probability of events that are not mutually exclusive.

In the general case when A and B are not necessarily mutually exclusive, how can we calculate the joint probability of A and B? From the general theory of probability, we can easily see two ways to accomplish this. First, we can use the classical approach. Both events (sets) A and B can be expressed in terms of atomic outcomes. We then write $A \cap B$ as the set of those atomic outcomes that is common to both and calculate the probabilities of each of these outcomes. Alternatively, we can use the relative frequency approach. Let $n_{A,B}$ be the number of times that A and B simultaneously occur in n trials. Then,

$$\Pr(A, B) = \lim_{n \to \infty} \frac{n_{A,B}}{n}. \tag{2.7}$$

EXAMPLE 2.11: A standard deck of playing cards has 52 cards that can be divided in several manners. There are four suits (spades, hearts, diamonds, and clubs), each of which has 13 cards (ace, 2, 3, 4, $\ldots$, 10, jack, queen, king). There are two red suits (hearts and diamonds) and two black suits (spades and clubs). Also, the jacks, queens, and kings are referred to as face cards, while the others are number cards. Suppose the cards are sufficiently shuffled (randomized) and one card is drawn from the deck. The experiment has 52 atomic outcomes corresponding to the 52 individual cards that could have been selected. Hence, each atomic outcome has a probability of $1/52$. Define the events: $A =$ {red card selected}, $B =$ {number card selected}, and $C =$ {heart selected}. Since the event A consists of 26 atomic outcomes (there are 26 red cards), then $\Pr(A) = 26/52 = 1/2$. Likewise, $\Pr(B) = 40/52 = 10/13$ and $\Pr(C) = 13/52 = 1/4$. Events A and B have 20 outcomes in common, hence $\Pr(A, B) = 20/52 = 5/13$. Likewise, $\Pr(A, C) = 13/52 = 1/4$ and $\Pr(B, C) = 10/52 = 5/26$. It is interesting to note that in this example, $\Pr(A, C) = \Pr(C)$. This is because $C \subset A$ and as a result $A \cap C = C$.

Often the occurrence of one event may be dependent upon the occurrence of another. In the previous example, the event $A = \{$a red card is selected$\}$ had a probability of $\Pr(A) = 1/2$. If it is known that event $C = \{$a heart is selected$\}$ has occurred, then the event A is now certain (probability equal to 1), since all cards in the heart suit are red. Likewise, if it is known that the event C did not occur, then there are 39 cards remaining, 13 of which are red (all the diamonds). Hence, the probability of event A in that case becomes $1/3$. Clearly, the probability of event A depends on the occurrence of event C. We say that the probability of A is conditional on C, and the probability of A given knowledge that the event C has occurred is referred to as the *conditional probability* of A given C. The shorthand notation $\Pr(A|C)$ is used to denote the probability of the event A given that the event C has occurred, or simply the probability of A given C.

DEFINITION 2.5: For two events A and B, the probability of A conditioned on knowing that B has occurred is

$$\Pr(A|B) = \frac{\Pr(A, B)}{\Pr(B)}. \tag{2.8}$$

The reader should be able to verify that this definition of conditional probability does indeed satisfy the axioms of probability (see Exercise 2.7).

We may find in some cases that conditional probabilities are easier to compute than the corresponding joint probabilities, and hence this formula offers a convenient way to compute joint probabilities:

$$\Pr(A, B) = \Pr(B|A)\Pr(A) = \Pr(A|B)\Pr(B). \tag{2.9}$$

This idea can be extended to more than two events. Consider finding the joint probability of three events, A, B, and C:

$$\Pr(A, B, C) = \Pr(C|A, B)\Pr(A, B) = \Pr(C|A, B)\Pr(B|A)\Pr(A). \tag{2.10}$$

In general, for M events, $A_1, A_2, \ldots, A_M$,

$$\Pr(A_1, A_2, \ldots, A_M) = \Pr(A_M|A_1, A_2, \ldots, A_{M-1})\Pr(A_{M-1}|A_1, \ldots, A_{M-2}) \cdots$$
$$\times \Pr(A_2|A_1)\Pr(A_1). \tag{2.11}$$

EXAMPLE 2.12: Return to the experiment of drawing cards from a deck as described in Example 2.11. Suppose now that we select two cards at random from the deck. When we select the second card, we do not return the first card to the deck. In this case, we say that we are selecting cards without replacement. As a result, the probabilities associated with selecting the second card are slightly different if we have knowledge of which card was drawn on the first selection. To illustrate this, let $A = \{$first card was a spade$\}$ and $B = \{$second card was a spade$\}$. The probability of the event A can be calculated as in the previous example to be $\Pr(A) = 13/52 = 1/4$. Likewise, if we have no knowledge of what was drawn on the first selection, the probability of the event B is the same, $\Pr(B) = 1/4$. To calculate the joint probability of A and B, we have to do some counting.

To begin, when we select the first card there are 52 possible outcomes. Since this card is not returned to the deck, there are only 51 possible outcomes for the second card. Hence, this experiment of selecting two cards from the deck has $52 * 51$ possible outcomes each of which is equally likely and has a probability of $1/52 * 51$. Similarly, there are $13 * 12$ outcomes that belong to the joint event $A \cap B$. Therefore, the joint probability for A and B is $\Pr(A, B) = (13 * 12)/(52 * 51) = 1/17$. The conditional probability of the second card being a spade given that the first card is a spade is then $\Pr(B|A) = \Pr(A, B)/\Pr(A) = (1/17)/(1/4) = 4/17$. However, calculating this conditional probability directly is probably easier than calculating the joint probability. Given that we know the first card selected was a spade, there are now 51 cards left in the deck, 12 of which are spades, thus $\Pr(B|A) = 12/51 = 4/17$. Once this is established, then the joint probability can be calculated as $\Pr(A, B) = \Pr(B|A)\Pr(A) = (4/17) * (1/4) = 1/17$.

EXAMPLE 2.13: In a game of poker, you are dealt five cards from a standard 52 card deck. What is the probability that you are dealt a flush in spades? (A flush is when you are dealt all five cards of the same suit.) What is the probability of a flush in any suit? To answer these questions requires a simple extension of the previous example. Let A_i be the event $\{i$th card dealt to us is a spade$\}$, $i = 1, 2, \ldots, 5$. Then

$$\Pr(A_1) = 1/4,$$

$$\Pr(A_1, A_2) = \Pr(A_2|A_1)\Pr(A_1) = (12/51) * (1/4) = 1/17,$$

$$\Pr(A_1, A_2, A_3) = \Pr(A_3|A_1, A_2)\Pr(A_1, A_2)$$

$$= (11/50) * (1/17) = 11/850,$$

$$\Pr(A_1, A_2, A_3, A_4) = \Pr(A_4|A_1, A_2, A_3)\Pr(A_1, A_2, A_3)$$
$$= (10/49) * (11/850) = 11/4165,$$
$$\Pr(A_1, A_2, A_3, A_4, A_5) = \Pr(A_5|A_1, A_2, A_3, A_4)\Pr(A_1, A_2, A_3, A_4)$$
$$= (9/48) * (11/4165) = 33/66,640.$$

To find the probability of being dealt a flush in any suit, we proceed as follows:

$$\Pr(\text{flush}) = \Pr(\{\text{flush in spades}\} \cup \{\text{flush in hearts}\}$$
$$\cup \{\text{flush in diamonds}\} \cup \{\text{flush in clubs}\})$$
$$= \Pr(\text{flush in spades}) + \Pr(\text{flush in hearts})$$
$$+ \Pr(\text{flush in diamonds}) + \Pr(\text{flush in clubs}).$$

Since all four events in the preceding expression have equal probability, then

$$\Pr(\text{flush}) = 4 * \Pr(\text{flush in spades}) = \frac{4 * 33}{66,640} = \frac{33}{16,660}.$$

2.5 Bayes's Theorem

In this section, we develop a few results related to the concept of conditional probability. While these results are fairly simple, they are so useful that we felt it was appropriate to devote an entire section to them. To start with, the following theorem was essentially proved in the previous section and is a direct result of the definition of conditional probability.

THEOREM 2.4: For any events A and B such that $\Pr(B) \neq 0$,

$$\Pr(A|B) = \frac{\Pr(B|A)\Pr(A)}{\Pr(B)}. \tag{2.12}$$

PROOF: From Definition 2.5,

$$\Pr(A, B) = \Pr(A|B)\Pr(B) = \Pr(B|A)\Pr(A). \tag{2.13}$$

Theorem 2.4 follows directly by dividing the preceding equations by $\Pr(B)$. ■

Theorem 2.4 is useful for calculating certain conditional probabilities since, in many problems, it may be quite difficult to compute $\Pr(A|B)$ directly, whereas calculating $\Pr(B|A)$ may be straightforward.

THEOREM 2.5 (Theorem of Total Probability): Let $B_1, B_2, \ldots, B_n$ be a set of mutually exclusive and exhaustive events. That is, $B_i \cap B_j = \emptyset$ for all $i \neq j$ and

$$\bigcup_{i=1}^{n} B_i = S \Rightarrow \sum_{i=1}^{n} \Pr(B_i) = 1. \tag{2.14}$$

Then

$$\Pr(A) = \sum_{i=1}^{n} \Pr(A|B_i)\Pr(B_i) \tag{2.15}$$

PROOF: As with Theorem 2.1, a Venn diagram (shown in Figure 2.2) is used here to aid in the visualization of our result. From the diagram, it can be seen that the event A can be written as

$$A = \{A \cap B_1\} \cup \{A \cap B_2\} \cup \cdots \cup \{A \cap B_n\} \tag{2.16}$$

$$\Rightarrow \Pr(A) = \Pr(\{A \cap B_1\} \cup \{A \cap B_2\} \cup \cdots \cup \{A \cap B_n\}) \tag{2.17}$$

Also, since the B_i are all mutually exclusive, then the $\{A \cap B_i\}$ are also mutually exclusive, so that

$$\Pr(A) = \sum_{i=1}^{n} \Pr(A, B_i) \qquad \text{(by Corollary 2.3)}, \tag{2.18}$$

$$= \sum_{i=1}^{n} \Pr(A|B_i)\Pr(B_i) \quad \text{(by Theorem 2.4).} \tag{2.19}$$

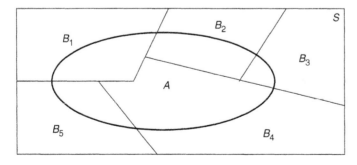

Figure 2.2 Venn diagram used to help prove the theorem of total probability.

Finally, by combining the results of Theorems 2.4 and 2.5, we get what has come to be known as Bayes's theorem.   ∎

THEOREM 2.6 (Bayes's Theorem): Let $B_1, B_2, \ldots, B_n$ be a set of mutually exclusive and exhaustive events. Then

$$\Pr(B_i|A) = \frac{\Pr(A|B_i)\Pr(B_i)}{\sum\limits_{i=1}^{n} \Pr(A|B_i)\Pr(B_i)}. \tag{2.20}$$

As a matter of nomenclature, $\Pr(B_i)$ is often referred to as the *a priori*[1] probability of event B_i, while $\Pr(B_i|A)$ is known as the *a posteriori*[2] probability of event B_i given A. Section 2.8 presents an engineering application showing how Bayes's theorem is used in the field of signal detection. We conclude here with a practical example showing how useful Bayes's theorem can be.

EXAMPLE 2.14: A certain auditorium has 30 rows of seats. Row 1 has 11 seats, while Row 2 has 12 seats, Row 3 has 13 seats, and so on to the back of the auditorium where Row 30 has 40 seats. A door prize is to be given away by randomly selecting a row (with equal probability of selecting any of the 30 rows) and then randomly selecting a seat within that row (with each seat in the row equally likely to be selected). Find the probability that Seat 15 was selected given that Row 20 was selected and also find the probability that Row 20 was selected given that Seat 15 was selected. The first task is straightforward. Given that Row 20 was selected, there are 30 possible seats in Row 20 that are equally likely to be selected. Hence, Pr(Seat 15|Row 20) $= 1/30$. Without the help of Bayes's theorem, finding the probability that Row 20 was selected given that we know Seat 15 was selected would seem to be a formidable problem. Using Bayes's theorem,

Pr(Row 20|Seat 15) = Pr(Seat 15|Row20)Pr(Row 20)/Pr(Seat 15).

---

[1]The term *a priori* is Latin and is literally translated "from the former." In this context, it refers to probabilities that are formed from self-evident or presupposed models.

[2]The term *a posteriori* is also Latin and is literally translated "from the latter." In this context, it refers to probabilities that are derived or calculated after observing certain events.

The two terms in the numerator on the right-hand side are both equal to $1/30$. The term in the denominator is calculated using the help of the theorem of total probability.

$$\Pr(\text{Seat } 15) = \sum_{k=5}^{30} \frac{1}{k+10} \frac{1}{30} = 0.0342.$$

With this calculation completed, the a posteriori probability of Row 20 being selected given seat 15 was selected is given by

$$\Pr(\text{Row } 20 | \text{Seat } 15) = \frac{\frac{1}{30}\frac{1}{30}}{0.0342} = 0.0325.$$

Note that the a priori probability that Row 20 was selected is $1/30 = 0.0333$. Therefore, the additional information that Seat 15 was selected makes the event that Row 20 was selected slightly less likely. In some sense, this may be counterintuitive, since we know that if Seat 15 was selected, there are certain rows that could not have been selected (i.e., Rows 1–4 have fewer than 15 seats) and, therefore, we might expect Row 20 to have a slightly higher probability of being selected compared to when we have no information about which seat was selected. To see why the probability actually goes down, try computing the probability that Row 5 was selected given that Seat 15 was selected. The event that Seat 15 was selected makes some rows much more probable, while it makes others less probable and a few rows now impossible.

2.6 Independence

In Example 2.14, it was seen that observing one event can change the probability of the occurrence of another event. In that particular case, the fact that it was known that Seat 15 was selected, lowered the probability that Row 20 was selected. We say that the event $A = \{$Row 20 was selected$\}$ is statistically dependent on the event $B = \{$Seat 15 was selected$\}$. If the description of the auditorium were changed so that each row had an equal number of seats (e.g., say all 30 rows had 20 seats each), then observing the event $B = \{$Seat 15 was selected$\}$ would not give us any new information about the likelihood of the event $A = \{$Row 20 was selected$\}$. In that case, we say that the events A and B are statistically independent.

Mathematically, two events A and B are independent if $\Pr(A|B) = \Pr(A)$. That is, the a priori probability of event A is identical to the a posteriori probability of A given B. Note that if $\Pr(A|B) = \Pr(A)$, then the following two conditions also hold (see Exercise 2.8):

$$\Pr(B|A) = \Pr(B) \tag{2.21}$$

$$\Pr(A, B) = \Pr(A)\Pr(B). \tag{2.22}$$

Furthermore, if $\Pr(A|B) \neq \Pr(A)$, then the other two conditions also do not hold. We can thereby conclude that any of these three conditions can be used as a test for independence and the other two forms must follow. We use the last form as a definition of independence since it is symmetric relative to the events A and B.

DEFINITION 2.6: Two events are statistically independent if and only if

$$\Pr(A, B) = \Pr(A)\Pr(B). \tag{2.23}$$

EXAMPLE 2.15: Consider the experiment of tossing two numbered dice and observing the numbers that appear on the two upper faces. For convenience, let the dice be distinguished by color, with the first die tossed being red and the second being white. Let $A = \{$number on the red die is less than or equal to 2$\}$, $B = \{$number on the white die is greater than or equal to 4$\}$, and $C = \{$the sum of the numbers on the two dice is 3$\}$. As mentioned in the preceding text, there are several ways to establish independence (or lack thereof) of a pair of events. One possible way is to compare $\Pr(A, B)$ with $\Pr(A)\Pr(B)$. Note that for the events defined here, $\Pr(A) = 1/3, \Pr(B) = 1/2, \Pr(C) = 1/18$. Also, of the 36 possible atomic outcomes of the experiment, six belong to the event $A \cap B$ and hence $\Pr(A, B) = 1/6$. Since $\Pr(A)\Pr(B) = 1/6$ as well, we conclude that the events A and B are independent. This agrees with intuition since we would not expect the outcome of the roll of one die to affect the outcome of the other. What about the events A and C? Of the 36 possible atomic outcomes of the experiment, two belong to the event $A \cap C$ and hence $\Pr(A, C) = 1/18$. Since $\Pr(A)\Pr(C) = 1/54$, the events A and C are not independent. Again, this is intuitive since whenever the event C occurs, the event A must also occur and so the two must be dependent. Finally, we look at the pair of events B and C. Clearly, B and C are mutually exclusive. If the white die shows a number greater than or equal to 4, there is no way the sum can be 3.

Hence, $\Pr(B,C) = 0$ and since $\Pr(B)\Pr(C) = 1/36$, these two events are also dependent.

The previous example brings out a point that is worth repeating. It is a common mistake to equate mutual exclusiveness with independence. Mutually exclusive events are not the same thing as independent events. In fact, for two events A and B for which $\Pr(A) \neq 0$ and $\Pr(B) \neq 0$, A and B can never be both independent and mutually exclusive. Thus, mutually exclusive events are necessarily statistically dependent.

A few generalizations of this basic idea of independence are in order. First, what does it mean for a set of three events to be independent? The following definition clarifies this and then we generalize the definition to any number of events.

DEFINITION 2.7: The events A, B, and C are mutually independent if each pair of events is independent; that is,

$$\Pr(A,B) = \Pr(A)\Pr(B), \tag{2.24a}$$

$$\Pr(A,C) = \Pr(A)\Pr(C), \tag{2.24b}$$

$$\Pr(B,C) = \Pr(B)\Pr(C), \tag{2.24c}$$

and in addition,

$$\Pr(A,B,C) = \Pr(A)\Pr(B)\Pr(C). \tag{2.24d}$$

DEFINITION 2.8: The events $A_1, A_2, \ldots, A_n$ are independent if any subset of $k < n$ of these events are independent, and in addition

$$\Pr(A_1, A_2, \ldots, A_n) = \Pr(A_1)\Pr(A_2) \ldots \Pr(A_n). \tag{2.25}$$

There are basically two ways in which we can use this idea of independence. As shown in Example 2.15, we can compute joint or conditional probabilities and apply one of the definitions as a test for independence. Alternatively, we can assume independence and use the definitions to compute joint or conditional probabilities that otherwise may be difficult to find. This latter approach is used extensively in engineering applications. For example, certain types of noise signals can be modeled in this way. Suppose we have some time waveform $X(t)$ which represents a noisy signal that we wish to sample at various points in time, $t_1, t_2, \ldots, t_n$. Perhaps we are interested in the probabilities that these samples might exceed some threshold, so we define the events $A_i = \Pr(X(t_i) > T)$, $i = 1, 2, \ldots, n$. How might we calculate the joint probability $\Pr(A_1, A_2, \ldots, A_n)$? In some cases, we have every reason to believe that the value of the noise at one point in time does not affect the value of the noise at another point in time. Hence, we assume that these events are independent and write $\Pr(A_1, A_2, \ldots, A_n) = \Pr(A_1)\Pr(A_2) \ldots \Pr(A_n)$.

2.7 Discrete Random Variables

Suppose we conduct an experiment, E, which has some sample space, S. Furthermore, let ξ be some outcome defined on the sample space, S. It is useful to define functions of the outcome ξ, $X = f(\xi)$. That is, the function f has as its domain all possible outcomes associated with the experiment, E. The range of the function f will depend upon how it maps outcomes to numerical values but in general will be the set of real numbers or some part of the set of real numbers. Formally, we have the following definition.

DEFINITION 2.9: A random variable is a real-valued function of the elements of a sample space, S. Given an experiment, E, with sample space, S, the random variable X maps each possible outcome, $\xi \in S$, to a real number $X(\xi)$ as specified by some rule. If the mapping $X(\xi)$ is such that the random variable X takes on a finite or countably infinite number of values, then we refer to X as a discrete random variable; whereas, if the range of $X(\xi)$ is an uncountably infinite number of points, we refer to X as a continuous random variable.

Since $X = f(\xi)$ is a random variable whose numerical value depends on the outcome of an experiment, we cannot describe the random variable by stating its value; rather, we must give it a probabilistic description by stating the probabilities that the variable X takes on a specific value or values (e.g., $\Pr(X = 3)$ or $\Pr(X > 8)$). For now, we will focus on random variables that take on discrete values and will describe these random variables in terms of probabilities of the form $\Pr(X = x)$. In the next chapter when we study continuous random variables, we will find this description to be insufficient and will introduce other probabilistic descriptions as well.

DEFINITION 2.10: The probability mass function (PMF), $P_X(x)$, of a random variable, X, is a function that assigns a probability to each possible value of the random variable, X. The probability that the random variable X takes on the specific value x is the value of the probability mass function for x. That is, $P_X(x) = \Pr(X = x)$. We use the convention that upper case variables represent random variables while lower case variables represent fixed values that the random variable can assume.

EXAMPLE 2.16: A discrete random variable may be defined for the random experiment of flipping a coin. The sample space of outcomes is $S = \{H, T\}$. We could define the random variable X to be $X(H) = 0$ and $X(T) = 1$. That is, the sample space H, T is mapped to the set $\{0, 1\}$ by the random variable X. Assuming a fair coin, the resulting probability mass function is $P_X(0) = 1/2$ and $P_X(1) = 1/2$. Note that the mapping is not unique and we could have just as easily mapped the sample space H, T to any other pair of real numbers (e.g., $\{1, 2\}$).

EXAMPLE 2.17: Suppose we repeat the experiment of flipping a fair coin n times and observe the sequence of heads and tails. A random variable, Y, could be defined to be the number of times tails occurs in n trials. It turns out that the probability mass function for this random variable is

$$P_Y(k) = \binom{n}{k} \left(\frac{1}{2}\right)^n, \quad k = 0, 1, \ldots, n.$$

The details of how this PMF is obtained will be deferred until later in this section.

EXAMPLE 2.18: Again, let the experiment be the flipping of a coin, and this time we will continue repeating the event until the first time a heads occurs. The random variable Z will represent the number of times until the first occurrence of a heads. In this case, the random variable Z can take on any positive integer value, $1 \leq Z < \infty$. The probability mass function of the random variable Z can be worked out as follows:

$$\Pr(Z = n) = \Pr(n - 1 \text{ tails followed by one heads})$$

$$= (\Pr(T))^{n-1} \Pr(H) = \left(\frac{1}{2}\right)^{n-1} \left(\frac{1}{2}\right) = 2^{-n}.$$

Hence,

$$P_Z(n) = 2^{-n}, \; n = 1, 2, 3, \ldots.$$

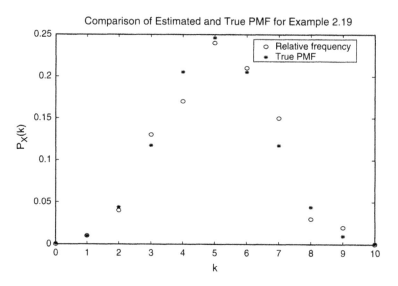

Figure 2.3   MATLAB simulation results from Example 2.19.

EXAMPLE 2.19:   In this example, we will estimate the PMF given in Example 2.17 via MATLAB simulation using the relative frequency approach. Suppose the experiment consists of tossing the coin $n = 10$ times and counting the number of tails. We then repeat this experiment a large number of times and count the relative frequency of each number of tails to estimate the PMF. The following MATLAB code can be used to accomplish this. Results of running this code are shown in Figure 2.3.

```
% Simulation code to estimate PMF of Example 2.17.
n=10;                           % Number of coin flips per
                                  experiment.
m=100;                          % Number of times to repeat
                                  experiment.
X=round(rand(n,m));             % Simulate coin flipping.
Y=sum(X);                       % Calculate number of tails per
                                  experiment.
Rel_Freq=hist(Y,[0:n])/m;       % Compute relative frequencies.
for k=0:n                       % Compute actual PMF.
PMF(k+1)=nchoosek(n,k)*(2^(-n));
end
% Plot Results
```

```
plot([0:n],Rel_Freq,'o',[0:n],PMF,'*')
legend('Relative frequency','True PMF')
xlabel('k')
ylabel('P_X(k)')
title('Comparison of Estimated and True PMF for Example 2.19')
```

Try running this code using a larger value for m. You should see more accurate relative frequency estimates as you increase m.

From the preceding examples, it should be clear that the probability mass function associated with a random variable, X, must obey certain properties. First, since $P_X(x)$ is a probability, it must be nonnegative and no greater than 1. Second, if we sum $P_X(x)$ over all x, then this is the same as the sum of the probabilities of all outcomes in the sample space, which must be equal to 1. Stated mathematically, we may conclude that

$$0 \le P_X(x) \le 1, \qquad (2.26a)$$

$$\sum_x P_X(x) = 1. \qquad (2.26b)$$

When developing the probability mass function for a random variable, it is useful to check that the PMF satisfies these properties.

In the paragraphs that follow, we describe some commonly used discrete random variables, along with their probability mass functions, and some real-world applications in which each might typically be used.

A. Bernoulli Random Variable   This is the simplest possible random variable and is used to represent experiments that have two possible outcomes. These experiments are called Bernoulli trials and the resulting random variable is called a *Bernoulli random variable*. It is most common to associate the values {0,1} with the two outcomes of the experiment. If X is a Bernoulli random variable, its probability mass function is of the form

$$P_X(0) = 1 - p, \quad P_X(1) = p. \qquad (2.27)$$

The coin tossing experiment would produce a Bernoulli random variable. In that case, we may map the outcome H to the value $X = 1$ and T to $X = 0$. Also, we would use the value $p = 1/2$ assuming that the coin is fair. Examples of engineering applications might include radar systems where the random variable could indicate the presence ($X = 1$) or absence ($X = 0$) of a target, or a digital communication system where $X = 1$ might indicate a bit was transmitted in error while $X = 0$

would indicate that the bit was received correctly. In these examples, we would probably expect that the value of p would be much smaller than $1/2$.

B. Binomial Random Variable   Consider repeating a Bernoulli trial n times, where the outcome of each trial is independent of all others. The Bernoulli trial has a sample space of $S = \{0, 1\}$ and we say that the repeated experiment has a sample space of $S_n = \{0, 1\}^n$, which is referred to as a *Cartesian space*. That is, outcomes of the repeated trials are represented as n element vectors whose elements are taken from S. Consider, for example, the outcome

$$\xi_k = \overbrace{(1, 1, \ldots, 1,}^{k \text{ times}} \overbrace{0, 0, \ldots, 0)}^{n - k \text{ times}} . \tag{2.28}$$

The probability of this outcome occurring is

$$\Pr(\xi_k) = \Pr(1, 1, \ldots, 1, 0, 0, \ldots, 0) = \Pr(1)\Pr(1) \ldots \Pr(1)\Pr(0)\Pr(0) \ldots \Pr(0)$$

$$= (\Pr(1))^k (\Pr(0))^{n-k} = p^k (1 - p)^{n-k}. \tag{2.29}$$

In fact, the order of the 1s and 0s in the sequence is irrelevant. Any outcome with exactly k 1s and $n - k$ 0s would have the same probability. Now let the random variable X represent the number of times the outcome 1 occurred in the sequence of n trials. This is known as a *binomial random variable* and takes on integer values from 0 to n. To find the probability mass function of the binomial random variable, let A_k be the set of all outcomes that have exactly k 1s and $n - k$ 0s. Note that all outcomes in this event occur with the same probability. Furthermore, all outcomes in this event are mutually exclusive. Then

$$P_X(k) = \Pr(A_k) = (\# \text{ of outcomes in } A_k) * (\text{probability of each outcome in } A_k)$$

$$= \binom{n}{k} p^k (1 - p)^{n-k}, \quad k = 0, 1, 2, \ldots, n. \tag{2.30}$$

The number of outcomes in the event A_k is just the number of combinations of n objects taken k at a time. This is the binomial coefficient, which is given by

$$\binom{n}{k} = \frac{n!}{k!(n-k)!}. \tag{2.31}$$

As a check, we verify that this probability mass function is properly normalized:

$$\sum_{k=0}^{n} \binom{n}{k} p^k (1 - p)^{n-k} = (p + 1 - p)^n = 1^n = 1. \tag{2.32}$$

In this calculation, we have used the binomial expansion

$$(a + b)^n = \sum_{k=0}^{n} \binom{n}{k} a^k b^{n-k}. \tag{2.33}$$

Binomial random variables occur in practice any time Bernoulli trials are repeated. For example, in a digital communication system, a packet of n bits may be transmitted and we might be interested in the number of bits in the packet that are received in error. Or, perhaps a bank manager might be interested in the number of tellers who are serving customers at a given point in time. Similarly, a medical technician might want to know how many cells from a blood sample are white and how many are red. In Example 2.17, the coin tossing experiment was repeated n times and the random variable Y represented the number of times heads occurred in the sequence of n tosses. This is a repetition of a Bernoulli trial, and hence the random variable Y should be a binomial random variable with $p = 1/2$ (assuming the coin is fair).

C. *Poisson Random Variable*   Consider a binomial random variable, X, where the number of repeated trials, n, is very large. In that case, evaluating the binomial coefficients can pose numerical problems. If the probability of success in each individual trial, p, is very small, then the binomial random variable can be well approximated by a *Poisson random variable*. That is, the Poisson random variable is a limiting case of the binomial random variable. Formally, let n approach infinity and p approach zero in such a way that $\lim_{n \to \infty} np = \alpha$. Then the binomial probability mass function converges to the form

$$P_X(m) = \frac{\alpha^m}{m!} e^{-\alpha}, \quad m = 0, 1, 2, \ldots, \tag{2.34}$$

which is the probability mass function of a Poisson random variable. We see that the Poisson random variable is properly normalized by noting that

$$\sum_{m=0}^{\infty} \frac{\alpha^m}{m!} e^{-\alpha} = e^{-\alpha} e^{\alpha} = 1. \tag{2.35}$$

see Appendix E, (E.14).

The Poisson random variable is extremely important as it describes the behavior of many physical phenomena. It is most commonly used in queuing theory and in communication networks. The number of customers arriving at a cashier in a store during some time interval may be well modeled as a Poisson random variable, as may the number of data packets arriving at a given node in a computer network. We will see increasingly in later chapters that the Poisson random variable plays a fundamental role in our development of a probabilistic description of noise.

D. Geometric Random Variable   Consider repeating a Bernoulli trial until the first occurrence of the outcome ξ_0. If X represents the number of times the outcome ξ_1 occurs before the first occurrence of ξ_0, then X is a *geometric random variable* whose probability mass function is

$$P_X(k) = (1 - p)p^k, \quad k = 0, 1, 2, \ldots . \tag{2.36}$$

We might also formulate the geometric random variable in a slightly different way. Suppose X counted the number of trials that were performed until the first occurrence of ξ_0. Then the probability mass function would take on the form

$$P_X(k) = (1 - p)p^{k-1}, \quad k = 1, 2, 3, \ldots . \tag{2.37}$$

The geometric random variable can also be generalized to the case where the outcome ξ_0 must occur exactly m times. That is, the generalized geometric random variable counts the number of Bernoulli trials that must be repeated until the mth occurrence of the outcome ξ_0. We can derive the form of the probability mass function for the generalized geometric random variable from what we know about binomial random variables. For the mth occurrence of ξ_0 to occur on the kth trial, then the first $k-1$ trials must have had $m-1$ occurrences of ξ_0 and $k-m$ occurrences of ξ_1. Then

$$P_X(k) = \Pr(\{((m-1) \text{ occurrences of } \xi_0 \text{ in } k-1) \text{ trials}\}$$
$$\cap \{\xi_0 \text{ occurs on the } k\text{th trial}\})$$
$$= \binom{k-1}{m-1} p^{k-m}(1-p)^{m-1}(1-p)$$
$$= \binom{k-1}{m-1} p^{k-m}(1-p)^m, \quad k = m, m+1, m+2, \ldots . \tag{2.38}$$

This generalized geometric random variable sometimes goes by the name of a Pascal random variable or the negative binomial random variable.

Of course, one can define many other random variables and develop the associated probability mass functions. We have chosen to introduce some of the more important discrete random variables here. In the next chapter, we will introduce some continuous random variables and the appropriate probabilistic descriptions of these random variables. However, to close out this chapter, we provide a section showing how some of the material covered herein can be used in at least one engineering application.

2.8 Engineering Application: An Optical Communication System

Figure 2.4 shows a simplified block diagram of an optical communication system. Binary data is transmitted by pulsing a laser or a light emitting diode (LED) that is coupled to an optical fiber. To transmit a binary 1 we turn on the light source for T seconds, while a binary 0 is represented by turning the source off for the same time period. Hence, the signal transmitted down the optical fiber is a series of pulses (or absence of pulses) of duration T seconds that represents the string of binary data to be transmitted. The receiver must convert this optical signal back into a string of binary numbers; it does this using a photodetector. The received light wave strikes a photoemissive surface, which emits electrons in a random manner. While the number of electrons emitted during a T second interval is random and thus needs to be described by a random variable, the probability mass function of that random variable changes according to the intensity of the light incident on the photoemissive surface during the T second interval. Therefore, we define a random variable X to be the number of electrons counted during a T second interval, and we describe this random variable in terms of two conditional probability mass functions $P_{X|0}(k) = \Pr(X = k|0 \text{ sent})$ and $P_{X|1}(k) = \Pr(X = k|1 \text{ sent})$. It can be shown through a quantum mechanical argument that these two probability mass functions should be those of Poisson random variables. When a binary 0 is sent, a relatively low number of electrons is typically observed; whereas, when a 1 is sent, a higher number of electrons is typically counted. In particular, suppose the two probability mass functions are given by

$$P_{X|0}(k) = \frac{R_0^k}{k!}e^{-R_0}, \quad k = 0, 1, 2, \dots, \tag{2.39a}$$

$$P_{X|1}(k) = \frac{R_1^k}{k!}e^{-R_1}, \quad k = 0, 1, 2, \dots. \tag{2.39b}$$

In these two PMFs, the parameters R_0 and R_1 are interpreted as the "average" number of electrons observed when a 0 is sent and when a 1 is sent, respectively.

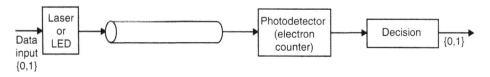

Figure 2.4   Block diagram of an optical communication system.

Also, it is assumed that $R_0 < R_1$, so when a 0 is sent we tend to observe fewer electrons than when a 1 is sent.

At the receiver, we count the number of electrons emitted during each T second interval and then must decide whether a 0 or 1 was sent during each interval. Suppose that during a certain bit interval, it is observed that k electrons are emitted. A logical decision rule would be to calculate $\Pr(0 \text{ sent}|X = k)$ and $\Pr(1 \text{ sent}|X = k)$ and choose according to whichever is larger. That is, we calculate the a posteriori probabilities of each bit being sent given the observation of the number of electrons emitted and choose the data bit that maximizes the a posteriori probability. This is referred to as a maximum a posteriori (MAP) decision rule, and we decide that a binary 1 was sent if

$$\Pr(1 \text{ sent}|X = k) > \Pr(0 \text{ sent}|X = k); \tag{2.40}$$

otherwise we decide a 0 was sent. Note that these desired a posteriori probabilities are backwards relative to how the photodetector was statistically described. That is, we know the probabilities of the form $\Pr(X = k|1 \text{ sent})$ but we want to know $\Pr(1 \text{ sent}|X = k)$. We call upon Bayes's theorem to help us convert what we know into what we desire to know. Using the theorem of total probability,

$$P_X(k) = \Pr(X = k) = P_{X|0}(k)\Pr(0 \text{ sent}) + P_{X|1}(k)\Pr(1 \text{ sent}). \tag{2.41}$$

The a priori probabilities $\Pr(0 \text{ sent})$ and $\Pr(1 \text{ sent})$ are taken to be equal (to $1/2$), so that

$$P_X(k) = \frac{1}{2}\frac{R_0^k}{k!}e^{-R_0} + \frac{1}{2}\frac{R_1^k}{k!}e^{-R_1}. \tag{2.42}$$

Therefore, applying Bayes's theorem,

$$\Pr(0 \text{ sent}|X = k) = \frac{P_{X|0}(k)\Pr(0 \text{ sent})}{P_X(k)} = \frac{\dfrac{1}{2}\dfrac{R_0^k}{k!}e^{-R_0}}{\dfrac{1}{2}\dfrac{R_0^k}{k!}e^{-R_0} + \dfrac{1}{2}\dfrac{R_1^k}{k!}e^{-R_1}}, \tag{2.43}$$

and

$$\Pr(1 \text{ sent}|X = k) = \frac{P_{X|1}(k)\Pr(1 \text{ sent})}{P_X(k)} = \frac{\dfrac{1}{2}\dfrac{R_1^k}{k!}e^{-R_1}}{\dfrac{1}{2}\dfrac{R_0^k}{k!}e^{-R_0} + \dfrac{1}{2}\dfrac{R_1^k}{k!}e^{-R_1}}, \tag{2.44}$$

Since the denominators of both a posteriori probabilities are the same, we decide that a 1 was sent if

$$\frac{1}{2}\frac{R_1^k}{k!}e^{-R_1} > \frac{1}{2}\frac{R_0^k}{k!}e^{-R_0}. \tag{2.45}$$

After a little algebraic manipulation, this reduces down to choosing in favor of a 1 if

$$k > \frac{R_1 - R_0}{\ln(R_1/R_0)};$$ (2.46)

otherwise, we choose in favor of 0. That is, the receiver for our optical communication system counts the number of electrons emitted and compares that number with a threshold. If the number of electrons emitted is above the threshold we decide that a 1 was sent; otherwise, we decide a 0 was sent.

We might also be interested in evaluating how often our receiver makes a wrong decision. Ideally, the answer is that errors are rare, but still we would like to quantify this. Toward that end, we note that errors can occur in two manners. First a 0 could be sent and the number of electrons observed could fall above the threshold, causing us to decide that a 1 was sent. Likewise, if a 1 is actually sent and the number of electrons observed is low, we would mistakenly decide that a 0 was sent. Again, invoking concepts of conditional probability, we see that

$$\Pr(\text{error}) = \Pr(\text{error}|0 \text{ sent})\Pr(0 \text{ sent}) + \Pr(\text{error}|1 \text{ sent})\Pr(1 \text{ sent}).$$ (2.47)

Let x_0 be the threshold with which we compare X to decide which data bit was sent. Specifically, let $x_0 = \lfloor (R_1 - R_0)/\ln(R_1/R_0) \rfloor$ so that we decide a 1 was sent if $X > x_0$, and we decide a 0 was sent if $X \le x_0$. Then

$$\Pr(\text{error} | 0 \text{ sent}) = \Pr(X > x_0|0 \text{ sent}) = \sum_{k=x_0+1}^{\infty} P_{X|0}(k)$$

$$= \sum_{k=x_0+1}^{\infty} \frac{R_0^k}{k!}e^{-R_0} = 1 - \sum_{k=0}^{x_0} \frac{R_0^k}{k!}e^{-R_0}.$$ (2.48)

Likewise,

$$\Pr(\text{error} | 1 \text{ sent}) = \sum_{k=0}^{x_0} P_{X|0}(k) = \sum_{k=0}^{x_0} \frac{R_1^k}{k!}e^{-R_1}.$$ (2.49)

Hence, the probability of error for our optical communication system is

$$\Pr(\text{error}) = \frac{1}{2} - \frac{1}{2}\sum_{k=0}^{x_0} \frac{R_0^k e^{-R_0} - R_1^k e^{-R_1}}{k!}.$$ (2.50)

Figure 2.5 shows a plot of the probability of error as a function of R_1 with R_0 as a parameter. The parameter R_0 is a characteristic of the photodetector used. We will see in later chapters that R_0 can be interpreted as the "average" number of electrons

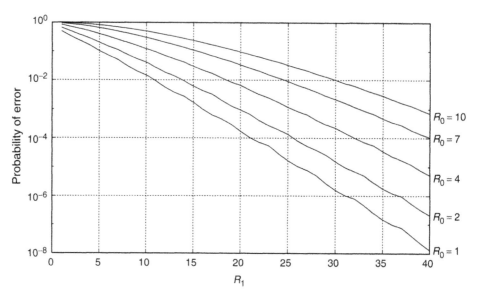

Figure 2.5 Probability of error curves for an optical communication system; curves are parameterized from bottom to top with $R_0 = 1, 2, 4, 7, 10$.

emitted during a bit interval when there is no signal incident of the photodetector. This is sometimes referred to as the "dark current." The parameter R_1 is controlled by the intensity of the incident light. Given a certain photodetector, the value of the parameter R_0 can be measured. The value of R_1 required to achieve a desired probability of error can be found from Figure 2.5 (or from Equation 2.50, which generated the figure). The intensity of the laser or LED can then be adjusted to produce the required value for the parameter R_1.

Exercises

2.1 Using mathematical induction, prove Corollary 2.1. Recall that Corollary 2.1 states that for M events $A_1, A_2, \ldots, A_M$ that are mutually exclusive (i.e., $A_i \cap A_j = \varnothing$ for all $i \neq j$),

$$\Pr\left(\bigcup_{i=1}^{M} A_i\right) = \sum_{i=1}^{M} \Pr(A_i).$$

2.2 Develop a careful proof of Theorem 2.1, which states that for any events A and B,

$$\Pr(A \cup B) = \Pr(A) + \Pr(B) - \Pr(A \cap B).$$

One way to approach this proof is to start by showing that the set $A \cup B$ can be written as the union of three mutually exclusive sets,

$$A \cup B = \{A \cap (\overline{A \cap B})\} \cup \{A \cap B\} \cup \{B \cap (\overline{A \cap B})\}$$

and hence by Corollary 2.1,

$$\Pr(A \cup B) = \Pr(A \cap (\overline{A \cap B})) + \Pr(A \cap B) + \Pr(B \cap (\overline{A \cap B})).$$

Next, show that

$$\Pr(A \cap (\overline{A \cap B})) = \Pr(A) - \Pr(A \cap B)$$

and likewise

$$\Pr(B \cap (\overline{A \cap B})) = \Pr(B) - \Pr(A \cap B).$$

(*Hint:* Recall DeMorgan's law) Put these results together to complete the desired proof.

2.3 Show that the above formula for the probability of the union of two events can be generalized to three events as follows:

$$\Pr(A \cup B \cup C) = \Pr(A) + \Pr(B) + \Pr(C)$$
$$- \Pr(A \cap B) - \Pr(A \cap C) - \Pr(B \cap C) + \Pr(A \cap B \cap C).$$

2.4 Prove Theorem 2.3, which states that if $A \subset B$ then $\Pr(A) \leq \Pr(B)$.

2.5 Formally prove the *union bound* which states that for any events $A_1, A_2, \ldots, A_M$ (not necessarily mutually exclusive),

$$\Pr\left(\bigcup_{i=1}^{M} A_i\right) \leq \sum_{i=1}^{M} \Pr(A_i).$$

2.6 (a) Demonstrate that the relative frequency approach to assigning probabilities satisfies the three axioms of probability.

(b) Demonstrate that the definition of conditional probability $\Pr(A|B) = \Pr(A, B)/\Pr(B)$ satisfies the three axioms of probability.

2.7 Prove that if $\Pr(B|A) = \Pr(B)$, then it follows that

(a) $\Pr(A, B) = \Pr(A)\Pr(B)$, and

(b) $\Pr(A|B) = \Pr(A)$.

Furthermore, show that if $\Pr(B|A) \neq \Pr(B)$, then the two conditions (a) and (b) do not hold as well.

2.8 We are given a number of darts. When we throw a dart at a target, we have a probability of 1/4 of hitting the target. What is the probability of obtaining at least one hit if three darts are thrown? Calculate this probability two ways. *Hint:* Construct the sample space. How many outcomes are in the sample space? Are all outcomes in the sample space equally likely?

2.9 A box of 30 diodes is known to contain five defective ones. If two diodes are selected at random without replacement, what is the probability that at least one of these diodes is defective?

2.10 Two balls are selected sequentially (without replacement) from an urn containing three red, four white, and five blue balls.

(a) What is the probability that the first is red and the second is blue?

(b) What is the probability of selecting a white ball on the second draw if the first ball is replaced before the second is selected?

(c) What is the probability of selecting a white ball on the second draw if the first ball is not replaced before the second is selected?

2.11 In pulse code modulation (PCM), a PCM word consists of a sequence of binary digits (bits) of 1s and 0s.

(a) Suppose the PCM word length is n bits long. How many distinct words are there?

(b) If each PCM word, three bits long, is equally likely to occur, what is the probability of a word with exactly two 1s occurring? Solve this problem in two ways. First, consider all words in a sample space. Second, suppose each bit is equally likely.

2.12 In pulse amplitude modulation (PAM), a PAM word consists of a sequence of pulses, where each pulse may take on a given number of amplitude levels. Suppose a PAM word is n pulses long and each pulse may take on m different levels.

(a) How many distinct PAM words are there?

(b) If each PAM word, four pulses long, is equally likely to occur and each pulse can have one of three levels, {0, 1, 2}, what is the probability of a PAM word occurring with exactly two pulses of level 2?

2.13 A balanced coin is tossed nine times. Find the probabilities of each of the following events:

(a) exactly 3 heads occurred;

(b) at least 3 heads occurred;

(c) at least 3 heads and at least 2 tails occurred.

2.14 Two six-sided (balanced) dice are thrown. Find the probabilities of each of the following events:

(a) a 5 does not occur on either throw;

(b) the sum is 7;

(c) a 5 and a 3 occur in any order;

(d) the first throw is a 5 and the second throw is a 5 or a 4;

(e) both throws are 5;

(f) either throw is a 6.

2.15 Two six-sided (balanced) dice are thrown. Find the probabilities of each of the following events:

(a) only 2, 3, or 4 appear on both dice;

(b) the value of the second roll subtracted from the value of the first roll is 2;

(c) the sum is 10 given that one roll is 6;

(d) the sum is 7 or 8 given that one roll is 5;

(e) one roll is a 4 given that the sum is 7.

2.16 Manufacturer X produces personal computers (PCs) at two different locations in the world. Fifteen percent of the PCs produced at location A are delivered defective to a retail outlet, while 5 percent of the PCs produced at location B are delivered defective to the same retail store. If the manufacturing plant at A produces 1,000,000 PCs per year and the plant at B produces 150,000 PCs per year, find the probability of purchasing a defective PC.

2.17 I deal myself 3 cards for a standard 52-card deck. Find the probabilities of each of the following events:

(a) 2 of a kind (e.g., 2 fives or 2 kings),

(b) 3 of a kind,

(c) 3 of the same suit (a.k.a. a flush, e.g., 3 hearts or 3 clubs),

(d) 3 cards in consecutive order (a.k.a. a straight, e.g., 2-3-4 or 10-J-Q).

2.18 I deal myself 5 cards for a standard 52-card deck. Find the probabilities of each of the following events:

  (a) 2 of a kind,
  (b) 3 of a kind,
  (c) 2 pair (e.g., 2 eights and 2 queens),
  (d) a flush (5 cards all of the same suit),
  (e) a full house (3 of one kind and 2 of another kind),
  (f) a straight (5 cards in consecutive order).

2.19 I deal myself 13 cards for a standard 52-card deck. Find the probabilities of each of the following events:

  (a) exactly one heart appears in my hand (of 13 cards);
  (b) at least 7 cards from a single suit appear in my hand;
  (c) my hand is void (0 cards) of at least one suit.

2.20 Cards are drawn from a standard 52-card deck until an ace is drawn. After each card is drawn, (if the card is not an ace), it is put back in the deck and the cards are reshuffled so that each card drawn is independent of all others.

  (a) Find the probability that the first ace is drawn on the 5th selection.
  (b) Find the probability that *at least* 5 cards are drawn *before* the first ace appears.
  (c) Repeat parts (a) and (b) if the cards are drawn without replacement. That is, after each card is drawn, (if it is not an ace) the card is set aside and not replaced in the deck.

2.21 Cards are drawn from a standard 52-card deck until the third club is drawn. After each card is drawn, it is put back in the deck and the cards are reshuffled so that each card drawn is independent of all others.

  (a) Find the probability that the 3rd club is drawn on the 8th selection.
  (b) Find the probability that *at least* 8 cards are drawn *before* the 3rd club appears.
  (c) Repeat parts (a) and (b) if the cards are drawn without replacement. That is, after each card is drawn, the card is set aside and not replaced in the deck.

2.22 A computer memory has the capability of storing 10^6 words. Due to outside forces, portions of the memory are often erased. Therefore, words are stored redundantly in various areas of the memory. If a particular word is stored in n different places in the memory, what is the probability that this word cannot be recalled if one-half of the memory is erased by electromagnetic radiation? *Hint:* Consider each word to be stored in a particular cell (or box). These cells (boxes) may be located anywhere, geometrically speaking, in memory. The contents of each cell may be either erased or not erased. Assume n is small compared to the memory capacity.

2.23 If two events A and B can occur and $Pr(A)$ is not zero and $Pr(B)$ is not zero, what combinations of independent (I), not independent (NI), mutually exclusive (M), and not mutually exclusive (NM) are permissible? In other words, which of the four combinations (I, M), (NI, M), (I, NM), and (NI, NM) are permissible? Construct an example for those combinations that are permissible.

2.24 A possible outcome of an experiment is the event A. The probability of this event is p. The experiment is performed n times, and the outcome of any trial is not affected by the results of the previous trials. Define a random variable X to be the number of times the event A occurs in n trials.

(a) What is the PMF $Pr(X = x)$?

(b) Show that the sum of the PMF over all x is 1.

(c) What is the name of this PMF?

2.25 For each of the following probability mass functions, find the value of the constant c.

(a) $P_X(k) = c(0.37)^k$, $k = 0, 1, 2, \ldots$

(b) $P_X(k) = c(0.82)^k$, $k = 0, 1, 2, 3, \ldots$

(c) $P_X(k) = c(0.41)^k$, $k = 0, 1, 2, \ldots, 24$

(d) $P_X(k) = c(0.91)^k$, $k = 0, 1, 2, 3, \ldots, 15$

(e) $P_X(k) = c(0.41)^k$, $k = 0, 2, 4, \ldots, 12$

2.26 Consider a Bernoulli trial where $Pr(1) = p$ and $Pr(0) = 1 - p$. Suppose this Bernoulli trial is repeated n times.

(a) Plot the probability mass function for a binomial random variable, X, with $p = 1/5$ and $n = 10$.

(b) Plot the corresponding probability mass function for a Poisson random variable X with $\alpha = np = 2$.

(c) Compare $\Pr(X \geq 5)$ as computed by both the binomial and Poisson random variables. Is the Poisson random variable a good approximation for the binomial random variable for this example?

2.27 Suppose the arrival of telephone calls at a switch can be modeled with a Poisson PMF. That is, if X is the number of calls that arrives in t minutes, then

$$\Pr(X = k) = \frac{(\lambda t)^k}{k!}e^{-\lambda t}, \ k = 0, 1, 2, \ldots,$$

where λ is the average arrival rate in calls/minute. Suppose that the average rate of calls is 10 per minute.

(a) What is the probability that fewer than three calls will be received in the first 6 seconds?

(b) What is the probability that fewer than three calls will be received in the first 6 minutes?

2.28 In a certain lottery, six numbers are randomly chosen from the set $\{0, 1, 2, \ldots, 49\}$ (without replacement). To win the lottery, a player must guess correctly all six numbers, but it is not necessary to specify in which order the numbers are selected.

(a) What is the probability of winning the lottery with only one ticket?

(b) Suppose in a given week, 6 million lottery tickets are sold. Suppose further that each player is equally likely to choose any of the possible number combinations and does so independently of the selections of all other players. What is the probability that exactly four players correctly select the winning combination?

(c) Again assuming 6 million tickets sold, what is the most probable number of winning tickets?

(d) Repeat parts (b) and (c) using the Poisson approximation to the binomial probability distribution. Is the Poisson distribution an accurate approximation in this example?

2.29 Imagine an audio amplifier contains six transistors. Harry has determined that two transistors are defective, but he does not know which two. Harry removes three transistors at random and inspects them. Let X be the number of defective transistors that Harry finds, where X may be 0, 1, or 2. Find the PMF for X.

2.30 A software manufacturer knows that one out of 10 software games that the company markets will be a financial success. The manufacturer selects 10 new games to market. What is the probability that exactly one game will be a financial success? What is the probability that at least two games will be a success?

2.31 Prove the following identities involving the binomial coefficient $\binom{n}{k} = \dfrac{n!}{k!(n-k)!}$.

(a) $\binom{n}{k} = \binom{n}{n-k}$

(d) $\displaystyle\sum_{k=0}^{n} \binom{n}{k}(-1)^k = 0$

(b) $\binom{n}{k} + \binom{n}{k+1} = \binom{n+1}{k+1}$

(e) $\displaystyle\sum_{k=1}^{n} \binom{n}{k}k = n2^{n-1}$

(c) $\displaystyle\sum_{k=0}^{n} \binom{n}{k} = 2^n$

(f) $\displaystyle\sum_{k=0}^{n} \binom{n}{k}k(-1)^k = 0$

2.32 In a digital communication system, a block of k data bits is mapped into an n bit codeword that typically contains the k information bits as well as $n-k$ redundant bits. This is known as an (n,k) block code. The redundant bits are included to provide error correction capability. Suppose that each transmitted bit in our digital communication system is received in error with probability p. Furthermore, assume that the decoder is capable of correcting any pattern of t or fewer errors in an n bit block. That is, if t or fewer bits in an n bit block are received in error, then the codeword will be decoded correctly, whereas if more than t errors occur, the decoder will decode the received word incorrectly. Assuming each bit is received in error with probability $p = 0.03$, find the probability of decoder error for each of the following codes.

(a) $(n,k) = (7,4)$, $t = 1$
(b) $(n,k) = (15,7)$, $t = 2$
(c) $(n,k) = (31,16)$, $t = 3$

MATLAB Exercises

2.33 Write the MATLAB code to produce a randomly generated number that is equally likely to produce any number from the set $\{0,1,2,\ldots,9\}$.

2.34 Write the MATLAB code to produce a randomly generated number that follows the Bernoulli distribution for an arbitrary parameter, p.

2.35 Modify the MATLAB code in Example 2.19 to produce a random variable that follows a binomial distribution for arbitrary parameters n, p.

2.36 Write the MATLAB code to simulate a random variable, Z, whose PMF is given by $P_z(k) = 2^{-k}$, $k = 1, 2, 3, \ldots$. *Hint:* See Example 2.18 for specifics on how this random variable arises and then follow the lead of Example 2.19.

2.37 (a) Write and execute a MATLAB program to calculate $n!$ for an arbitrary n. Use your program to calculate 64!.

(b) What is the largest integer n for which your program gives a finite answer?

(c) Sterling's approximation for the factorial function is given by

$$n! \approx \sqrt{2\pi} \left(n^{n+\frac{1}{2}} \right) e^{-n} \left(1 - \frac{1}{12n} \right).$$

Use your program to compare the true value of $n!$ with Sterling's approximation. For what ranges of n is the approximation within 1 percent of the true value?

2.38 Write your own program to evaluate the binomial coefficient $\binom{n}{k} = \frac{n!}{k!(n-k)!}$. Create your program in such a way that it need not directly evaluate $n!$. That way the program will not crash if you use it to evaluate a binomial coefficient n greater than the value you found in Exercise 2.37b. Use your program to evaluate $\binom{384}{15}$.

Random Variables, Distributions, and Density Functions

At the end of the last chapter, we introduced the concept of a random variable and gave several examples of common discrete random variables. These random variables were described by their probability mass functions. While this description works fine for discrete random variables, it is inadequate to describe random variables that take on a continuum of values. We will illustrate through an example shortly. In this chapter, we introduce the cumulative distribution function as an alternative description of random variables that is appropriate for describing continuous as well as discrete random variables. A related function, the probability density function is also covered. With these tools in hand, the concepts of random variables can be fully developed. Several examples of commonly used continuous random variables are also discussed.

To show the need for an alternative to the probability mass function, consider a discrete random variable, X, that takes on values from the set $\{0, 1/N, 2/N, \ldots, (N-1)/N\}$ with equal probability. That is, the probability mass function of X is

$$P_X\left(\frac{k}{N}\right) = \frac{1}{N}, \quad k = 0, 1, 2, \ldots, N-1. \tag{3.1}$$

This is the type of random variable that is produced by "random" number generators in high-level languages, such as Fortran and C, and in math packages such as MATLAB, MathCAD, and Mathematica. In these cases, N is taken to be a fairly large number so that it appears that the random number can be anything in the continuous range $[0, 1)$. The reader is referred to Chapter 12, Simulation Techniques, for more details on how computer-generated random numbers work. For now, consider the limiting case as $N \to \infty$ so that the random variable can truly fall anywhere in the interval $[0,1)$. One curious result of passing to the limit is that now

$$P_X\left(\frac{k}{N}\right) = \lim_{N \to \infty} \frac{1}{N} = 0. \tag{3.2}$$

That is, each point has zero probability of occurring. Yet, something has to occur! This problem is common to continuous random variables, and it is clear that the probability mass function is not a suitable description for such a random variable. The next sections develop two alternative descriptions for continuous random variables, which will be used extensively throughout the rest of the text.

3.1 The Cumulative Distribution Function

Since a continuous random variable will typically have a zero probability of taking on a specific value, we avoid talking about such probabilities. Instead, events of the form $\{X \le x\}$ can be considered.

DEFINITION 3.1: The cumulative distribution function (CDF) of a random variable, X, is

$$F_X(x) = \Pr(X \le x). \tag{3.3}$$

From this definition, several properties of the CDF can be inferred. First, since the CDF is a probability, it must take on values between 0 and 1. Since random variables are real-valued, it is easy to conclude that $F_X(-\infty) = 0$ and $F_X(\infty) = 1$. That is, a real number cannot be less than $-\infty$ and must be less than ∞. Next, if we consider two fixed values, x_1 and x_2, such that $x_1 < x_2$, then the event $\{X \le x_1\}$ is a subset of $\{X \le x_2\}$. Hence, $F_x(x_1) \le F_x(x_2)$. This implies that the CDF is a monotonic nondecreasing function. Also, we can break the event $\{X \le x_2\}$ into the union of two mutually exclusive events, $\{X \le x_2\} = \{X \le x_1\} \cup \{x_1 < X \le x_2\}$. Hence, $F_X(x_2) = F_X(x_1) + \Pr(x_1 < X \le x_2)$ or, equivalently, $\Pr(x_1 < X \le x_2) = F_X(x_2) - F_X(x_1)$. Thus, the CDF can also be used to measure the probability that a random variable takes on a value in a certain interval. These properties of cumulative distribution functions are summarized as follows:

(1) $F_X(-\infty) = 0, F_X(\infty) = 1,$ \hfill (3.4a)

(2) $0 \le F_X(x) \le 1,$ \hfill (3.4b)

(3) For $x_1 < x_2, F_X(x_1) \le F_X(x_2),$ \hfill (3.4c)

(4) For $x_1 < x_2, \Pr(x_1 < X \le x_2) = F_X(x_2) - F_X(x_1).$ \hfill (3.4d)

EXAMPLE 3.1: Which of the following mathematical functions could be the CDF of some random variable?

(a) $F_X(x) = \dfrac{1}{2} + \dfrac{1}{\pi} \tan^{-1}(x),$

(b) $F_X(x) = [1 - e^{-x}]u(x),$ ($u(x)$ is the unit step function),

(c) $F_X(x) = e^{-x^2},$

(d) $F_X(x) = x^2 u(x).$

To determine this, we need to check that the function starts at 0 when $x = -\infty$, ends at 1 when $x = \infty$, and is monotonic increasing in between. The first two functions satisfy these properties and thus are valid CDFs, while the last two do not. The function in (c) is decreasing for positive values of x, while the function in (d) takes on values greater than 1 and $F_X(\infty) \ne 1$.

To more carefully illustrate the behavior of the CDF, let us return to the computer random number generator that generates N possible values from the set $\{0, 1/N, 2/N, \ldots, (N-1)/N\}$ with equal probability. The CDF for this particular random variable can be described as follows. First, $F_X(x) = 0$ for all $x < 0$, since the random variable cannot take on negative values. Similarly, $F_X(x) = 1$ for all $x \ge (N-1)/N$ since the random variable cannot be greater than $(N-1)/N$. Next, consider a value of x in the range $0 \le x < 1/N$. In this case, $\Pr(X \le x) = \Pr(X = 0)$ since the only value in the specified range that this random variable can take on is $X = 0$. Hence, $F_X(x) = \Pr(X = 0) = 1/N$ for $0 \le x < 1/N$. Similarly, for $1/N \le x < 2/N, F_X(x) = \Pr(X = 0) + \Pr(X = 1/2N) = 2/N$.

Following this same reasoning, it is seen that, in general, for an integer k such that $0 < k < N$ and $(k-1)/N \le x < k/N$, $F_X(x) = k/N$. A plot of $F_X(x)$ as a function of x would produce the general staircase type function shown in Figure 3.1. In Figures 3.2(a) and 3.2(b), the CDF is shown for specific values of $N = 10$ and $N = 50$, respectively. It should be clear from these plots that in the limit as N passes to infinity, the CDF of Figure 3.2(c) results. The functional form of this CDF is

$$F_X(x) = \begin{cases} 0 & x \le 0 \\ x & 0 < x \le 1. \\ 1 & x > 1 \end{cases} \qquad (3.5)$$

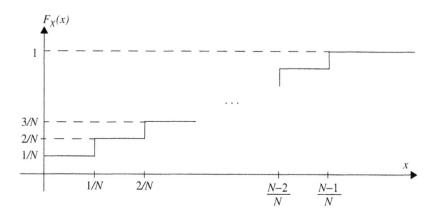

Figure 3.1   General CDF of the random variable X.

In this limiting case, the random variable X is a continuous random variable and takes on values in the range [0, 1) with equal probability. Later in the chapter, this will be referred to as a uniform random variable. Note that when the random variable was discrete, the CDF was discontinuous and had jumps at the specific values that the random variable could take on; whereas, for the continuous random variable, the CDF was a continuous function (although its derivative was not always continuous). This last observation turns out to be universal in that continuous random variables have a continuous CDF, while discrete random variables have a discontinuous CDF with a staircase type of function. Occasionally, one also needs to work with a random variable whose CDF is continuous in some ranges and yet also has some discontinuities. Such a random variable is referred to as a *mixed random variable*.

EXAMPLE 3.2: Suppose we are interested in observing the occurrence of certain events and noting the time of first occurrence. The event might be the emission of a photon in our optical photo detector at the end of Chapter 2, the arrival of a message at a certain node in a computer communications network, or perhaps the arrival of a customer in a store. Let X be a random variable that represents the time that the event first occurs. We would like to find the CDF of such a random variable, $F_X(t) = \Pr(X \le t)$. Since the event could happen at any point in time and time is continuous, we expect X to be a continuous random variable. To formulate a reasonable CDF for this random variable, suppose we divide the time interval $(0, t]$ into many, tiny nonoverlapping time intervals of

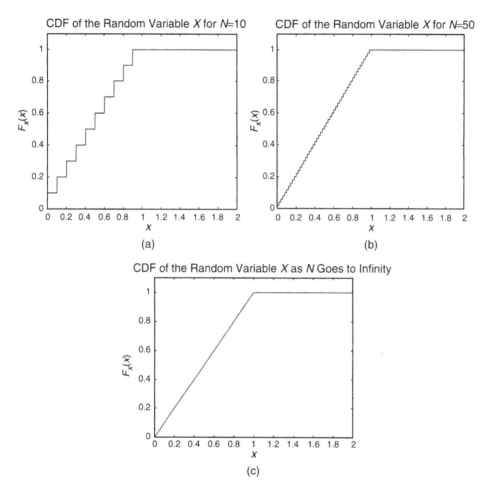

Figure 3.2 CDF of the random variable X for (a) $N = 10$, (b) $N = 50$, and (c) $N \to \infty$.

length Δt. Assume that the probability that our event occurs in a time interval of length Δt is proportional to Δt and take λ to be the constant of proportionality. That is

$$\Pr(\text{event occurs in } (k\Delta t, \ (k + 1)\Delta t)) = \lambda \Delta t.$$

We also assume that the event occurring in one interval is independent of the event occurring in another nonoverlapping time interval. With these rather simple assumptions, we can develop the CDF of the random variable X as follows:

$$F_X(t) = \Pr(X \le t) = 1 - \Pr(X > t),$$

$$\Pr(X > t) = \Pr(X \notin (0, t]) = \Pr(\{X \notin (0, \Delta t]\} \cap \{X \notin (\Delta t, 2\Delta t]\}$$

$$\cap \cdots \cap \{X \notin ((k-1)\Delta t, k\Delta t]\}).$$

In this equation, it is assumed that the time interval $(0, t]$ has been divided into k intervals of length Δt. Since each of the events in the expression are independent, the probability of the intersection is just the product of the probabilities, so that

$$\Pr(X > t) = \Pr(X \notin (0, \Delta t]) \Pr(X \notin (\Delta t, 2\Delta t]) \cdots \Pr(X \notin ((k-1)\Delta t, k\Delta t])$$

$$= (1 - \lambda \Delta t)^k = \left(1 - \frac{\lambda t}{k}\right)^k.$$

Finally, we pass to the limit as $\Delta t \to 0$ or, equivalently, $k \to \infty$ to produce

$$\Pr(X > t) = e^{-\lambda t} u(t) \Rightarrow F_X(t) = (1 - e^{-\lambda t}) u(t).$$

EXAMPLE 3.3: Suppose a random variable has a CDF given by $F_X(x) = (1 - e^{-x})u(x)$. Find the following quantities:

(a) $\Pr(X > 5)$,

(b) $\Pr(X < 5)$,

(c) $\Pr(3 < X < 7)$,

(d) $\Pr(X > 5 | X < 7)$.

For part (a), we note that $\Pr(X > 5) = 1 - \Pr(X \le 5) = 1 - F_X(5) = e^{-5}$. In part (b), we note that $F_X(5)$ gives us $\Pr(X \le 5)$, which is not quite what we want. However, we note that

$$F_X(5) = \Pr(\{X < 5\} \cup \{X = 5\}) = \Pr(X < 5) + \Pr(X = 5).$$

Hence,

$$\Pr(X < 5) = F_X(5) - \Pr(X = 5).$$

In this case, since X is a continuous random variable, $\Pr(X = 5) = 0$ and so there is no need to make a distinction between $\Pr(X < 5)$ and

$\Pr(X \le 5)$; however, for discrete random variables we would need to be careful. Accordingly, $\Pr(X < 5) = F_X(5) = 1 - \exp(-5)$. For part (c), we note that in general $F_X(7) - F_X(3) = \Pr(3 < X \le 7)$. Again, for this continuous random variable, $\Pr(X = 7) = 0$, so we can also write $\Pr(3 < X < 7) = F_X(7) - F_X(3) = e^{-3} - e^{-7}$. Finally, for part (d) we invoke the definition of conditional probability to write the required quantity in terms of the CDF of X:

$$\Pr(X > 5|X < 7) = \frac{\Pr(\{X > 5\} \cap \{X < 7\})}{\Pr(X < 7)} = \frac{\Pr(5 < X < 7)}{\Pr(X < 7)}$$

$$= \frac{F_X(7) - F_X(5)}{F_X(7)} = \frac{e^{-5} - e^{-7}}{1 - e^{-7}} .$$

For discrete random variables, the CDF can be written in terms of the probability mass function defined in Chapter 2. Consider a general random variable, X, which can take on values from the discrete set $\{x_1, x_2, x_3, \dots\}$. The CDF for this random variable is

$$F_X(x) = \sum_{i=1}^{k} P_X(x_i), \quad \text{for } x_k \le x < x_{k+1}. \tag{3.6}$$

The constraint in this equation can be incorporated using unit step functions, in which case the CDF of a discrete random variable can be written as

$$F_X(x) = \sum_{i=1}^{k} P_X(x_i) u(x - x_i). \tag{3.7}$$

In conclusion, if we know the PMF of a discrete random variable, we can easily construct its CDF.

3.2 The Probability Density Function

While the CDF introduced in the last section represents a mathematical tool to statistically describe a random variable, it is often quite cumbersome to work with CDFs. For example, we will see later in this chapter that the most important and commonly used random variable, the Gaussian random variable, has a CDF that cannot be expressed in closed form. Furthermore, it can often be difficult to infer various properties of a random variable from its CDF. To help circumvent these problems, an alternative and often more convenient description known as the *probability density function* (PDF) is often used.

DEFINITION 3.2: The probability density function (PDF) of the random variable X evaluated at the point x is

$$f_X(x) = \lim_{\varepsilon \to 0} \frac{\Pr(x \le X < x + \varepsilon)}{\varepsilon}. \tag{3.8}$$

As the name implies, the probability density function is the probability that the random variable X lies in an infinitesimal interval about the point $X = x$, normalized by the length of the interval.

Note that the probability of a random variable falling in an interval can be written in terms of its CDF as specified in Equation 3.4d. For continuous random variables,

$$\Pr(x \le X < x + \varepsilon) = F_X(x + \varepsilon) - F_X(x) \tag{3.9}$$

so that

$$f_X(x) = \lim_{\varepsilon \to 0} \frac{F_X(x + \varepsilon) - F_X(x)}{\varepsilon} = \frac{dF_X(x)}{dx}. \tag{3.10}$$

Hence, it is seen that the PDF of a random variable is the derivative of its CDF. Conversely, the CDF of a random variable can be expressed as the integral of its PDF. This property is illustrated in Figure 3.3. From the definition of the PDF in Equation 3.8, it is apparent that the PDF is a nonnegative function, although it is not restricted to be less than unity as with the CDF. From the properties of the CDFs, we can also infer several important properties of PDFs as well. Some properties of PDFs are

(1) $f_X(x) \ge 0;$ $\hspace{6cm}$ (3.11a)

(2) $f_X(x) = \dfrac{dF_X(x)}{dx};$ $\hspace{5cm}$ (3.11b)

(3) $F_X(x) = \displaystyle\int_{-\infty}^{x} f_X(y)\,dy;$ $\hspace{5cm}$ (3.11c)

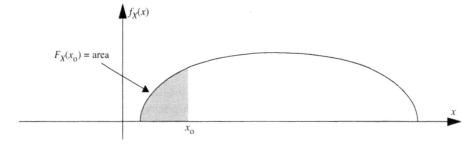

Figure 3.3   Relationship between the PDF and CDF of a random variable.

(4) $\displaystyle\int_{-\infty}^{\infty} f_X(x)\,dx = 1;$ (3.11d)

(5) $\displaystyle\int_{a}^{b} f_X(x)\,dx = \Pr(a < X \le b).$

(3.11e)

EXAMPLE 3.4: Which of the following are valid probability density functions?

(a) $f_X(x) = e^{-x}u(x);$

(b) $f_X(x) = e^{-|x|};$

(c) $f_X(x) = \begin{cases} \frac{3}{4}(x^2 - 1) & |x| < 2 \\ 0 & \text{otherwise} \end{cases};$

(d) $f_X(x) = \begin{cases} 1 & 0 \le x < 1 \\ 0 & \text{otherwise} \end{cases};$

(e) $f_X(x) = 2xe^{-x^2}u(x).$

To verify the validity of a potential PDF, we need to verify only that the function is nonnegative and normalized so that the area underneath the function is equal to unity. The function in part (c) takes on negative values while the function in part (b) is not properly normalized, and hence these are not valid PDFs. The other three functions are valid PDFs.

EXAMPLE 3.5: A random variable has a CDF given by $F_X(x) = \left(1 - e^{-\lambda x}\right)u(x)$. Its PDF is then given by

$$f_X(x) = \frac{dF_X(x)}{dx} = \lambda e^{-\lambda x}u(x).$$

Likewise, if a random variable has a PDF given by $f_X(x) = 2xe^{-x^2}u(x)$, then its CDF is given by

$$F_X(x) = \int_{-\infty}^{x} f_X(y)\,dy = \int_{-\infty}^{x} 2ye^{-y^2}u(y)\,dy$$

$$= \int_{0}^{x} 2ye^{-y^2}\,dy\,u(x) = (1 - e^{-x^2})u(x).$$

EXAMPLE 3.6: The MATLAB function rand generates random numbers that are uniformly distributed in the interval (0,1) using an algorithm to be discussed in Chapter 12, Simulation Techniques. For the present, consider the algorithm to select a number from a table in a random manner. To construct a histogram for the random numbers generated by rand, we write a script that calls rand repeatedly. Since we can do this only a finite number of times, we quantize the range of the random numbers into increments of 0.1. We then calculate the number of times a random number falls in each quantized interval and divide by the total number of numbers generated for the example. If we plot this ratio of relative frequencies using a bar graph, the resulting plot is called a histogram. The MATLAB script for this example follows and the histogram is shown in Figure 3.4, where the total number of values generated is 10,000. Try changing the value of *N* or the number and width of the bins in this example to see how results vary.

```
N=10,000;                    % Do N times
x=rand(1,N);                 % Produce N random numbers.
bins=[0.05:0.1:0.95];        % Create 10 bins,
                             % with centers at 0.05,
                             % 0.15, ....
```

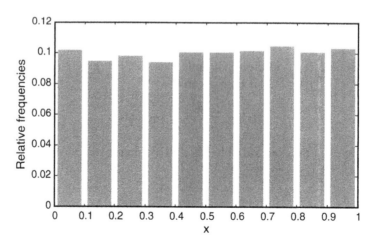

Figure 3.4   Histogram of relative frequencies for uniform random variable generated by MATLAB's rand function using 10,000 trials.

```
[yvalues,xvalues]=hist(x,bins);    % Define xvalues and
                                   yvalues.
yvalues=yvalues/N;                 % Normalize to produce
                                   % relative frequencies.

bar(xvalues,yvalues);              % Plot bar graph.
xlabel('x')
ylabel('Relative Frequencies')
```

3.3 The Gaussian Random Variable

In the study of random variables, the Gaussian random variable is clearly the most commonly used and of most importance. As we will see later in the text, many physical phenomena can be modeled as Gaussian random variables, including the thermal noise encountered in electronic circuits. Although many students may not realize it, they are probably quite familiar with the Gaussian random variable, for it is this random variable that leads to the so-called curve on which many students are graded.

DEFINITION 3.3: A Gaussian random variable is one whose probability density function can be written in the general form

$$f_X(x) = \frac{1}{\sqrt{2\pi\sigma^2}} \exp\left(-\frac{(x-m)^2}{2\sigma^2}\right). \tag{3.12}$$

The PDF of the Gaussian random variable has two parameters, m and σ, which have the interpretation of the mean and standard deviation respectively.[1] The parameter σ^2 is referred to as the variance.

An example of a Gaussian PDF is shown in Figure 3.5. In general, the Gaussian PDF is centered about the point $x = m$ and has a width that is proportional to σ.

It should be pointed out that in the mathematics and statistics literature, this random variable is referred to as a "normal" random variable. Furthermore, for the special case when $m = 0$ and $\sigma = 1$, it is called a "standard normal"

---

[1] The terms *mean*, *standard deviation*, and *variance* will be defined and explained carefully in the next chapter.

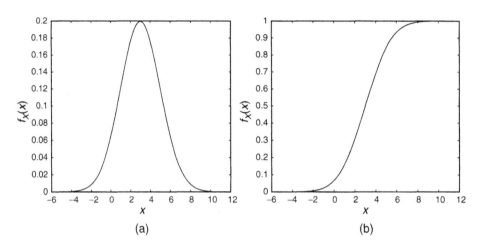

Figure 3.5   PDF (a) and CDF (b) of a Gaussian random variable with $m = 3$ and $\sigma = 2$.

random variable. However, in the engineering literature the term *Gaussian* is much more common, so this nomenclature will be used throughout the text.

Because Gaussian random variables are so commonly used in such a wide variety of applications, it is standard practice to introduce a shorthand notation to describe a Gaussian random variable, $X \sim N(m, \sigma^2)$. This is read "X is distributed normally (or Gaussian) with mean, m, and variance, σ^2."

The first goal to be addressed in the study of Gaussian random variables is to find its CDF. The CDF is required whenever we want to find the probability that a Gaussian random variable lies above or below some threshold or in some interval. Using the relationship in Equation 3.11c, the CDF of a Gaussian random variable is written as

$$F_X(x) = \int_{-\infty}^{x} \frac{1}{\sqrt{2\pi\sigma^2}} \exp\left(-\frac{(y-m)^2}{2\sigma^2}\right) dy. \tag{3.13}$$

It can be shown that it is impossible to express this integral in closed form. While this is unfortunate, it does not stop us from extensively using the Gaussian random variable. Two approaches can be taken to deal with this problem. As with other important integrals that cannot be expressed in closed form (e.g., Bessel functions), the Gaussian CDF has been extensively tabulated, and one can always look up values of the required CDF in a table (such as the one provided in Appendix E). However, it is often a better option to use one of several numerical routines that can approximate the desired integral to any desired accuracy.

The same sort of situation exists with many more commonly known mathematical functions. For example, what if a student was asked to find the tangent of 1.23 radians? While the student could look up the answer in a table of trig functions, that seems like a rather archaic approach. Any scientific calculator, high-level programming language, or math package will have internally generated functions to evaluate such standard mathematical functions. While not all scientific calculators and high-level programming languages have internally generated functions for evaluating Gaussian CDFs, most mathematical software packages do, and in any event, it is a fairly simple thing to write a short program to evaluate the required function. Some numerical techniques for evaluating functions related to the Gaussian CDF will be discussed specifically in Appendix F.

Whether the Gaussian CDF is to be evaluated by using a table or a program, the required CDF must be converted into one of a few commonly used standard forms. A few of these common forms are

- error function integral, $\mathrm{erf}(x) = \dfrac{2}{\sqrt{\pi}} \displaystyle\int_0^x \exp\left(-t^2\right) dt$,

- complementary error function integral, $\mathrm{erfc}(x) = 1 - \mathrm{erf}(x) = \dfrac{2}{\sqrt{\pi}} \displaystyle\int_x^\infty \exp(-t^2)\, dt$,

- Φ-function, $\Phi(x) = \dfrac{1}{\sqrt{2\pi}} \displaystyle\int_{-\infty}^x \exp\left(-\dfrac{t^2}{2}\right) dt$,

- Q-function, $Q(x) = \dfrac{1}{\sqrt{2\pi}} \displaystyle\int_x^\infty \exp\left(-\dfrac{t^2}{2}\right) dt$.

The error function and its complement are most commonly used in the mathematics community; however, in this text we will primarily use the Q-function. Nevertheless, students at least need to be familiar with the relationship between all of these functions because most math packages will have internally defined routines for the error function integral and perhaps its complement as well, but usually not for the Φ-function or the Q-function. So, why not just use error functions? The reason is that if one compares the integral expression for the Gaussian CDF in Equation 3.13 with the integral functions defined in our list, it is a more straightforward thing to express the Gaussian CDF in terms of a Φ-function or a Q-function. Also, the Q-function seems to be enjoying the most common usage in the engineering literature in recent years.

Perhaps the advantage is clearer if we note that the Φ-function is simply the CDF of a standard normal random variable. For general Gaussian random variables that are not in the normalized form, the CDF can be expressed in terms of a Φ-function using a simple transformation. Starting with the Gaussian CDF in Equation 3.13, make the transformation $t = (y - m)/\sigma$, resulting in

$$F_X(x) = \int_{-\infty}^x \frac{1}{\sqrt{2\pi\sigma^2}} \exp\left(-\frac{(y-m)^2}{2\sigma^2}\right) dy = \int_{-\infty}^{\frac{x-m}{\sigma}} \frac{1}{\sqrt{2\pi}} \exp\left(-\frac{t^2}{2}\right) dt = \Phi\left(\frac{x-m}{\sigma}\right).$$

$$(3.14)$$

Hence, to evaluate the CDF of a Gaussian random variable, we just evaluate the Φ-function at the points $(x - m)/\sigma$.

The Q-function is more natural for evaluating probabilities of the form $\Pr(X > x)$. Following a line of reasoning identical to the previous paragraph, it is seen that if $X \sim N(m, \sigma^2)$, then

$$\Pr(X > x) = \int_{\frac{x-m}{\sigma}}^{\infty} \frac{1}{\sqrt{2\pi}} \exp\left(-\frac{t^2}{2}\right) dt = Q\left(\frac{x - m}{\sigma}\right). \qquad (3.15)$$

Furthermore, since we have shown that $\Pr(X > x) = Q((x-m)/\sigma)$ and $\Pr(X \leq x) = \Phi((x - m)/\sigma)$, it is apparent that the relationship between the Φ-function and the Q-function is

$$Q(x) = 1 - \Phi(x). \qquad (3.16)$$

This and other symmetry relationships can be visualized using the graphical definitions of the Φ-function (phi function) and the Q-function shown in Figure 3.6. Note that the CDF of a Gaussian random variable can be written in terms of a Q-function as

$$F_X(x) = 1 - Q\left(\frac{x - m}{\sigma}\right). \qquad (3.17)$$

Once the desired CDF has been expressed in terms of a Q-function, the numerical value can be looked up in a table or calculated with a numerical program.

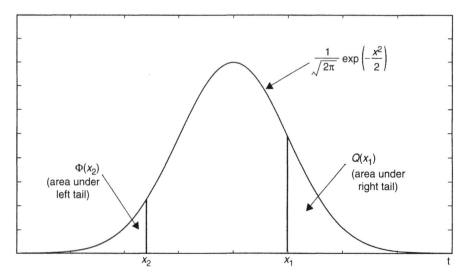

Figure 3.6   Standardized integrals related to the Gaussian CDF: the $\Phi(\cdot)$ and $Q(\cdot)$ functions.

Other probabilities can be found in a similar manner as shown in the next example. It should be noted that some internally defined programs for Q and related functions may expect a positive argument. If it is required to evaluate the Q- function at a negative value, the relationship $Q(x) = 1 - Q(-x)$ can be used. That is, to evaluate $Q(-2)$ for example, $Q(2)$ can be evaluated and then use $Q(-2) = 1 - Q(2)$.

EXAMPLE 3.7: A random variable has a PDF given by

$$f_X(x) = \frac{1}{\sqrt{8\pi}} \exp\left(-\frac{(x+3)^2}{8}\right).$$

Find each of the following probabilities and express the answers in terms of Q-functions.

(a) $\Pr(X \leq 0)$,

(b) $\Pr(X > 4)$,

(c) $\Pr(|X + 3| < 2)$,

(d) $\Pr(|X - 2| > 1)$.

For the given Gaussian pdf, $m = -3$ and $\sigma = 2$. For part (a),

$$\Pr(X \leq 0) = \Phi((0 - (-3))/2) = \Phi(1.5).$$

This can be rewritten in terms of a Q-function as

$$\Pr(X \leq 0) = 1 - Q(1.5).$$

The probability in part (b) is easier to express directly in terms of a Q-function.

$$\Pr(X > 4) = Q((4 - (-3))/2) = Q(3.5).$$

In part (c), the probability of the random variable X falling in an interval is required. This event can be rewritten as

$$\Pr(|X + 3| < 2) = \Pr(-5 < X < -1) = \Pr(X > -5) - \Pr(X > -1)$$
$$= Q(-1) - Q(1) = 1 - 2Q(1).$$

We can proceed in a similar manner for part (d).

$$\Pr(|X - 2| > 1) = \Pr(X < 1) + \Pr(X > 3) = \Phi(2) + Q(3) = 1 - Q(2) + Q(3).$$

EXAMPLE 3.8: While MATLAB does not have a built-in function to evaluate the Q-function, it does have built-in functions for the `erf` and the `erfc` functions. Hence, it is very simple to write your own MATLAB function to evaluate the Q-function. Such a one-line function is shown in the MATLAB code that follows.

```
function output=Q(x)
% Computes the Q-function.
output=erfc(x/sqrt(2))/2;
```

EXAMPLE 3.9: MATLAB also has a built-in function, `randn`, which generates random variables according to a Gaussian or normal distribution. In particular, `randn(k,n)` creates a $k \times n$ matrix whose elements are randomly chosen according to a standard normal distribution. This example constructs a histogram of the numbers generated by the `randn` function similar to what was done in Example 3.6 using the `rand` function. Note that by multiplying the output of the `randn` function by σ and adding m, the Gaussian random variable produced by `randn` now has mean m and variance σ^2. We will elaborate on such transformations and others in the next chapter. Note that the MATLAB code that follows is similar to that of Example 3.6 with the exception that we are now using `randn` instead of `rand`. Also the Gaussian PDF has a domain that is infinite and, thus, in principle we would need an infinite number of bins in our histogram. Since this is impractical, we choose a sufficiently large number of bins such that those not included would represent relatively insignificant values. Note also that in this histogram we are plotting probability densities rather than relative frequencies so that a direct comparison can be made between the histogram and the true PDF. The histogram obtained using the following code is shown in Figure 3.7.

```
N=10,000;                        % Do N times.
m=5; sigma=2;                    % Set mean and variance.
x=m+sigma*randn(1,N);           % Produce N random numbers.
left=-4.5; width=1; right=14.5; % Set bin parameters.
bins=[left:width:right];        % Create bins with centers at
                                % left, left+width,..., right.
[yvalues,xvalues]=hist(x,bins); % Define xvalues and yvalues.
yvalues=yvalues/(N*width);      % Normalize to produce
                                % probability densities.
```

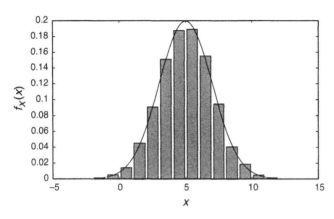

Figure 3.7 Histogram of random numbers produced by randn along with a Gaussian PDF, where $m = 5, \sigma = 2$.

```
bar(xvalues,yvalues);              % Plot bar graph.
z=[left-width/2:width/10:right+width/2];
pdf=exp(-(z-m).^2/(2*sigma^2));    % Compute true PDF
pdf=pdf/sqrt(2*pi*sigma^2);
hold on                            % Place plot of true PDF on
plot(z,pdf)                        % top of histogram.
xlabel('x')
ylabel('f_X(x)')
```

3.4 Other Important Random Variables

This section provides a summary of other important random variables that find use in engineering applications. For each random variable, an indication is given as to the sorts of applications that might best employ that function.

3.4.1 Uniform Random Variable

The uniform probability density function is constant over an interval $[a, b)$. The PDF and its corresponding CDF are

$$f_X(x) = \begin{cases} \dfrac{1}{b-a} & a \le x < b, \\ 0 & \text{elsewhere.} \end{cases} \tag{3.18a}$$

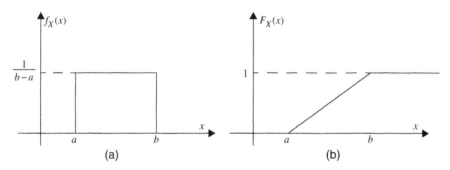

Figure 3.8   Probability density function (a) and cumulative distribution function (b) of a uniform random variable.

$$F_X(x) = \begin{cases} 0 & x < a, \\ \dfrac{x-a}{b-a} & a \le x < b, \\ 1 & x \ge b. \end{cases} \tag{3.18b}$$

Since this is a continuous random variable, the interval over which the PDF is nonzero can be open or closed on either end. A plot of the PDF and CDF of a uniform random variable is shown in Figure 3.8. Most computer random number generators will generate a random variable that closely approximates a uniform random variable over the interval (0,1). We will see in the next chapter that by performing a transformation on this uniform random variable, we can create many other random variables of interest. An example of a uniform random variable would be the phase of a radio frequency sinusoid in a communications system. Although the transmitter knows the phase of the sinusoid, the receiver may have no information about the phase. In this case, the phase at the receiver could be modeled as a random variable uniformly distributed over the interval $[0, 2\pi)$.

3.4.2 Exponential Random Variable

The exponential random variable has a probability density function and cumulative distribution function given (for any $b > 0$) by

$$f_X(x) = \frac{1}{b} \exp\left(-\frac{x}{b}\right) u(x), \tag{3.19a}$$

$$F_X(x) = \left[1 - \exp\left(-\frac{x}{b}\right)\right] u(x). \tag{3.19b}$$

A plot of the PDF and the CDF of an exponential random variable is shown in Figure 3.9. The parameter b is related to the width of the PDF and the PDF has

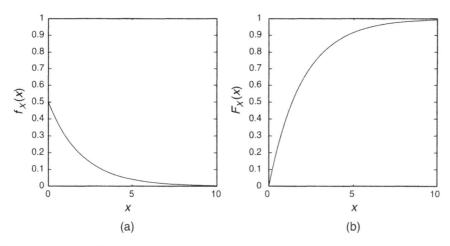

Figure 3.9 Probability density function (a) and cumulative distribution function (b) of an exponential random variable, $b = 2$.

a peak value of $1/b$ which occurs at $x = 0$. The PDF and CDF are nonzero over the semi-infinite interval $(0, \infty)$, which may be either open or closed on the left endpoint.

Exponential random variables are commonly encountered in the study of queueing systems. The time between arrivals of customers at a bank, for example, is commonly modeled as an exponential random variable, as is the duration of voice conversations in a telephone network.

3.4.3 Laplace Random Variable

A Laplace random variable has a PDF which takes the form of a two-sided exponential. The functional forms of the PDF and CDF are given (for any $b > 0$) by

$$f_X(x) = \frac{1}{2b} \exp\left(-\frac{|x|}{b}\right),$$ (3.20a)

$$F_X(x) = \begin{cases} \dfrac{1}{2} \exp\left(\dfrac{x}{b}\right) & x < 0 \\ 1 - \dfrac{1}{2} \exp\left(-\dfrac{x}{b}\right) & x \geq 0 \end{cases}.$$ (3.20b)

A plot of these functions is shown in Figure 3.10. The width of the PDF is determined by the parameter b while the peak value of the PDF is $1/2b$. Note that this peak value is one-half of what it is in the case of the (one-sided) exponential

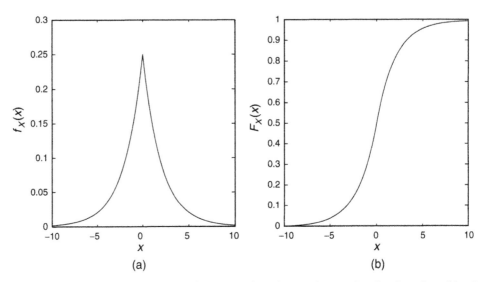

Figure 3.10   Probability density function (a) and cumulative density function (b) of a Laplace random variable, $b = 2$.

shown in Figure 3.9. This makes sense since the Laplace random variable has two sides and the area under the curve must remain constant (and equal to unity). The Laplace random variable has been used to model the probability distribution of a speech (voice) signal.

3.4.4 Gamma Random Variable

A random variable that follows a gamma distribution has a PDF and CDF given (for any $b > 0$ and any $c > 0$) by

$$f_X(x) = \frac{(x/b)^{c-1} \exp\left(-\frac{x}{b}\right)}{b\,\Gamma(c)} u(x), \tag{3.21a}$$

$$F_X(x) = \frac{\gamma(c, x/b)}{\Gamma(c)} u(x). \tag{3.21b}$$

In these two equations, the gamma function is a generalization of the factorial function defined by

$$\Gamma(\alpha) = \int_0^\infty e^{-t} t^{\alpha-1}\, dt, \tag{3.22}$$

and the incomplete gamma function is given by

$$\gamma(\alpha, \beta) = \int_0^\beta e^{-t} t^{\alpha-1} \, dt. \tag{3.23}$$

The gamma random variable is used in queueing theory and has several other random variables as special cases. If the parameter c is an integer, the resulting random variable is also known as an Erlang random variable; whereas, if $b = 2$ and c is a half integer, a chi-squared (χ^2) random variable results. Finally, if $c = 1$, the gamma random variable reduces to an exponential random variable.

3.4.5 Erlang Random Variable

As we've mentioned, the Erlang random variable is a special case of the gamma random variable. The PDF and CDF are given (for positive integer n and any $b > 0$) by

$$f_X(x) = \frac{(x/b)^{n-1} \exp\left(-\frac{x}{b}\right)}{b(n-1)!} u(x), \tag{3.24}$$

$$F_X(x) = \left[1 - \exp\left(-\frac{x}{b}\right) \sum_{m=0}^{n-1} \frac{(x/b)^m}{m!}\right] u(x). \tag{3.25}$$

The Erlang distribution plays a fundamental role in the study of wireline telecommunication networks. In fact, this random variable plays such an important role in the analysis of trunked telephone systems that the amount of traffic on a telephone line is measured in Erlangs.

3.4.6 Chi-Squared Random Variable

Another special case of the gamma random variable, the chi-squared (χ^2) random variable has a PDF and CDF given (for positive integer or half-integer values of c) by

$$f_X(x) = \frac{x^{c-1} \exp\left(-\frac{x}{2}\right)}{2^c \Gamma(c)} u(x), \tag{3.26}$$

$$F_X(x) = \frac{\gamma(c, x/2)}{\Gamma(c)} u(x). \tag{3.27}$$

Many engineering students are probably familiar with the χ^2 random variable from previous studies of statistics. It also commonly appears in various detection problems.

3.4.7 Rayleigh Random Variable

A Rayleigh random variable, like the exponential random variable, has a one-sided PDF. The functional form of the PDF and CDF are given (for any $\sigma > 0$) by

$$f_X(x) = \frac{x}{\sigma^2} \exp\left(-\frac{x^2}{2\sigma^2}\right) u(x), \tag{3.28a}$$

$$F_X(x) = \left(1 - \exp\left(-\frac{x^2}{2\sigma^2}\right)\right) u(x). \tag{3.28b}$$

Plots of these functions are shown in Figure 3.11. The Rayleigh distribution is described by a single parameter, σ^2, which is related to the width of the Rayleigh PDF. In this case, the parameter σ^2 is not to be interpreted as the variance of the Rayleigh random variable. It will be shown later (Example 5.23) that the Rayleigh distribution arises when studying the magnitude of a complex number whose real and imaginary parts both follow a zero-mean Gaussian distribution. The Rayleigh distribution arises often in the study of noncoherent communication systems and also in the study of land mobile communication channels, where the phenomenon known as fading is often modeled using Rayleigh random variables.

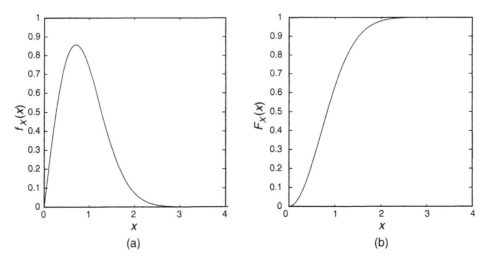

(a)   (b)

Figure 3.11 Probability density function (a) and cumulative density function (b) of a Rayleigh random variable, $\sigma^2 = 1/2$.

3.4.8 Rician Random Variable

A Rician random variable is closely related to the Rayleigh random variable (in fact, the Rayleigh distribution is a special case of the Rician distribution). The functional form of the PDF for a Rician random variable is given (for any $a > 0$ and any $\sigma > 0$) by

$$f_X(x) = \frac{x}{\sigma^2} \exp\left(-\frac{x^2 + a^2}{2\sigma^2}\right) I_0\left(\frac{ax}{\sigma^2}\right) u(x). \tag{3.29}$$

In this expression, the function $I_0(x)$ is the modified Bessel function of the first kind of order zero, which is defined by

$$I_0(x) = \frac{1}{2\pi} \int_0^{2\pi} e^{x \cos(\theta)} \, d\theta. \tag{3.30}$$

Like the Gaussian random variable, the CDF of a Rician random variable cannot be written in closed form. Similar to the Q-function used to describe the Gaussian CDF, there is another function known as *Marcum's Q-function* which describes the CDF of a Rician random variable. It is defined by

$$Q(\alpha, \beta) = \int_\beta^\infty z \exp\left(-\frac{(z^2 + \alpha^2)}{2}\right) I_0(\alpha z) \, dz. \tag{3.31}$$

The CDF of the Rician random variable is then given by

$$F_X(x) = 1 - Q\left(\frac{a}{\sigma}, \frac{x}{\sigma}\right). \tag{3.32}$$

Tables of the Marcum Q-function can be found as well as efficient numerical routines for calculating it. A plot of the Rician PDF is shown in Figure 3.12. The Rician distribution is described by two parameters, a and σ^2, which are related to the center and width, respectively, of the PDF. As with the Rayleigh random variable, the parameter σ^2 is not to be interpreted as the variance of the Rician random variable. The Rician distribution arises in the study of noncoherent communication systems and also in the study of satellite communication channels, where fading is modeled using Rician random variables.

3.4.9 Cauchy Random Variable

The Cauchy random variable has a PDF and CDF given (for any a and any $b > 0$) by

$$f_X(x) = \frac{b/\pi}{b^2 + (x - a)^2}, \tag{3.33}$$

$$F_X(x) = \frac{1}{2} + \frac{1}{\pi} \tan^{-1}\left(\frac{x - a}{b}\right). \tag{3.34}$$

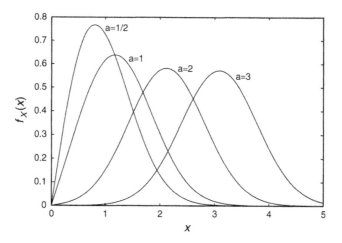

Figure 3.12   PDF of a Rician random variable, where $\sigma^2 = 1/2, a = 1/2, 1, 2, 3$.

The Cauchy random variable occurs when observing the tangent of a random variable which is uniformly distributed over $[0, 2\pi)$. The PDF is centered around $x = a$ and its width is determined by the parameter b. Unlike most of the other random variables where the PDFs decrease exponentially in the tails, the Cauchy PDF decays quadratically as $|x - a|$ increases. Hence, there is a greater amount of probability in the tails of the Cauchy PDF than in many of the other commonly used random variables. We say that this type of distribution is "heavy-tailed."

EXAMPLE 3.10: One can construct many new types of random variables by making functions of random variables. In this example, we construct a random variable that is the sine of a uniform random phase. That is, we construct a random variable Θ which is uniformly distributed over $[0, 2\pi)$, and then form a new random variable according to $X = \sin(\Theta)$. In the next chapter, we will develop the tools to analytically determine what the distribution of X should be, but for now we will simply observe its PDF by plotting a histogram. The MATLAB code that follows was used to accomplish the plot and the results are illustrated in Figure 3.13.

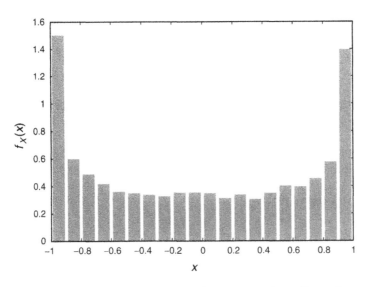

Figure 3.13 Histogram from Example 3.10: sine of a uniform phase.

```
N=10000;
Theta=2*pi*rand(1,N);              % Uniform phase.
X=sin(Theta);                      % sinusoidal transformation.
bins=[-0.95:0.1:0.95];             % Histogram bins.
[yvalues,xvalues]=hist(X,bins);    % Histogram values.
pdf_estimate=yvalues/(N*0.1);      % Normalize probability
                                   densities.

bar(xvalues,pdf_estimate)          % Plot PDF histogram.
xlabel('x'); ylabel('f_X(x)')      % label plot.
```

3.5 Conditional Distribution and Density Functions

In Chapter 2, we defined the notion of conditional probability. In a similar manner, it is quite common to talk about the distribution or density of a random variable conditioned on some event, A. As with the initial study of random variables in the beginning of this chapter, it is convenient to start with the notion of a conditional CDF.

DEFINITION 3.4: The conditional cumulative distribution function of a random variable, X, conditioned on the event A having occurred is

$$F_{X|A}(x) = \Pr(X \le x|A) = \frac{\Pr(\{X \le x\}, A)}{\Pr(A)}. \tag{3.35}$$

Naturally, this definition requires the caveat that the probability of the event A must not be zero.

The properties of CDFs listed in Equations 3.4a–3.4d also apply to conditional CDFs, resulting in the following properties of conditional CDFs:

(1) $F_{X|A}(-\infty) = 0, F_{X|A}(\infty) = 1,$ \hfill (3.36a)

(2) $0 \le F_{X|A}(x) \le 1,$ \hfill (3.36b)

(3) For $x_1 < x_2, F_{X|A}(x_1) \le F_{X|A}(x_2),$ \hfill (3.36c)

(4) For $x_1 < x_2, \Pr(x_1 < X \le x_2|A) = F_{X|A}(x_2) - F_{X|A}(x_1).$ \hfill (3.36d)

It is left as an exercise for the reader (see Exercise 3.13) to prove that these properties of CDFs do indeed apply to conditional CDFs as well.

EXAMPLE 3.11: Suppose a random variable X is uniformly distributed over the interval $[0, 1)$ so that its CDF is given by

$$F_X(x) = \begin{cases} 0 & x < 0 \\ x & 0 \le x \le 1 \\ 1 & x > 1 \end{cases}.$$

Suppose further that we want to find the conditional CDF of X given that $X < 1/2$. Here the event $A = \{X < 1/2\}$ is related to a numerical condition on the random variable itself. From the definition of a conditional CDF,

$$F_{X|\{X<1/2\}}(x) = \frac{\Pr(X \le x, X < 1/2)}{\Pr(X < 1/2)}.$$

For $x < 0$, the event $X \le x$ has probability zero and hence $F_{X|\{X<1/2\}}(x) = 0$ for $x < 0$. When $0 \le x \le 1/2$, the intersection of the events $X \le x$ and $X < 1/2$ is simply the event $X \le x$, so that

$$F_{X|\{X<1/2\}}(x) = \frac{\Pr(X \le x)}{\Pr(X < 1/2)} = \frac{x}{1/2} = 2x \quad \text{for } 0 \le x \le 1/2.$$

Finally, for $x > 1/2$, the intersection of the events $X \le x$ and $X < 1/2$ is simply the event $X < 1/2$ and the conditional CDF reduces to 1. Putting this all together, the desired conditional CDF is

$$F_X(x) = \begin{cases} 0 & x < 0 \\ 2x & 0 \le x \le 1/2 \, . \\ 1 & x > 1/2 \end{cases}$$

In order to generalize the result of the previous example, suppose that for some arbitrary random variable X, the conditioning event is of the form $A = a < X \le b$ for some constants $a < b$. Then

$$F_{X|\{a < X \le b\}}(x) = \frac{\Pr(X \le x, a < X \le b)}{\Pr(a < X \le b)}. \tag{3.37}$$

If $x \le a$, then the events $\{X \le x\}$ and $a < X \le b$ are mutually exclusive and the conditional CDF is zero. For $x > b$ the event $a < X \le b$ is a subset of $\{X \le x\}$ and hence $\Pr(X \le x, a < X \le b) = \Pr(a < X \le b)$ so that the conditional CDF is 1. When $a < x \le b$, then $\{X \le x\} \cap \{a < X \le b\} = a < X \le x$ and $\Pr(X \le x, a < X \le b) = \Pr(a < X \le x)$. This can be written in terms of the CDF (unconditional) of X as $\Pr(a < X \le x) = F_X(x) - F_X(a)$. Likewise, $\Pr(a < X \le b) = F_X(b) - F_X(a)$. Putting these results together gives

$$F_{X|\{a \le X < b\}}(x) = \begin{cases} 0 & x < a \\ \dfrac{F_X(x) - F_X(a)}{F_X(b) - F_X(a)} & a < x \le b \, . \\ 1 & x > b \end{cases} \tag{3.38}$$

This result could also be extended to conditioning events where X is conditioned on being in more extravagant regions.

As with regular random variables, it is often more convenient to work with a conditional PDF rather than a conditional CDF. The definition of the conditional PDF is a straightforward extension of the previous definition given for a PDF.

DEFINITION 3.5: The conditional probability density function of a random variable X conditioned on some event A is

$$f_{X|A}(x) = \lim_{\varepsilon \to 0} \frac{\Pr(x \le X < x + \varepsilon | A)}{\varepsilon}. \tag{3.39}$$

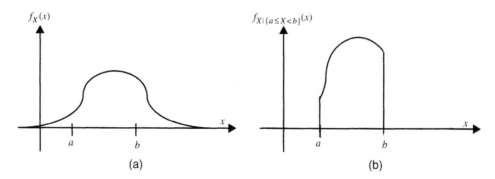

Figure 3.14   A PDF (a) and the corresponding conditional PDF (b).

As with the conditional CDF, it is not difficult to show that all of the properties of regular PDFs apply to conditional PDFs as well. In particular,

(1) $f_{X|A}(x) \geq 0.$ (3.40a)

(2) $f_{X|A}(x) = \frac{dF_{X|A}(x)}{dx}.$ (3.40b)

(3) $F_{X|A}(x) = \int_{-\infty}^{x} f_{X|A}(y)\, dy.$ (3.40c)

(4) $\int_{-\infty}^{\infty} f_{X|A}(x)\, dx = 1.$ (3.40d)

(5) $\int_{a}^{b} f_{X|A}(x)\, dx = \Pr(a < X \leq b|A).$ (3.40e)

Furthermore, the result in Equation 3.38 can be extended to the conditional PDF by applying Equation 3.40b. This results in the following general formula for the conditional PDF of a random variable, X, when the conditioning event is of the nature $A = \{a \leq X < b\}$:

$$f_{X|\{a \leq X < b\}}(x) = \begin{cases} \dfrac{f_X(x)}{\Pr(a \leq X < b)} & a \leq x < b \\ 0 & \text{otherwise} \end{cases}.$$ (3.41)

To summarize, the conditional PDF takes on the same functional form (but is scaled by the probability of the conditioning event) over the range of x where the condition is satisfied, and the conditional PDF is zero wherever the conditioning event is not true. This result is illustrated in Figure 3.14.

EXAMPLE 3.12: Let X be a random variable representing the length of time we spend waiting in the grocery store checkout line. Suppose the random variable X has an exponential PDF given by $f_X(x) = (1/c)\exp(-x/c)u(x)$, where $c = 3$ minutes. What is the PDF for the amount of time we spend waiting in line given that we have already been waiting for 2 minutes? Here the conditioning event is of the form $X > 2$. We can use the result in Equation 3.41 by letting $a = 2$ and $b = \infty$. The probability of the conditioning event is $\Pr(X > 2) = 1 - F_X(2) = \exp(-2/3)$. Therefore, the conditional PDF is

$$f_{X|\{X>2\}}(x) = \exp(2/3)f_X(x)u(x-2) = \frac{1}{3}\exp\left(-\frac{x-2}{3}\right)u(x-2).$$

It is curious to note that for this example, $f_{X|\{X>2\}}(x) = f_X(x-2)$. That is, given that we have been waiting in line for 2 minutes, the PDF of the total time we must wait in line is simply shifted by 2 minutes. This interesting behavior is unique to the exponential distribution and we might not have seen the same result if we had started with a different distribution. For example, try working the same problem starting with a Rayleigh distribution.

Up to this point, we have primarily looked at conditioning events that impose a numerical constraint. It is also common to consider conditioning events of a qualitative, or nonnumerical, nature. Consider, for example, a random variable X that represents a student's score on a certain standardized test (e.g., the SAT or GRE). We might be interested in determining if there is any gender bias in the test. To do so we could compare the distribution of the variable X given that the student is female, $F_{X|F}(x)$, with the distribution of the same variable given that the student is male, $F_{X|M}(x)$. If these distributions are substantially different, then we might conclude a gender bias exists and work to fix the bias in the exam. Naturally, we could work with conditional PDFs $f_{X|F}(x)$ and $f_{X|M}(x)$ as well. Here, the conditioning event is a characteristic of the experiment that may affect the outcome rather than a restriction on the outcome itself.

In general, consider a set of mutually exclusive and exhaustive conditioning events, $A_1, A_2, \ldots, A_N$. Suppose we had access to the conditional CDFs, $F_{X|A_n}(x)$, where $n = 1, 2, \ldots, N$, and wanted to find the unconditional CDF, $F_X(x)$. According to the theorem of total probability (Theorem 2.6),

$$F_X(x) = \Pr(X \leq x) = \sum_{n=1}^{N} \Pr(X \leq x|A_n)\Pr(A_n) = \sum_{n=1}^{N} F_{X|A_n}(x)\Pr(A_n). \qquad (3.42)$$

Hence, the CDF of X (unconditional) can be found by forming a weighted sum of the conditional CDFs with the weights determined by the probabilities that each of the conditioning events is true. By taking derivatives of both sides of the previous equation, a similar result is obtained for conditional PDFs, namely,

$$f_X(x) = \sum_{n=1}^{N} f_{X|A_n}(x) \Pr(A_n). \tag{3.43}$$

We might also be interested in looking at things in the reverse direction. That is, suppose we observe that the random variable has taken on a value of $X = x$. Does the probability of the event A_n change? To answer this, we need to compute $\Pr(A_n|X = x)$. If X were a discrete random variable, we could do this by invoking the results of Theorem 2.5

$$\Pr(A_n|X = x) = \frac{\Pr(X = x|A_n) \Pr(A_n)}{\Pr(X = x)}. \tag{3.44}$$

In the case of continuous random variables, greater care must be taken since both $\Pr(X = x|A_n)$ and $\Pr(X = x)$ will be zero, resulting in an indeterminate expression. To avoid that problem, rewrite the event $\{X = x\}$ as $\{x \le X < x + \varepsilon\}$ and consider the result in the limit as $\varepsilon \to 0$:

$$\Pr(A_n|x \le X < x + \varepsilon) = \frac{\Pr(x \le X < x + \varepsilon|A_n) \Pr(A_n)}{\Pr(x \le X < x + \varepsilon)}. \tag{3.45}$$

Note that for infinitesimal ε, $\Pr(x \le X < x + \varepsilon) = f_X(x)\varepsilon$ and similarly $\Pr(x \le X < x + \varepsilon|A_n) = f_{X|A_n}(x)\varepsilon$. Hence,

$$\Pr(A_n|x \le X < x + \varepsilon) = \frac{f_{X|A_n}(x)\varepsilon \Pr(A_n)}{f_X(x)\varepsilon} = \frac{f_{X|A_n}(x) \Pr(A_n)}{f_X(x)}. \tag{3.46}$$

Finally, passing to the limit as $\varepsilon \to 0$ gives the desired result:

$$\Pr(A_n|X = x) = \lim_{\varepsilon \to 0} \Pr(A_n|x \le X < x + \varepsilon) = \frac{f_{X|A_n}(x) \Pr(A_n)}{f_X(x)}. \tag{3.47}$$

We could also combine this result with Equation 3.43 to produce an extension to Bayes's theorem:

$$\Pr(A_n|X = x) = \frac{f_{X|A_n}(x) \Pr(A_n)}{\sum_{n=1}^{N} f_{X|A_n}(x) \Pr(A_n)}. \tag{3.48}$$

EXAMPLE 3.13: In a certain junior swimming competition, swimmers are placed in one of two categories based on their previous times so that all children can compete against others of their own abilities. The fastest swimmers are placed in the A category, while the slower swimmers are put in the B group. Let X be a random variable representing a child's time (in seconds) in the 50-meter, freestyle race. Suppose that it is determined that for those swimmers in group A, the PDF of a child's time is given by $f_{X|A}(x) = (4\pi)^{-1/2} \exp(-(x - 40)^2/4)$, while for those in the B group the PDF is given by $f_{X|B}(x) = (4\pi)^{-1/2} \exp(-(x - 45)^2/4)$. Furthermore, assume that 30 percent of the swimmers are in the A group and 70 percent are in the B group. If a child swims the race with a time of 42 seconds, what is the probability that the child was in the B group? Applying Equation 3.48, we get

$$\Pr(B|X = 42) = \frac{0.7 f_{X|B}(42)}{0.3 f_{X|A}(42) + 0.7 f_{X|B}(42)}$$

$$= \frac{0.7 \exp(-9/4)}{0.3 \exp(-1) + 0.7 \exp(-9/4)} = 0.4007.$$

Naturally, the probability of the child being from group A must then be $\Pr(A|X = 42) = 1 - \Pr(B|X = 42) = 0.5993$.

3.6 Engineering Application: Reliability and Failure Rates

The concepts of random variables presented in this chapter are used extensively in the study of system reliability. Consider an electronic component that is to be assembled with other components as part of a larger system. Given a probabilistic description of the lifetime of such a component, what can we say about the lifetime of the system itself? The concepts of reliability and failure rates are introduced in this section to provide tools to answer such questions.

DEFINITION 3.6: Let X be a random variable that represents the lifetime of a device. That is, if the device is turned on at time zero, X would represent the time at which the device fails. The *reliability function* of the device, $R_X(t)$, is simply the probability that the device is still functioning at time t:

$$R_X(t) = \Pr(X > t). \tag{3.49}$$

Note that the reliability function is just the complement of the CDF of the random variable. That is, $R_X(t) = 1 - F_X(t)$. As it is often more convenient to work with

PDFs rather than CDFs, we note that the derivative of the reliability function can be related to the PDF of the random variable X by $R'_X(t) = -f_X(t)$.

With many devices, the reliability changes as a function of how long the device has been functioning. Suppose we observe that a particular device is still functioning at some point in time, t. The remaining lifetime of the device may behave (in a probabilistic sense) very differently from when it was first turned on. The concept of failure rate is used to quantify this effect.

DEFINITION 3.7: Let X be a random variable that represents the lifetime of a device. The *failure rate function* is

$$r(t) = f_{X|\{X>t\}}(x)|_{x=t} . \tag{3.50}$$

To give this quantity some physical meaning, we note that $\Pr(t < X < t + dt | X > t) = r(t)dt$. Thus, $r(t)dt$ is the probability that the device will fail in the next time instant of length dt, given that the device has survived up to now (time t). Different types of "devices" have failure rates that behave in different manners. Our pet goldfish, Elvis, might have an increasing failure rate function (as do most biological creatures). That is, the chances of Elvis "going belly up" in the next week is greater when Elvis is six months old than when he is just one month old. We could also imagine devices that have a decreasing failure rate function (at least for part of their lifetime). For example, an integrated circuit might be classified into one of two types, those fabricated correctly with expected long lifetimes and those with defects which generally fail fairly quickly. When we select an IC, we may not know which type it is. Once the device lives beyond that initial period when the defective ICs tend to fail, the failure rate may go down (at least for a while). Finally, there may be some devices whose failure rates remain constant with time.

The failure rate of a device can be related to its reliability function. From Equation 3.41, it is noted that

$$f_{X|\{X>t\}}(x) = \frac{f_X(x)u(x-t)}{1 - F_x(t)} . \tag{3.51}$$

The denominator in this expression is the reliability function, $R_X(t)$, while the PDF in the numerator is simply $-R'_X(x)$. Evaluating at $x = t$ produces the failure rate function

$$r(t) = \frac{-R'_X(t)}{R_X(t)} . \tag{3.52}$$

Conversely, given a failure rate function, $r(t)$, one can solve for the reliability function by solving the first order differential equation:

$$\frac{d}{dt} R_x(t) = -r(t) R_X(t). \tag{3.53}$$

The general solution to this differential equation (subject to the initial condition $R_X(0) = 1$) is

$$R_X(t) = \exp\left[-\int_0^t r(u)du\right]u(t). \tag{3.54}$$

It is interesting to note that a failure rate function completely specifies the PDF of a device's lifetime:

$$f_X(t) = -R_X'(t) = r(t)\exp\left(-\int_0^t r(u)du\right)u(t). \tag{3.55}$$

For example, suppose a device had a constant failure rate function, $r(t) = \lambda$. The PDF of the device's lifetime would then follow an exponential distribution, $f_X(t) = \lambda\exp(-\lambda t)u(t)$. The corresponding reliability function would also be exponential, $R_X(t) = \exp(-\lambda t)u(t)$. We say that the exponential random variable has the memoryless property. That is, it does not matter how long the device has been functioning, the failure rate remains the same.

EXAMPLE 3.14: Suppose the lifetime of a certain device follows a Rayleigh distribution given by $f_X(t) = 2bt\exp(-bt^2)u(t)$. What are the reliability function and the failure rate function? The reliability function is given by

$$R_X(t) = \Pr(X > t) = \left[\int_t^\infty 2bu\exp(-bu^2)du\right]u(t) = \exp(-bt^2)u(t).$$

A straightforward application of Equation 3.52 produces the failure rate function, $r(t) = 2bt\,u(t)$. In this case, the failure rate is linearly increasing in time.

Next, suppose we have a system which consists of N components, each of which has a lifetime described by the random variable X_n, $n = 1, 2, \ldots, N$. Furthermore, assume that for the system to function, all N components must be functioning. In other words, if any of the individual components fails, the whole system fails. This is usually referred to as a series connection of components. If we can characterize the reliability and failure rate functions of each individual component, can we calculate the same functions for the entire system? The answer is yes, under some mild assumptions. Define X to be the random variable representing the lifetime of the system. Then

$$X = \min(X_1, X_2, \ldots, X_N). \tag{3.56}$$

Furthermore,

$$R_X(t) = \Pr(X > t)$$

$$= \Pr(\{X_1 > t\} \cap \{X_2 > t\} \cap \cdots \cap \{X_N > t\}). \tag{3.57}$$

We assume that all of the components fail independently. That is, the event $\{X_i > t\}$ is taken to be independent of $\{X_j > t\}$ for all $i \neq j$. Under this assumption,

$$R_X(t) = \Pr(X_1 > t)\Pr(X_2 > t) \cdots \Pr(X_N > t)$$

$$= R_{X_1}(t) R_{X_2}(t) \cdots R_{X_N}(t). \tag{3.58}$$

Furthermore, application of Equation 3.52 provides an expression for the failure rate function:

$$r(t) = \frac{-R'_X(t)}{R_X(t)} = -\frac{\dfrac{d}{dt}[R_{X_1}(t)R_{X_2}(t) \cdots R_{X_N}(t)]}{R_{X_1}(t)R_{X_2}(t) \cdots R_{X_N}(t)} \tag{3.59}$$

$$= -\sum_{n=1}^{N} \frac{R'_{X_n}(t)}{R_{X_n}(t)} = \sum_{n=1}^{N} r_n(t), \tag{3.60}$$

where $r_n(t)$ is the failure rate function of the nth component. We have shown that for a series connection of components, the reliability function of the system is the product of the reliability functions of each component and the failure rate function of the system is the sum of the failure rate functions of the individual components.

We may also consider a system that consists of a parallel interconnection of components. That is, the system will be functional as long as any of the components are functional. We can follow a similar derivation to compute the reliability and failure rate functions for the parallel interconnection system. First, the reliability function is written as

$$R_X(t) = \Pr(\{X_1 > t\} \cup \{X_2 > t\} \cup \cdots \cup \{X_N > t\}). \tag{3.61}$$

In this case, it is easier to work with the complement of the reliability function (the CDF of the lifetime). Since the reliability function represents the probability that the system is still functioning at time t, the complement of the reliability function represents the probability that the system is not working at time t. With the parallel interconnections, the system will fail only if all the individual components fail. Hence,

$$1 - R_X(t) = \Pr(X \leq t) = \Pr(\{X_1 \leq t\} \cap \{X_2 \leq t\} \cap \cdots \cap \{X_N \leq t\})$$

$$= \Pr(X_1 \leq t)\Pr(X_2 \leq t) \cdots \Pr(X_N \leq t)$$

$$= (1 - R_{X_1}(t))(1 - R_{X_2}(t)) \cdots (1 - R_{X_N}(t)). \tag{3.62}$$

As a result, the reliability function of the parallel interconnection system is given by

$$R_X(t) = 1 - \prod_{n=1}^{N}(1 - R_{X_n}(t)). \tag{3.63}$$

Unfortunately, the general formula for the failure rate function is not as simple as in the serial interconnection case. Application of Equation 3.52 to our preceding equation gives (after some straightforward manipulations)

$$r(t) = -\frac{1 - R_X(t)}{R_X(t)} \sum_{n=1}^{N} \frac{R'_{X_n}(t)}{1 - R_{X_n}(t)}, \tag{3.64}$$

or, equivalently,

$$r(t) = \left[\frac{1}{R_X(t)} - 1\right] \sum_{n=1}^{N} \frac{r_n(t)}{\dfrac{1}{R_{X_n}(t)} - 1}. \tag{3.65}$$

EXAMPLE 3.15: Suppose a system consists of N components each with a constant failure rate, $r_n(t) = \lambda_n$, $n = 1, 2, \ldots, N$. Find the reliability and failure rate functions for a series interconnection. Then find the same functions for a parallel interconnection. It was shown previously that a constant failure rate function corresponds to an exponential reliability function. That is, $R_{X_n}(t) = \exp(-\lambda_n t)u(t)$. For the serial interconnection, we then have

$$R_X(t) = \prod_{n=1}^{N} R_{X_n}(t) = \prod_{n=1}^{N} \exp(-\lambda_n t)u(t) = \exp\left(-\left[\sum_{n=1}^{N}\lambda_n\right]t\right)u(t),$$

$$r(t) = \sum_{n=1}^{N} r_n(t) = \sum_{n=1}^{N}\lambda_n.$$

For the parallel interconnection,

$$R_X(t) = \left\{1 - \prod_{n=1}^{N}[1 - \exp(-\lambda_n t)]\right\}u(t),$$

$$r(t) = \frac{\displaystyle\prod_{n=1}^{N}[1 - \exp(-\lambda_n t)]}{1 - \displaystyle\prod_{n=1}^{N}[1 - \exp(-\lambda_n t)]} \sum_{n=1}^{N} \frac{\lambda_n}{\exp(\lambda_n t) - 1}.$$

Exercises

3.1 Suppose a random variable is equally likely to fall anywhere in the interval $[a, b]$. Then the PDF is of the form

$$f_X(x) = \begin{cases} \dfrac{1}{b-a} & a \leq x \leq b \\ 0 & \text{otherwise} \end{cases}.$$

Find and sketch the corresponding CDF.

3.2 Find and plot the CDFs corresponding to each of the following PDFs:

(a) $f_X(x) = \begin{cases} 1 & 0 \leq x < 1 \\ 0 & \text{otherwise} \end{cases}.$

(b) $f_X(x) = \begin{cases} x & 0 \leq x < 1 \\ 2 - x & 1 \leq x < 2 \\ 0 & \text{otherwise} \end{cases}.$

3.3 A random variable has the following exponential PDF:

$$f_X(x) = \begin{cases} a^{-bx} & x \leq 0 \\ 0 & \text{otherwise} \end{cases},$$

where a and b are constants.

(a) Determine the required relationship between a and b.
(b) Determine the corresponding CDF.

3.4 A certain random variable has a probability density function of the form $f_X(x) = ce^{-2x}u(x)$. Find the following:

(a) the constant c,
(b) $\Pr(X > 2)$,
(c) $\Pr(X < 3)$,
(d) $\Pr(X < 3 | X > 2)$.

3.5 Repeat Problem 3.4 using the PDF $f_X(x) = \dfrac{c}{x^2 + 4}$.

3.6 Repeat Problem 3.4 using the PDF $f_X(x) = \dfrac{c}{\sqrt{25 - x^2}}$, $-5 < x < 5$.

3.7 The voltage of communication signal S is measured. However, the measurement procedure is corrupted by noise resulting in a random measurement with the PDF shown in the accompanying diagram. Find the probability that for any particular measurement, the error will exceed ± 0.75 percent of the correct value if this correct value is 10 volts.

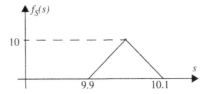

3.8 Prove the integral identity, $I = \displaystyle\int_{-\infty}^{\infty} \exp\left(-\dfrac{x^2}{2}\right) dx = \sqrt{2\pi}$. *Hint:* It may be easier to show that $I^2 = 2\pi$.

3.9 Using the normalization integral for a Gaussian random variable, find an analytical expression for the following integral:

$$I = \int_{-\infty}^{\infty} \exp(-(ax^2 + bx + c))\, dx,$$

where $a > 0$, b, and c are constants.

3.10 A Gaussian random variable has a probability density function of the form

$$f_X(x) = c \exp(-(2x^2 + 3x + 1)).$$

(a) Find the value of the constant c.
(b) Find the values of the parameters m and σ for this Gaussian random variable.

3.11 A Gaussian random variable has a PDF of the form

$$f_X(x) = \frac{1}{\sqrt{50\pi}} \exp\left(-\frac{(x - 10)^2}{50}\right).$$

Write each of the following probabilities in terms of Q-functions (with positive arguments) and also give numerical evaluations.

(a) $\Pr(X > 17)$	(e) $\Pr(|X - 10| > 7)$
(b) $\Pr(X > 4)$	(f) $\Pr(|X - 10| < 3)$
(c) $\Pr(X < 15)$	(g) $\Pr(|X - 7| > 5)$
(d) $\Pr(X < -2)$	(h) $\Pr(|X - 4| < 7)$

3.12 Prove the following properties of the Gamma function.

(a) $\Gamma(n) = (n - 1)!$, for $n = 1, 2, 3, \ldots$

(b) $\Gamma(x + 1) = x\Gamma(x)$

(c) $\Gamma(1/2) = \sqrt{\pi}$

3.13 Prove the following properties of conditional CDFs.

(a) $F_{X|A}(-\infty) = 0$, $F_{X|A}(\infty) = 1$

(b) $0 \leq F_{X|A}(x) \leq 1$

(c) For $x_1 < x_2$, $F_{X|A}(x_1) \leq F_{X|A}(x_2)$

(d) For $x_1 < x_2$, $\Pr(x_1 < X \leq x_2|A) = F_{X|A}(x_2) - F_{X|A}(x_1)$

3.14 Let X be a Gaussian random variable such that $X \sim N(0, \sigma^2)$. Find and plot the following conditional PDFs.

(a) $f_{X|X>0}(x)$

(b) $f_{X||X|<3}(x)$

(c) $f_{X||X|>3}(x)$

3.15 Mr. Hood is a good archer. He can regularly hit a target having a 3-ft. diameter and often hits the bull's-eye, which is 0.5 ft. in diameter, from 50 ft. away. Suppose the miss is measured as the radial distance from the center of the target and, further, that the radial miss distance is a Rayleigh random variable with the constant in the Rayleigh PDF being $\sigma^2 = 4$(sq. ft.).

(a) Determine the probability of Mr. Hood's hitting the target.

(b) Determine the probability of Mr. Hood's hitting the bull's-eye.

(c) Determine the probability of Mr. Hood's hitting the bull's-eye given that he hits the target.

3.16 A digital communication system sends two messages, $M = 0$ or $M = 1$, with equal probability. A receiver observes a voltage which can be modeled as a Gaussian random variable, X, whose PDFs conditioned on the transmitted message are given by

$$f_{X|M=0}(x) = \frac{1}{\sqrt{2\pi\sigma^2}} \exp\left(-\frac{x^2}{2\sigma^2}\right) \quad \text{and}$$

$$f_{X|M=1}(x) = \frac{1}{\sqrt{2\pi\sigma^2}} \exp\left(-\frac{(x-1)^2}{2\sigma^2}\right).$$

(a) Find and plot $Pr(M = 0|X = x)$ as a function of x for $\sigma^2 = 1$. Repeat for $\sigma^2 = 5$.

(b) Repeat part (a) assuming that the a priori probabilities are $Pr(M = 0) = 1/4$ and $Pr(M = 1) = 3/4$.

3.17 In Problem 3.16, suppose our receiver must observe the random variable X and then make a decision as to what message was sent. Furthermore, suppose the receiver makes a three-level decision as follows:

(1) Decide 0 was sent if $Pr(M = 0|X = x) \geq 0.9$.

(2) Decide 1 was sent if $Pr(M = 1|X = x) \geq 0.9$.

(3) Erase the symbol (decide not to decide) if both $Pr(M = 0|X = x) < 0.9$ and $Pr(M = 1|X = x) < 0.9$.

Assuming the two messages are equally probable, $Pr(M = 0) = Pr(M = 1) = 1/2$, and that $\sigma^2 = 1$, find

(a) the range of x over which each of the three decisions should be made,

(b) the probability that the receiver erases a symbol,

(c) the probability that the receiver makes an error (i.e., decides a 0 was sent when a 1 was actually sent, or vice versa).

3.18 Recalling Example 3.15, suppose that a serial connection system has 10 components and the failure rate function is the same constant for all components and is 1 per 100 days.

(a) Determine the probability that the lifetime of the system exceeds 10 days.

(b) What is the probability that the lifetime of one component exceeds 10 days?

(c) What is the reliability function of each component and of the system as a whole?

MATLAB Exercises

3.19 Write a MATLAB program to calculate the probability $\Pr(x_1 \leq X \leq x_2)$ if X is a Gaussian random variable for an arbitrary x_1 and x_2. Note that you will have to specify the mean and variance of the Gaussian random variable.

3.20 Write a MATLAB program to calculate the probability $\Pr(|X - a| < b)$ if X is a Gaussian random variable for an arbitrary a and $b > 0$. Note that you will have to specify the mean and variance of the Gaussian random variable.

3.21 Use the MATLAB `rand` function to create a random variable X uniformly distributed over $(0, 1)$. Then create a new random variable according to $Y = -\ln(X)$. Repeat this procedure many times to create a large number of realizations of Y. Using these samples, estimate and plot the probability density function of Y. Find an analytical model that seems to fit your estimated PDF.

3.22 Use the MATLAB `randn` function to create a Gaussian distributed random variable X. Repeat this procedure and form a new random variable Y. Finally, form a random variable Z according to $Z = \sqrt{X^2 + Y^2}$. Repeat this procedure many times to create a large number of realizations of Z. Using these samples, estimate and plot the probability density function of Z. Find an analytical model that seems to fit your estimated PDF.

3.23 Use the MATLAB `randn` function to generate a large number of samples according to a Gaussian distribution. Let A be the event $A = \{$the sample is greater than 1.5$\}$. Of those samples that are members of the event A, what proportion (relative frequency) is greater than 2. By computing this proportion you will have estimated the conditional probability $\Pr(X > 2|X > 1.5)$. Calculate the exact conditional probability analytically and compare it with the numerical results obtained through your MATLAB program.

Operations on a Single Random Variable

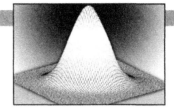

In our study of random variables we use the probability density function or the cumulative distribution function to provide a complete statistical description of the random variable. From these functions we could, in theory, determine just about anything we might want to know about the random variable. In many cases, it is of interest to distill this information down to a few parameters that describe some of the important features of the random variable. For example, we saw in Chapter 3 that the Gaussian random variable is described by two parameters, which were referred to as the mean and variance. In this chapter, we will look at these parameters as well as several others that describe various characteristics of random variables. We will see that these parameters can be viewed as the results of performing various operations on a random variable.

4.1 Expected Value of a Random Variable

To begin, we introduce the idea of an average or expected value of a random variable. This is perhaps the single most important characteristic of a random variable and is also a concept very familiar to most students. After taking a test, one of the most common questions a student will ask after they see their grade is, What was the average? On the other hand, how often does a student ask, What was the probability density function of the exam scores? While the answer to the second question would provide the student with more information about how the class performed, the student may not want all that information. Just knowing the average may be sufficient to tell the student how he or she performed relative to the rest of the class.

DEFINITION 4.1: The *expected value* of a random variable X which has a PDF, $f_X(x)$, is

$$E[X] = \int_{-\infty}^{\infty} x f_X(x)\, dx. \tag{4.1}$$

The terms *average, mean, expectation,* and *first moment* are all alternative names for the concept of expected value and will be used interchangeably throughout the text. Furthermore, an overbar is often used to denote expected value so that the symbol $\overline{X}$ is to be interpreted as meaning the same thing as $E[X]$. Another commonly used notation is to write $\mu_X = E[X]$.

For discrete random variables, the PDF can be written in terms of the probability mass function,

$$f_X(x) = \sum_k P_X(x_k)\delta(x - x_k). \tag{4.2}$$

In this case, using the properties of delta functions, the definition of expected values for discrete random variables reduces to

$$E[X] = \sum_k x_k P_X(x_k). \tag{4.3}$$

Hence, the expected value of a discrete random variable is simply a weighted average of the values that the random variable can take on, weighted by the probability mass of each value. Naturally, the expected value of a random variable exists only if the integral in Equation 4.1 or the series in Equation 4.3 converges. One can dream up many random variables for which the integral or series does not converge and thus their expected values don't exist (or less formally, their expected value is infinite). To gain some physical insight into this concept of expected value, we may think of $f_X(x)$ as a mass distribution of an object along the x-axis; then Equation 4.1 calculates the centroid or center of gravity of the mass.

EXAMPLE 4.1: Consider a random variable that has an exponential PDF given by $f_X(x) = \dfrac{1}{b}\exp\left(-\dfrac{x}{b}\right)u(x)$. Its expected value is calculated as follows:

$$E[X] = \int_0^{\infty} \frac{x}{b}\exp\left(-\frac{x}{b}\right)dx = b\int_0^{\infty} y\exp(-y)dy = b.$$

The last equality in the series is obtained by using integration by parts once. It is seen from this example that the parameter b that appears in this exponential distribution is, in fact, the mean (or expected value) of the random variable.

EXAMPLE 4.2: Next, consider a Poisson random variable whose probability mass function is given by $P_X(k) = \alpha^k e^{-\alpha}/k!$, $k = 0, 1, 2, \ldots$. Its expected value is found in a similar manner.

$$E[X] = \sum_{k=0}^{\infty} k \frac{\alpha^k e^{-\alpha}}{k!} = e^{-\alpha} \sum_{k=1}^{\infty} \frac{\alpha^k}{(k-1)!} = \alpha e^{-\alpha} \sum_{k=1}^{\infty} \frac{\alpha^{k-1}}{(k-1)!}$$

$$= \alpha e^{-\alpha} \sum_{m=0}^{\infty} \frac{\alpha^m}{m!} = \alpha e^{-\alpha} e^{\alpha} = \alpha$$

Once again, we see that the parameter α in the Poisson distribution is equal to the mean.

EXAMPLE 4.3: In the last two examples, we saw that a random variable whose PDF or PMF was described by a single parameter in both cases turned out to be the mean. We work one more example here to show that this does not always have to be the case. Consider a Rayleigh random variable with PDF

$$f_X(x) = \frac{x}{\sigma^2} \exp\left(-\frac{x^2}{2\sigma^2}\right) u(x).$$

The mean is calculated as follows:

$$E[X] = \int_0^{\infty} \frac{x^2}{\sigma^2} \exp\left(-\frac{x^2}{2\sigma^2}\right) dx = \sqrt{2}\sigma \int_0^{\infty} y^{1/2} \exp(-y) dy$$

$$= \sqrt{2}\sigma \Gamma(3/2) = \sqrt{\frac{\pi}{2}} \sigma.$$

The last equality is obtained using the fact that $\Gamma(3/2) = \sqrt{\pi}/2$. Alternatively (for those students not familiar with the properties of gamma functions), one could obtain this result using integration by parts once on the original integral (setting $u = x$ and $dv = (x/\sigma^2) \exp(-x^2/(2\sigma^2))$). In this case, neither the parameter σ nor σ^2 is equal to the expected value of the random variable. However, since the mean is proportional to σ, we could, if we wanted to, rewrite the Rayleigh PDF in terms of its expected value, μ_X, as follows:

$$f_X(x) = \frac{\pi x}{2\mu_X^2} \exp\left(-\frac{\pi x^2}{4\mu_X^2}\right) u(x).$$

4.2 Expected Values of Functions of Random Variables

The concept of expectation can be applied to functions of random variables as well as to the random variable itself. This will allow us to define many other parameters that describe various aspects of a random variable.

DEFINITION 4.2: Given a random variable X with PDF $f_X(x)$, the expected value of a function, $g(X)$, of that random variable is given by

$$E[g(X)] = \int_{-\infty}^{\infty} g(x) f_X(x)\, dx. \tag{4.4}$$

For a discrete random variable, this definition reduces to

$$E[g(X)] = \sum_k g(x_k) P_X(x_k). \tag{4.5}$$

To start with, we demonstrate one extremely useful property of expectations in the following theorem.

THEOREM 4.1: For any constants a and b,

$$E[aX + b] = aE[X] + b. \tag{4.6}$$

Furthermore, for any function $g(x)$ that can be written as a sum of several other functions (i.e., $g(x) = g_1(x) + g_2(x) + \cdots + g_N(x)$),

$$E\left[\sum_{k=1}^{N} g_k(X)\right] = \sum_{k=1}^{N} E[g_k(X)] \tag{4.7}$$

In other words, expectation is a linear operation and the expectation operator can be exchanged (in order) with any other linear operation.

PROOF: The proof follows directly from the linearity of the integration operator.

$$E[aX + b] = \int_{-\infty}^{\infty} (ax + b) f_X(x)\, dx$$

$$= a \int_{-\infty}^{\infty} x f_X(x)\, dx + b = aE[X] + b. \tag{4.8}$$

The second part of the theorem is proved in an identical manner:

$$E\left[\sum_{k=1}^{N} g_k(X)\right] = \int_{-\infty}^{\infty} \left[\sum_{k=1}^{N} g_k(x)\right] f_X(x)\, dx = \sum_{k=1}^{N} \int_{-\infty}^{\infty} g_k(x) f_X(x)\, dx = \sum_{k=1}^{N} E[g_k(X)]. \tag{4.9}$$

∎

Table 4.1   Expected Values of Various Functions of Random Variables

Name	Function of X	Expected value, notation
Mean, average, expected value, expectation, first moment	$g(x) = x$	$\mu_X = \overline{X} = E[X]$
nth moment	$g(x) = x^n$	$\overline{X^n} = E[X^n]$
nth central moment	$g(x) = (x - \mu_X)^n$	$\overline{(X - \mu_X)^n} = E[(X - \mu_X)^n]$
Variance	$g(x) = (x - \mu_X)^2$	$\sigma_X^2 = E[(X - \mu_X)^2]$
Coefficient of skewness	$g(x) = \left(\dfrac{x - \mu_X}{\sigma_X}\right)^3$	$c_s = E\left[\left(\dfrac{X - \mu_X}{\sigma_X}\right)^3\right]$
Coefficient of kurtosis	$g(x) = \left(\dfrac{x - \mu_X}{\sigma_X}\right)^4$	$c_k = E\left[\left(\dfrac{X - \mu_X}{\sigma_X}\right)^4\right]$
Characteristic function	$g(x) = e^{j\omega x}$	$\Phi_X(\omega) = E[e^{j\omega X}]$
Moment generating function	$g(x) = e^{sx}$	$M_X(s) = E[e^{sX}]$
Probability generating function	$g(x) = z^x$	$H_X(z) = E[z^X]$

Different functional forms of $g(X)$ lead to various different parameters that describe the random variable and are known by special names. A few of the more common ones are listed in Table 4.1. In the following sections, selected parameters will be studied in more detail.

4.3 Moments

DEFINITION 4.3: The nth *moment* of a random variable X is defined as

$$E[X^n] = \int_{-\infty}^{\infty} x^n f_X(x)\, dx. \tag{4.10}$$

For a discrete random variable, this definition reduces to

$$E[X^n] = \sum_k x_k^n P_X(x_k). \tag{4.11}$$

The zeroth moment is simply the area under the PDF and hence must be 1 for any random variable. The most commonly used moments are the first and second moments. The first moment is what we previously referred to as the mean, while the second moment is the mean squared value. For some random variables, the second moment might be a more meaningful characterization than the first. For example, suppose X is a sample of a noise waveform. We might expect that the distribution of the noise is symmetric about zero (i.e., just as likely to be positive as negative) and hence the first moment will be zero. So if we are told that X has a zero mean, this merely says that the noise does not have a bias. On the other hand, the second moment of the random noise sample is in some sense a measure of the strength of the noise. In fact, in a later Chapter 10, we will associate the second moment of a noise process with the power in the process. Hence, specifying the second moment can give us some useful physical insight into the noise process.

EXAMPLE 4.4: Consider a discrete random variable that has a binomial distribution. Its probability mass function is given by

$$P_X(k) = \binom{n}{k} p^k (1-p)^{n-k}, \quad k = 0, 1, 2, \ldots, n.$$

The first moment is calculated as follows:

$$E[X] = \sum_{k=0}^{n} k \binom{n}{k} p^k (1-p)^{n-k} = \sum_{k=1}^{n} \frac{kn!}{k!(n-k)!} p^k (1-p)^{n-k}$$

$$= \sum_{k=1}^{n} \frac{n!}{(k-1)!(n-k)!} p^k (1-p)^{n-k} = np \sum_{k=1}^{n} \frac{(n-1)!}{(k-1)!(n-k)!} p^{k-1} (1-p)^{n-k}$$

$$= np \sum_{k=1}^{n} \binom{n-1}{k-1} p^{k-1} (1-p)^{n-k} = np \sum_{m=0}^{n-1} \binom{n-1}{m} p^m (1-p)^{n-1-m}.$$

In this last expression, the summand is a valid probability mass function (i.e., that of a binomial random variable with parameters p and $n-1$) and hence must sum to unity. Therefore, $E[X] = np$. To calculate the second moment, we employ a helpful little trick. Note that we can write $k^2 = k(k-1) + k$. Then

$$E[X^2] = \sum_{k=0}^{n} k^2 \binom{n}{k} p^k (1-p)^{n-k}$$

$$= \sum_{k=0}^{n} k(k-1) \binom{n}{k} p^k (1-p)^{n-k} + \sum_{k=0}^{n} k \binom{n}{k} p^k (1-p)^{n-k}.$$

The second sum is simply the first moment, which has already been calculated. The first sum is evaluated in a manner similar to that used to calculate the mean.

$$\sum_{k=0}^{n} k(k-1)\binom{n}{k}p^k(1-p)^{n-k} = \sum_{k=2}^{n} \frac{n!}{(k-2)!(n-k)!}p^k(1-p)^{n-k}$$

$$= n(n-1)p^2 \sum_{k=2}^{n} \frac{(n-2)!}{(k-2)!(n-k)!}p^{k-2}(1-p)^{n-k}$$

$$= n(n-1)p^2 \sum_{k=2}^{n} \binom{n-2}{k-2}p^{k-2}(1-p)^{n-k}$$

$$= n(n-1)p^2 \sum_{m=0}^{n-2} \binom{n-2}{m}p^m(1-p)^{n-2-m} = n(n-1)p^2$$

Putting these two results together gives

$$E[X^2] = n(n-1)p^2 + np = n^2p^2 + np(1-p).$$

EXAMPLE 4.5: Consider a random variable with a uniform probability density function given as

$$f_X(x) = \begin{cases} 1/a & 0 \le x \le a \\ 0 & \text{otherwise} \end{cases}$$

The mean is given by

$$E[X] = \int_0^a \frac{x}{a}dx = \frac{x^2}{2a}\Big|_0^a = \frac{a}{2},$$

while the second moment is

$$E[X^2] = \int_0^a \frac{x^2}{a}dx = \frac{x^3}{3a}\Big|_0^a = \frac{a^2}{3}.$$

In fact, it is not hard to see that in general, the nth moment of this uniform random variable is given by

$$E[X^n] = \int_0^a \frac{x^n}{a}dx = \frac{x^{n+1}}{(n+1)a}\Big|_0^a = \frac{a^n}{n+1}.$$

4.4 Central Moments

Consider a random variable Y which could be expressed as the sum, $Y = a + X$ of a deterministic (i.e., not random) part a and a random part X. Furthermore, suppose that the random part tends to be very small compared to the fixed part. That is, the random variable Y tends to take small fluctuations about a constant value, a. Such might be the case in a situation where there is a fixed signal corrupted by noise. In this case, we might write $Y^n = (a + X)^n \approx a^n$. In this case, the nth moment of Y would be dominated by the fixed part. That is, it's difficult to characterize the randomness in Y by looking at the moments. To overcome this, we can use the concept of central moments.

DEFINITION 4.4: The nth *central moment* of a random variable X is defined as

$$E[(X - \mu_X)^n] = \int_{-\infty}^{\infty} (x - \mu_X)^n f_X(x)\, dx. \tag{4.12}$$

In this equation, μ_X is the mean (first moment) of the random variable. For discrete random variables, this definition reduces to

$$E[(X - \mu_X)^n] = \sum_k (x_k - \mu_X)^k P_X(x_k). \tag{4.13}$$

With central moments, the mean is subtracted from the variable before the moment is taken in order to remove the bias in the higher moments due to the mean. Note that, like regular moments, the zeroth central moment is $E[(X - \mu_X)^0] = E[1] = 1$. Furthermore, the first central moment is $E[X - \mu_X] = E[X] - \mu_X = \mu_X - \mu_X = 0$. Therefore, the lowest central moment of any real interest is the second central moment. This central moment is given a special name, the *variance*, and we quite often use the notation σ_X^2 to represent the variance of the random variable X. Note that

$$\sigma_X^2 = E[(X - \mu_X)^2] = E[X^2 - 2\mu_X X + \mu_X^2] = E[X^2] - 2\mu_X E[X] + \mu_X^2$$
$$= E[X^2] - \mu_X^2. \tag{4.14}$$

In many cases, the best way to calculate the variance of a random variable is to calculate the first two moments and then form the second moment minus the first moment squared.

EXAMPLE 4.6: For the binomial random variable in Example 4.4, recall that the mean was $E[X] = np$ and the second moment was $E[X^2] = n^2 p^2 + np(1 - p)$. Therefore, the variance is given by $\sigma_X^2 = np(1 - p)$.

Similarly, for the uniform random variable in Example 4.5, $E[X] = a/2$, $E[X^2] = a^2/3$, and hence $\sigma_X^2 = a^2/3 - a^2/4 = a^2/12$. Note that if the moments have not previously been calculated, it may be just as easy to compute the variance directly. In the case of the uniform random variable, once the mean has been calculated, the variance can be found as

$$\sigma_X^2 = \int_0^a (x - a/2)^2 \frac{1}{a} dx = \int_{-a/2}^{a/2} \frac{x^2}{a} dx = \frac{x^3}{3a}\bigg|_{-a/2}^{a/2} = \frac{a^2}{12}.$$

Another common quantity related to the second central moment of a random variable is the *standard deviation*, which is defined as the square root of the variance,

$$\sigma_X = \sqrt{E[(X - \mu_X)^2]}.$$

Both the variance and the standard deviation serve as a measure of the width of the PDF of a random variable. Some of the higher order central moments also have special names, although they are much less frequently used. The third central moment is known as the *skewness* and is a measure of the symmetry of the PDF about the mean. The fourth central moment is called the *kurtosis* and is a measure of the peakedness of a random variable near the mean. Note that not all random variables have finite moments and/or central moments. We give an example of this later for the Cauchy random variable. Some quantities related to these higher order central moments are given in Definition 4.5.

DEFINITION 4.5: The *coefficient of skewness* is

$$c_s = \frac{E[(X - \mu_X)^3]}{\sigma_X^3}. \tag{4.15}$$

This is a dimensionless quantity that is positive if the random variable has a PDF skewed to the right and negative if skewed to the left. The *coefficient of kurtosis* is also dimensionless and is given as

$$c_k = \frac{E[(X - \mu_X)^4]}{\sigma_X^4}. \tag{4.16}$$

The more the PDF is concentrated near its mean, the larger the coefficient of kurtosis. In other words, a random variable with a large coefficient of kurtosis will have a large peak near the mean.

EXAMPLE 4.7: An exponential random variable has a PDF given by

$$f_X(x) = b \exp(-bx)u(x).$$

The mean value of this random variable is $\mu_X = 1/b$. The nth central moment is given by

$$E[(X - \mu_X)^n] = \int_0^\infty (x - 1/b)^n b \exp(-bx) dx$$

$$= \frac{b}{e} \int_{-\frac{1}{b}}^\infty y^n \exp(-by) dy$$

$$= \frac{1}{b^n} \sum_{m=0}^n \frac{n!}{m!} (-1)^m.$$

In the preceding expression, it is understood that $0! = 1$. As expected, it is easily verified from the expression that the zeroth central moment is 1 and the first central moment is 0. Beyond these, the second central moment is $\sigma_X^2 = 1/b^2$, the third central moment is $E[(X - 1/b)^3] = -2/b^3$, and the fourth central moment is $E[(X - 1/b)^4] = 9/b^4$. The coefficients of skewness and kurtosis are given by

$$c_s = \frac{E[(X - \mu_X)^3]}{\sigma_X^3} = \frac{-2/b^3}{1/b^3} = -2,$$

$$c_k = \frac{E[(X - \mu_X)^4]}{\sigma_X^4} = \frac{9/b^4}{1/b^4} = 9.$$

The fact that the coefficient of skewness is negative shows that the exponential PDF is skewed to the left of its mean.

EXAMPLE 4.8: Next consider a Laplace (two-sided exponential) random variable with a PDF given by

$$f_X(x) = \frac{b}{2} \exp(-b|x|).$$

Since this PDF is symmetric about zero, its mean is zero, and hence in this case the central moments are simply the moments

$$E[(X - \mu_X)^n] = E[X^n] = \int_{-\infty}^\infty \frac{bx^n}{2} \exp(-b|x|) dx.$$

Since the two-sided exponential is an even function and x^n is an odd function for any odd n, the integrand is then an odd function for any odd n. The integral of any odd function over an interval symmetric about zero is equal to zero, and hence all odd moments of the Laplace random variable are zero. The even moments can be calculated individually:

$$\sigma_X^2 = E[X^2] = \int_{-\infty}^{\infty} \frac{bx^2}{2} \exp(-b|x|)\,dx = \int_0^{\infty} bx^2 \exp(-bx)\,dx = \frac{2}{b^2},$$

$$E[(X-\mu_X)^4] = E[X^4] = \int_{-\infty}^{\infty} \frac{bx^4}{2} \exp(-b|x|)\,dx = \int_0^{\infty} bx^4 \exp(-bx)\,dx = \frac{24}{b^4}.$$

The coefficient of skewness is zero (since the third central moment is zero) and the coefficient of kurtosis is

$$c_k = \frac{E[(X-\mu_X)^4]}{\sigma_X^4} = \frac{24/b^4}{4/b^4} = 6.$$

Note that the Laplace distribution has a sharp peak near its mean as evidenced by a large coefficient of kurtosis. The fact that the coefficient of skewness is zero is consistent with the fact that the distribution is symmetric about its mean.

EXAMPLE 4.9: It is often the case that the PDF of random variables of practical interest may be too complicated to allow us to compute various moments and other important parameters of the distribution in an analytic fashion. In those cases, we can use a computer to calculate the needed quantities numerically. Take, for example, a Rician random variable whose PDF is given by

$$f_X(x) = \frac{x}{\sigma^2} \exp\left(-\frac{x^2 + a^2}{2\sigma^2}\right) I_o\left(\frac{ax}{\sigma^2}\right) u(x).$$

Suppose we wanted to know the mean of this random variable. This requires us to evaluate:

$$\mu_X = \int_0^{\infty} \frac{x^2}{\sigma^2} \exp\left(-\frac{x^2 + a^2}{2\sigma^2}\right) I_o\left(\frac{ax}{\sigma^2}\right) dx.$$

Note that the parameter σ^2 that shows up in the Rician PDF is *not* the variance of the Rician random variable. While analytical evaluation of this integral looks formidable, given numerical values for a and σ^2, this integral can be evaluated (at least approximately) using standard numerical integration techniques. In order to use the numerical integration routines built into MATLAB, we must first write a function that evaluates the integrand. For evaluating the mean of the Rician PDF, this can be accomplished as follows: (see Appendix D, (D.52) for the analytic expression for the mean.)

```
function pdf=Rician_mean(x,a,sigma)
% Evaluate the integrand needed for calculating the mean of a
% Rician random variable with parameters a and sigma.
pdf=(x./sigma).^2.*exp(-(x.^2+a^2)/(2*sigma^2));
pdf=pdf.*besseli(0,a*x/sigma^2);
```

Once this function is defined, the MATLAB function quad8 can be called upon to perform the numerical integration. For example, if $a = 2$ and $\sigma = 3$, the mean could be calculated as follows:

```
a=2; sigma=3;                    % set parameters
limit1=0; limit2=20;             % set limits of integration.
mean=quad8('Rician_mean',limit1,limit2,[],[],a,sigma);
```

Executing this code produced an answer of $\mu_X = 4.1665$. Note that in order to calculate the mean, the upper limit of the integral should be infinite. However, using limit2=Inf in the preceding code would have led MATLAB to produce a result of NaN ("not a number"). Instead, we must use an upper limit sufficiently large that for the purposes of evaluating the integral it is essentially infinite. This can be done by observing the integrand and seeing at what point the integrand dies off. The reader is encouraged to execute the code in this example using different values of the upper integration limit to see how large the upper limit must be to produce accurate results.

4.5 Conditional Expected Values

Another important concept involving expectation is that of conditional expected value. As specified in Definition 4.6, the conditional expected value of a random variable is a weighted average of the values the random variable can take on, weighted by the conditional PDF of the random variable.

DEFINITION 4.6: The expected value of a random variable X, conditioned on some event A is

$$E[X|A] = \int_{-\infty}^{\infty} x f_{X|A}(x)\, dx. \tag{4.17}$$

For a discrete random variable, this definition reduces to

$$E[X|A] = \sum_k x_k P_{X|A}(x_k). \tag{4.18}$$

Similarly, the conditional expectation of a function, $g(\cdot)$, of a random variable, conditioned on the event A is

$$E[g(X)|A] = \int_{-\infty}^{\infty} g(x) f_{X|A}(x)\, dx \ \text{ or } \ E[g(X)|A] = \sum_k g(x_k) P_{X|A}(x_k), \tag{4.19}$$

depending on whether the random variable is continuous or discrete.

Conditional expected values are computed in the same manner as regular expected values with the PDF or PMF replaced by a conditional PDF or conditional PMF.

EXAMPLE 4.10: Consider a Gaussian random variable of the form

$$f_X(x) = \frac{1}{\sqrt{2\pi}} \exp\left(-\frac{x^2}{2}\right).$$

Suppose the event A is that the random variable X is positive, $A = \{X > 0\}$. Then

$$f_{X|A}(x) = \frac{f_X(x)}{\Pr(X > 0)} u(x) = \sqrt{\frac{2}{\pi}} \exp\left(-\frac{x^2}{2}\right) u(x).$$

The conditional expected value of X given that $X > 0$ is then

$$E[X|X > 0] = \int_{-\infty}^{\infty} x f_{X|X>0}(x)\, dx = \sqrt{\frac{2}{\pi}} \int_0^{\infty} x \exp\left(-\frac{x^2}{2}\right) dx$$

$$= -\sqrt{\frac{2}{\pi}} \exp\left(-\frac{x^2}{2}\right)\Big|_0^{\infty} = \sqrt{\frac{2}{\pi}}.$$

4.6 Transformations of Random Variables

Consider a random variable X with a PDF and CDF given by $f_X(x)$ and $F_X(x)$, respectively. Define a new random variable Y such that $Y = g(X)$ for some function $g(\cdot)$. What is the PDF, $f_Y(y)$ (or CDF), of the new random variable? This problem is often encountered in the study of systems where the PDF for the input random variable X is known and the PDF for the output random variable Y needs to be determined. In such a case, we say that the input random variable has undergone a transformation.

A. Monotonically Increasing Functions   To begin our exploration of transformations of random variables, let's assume that the function is continuous, one-to-one, and monotonically increasing. A typical function of this form is illustrated in Figure 4.1(a). This assumption will be lifted later when we consider more general functions, but for now this simpler case applies. Under these assumptions, the inverse function, $X = g^{-1}(Y)$, exists and is well behaved. In order to obtain the PDF of Y, we first calculate the CDF. Recall that $F_Y(y) = \Pr(Y \leq y)$. Since there is a one-to-one relationship between values of Y and their corresponding values of X, this CDF can be written in terms of X according to

$$F_Y(y) = \Pr(g(X) \leq y) = \Pr(X \leq g^{-1}(y)) = F_X(g^{-1}(y)). \qquad (4.20)$$

Note that this can also be written as

$$F_X(x) = F_Y(g(x)). \qquad (4.21)$$

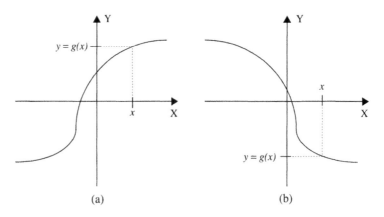

(a)   (b)

Figure 4.1   A monotonic increasing function (a) and a monotonic decreasing function (b).

Differentiating Equation 4.20 with respect to y produces

$$f_Y(y) = f_X(g^{-1}(y))\frac{dg^{-1}(y)}{dy} = f_X(x)\frac{dx}{dy}\bigg|_{x=g^{-1}(y)},$$ (4.22)

while differentiating Equation 4.21 with respect to x gives

$$f_X(x) = f_Y(g(x))\frac{dy}{dx} \Rightarrow f_Y(y) = \frac{f_X(x)}{\frac{dy}{dx}}\bigg|_{x=g^{-1}(y)}$$ (4.23)

Either Equation 4.22 or 4.23 can be used (whichever is more convenient) to compute the PDF of the new random variable.

EXAMPLE 4.11: Suppose X is a Gaussian random variable with mean, μ, and variance, σ^2. A new random variable is formed according to $Y = aX + b$, where $a > 0$ (so that the transformation is monotonically increasing). Since $dy/dx = a$, then applying Equation 4.23 produces

$$f_Y(y) = \frac{1}{a} f_X\left(\frac{y-b}{a}\right).$$

Furthermore, plugging in the Gaussian PDF of X results in

$$f_Y(y) = \frac{1}{a\sqrt{2\pi\sigma^2}}\exp\left(-\frac{\left(\frac{y-b}{a}-\mu\right)^2}{2\sigma^2}\right)$$

$$= \frac{1}{\sqrt{2\pi(a\sigma^2)}}\exp\left(-\frac{(y-(b+a\mu))^2}{2(a\sigma^2)}\right).$$

Note that the PDF of Y still has a Gaussian form. In this example, the transformation did not change the form of the PDF; it merely changed the mean and variance.

EXAMPLE 4.12: Let X be an exponential random variable with $f_X(x) = 2e^{-2x}u(x)$ and let the transformation be $Y = X^3$. Then $dy/dx = 3x^2$ and hence,

$$f_Y(y) = \frac{f_X(x)}{3x^2}\bigg|_{x=\sqrt[3]{y}} = \frac{2}{3}y^{-2/3}\exp(-2y^{1/3})u(y).$$

EXAMPLE 4.13: Suppose a phase angle Θ is uniformly distributed over $(-\pi/2, \pi/2)$, and the transformation is $Y = \sin(\Theta)$. Note that in general, $y = \sin(\theta)$ is not a monotonic transformation, but under the restriction $-\pi/2 < \theta < \pi/2$, this transformation is indeed monotonically increasing. Also note that with this transformation the resulting random variable, Y, must take on values in the range $(-1, 1)$. Therefore, whatever PDF is obtained for Y, it must be understood that the PDF is zero outside $(-1, 1)$. Applying Equation 4.23 gives

$$f_Y(y) = \frac{f_\Theta(\theta)}{\cos(\theta)}\bigg|_{\theta=\sin^{-1}(y)} = \frac{1}{\pi \cos(\sin^{-1}(y))} = \frac{1}{\pi\sqrt{1-y^2}}, \quad -1 < y < 1.$$

This is known as an *arcsine distribution*.

B. Monotonically Decreasing Functions If the transformation is monotonically decreasing rather than increasing, a simple modification to the previous derivations can lead to a similar result. First, note that for monotonic decreasing functions, the event $\{Y \leq y\}$ is equivalent to the event $X \geq g^{-1}(y)$, giving us

$$F_Y(y) = \Pr(Y \leq y) = \Pr(X \geq g^{-1}(y)) = 1 - F_X(g^{-1}(y)). \tag{4.24}$$

Differentiating with respect to y gives

$$f_Y(y) = -f_X(x)\frac{dx}{dy}\bigg|_{x=g^{-1}(y)}. \tag{4.25}$$

Similarly, writing $F_Y(g(x)) = 1 - F_X(x)$ and differentiating with respect to x results in

$$f_Y(y) = -\frac{f_X(x)}{\dfrac{dy}{dx}}\bigg|_{x=g^{-1}(y)}. \tag{4.26}$$

Equations 4.22, 4.23, 4.25, and 4.26 can be consolidated into the following compact form:

$$f_Y(y) = f_X(x)\left|\frac{dx}{dy}\right|\bigg|_{x=g^{-1}(y)} = \frac{f_X(x)}{\left|\dfrac{dy}{dx}\right|}\bigg|_{x=g^{-1}(y)}, \tag{4.27}$$

where now the sign differences have been accounted for by the absolute value operation. This equation is valid for any monotonic function, either monotonic increasing or monotonic decreasing.

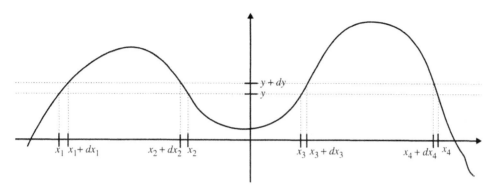

Figure 4.2   A nonmonotonic function; the inverse function may have multiple roots.

C. Nonmonotonic Functions   Finally, we consider a general function that is not necessarily monotonic. Figure 4.2 illustrates one such example. In this case, we cannot associate the event $\{Y \leq y\}$ with events of the form $\{X \leq g^{-1}(y)\}$ or $\{X \geq g^{-1}(y)\}$ because the transformation is not monotonic. To avoid this problem, we calculate the PDF of Y directly, rather than first finding the CDF. Consider an event of the form $\{y \leq Y < y + dy\}$ for an infinitesimal dy. The probability of this event is $\Pr(y \leq Y < y + dy) = f_Y(y)dy$. In order to relate the PDF of Y to the PDF of X, we relate the event $\{y \leq Y < y + dy\}$ to events involving the random variable X. Because the transformation is not monotonic, there may be several values of x that map to the same value of y. These are the roots of the equation $x = g^{-1}(y)$. Let us refer to these roots as $x_1, x_2, \ldots, x_N$. Furthermore, let X^+ be the subset of these roots at which the function $g(x)$ has a positive slope, and similarly let X^- be the remaining roots for which the slope of the function is negative. Then

$$\{y \leq Y < y + dy\} = \left[\bigcup_{i:x_i \in X^+} \{x_i \leq X < x_i + dx_i\} \right] \cup \left[\bigcup_{i:x_i \in X^-} \{x_i + dx_i < X \leq x_i\} \right].$$

(4.28)

Since each of the events on the right-hand side is mutually exclusive, the probability of the union is simply the sum of the probabilities, so that

$$f_Y(y)\, dy = \sum_{x_i \in X^+} \Pr(x_i \leq X < x_i + dx_i) + \sum_{x_i \in X^-} \Pr(x_i + dx_i < X \leq x_i)$$

$$= \sum_{x_i \in X^+} f_X(x_i)dx_i + \sum_{x_i \in X^-} f_X(x_i)(-dx_i)$$

(4.29)

Again, invoking absolute value signs to circumvent the need to have two separate sums and dividing by dy, the following result is obtained:

$$f_Y(y) = \sum_{x_i} f_X(x) \left| \frac{dx}{dy} \right| \Big|_{x_i = g^{-1}(y)}. \tag{4.30}$$

When it is more convenient, the equivalent expression

$$f_Y(y) = \sum_{x_i} \frac{f_X(x)}{\left| \dfrac{dy}{dx} \right|} \Bigg|_{x_i = g^{-1}(y)}, \tag{4.31}$$

can also be used. The following theorem summarizes the general formula for transformations of random variables.

THEOREM 4.2: Given a random variable X with known PDF, $f_X(x)$, and a transformation $Y = g(X)$, the PDF of Y is

$$f_Y(y) = \sum_{x_i} f_X(x) \left| \frac{dx}{dy} \right| \Big|_{x_i = g^{-1}(y)} = \sum_{x_i} \frac{f_X(x)}{\left| \dfrac{dy}{dx} \right|} \Bigg|_{x_i = g^{-1}(y)}, \tag{4.32}$$

where the x_i are the roots of the equation $y = g(x)$. The proof precedes the theorem.

EXAMPLE 4.14: Suppose X is a Gaussian random variable with zero mean and variance σ^2 together with a quadratic transformation, $Y = X^2$. For any positive value of y, $y = x^2$ has two real roots, namely $x = \pm\sqrt{y}$ (for negative values of y, there are no real roots). Application of Equation 4.32 gives

$$f_Y(y) = \left[\frac{f_X(+\sqrt{y})}{2|+\sqrt{y}|} + \frac{f_X(-\sqrt{y})}{2|-\sqrt{y}|} \right] u(y) = \frac{f_X(+\sqrt{y}) + f_X(-\sqrt{y})}{2\sqrt{y}} u(y).$$

For a zero mean Gaussian PDF, $f_X(x)$ is an even function so that $f_X(+\sqrt{y}) = f_X(-\sqrt{y})$. Therefore,

$$f_Y(y) = \frac{1}{\sqrt{y}} f_X(\sqrt{y}) u(y) = \sqrt{\frac{1}{2\pi y \sigma^2}} \exp\left(-\frac{y}{2\sigma^2} \right) u(y).$$

Hence, Y is a gamma random variable.

EXAMPLE 4.15: Suppose the same Gaussian random variable from the previous example is passed through a half-wave rectifier which is described by the input-output relationship

$$y = g(x) = \begin{cases} x & x \geq 0 \\ 0 & x \leq 0 \end{cases}.$$

For $x > 0$, $dy/dx = 1$ so that $f_Y(y) = f_X(y)$. However, when $x < 0$, $dy/dx = 0$, which will create a problem if we try to insert this directly into Equation 4.32. To treat this case, we note that the event $X < 0$ is equivalent to the event $Y = 0$; hence $\Pr(Y = 0) = \Pr(X < 0)$. Since the input Gaussian PDF is symmetric about zero, $\Pr(X < 0) = 1/2$. Basically, the random variable Y is a mixed random variable. It has a continuous part over the region $y > 0$ and a discrete part at $y = 0$. Using a delta function, we can write the PDF of Y as

$$f_Y(y) = \frac{1}{\sqrt{2\pi\sigma^2}} \exp\left(-\frac{y^2}{2\sigma^2}\right) u(y) + \frac{1}{2}\delta(y).$$

Example 4.15 illustrates how to deal with transformations that are flat over some interval of nonzero length. In general, suppose the transformation $y = g(x)$ is such that $g(x) = y_o$ for any x in the interval $x_1 \leq x \leq x_2$. Then the PDF of Y will include a discrete component (a delta function) of height $\Pr(Y = y_o) = \Pr(x_1 \leq x \leq x_2)$ at the point $y = y_o$. One often encounters transformations that have several different flat regions. One such "staircase" function is shown in Figure 4.3. Here, a random variable X that may be continuous will be converted into a discrete random variable. The classical example of this is analog-to-digital conversion of signals. Suppose the transformation is of a general staircase form,

$$y = \begin{cases} y_0 & x < x_1 \\ y_i & x_i \leq x < x_{i+1}, \quad i = 1, 2, \ldots, N-1 \\ y_N & x \geq x_N \end{cases}. \tag{4.33}$$

Then Y will be a discrete random variable whose PMF is

$$P(Y = y_i) = \begin{cases} \Pr(X < x_1) & i = 0 \\ \Pr(x_i \leq x < x_{i+1}) & i = 1, 2, \ldots, N-1 \\ \Pr(X \geq x_N) & i = N \end{cases}. \tag{4.34}$$

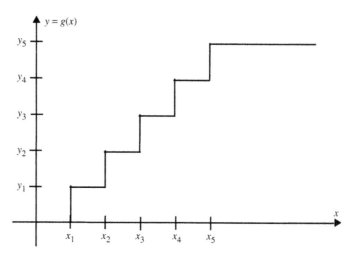

Figure 4.3 A staircase (quantizer) transformation: a continuous random variable will be converted into a discrete random variable.

EXAMPLE 4.16: Suppose X is an exponential random variable with a PDF $f_X(x) = \exp(-x)u(x)$ and we form a new random variable Y by rounding X down to the nearest integer. That is,

$$Y = g(X) = \text{floor}(X) = k, \quad k \le X < k + 1.$$

Then, the PMF of Y is

$$P(Y = k) = \Pr(k \le X < k + 1) = \int_k^{k+1} e^{-x} dx = e^{-k} - e^{-(k+1)}$$

$$= e^{-k}(1 - 1/e), \quad k = 0, 1, 2, \ldots.$$

Hence, quantization of an exponential random variable produces a geometric random variable.

EXAMPLE 4.17: Let X be a random variable uniformly distributed over $(-a/2, a/2)$. Accordingly, its PDF is of the form

$$f_X(x) = \frac{1}{a}\left(u\left(x + \frac{a}{2}\right) - u\left(x - \frac{a}{2}\right)\right).$$

A new random variable is to be formed according to the square law transformation $Y = X^2$. By applying the theory developed in this section you should be able to demonstrate that the new random variable has a PDF given by

$$f_Y(y) = \frac{1}{a\sqrt{y}}(u(y) - u(y - a^2/4)).$$

Using MATLAB, we create a large number of samples of the uniform random variable, pass these samples through the square law transformation, and then construct a histogram of the resulting probability densities. The MATLAB code to do so follows and the results of running this code are shown in Figure 4.4. These are then compared with the analytically determined PDF.

```
clear
N=10000;
a=5; ymax=a^2/4;            % Set parameters.
x=a*(rand(1,N)-0.5);        % Generate uniform RVs.
y=x.^2;                     % Square law transformation.
bw=0.25;                    % Bin width.
bins=[bw/2:bw:ymax];        % Histogram bins.
[yvals,xvals]=hist(y,bins); % Compute histogram values.
pdf_est=yvals/(N*bw);       % Convert to probability densities.
bar(xvals,pdf_est)          % Plot histogram.
```

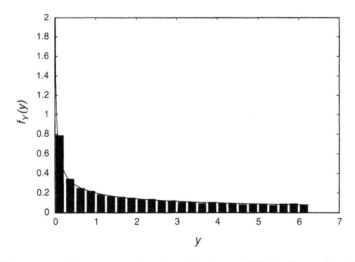

Figure 4.4 Comparison of estimated and true PDF for Example 4.17.

```
% Compare true PDF with histogram.
y=[0.01:0.01:ymax];
pdf=1./(a*sqrt(y));
hold on
plot(y,pdf)
xlabel('y'); ylabel('f_Y(y)')
hold off
```

4.7 Characteristic Functions

In this section we introduce the concept of a characteristic function. The characteristic function of a random variable is closely related to the Fourier transform of the PDF of that random variable. Thus, the characteristic function provides a sort of "frequency domain" representation of a random variable, although in this context there is no connection between our frequency variable ω and any physical frequency. In studies of deterministic signals, it was found that the use of Fourier transforms greatly simplified many problems, especially those involving convolutions. We will see in future chapters the need for performing convolution operations on PDFs of random variables and hence frequency domain tools will become quite useful. Furthermore, we will find that characteristic functions have many other uses. For example, the characteristic function is quite useful for finding moments of a random variable. In addition to the characteristic function, two other related functions, namely, the moment-generating function (analogous to the Laplace transform) and the probability-generating function (analogous to the z-transform), will also be studied in the following sections.

DEFINITION 4.7: The *characteristic function* of a random variable, X, is given by

$$\Phi_X(\omega) = E[e^{j\omega X}] = \int_{-\infty}^{\infty} e^{j\omega x} f_X(x)\, dx. \tag{4.35}$$

Note the similarity between this integral and the Fourier transform. In most of the electrical engineering literature, the Fourier transform of the function $f_X(x)$ would be $\Phi(-\omega)$. Given this relationship between the PDF and the characteristic function, it should be clear that one can get the PDF of a random variable from its characteristic function through an inverse Fourier transform operation:

$$f_X(x) = \frac{1}{2\pi} \int_{-\infty}^{\infty} e^{-j\omega x} \Phi_X(\omega)\, d\omega. \tag{4.36}$$

The characteristic functions associated with various random variables can be easily found using tables of commonly used Fourier transforms, but one must be careful since the Fourier integral used in Equation 4.35 may be different from the definition used to generate common tables of Fourier transforms. In addition, various properties of Fourier transforms can also be used to help calculate characteristic functions as shown in the following example.

EXAMPLE 4.18: An exponential random variable has a PDF given by $f_X(x) = \exp(-x)u(x)$. Its characteristic function is found to be

$$\Phi_X(\omega) = \int_{-\infty}^{\infty} e^{j\omega x} f_X(x)\, dx = \int_0^{\infty} e^{j\omega x} e^{-x} dx = -\frac{e^{-(1-j\omega)x}}{1-j\omega}\Big|_0^{\infty} = \frac{1}{1-j\omega}.$$

This result assumes that ω is a real quantity. Now suppose another random variable Y has a PDF given by $f_Y(y) = a\exp(-ay)u(y)$. Note that $f_Y(y) = a f_X(ay)$, thus using the scaling property of Fourier transforms, the characteristic function associated with the random variable Y is given by

$$\Phi_Y(\omega) = a\frac{1}{|a|}\Phi_X\left(\frac{\omega}{a}\right) = \frac{1}{1-j\omega/a} = \frac{a}{a-j\omega},$$

assuming a is a positive constant (which it must be for Y to have a valid PDF). Finally, suppose that Z has a PDF given by $f_Z(z) = a\exp(-a(z-b))u(z-b)$. Since $f_Z(z) = f_Y(z-b)$, the shifting property of Fourier transforms can be used to help find the characteristic function associated with the random variable Z:

$$\Phi_Z(\omega) = \Phi_Y(\omega)e^{-j\omega b} = \frac{ae^{-j\omega b}}{a-j\omega}.$$

The next example demonstrates that the characteristic function can also be computed for discrete random variables. In Section 4.8, the probability-generating function will be introduced and is preferred by some when dealing with discrete random variables.

EXAMPLE 4.19: A binomial random variable has a PDF that can be expressed as

$$f_X(x) = \sum_{k=0}^{n} \binom{n}{k} p^k (1-p)^{n-k} \delta(x-k).$$

Its characteristic function is computed as follows:

$$\Phi_X(\omega) = \int_{-\infty}^{\infty} e^{j\omega x} \left(\sum_{k=0}^{n} \binom{n}{k} p^k (1-p)^{n-k} \delta(x-k) \right) dx$$

$$= \sum_{k=0}^{n} \binom{n}{k} p^k (1-p)^{n-k} \int_{-\infty}^{\infty} \delta(x-k) e^{j\omega x} dx = \sum_{k=0}^{n} \binom{n}{k} p^k (1-p)^{n-k} e^{j\omega k}$$

$$= \sum_{k=0}^{n} \binom{n}{k} (pe^{j\omega})^k (1-p)^{n-k} = (1-p+pe^{j\omega})^n.$$

Since the Gaussian random variable plays such an important role in so many studies, we derive its characteristic function in Example 4.20. We recommend that the student commit the result of this example to memory. The techniques used to arrive at this result are also important and should be carefully studied and understood.

EXAMPLE 4.20: For a standard normal random variable, the characteristic function can be found as follows:

$$\Phi_X(\omega) = \int_{-\infty}^{\infty} \frac{1}{\sqrt{2\pi}} e^{-\frac{x^2}{2}} e^{j\omega x} dx = \int_{-\infty}^{\infty} \frac{1}{\sqrt{2\pi}} \exp\left(-\frac{(x^2 - 2j\omega x)}{2} \right) dx.$$

To evaluate this integral, we complete the square in the exponent.

$$\Phi_X(\omega) = \exp\left(-\frac{\omega^2}{2} \right) \int_{-\infty}^{\infty} \frac{1}{\sqrt{2\pi}} \exp\left(-\frac{(x^2 - 2j\omega x - \omega^2)}{2} \right) dx$$

$$= \exp\left(-\frac{\omega^2}{2} \right) \int_{-\infty}^{\infty} \frac{1}{\sqrt{2\pi}} \exp\left(-\frac{(x - j\omega)^2}{2} \right) dx$$

The integrand in this expression looks like the properly normalized PDF of a Gaussian random variable, and since the integral is over all values of x, the integral must be unity. However, close examination of the integrand reveals that the "mean" of this Gaussian integrand is complex. It is left to the student to rigorously verify that this integral still evaluates to unity even though the integrand is not truly a Gaussian PDF (since it is a complex function and hence not a PDF at all). The resulting characteristic function is then

$$\Phi_X(\omega) = \exp\left(-\frac{\omega^2}{2}\right).$$

For a Gaussian random variable whose mean is not zero or whose standard deviation is not unity (or both), the shifting and scaling properties of Fourier transforms can be used to show that

$$f_X(x) = \frac{1}{\sqrt{2\pi\sigma^2}}e^{-\frac{(x-\mu)^2}{2\sigma^2}} \leftrightarrow \Phi_X(\omega) = \exp\left(j\mu\omega - \frac{\omega^2\sigma^2}{2}\right).$$

THEOREM 4.3: For any random variable whose characteristic function is differentiable at $\omega = 0$,

$$E[X] = -j\frac{d}{d\omega}\Phi_X(\omega)\Big|_{\omega=0}. \tag{4.37}$$

PROOF: The proof follows directly from the fact that the expectation and differentiation operations are both linear and consequently the order of these operations can be exchanged.

$$\frac{d}{d\omega}\Phi_X(\omega) = \frac{d}{d\omega}E\left[e^{j\omega X}\right] = E\left[\frac{d}{d\omega}e^{j\omega X}\right] = E\left[jXe^{j\omega X}\right] = jE\left[Xe^{j\omega X}\right].$$

Multiplying both sides by $-j$ and evaluating at $\omega = 0$ produces the desired result. ■

Theorem 4.3 demonstrates a very powerful use of the characteristic function. Once the characteristic function of a random variable has been found, it is generally a very straightforward thing to produce the mean of the random variable. Furthermore, by taking the kth derivative of the characteristic function and evaluating at $\omega = 0$, an expression proportional to the kth moment of the random variable is produced. In particular,

$$E[X^k] = (-j)^k\frac{d^k}{d\omega^k}\Phi_X(\omega)\Big|_{\omega=0}. \tag{4.38}$$

Hence, the characteristic function represents a convenient tool to easily determine the moments of a random variable.

EXAMPLE 4.21: Consider the exponential random variable of Example 4.18 where $f_Y(y) = a \exp(-ay)u(y)$. The characteristic function was found to be

$$\Phi_Y(\omega) = \frac{a}{a - j\omega}.$$

The derivative of the characteristic function is

$$\frac{d}{d\omega}\Phi_Y(\omega) = \frac{ja}{(a - j\omega)^2},$$

and thus the first moment of Y is

$$E[Y] = -j\frac{d}{d\omega}\Phi_Y(\omega)\bigg|_{\omega=0} = \frac{a}{(a - j\omega)^2}\bigg|_{\omega=0} = \frac{1}{a}.$$

For this example, it is not difficult to show that the kth derivative of the characteristic function is

$$\frac{d^k}{d\omega^k}\Phi_Y(\omega) = \frac{j^k k! a}{(a - j\omega)^{k+1}},$$

and from this, the kth moment of the random variable is found to be

$$E[Y^k] = (-j)^k \frac{d^k}{d\omega^k}\Phi_Y(\omega)\bigg|_{\omega=0} = \frac{k! a}{(a - j\omega)^{k+1}}\bigg|_{\omega=0} = \frac{k!}{a^k}.$$

For random variables that have a more complicated characteristic function, evaluating the kth derivative in general may not be an easy task. However, Equation 4.38 only calls for the kth derivative evaluated at a single point ($\omega = 0$), which can be extracted from the Taylor series expansion of the characteristic function. To see this, note that from Taylor's Theorem, the characteristic function can be expanded in a power series as

$$\Phi_X(\omega) = \sum_{k=0}^{\infty} \frac{1}{k!}\left(\frac{d^k}{d\omega^k}\Phi_X(\omega)\bigg|_{\omega=0}\right)\omega^k. \tag{4.39}$$

If one can obtain a power series expansion of the characteristic function, then the required derivatives are proportional to the coefficients of the power series. Specifically, suppose an expansion of the form

$$\Phi_X(\omega) = \sum_{k=0}^{\infty} \phi_k \omega^k \tag{4.40}$$

is obtained. Then the derivatives of the characteristic function are given by

$$\frac{d^k}{d\omega^k}\Phi_X(\omega)\bigg|_{\omega=0} = k!\phi_k.$$
(4.41)

The moments of the random variable are then given by

$$E[X^k] = (-j)^k k!\phi_k.$$
(4.42)

This procedure is illustrated using a Gaussian random variable in the next example.

EXAMPLE 4.22: Consider a Gaussian random variable with a mean of $\mu = 0$ and variance σ^2. Using the result of Example 4.20, the characteristic function is $\Phi_X(\omega) = \exp(-\omega^2\sigma^2/2)$. Using the well-known Taylor series expansion of the exponential function, the characteristic function is expressed as

$$\Phi_X(\omega) = \sum_{n=0}^{\infty} \frac{(-\omega^2\sigma^2/2)^n}{n!} = \sum_{n=0}^{\infty} \frac{(-1)^n\sigma^{2n}}{2^n n!}\omega^{2n}.$$

The coefficients of the general power series as expressed in Equation 4.40 are given by

$$\phi_k = \begin{cases} \dfrac{j^k(\sigma/\sqrt{2})^k}{(k/2)!} & k \text{ even} \\ 0 & k \text{ odd} \end{cases}.$$

Hence the moments of the zero-mean Gaussian random variable are

$$E[X^k] = \begin{cases} \dfrac{k!}{(k/2)!}\left(\dfrac{\sigma}{\sqrt{2}}\right)^k & k \text{ even} \\ 0 & k \text{ odd} \end{cases}.$$

As expected, $E[X^0] = 1$, $E[X] = 0$ (since it was specified that $\mu = 0$), and $E[X^2] = \sigma^2$ (since in the case of zero-mean variables, the second moment and variance are one and the same). Now, we also see that $E[X^3] = 0$ (i.e., all odd moments are equal to zero), $E[X^4] = 3\sigma^4$, $E[X^6] = 15\sigma^6$, and so on. We can also conclude from this that for Gaussian random variables, the coefficient of skewness is $c_s = 0$, while the coefficient of kurtosis is $c_k = 3$.

In many cases of interest, the characteristic function has an exponential form. The Gaussian random variable is a typical example. In such cases, it is convenient to deal with the natural logarithm of the characteristic function.

DEFINITION 4.8: In general, we can write a series expansion of $\ln[\Phi_X(\omega)]$ as

$$\ln[\Phi_X(\omega)] = \sum_{n=1}^{\infty} \lambda_n \frac{(j\omega)^n}{n!}. \tag{4.43}$$

where the coefficients, λ_n, are called the *cumulants* and are given as

$$\lambda_n = \frac{d^n}{d(j\omega)^n} \{\ln[\Phi_X(\omega)]\}\Big|_{\omega=0}, \quad n = 1,2,3,\ldots. \tag{4.44}$$

The cumulants are related to the moments of the random variable. By taking the derivatives specified in Equation 4.44 we obtain

$$\lambda_1 = \mu_X, \tag{4.45}$$

$$\lambda_2 = E[X^2] - \mu_X^2 = \sigma_X^2, \tag{4.46}$$

$$\lambda_3 = E[X^3] - 3\mu_X E[X^2] + 2\mu_X^3 = E[(X - \mu_X)^3]. \tag{4.47}$$

Thus, λ_1 is the mean, λ_2 is the second central moment (or the variance), and λ_3 is the third central moment. However, higher order cumulants are not as simply related to the central moments.

4.8 Probability Generating Functions

In the world of signal analysis, we often use Fourier transforms to describe continuous time signals, but when we deal with discrete time signals, it is common to use a z-transform instead. In the same way, the characteristic function is a useful tool for working with continuous random variables, but when discrete random variables are concerned, it is often more convenient to use a device similar to the z-transform which is known as the probability generating function.

DEFINITION 4.9: For a discrete random variable with a probability mass function, $P_X(k)$, defined on the nonnegative integers[1], $k = 0, 1, 2, \ldots$, the *probability*

---

[1]Note that this definition assumes that the discrete random variable, X, is defined on nonnegative integer values, k. One could also define a probability generating function based on a bilateral z-transform that would allow for random variables that can take on negative integer values as well. However, since this is less common, we do not consider it further here.

generating function, $H_X(z)$, is defined as

$$H_X(z) = \sum_{k=0}^{\infty} P_X(k)z^k. \tag{4.48}$$

Note the similarity between the probability generating function and the unilateral z-transform of the probability mass function.

Since the PMF is seen as the coefficients of the Taylor series expansion of $H_X(z)$, it should be apparent that the PMF can be obtained from the probability generating function through

$$P_X(k) = \frac{1}{k!} \frac{d^k}{dz^k} H_X(z)\Big|_{z=0}. \tag{4.49}$$

The derivatives of the probability generating function evaluated at zero return the PMF and not the moments as with the characteristic function. However, the moments of the random variable can be obtained from the derivatives of the probability generating function at $z = 1$.

THEOREM 4.4: The mean of a discrete random variable can be found from its probability generating function according to

$$E[X] = \frac{d}{dz} H_X(z)\Big|_{z=1}. \tag{4.50}$$

Furthermore, the higher order derivatives of the probability generating function evaluated at $z = 1$ lead to quantities known as the *factorial moments*,

$$h_k = \frac{d^k}{dz^k} H_X(z)\Big|_{z=1} = E[X(X-1)(X-2)\cdots(X-k+1)]. \tag{4.51}$$

PROOF: The result follows directly from differentiating Equation 4.48. The details are left to the reader. ∎

It is a little unfortunate that these derivatives don't produce the moments directly, but the moments can be calculated from the factorial moments. For example,

$$h_2 = E[X(X-1)] = E[X^2] - E[X] = E[X^2] - h_1,$$

$$\Rightarrow E[X^2] = h_2 + h_1. \tag{4.52}$$

Hence, the second moment is simply the sum of the first two factorial moments. If we were interested in the variance as well, we would obtain

$$\sigma_X^2 = E[X^2] - \mu_X^2 = h_2 + h_1 - h_1^2. \tag{4.53}$$

EXAMPLE 4.23: Consider the binomial random variable of Example 4.4 whose PMF is

$$P_X(k) = \binom{n}{k} p^k (1-p)^{n-k}, \quad k = 0, 1, 2, \ldots, n.$$

The corresponding probability generating function is

$$H_X(z) = \sum_{k=0}^{n} \binom{n}{k} p^k (1-p)^{n-k} z^k = \sum_{k=0}^{n} \binom{n}{k} (pz)^k (1-p)^{n-k} = (1-p+pz)^n.$$

Evaluating the first few derivatives at $z = 1$ produces

$$h_1 = \frac{d}{dz} H_X(z)\bigg|_{z=1} = np(1-p+pz)^{n-1}\bigg|_{z=1} = np,$$

$$h_2 = \frac{d^2}{dz^2} H_X(z)\bigg|_{z=1} = n(n-1)p^2(1-p+pz)^{n-2}\bigg|_{z=1} = n(n-1)p^2.$$

From these factorial moments, we calculate the mean, second moment, and variance of a binomial random variable as

$$\mu_X = h_1 = np, \ E[X^2] = h_1 + h_2 = (np)^2 + np(1-p), \ \sigma_X^2 = h_2 + h_1 - h_1^2$$
$$= np(1-p).$$

In order to gain an appreciation for the power of these "frequency domain" tools, compare the amount of work used to calculate the mean and variance of the binomial random variable using the probability generating function in Example 4.23 with the direct method used in Example 4.4.

As was the case with the characteristic function, we can compute higher order factorial moments without having to take many derivatives by expanding the probability generating function into a Taylor series. In this case, the Taylor series must be about the point $z = 1$.

$$H_X(z) = \sum_{k=0}^{\infty} \frac{1}{k!} \left(\frac{d^k}{dz^k} H_X(z) \bigg|_{z=1} \right) (z-1)^k = \sum_{k=0}^{\infty} \frac{1}{k!} h_k (z-1)^k \qquad (4.54)$$

Once this series is obtained, one can easily identify all of the factorial moments. This is illustrated using a geometric random variable in Example 4.24.

EXAMPLE 4.24: A geometric random variable has a PMF given by $P_X(k) = (1-p)p^k, k = 0, 1, 2, \ldots$. The probability generating function is found to be

$$H_X(z) = \sum_{k=0}^{\infty} (1-p)(pz)^k = \frac{1-p}{1-pz}.$$

In order to facilitate forming a Taylor series expansion of this function about the point $z = 1$, it is written explicitly as a function of $z - 1$. From there, the power series expansion is fairly simple:

$$H_X(z) = \frac{1-p}{1-p-p(z-1)} = \frac{1}{1-\dfrac{p}{1-p}(z-1)} = \sum_{k=0}^{\infty} \left(\frac{p}{1-p} \right)^k (z-1)^k.$$

Comparing the coefficients of this series with the coefficients given in Equation 4.54 leads to immediate identification of the factorial moments,

$$h_k = \frac{k! p^k}{(1-p)^k}.$$

4.9 Moment Generating Functions

In many problems, the random quantities we are studying are often inherently nonnegative. Examples include the frequency of a random signal, the time between arrivals of successive customers in a queueing system, or the number of points scored by your favorite football team. The resulting PDFs of these quantities are naturally one-sided. For such one-sided waveforms, it is common to use Laplace transforms as a frequency domain tool. The moment generating function is the equivalent tool for studying random variables.

DEFINITION 4.10: The *moment generating function*, $M_X(u)$, of a nonnegative[2] random variable, X, is

$$M_X(u) = E[e^{uX}] = \int_0^\infty f_X(x)e^{ux}\,dx. \tag{4.55}$$

Note the similarity between the moment generating function and the Laplace transform of the PDF.

The PDF can in principle be retrieved from the moment generating function through an operation similar to an inverse Laplace transform,

$$f_X(x) = \frac{1}{2\pi j}\int_{c-j\infty}^{c+j\infty} M_X(u)e^{-ux}\,du. \tag{4.56}$$

Because the sign in the exponential term in the integral in Equation 4.55 is the opposite of the traditional Laplace transform, the contour of integration (the so-called Bromwich contour) in the integral specified in Equation 4.56 must now be placed to the left of all poles of the moment generating function. As with the characteristic function, the moments of the random variable can be found from the derivatives of the moment generating function (hence, its name) according to

$$E[X^k] = \frac{d^k}{du^k}M_X(u)\bigg|_{u=0}. \tag{4.57}$$

It is also noted that if the moment generating function is expanded in a power series of the form

$$M_X(u) = \sum_{k=0}^\infty m_k u^k, \tag{4.58}$$

then the moments of the random variable are given by $E[X^k] = k!m_k$.

EXAMPLE 4.25: Consider an Erlang random variable with a PDF of the form

$$f_X(x) = \frac{x^{n-1}\exp(-x)u(x)}{(n-1)!}.$$

---

[2]One may also define a moment generating function for random variables that are not necessarily nonnegative. In that case, a two-sided Laplace transform would be appropriate. This would be identical to the characteristic function with the association $u = j\omega$.

The moment generating function is calculated according to

$$M_X(u) = \int_0^\infty f_X(x)e^{ux}dx = \int_0^\infty \frac{x^{n-1}\exp(-(1-u)x)}{(n-1)!}dx.$$

To evaluate this function, we note that the integral looks like the Laplace transform of the function $x^{n-1}/(n-1)!$ evaluated at $s = 1 - u$. Using standard tables of Laplace transforms (or using integration by parts several times), we get

$$M_X(u) = \frac{1}{(1-u)^n}.$$

The first two moments are then found as follows:

$$E[X] = \frac{d}{du}(1-u)^{-n}\bigg|_{u=0} = n(1-u)^{-(n+1)}\bigg|_{u=0} = n,$$

$$E[X^2] = \frac{d^2}{du^2}(1-u)^{-n}\bigg|_{u=0} = n(n+1)(1-u)^{-(n+2)}\bigg|_{u=0} = n(n+1).$$

From this, we could also infer that the variance is $\sigma_X^2 = n(n+1)-n^2 = n$. If we wanted a general expression for the kth moment, it is not hard to see that

$$E[X^k] = \frac{d^k}{du^k}(1-u)^{-n}\bigg|_{u=0} = n(n+1)\cdots(n+k-1) = \frac{(n+k-1)!}{(n-1)!}.$$

4.10 Evaluating Tail Probabilities

A common problem encountered in a variety of applications is the need to compute the probability that a random variable exceeds a threshold, $\Pr(X > x_o)$. Alternatively, we might want to know, $\Pr(|X - \mu_X| > x_o)$. These quantities are referred to as *tail probabilities*. That is, we are asking, what is the probability that the random variable takes on a value that is in the tail of the distribution? While this can be found directly from the CDF of the random variable, quite often the CDF may be difficult or even impossible to find. In those cases, one can always resort to numerical integration of the PDF. However, this involves a numerical integration over a semi-infinite region, which in some cases may be problematic. Then, too, in some situations we might not even have the PDF of the random variable, but rather the random variable may be described in some other fashion. For example, we may know only the mean or the mean and variance, or the random variable may be described by one of the frequency domain functions described in the previous sections.

Obviously, if we are given only partial information about the random variable, we would not expect to be able to perfectly evaluate the tail probabilities, but we can obtain bounds on these probabilities. In this section, we present several techniques for obtaining various bounds on tail probabilities based on different information about the random variable. We then conclude the section by showing how to exactly evaluate the tail probabilities directly from one of the frequency domain descriptions of the random variables.

THEOREM 4.5 (Markov's Inequality): Suppose that X is a nonnegative random variable (i.e., one whose PDF is nonzero only over the range $[0, \infty)$). Then

$$\Pr(X \geq x_o) \leq \frac{E[X]}{x_o}. \tag{4.59}$$

PROOF: For nonnegative random variables, the expected value is

$$E[X] = \int_0^\infty x f_X(x) \, dx = \int_0^{x_o} x f_X(x) \, dx + \int_{x_o}^\infty x f_X(x) \, dx \geq \int_{x_o}^\infty x f_X(x) \, dx$$

$$\geq x_o \int_{x_o}^\infty f_X(x) \, dx. \tag{4.60}$$

Dividing both sides by x_o gives the desired result.

Markov's inequality provides a bound on the tail probability. The bound requires only knowledge of the mean of the random variable. Because the bound uses such limited information, it has the potential of being very loose. In fact, if $x_o < E[X]$, then the Markov inequality states that $\Pr(X \geq x_o)$ is bounded by a number that is greater than 1. While this is true, in this case the Markov inequality gives us no useful information. Even in less extreme cases, the result can still be very loose as shown by the next example.   ∎

EXAMPLE 4.26: Suppose the average life span of a person was 75 years. The probability of a human living to be 110 years would then be bounded by

$$\Pr(X \geq 110) \leq \frac{75}{110} = 0.6818.$$

Of course, we know that in fact very few people live to be 110 years old, and hence this bound is almost useless to us.

If we know more about the random variable than just its mean, we can obtain a more precise estimate of its tail probability. In Example 4.26, we know that the bound given by the Markov inequality is ridiculously loose because we know something about the variability of the human life span. The next result allows us to use the variance as well as the mean of a random variable to form a different bound on the tail probability.

THEOREM 4.6 (Chebyshev's Inequality): Suppose that X is a random variable with mean μ_X and variance σ_X^2. The probability that the random variable takes on a value that is removed from the mean by more than x_o is given by

$$\Pr(|X - \mu_X| \geq x_o) \leq \frac{\sigma_X^2}{x_o^2}. \tag{4.61}$$

PROOF: Chebyshev's inequality is a direct result of Markov's inequality. First, note that the event $\{|X - \mu_X| \geq x_o\}$ is equivalent to the event $\{(X - \mu_X)^2 \geq x_o^2\}$. Applying Markov's inequality to the later event results in

$$\Pr((X - \mu_X)^2 \geq x_o^2) \leq \frac{E[(X - \mu_X)^2]}{x_o^2} = \frac{\sigma_X^2}{x_o^2}. \tag{4.62}$$

Note that Chebyshev's inequality gives a bound on the two-sided tail probability, whereas the Markov inequality applies to the one-sided tail probability. Also, the Chebyshev inequality can be applied to any random variable, not just to those that are nonnegative. ∎

EXAMPLE 4.27: Continuing the previous example, suppose that in addition to a mean of 75 years, the human life span had a standard deviation of 5 years. In this case,

$$\Pr(X \geq 110) \leq \Pr(X \geq 110) + \Pr(X \leq 40) = \Pr(|X - 75| \geq 35).$$

Now the Chebyshev inequality can be applied to give

$$\Pr(|X - 75| \geq 35) \leq \left(\frac{5}{35}\right)^2 = \frac{1}{49}.$$

While this result may still be quite loose, by using the extra piece of information provided by the variance, a better bound is obtained.

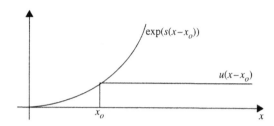

Figure 4.5   The unit step function and an exponential upper bound.

THEOREM 4.7 (Chernoff Bound): Suppose X is a random variable whose moment generating function is $M_X(s)$. Then

$$\Pr(X \geq x_0) \leq \min_{s \geq 0} e^{-s x_0} M_X(s). \tag{4.63}$$

PROOF: First, note that

$$\Pr(X \geq x_0) = \int_{x_0}^{\infty} f_X(x)\, dx = \int_{-\infty}^{\infty} f_X(x) u(x - x_0)\, dx. \tag{4.64}$$

Next, upper bound the unit step function in the integrand by an exponential function of the form $u(x - x_0) \leq \exp(s(x - x_0))$. This bound is illustrated in Figure 4.5. Note that the bound is valid for any real $s \geq 0$. The tail probability is then upper bounded by

$$\Pr(X \geq x_0) \leq e^{-s x_0} \int_{-\infty}^{\infty} f_X(x) e^{s x} dx = e^{-s x_0} M_X(s). \tag{4.65}$$

Since this bound is valid for any $s \geq 0$, it can be tightened by finding the value of s that minimizes the right-hand side. In this expression, a two-sided Laplace transform must be used to obtain the moment generating function if the random variable is not nonnegative (see footnote 2 associated with Definition 4.10).   ■

EXAMPLE 4.28: Consider a standard normal random variable whose moment generating function is given by $M_X(u) = \exp(u^2/2)$ (see result of Example 4.20, where the characteristic function is found, and replace ω with $-ju$). The tail probability, $\Pr(X \geq x_0)$, in this case is simply the

Q-function, $Q(x_o)$. According to Equation 4.65, this tail probability can be bounded by

$$Q(x_o) \leq \exp\left(-ux_o + \frac{u^2}{2}\right)$$

for any $u \geq 0$. Minimizing with respect to u, we get

$$\frac{d}{du} \exp\left(-ux_o + \frac{u^2}{2}\right) = (-x_o + u) \exp\left(-ux_o + \frac{u^2}{2}\right) = 0 \Rightarrow u = x_o.$$

Hence, the Chernoff bound on the tail probability for a standard normal random variable is

$$Q(x_o) \leq \exp\left(-\frac{x_o^2}{2}\right).$$

The result of this example provides a convenient upper bound on the Q-function.

Evaluating the Chernoff bound requires knowledge of the moment generating function of the random variable. This information is sufficient to calculate the tail probability exactly since, in theory, one can obtain the PDF from the moment generating function, and from there the exact tail probability can be obtained. However, in cases where the moment generating function is of a complicated analytical form, determining the PDF may be exceedingly difficult. Indeed, in some cases, it may not be possible to express the tail probability in closed form (as with the Gaussian random variable). In these cases, the Chernoff bound will often provide an analytically tractable expression that can give a crude bound on the tail probability. If a precise expression for the tail probability is required, the result of Theorem 4.8 will show how this can be obtained directly from the moment generating function (without having to explicitly find the PDF).

THEOREM 4.8: For a random variable, X, with a moment generating function, $M_X(u)$, an exact expression for the tail probability, $\Pr(X \geq x_o)$, is given by

$$\Pr(X \geq x_o) = \frac{1}{2\pi j} \int_{c-j\infty}^{c+j\infty} \frac{M_X(u)}{u} e^{-ux_o} du, \qquad (4.66)$$

where the contour of integration is to the right of the origin but to the left of all singularities of the moment generating function in the right half-plane.

PROOF: The right tail probability is given in general by

$$\Pr(X \geq x_o) = \int_{x_o}^{\infty} f_X(x)\,dx. \tag{4.67}$$

Then, replace the PDF in this integral with an inverse transform of the moment generating function as specified in Equation 4.56.

$$\Pr(X \geq x_o) = \int_{x_o}^{\infty} \frac{1}{2\pi j} \int_{c-j\infty}^{c+j\infty} M_X(u) e^{-ux}\,du\,dx$$

$$= \frac{1}{2\pi j} \int_{c-j\infty}^{c+j\infty} M_X(u) \int_{x_o}^{\infty} e^{-ux}\,dx\,du \tag{4.68}$$

Evaluating the inner integral results in

$$\Pr(X \geq x_o) = \frac{1}{2\pi j} \int_{c-j\infty}^{c+j\infty} \frac{M_X(u)}{u} e^{-ux_o}\,du, \quad \text{for } Re[u] > 0. \tag{4.69}$$

The integral specified in Equation 4.66 can be evaluated numerically or, when convenient to do so, it can also be evaluated by computing the appropriate residues[3]. For $x_o > 0$, the contour of integration can be closed to the right. The resulting closed contour will encompass all the singularities of the moment generating function in the right half-plane. According to Cauchy's residue theorem, the value of the integral will then be $-2\pi j$ times the sum of the residues of all the singularities encompassed by the contour. Hence,

$$\Pr(X \geq x_o) = \sum_{\substack{\text{right half-plane}}} \text{Residues} \left\{ \frac{M_X(u)}{u} e^{-ux_o} \right\}. \tag{4.70}$$

If a precise evaluation of the tail probability is not necessary, several approximations to the integral in Equation 4.66 are available. Perhaps the simplest and most useful is known as the *saddle point approximation*. To develop the saddle point approximation, define $\psi(u) = \ln(M_X(u))$ and

$$\lambda(u) = \ln\left(\frac{M_X(u)}{u} e^{-ux_o} \right) = \psi(u) - ux_o - \ln(u). \tag{4.71}$$

Furthermore, consider a Taylor series expansion of the function $\lambda(u)$ about some point $u = u_o$,

$$\lambda(u) = \lambda(u_o) + \lambda'(u_o)(u - u_o) + \frac{1}{2}\lambda''(u_o)(u - u_o)^2 + \cdots. \tag{4.72}$$

---

[3]The remainder of this section assumes a familiarity with the concepts of contour integration and residue calculus. For those unfamiliar with these topics, the remainder of this section can be skipped without any loss in continuity.

In particular, if u_o is chosen so that $\lambda'(u_o) = 0$, then near the point $u = u_o$, the integrand in Equation 4.66 behaves approximately like

$$\frac{M_X(u)}{u} e^{-ux_o} = e^{\lambda(u)} \approx \exp(\lambda(u_o)) \exp\left(\frac{1}{2}\lambda''(u_o)(u - u_o)^2\right). \qquad (4.73)$$

In general, the point $u = u_o$ will be a minima of the integrand as viewed along the real axis. This follows from the fact that the integrand is a concave function of u. A useful property of complex (analytic) functions tells us that if the function has a minima at some point u_o as u passes through it in one direction, the function will also have a maxima as u passes through the same point in the orthogonal direction. Such a point is called a "saddle point" since the shape of the function resembles a saddle near that point. If the contour of integration is selected to pass through the saddle point, the integrand will reach a local maximum at the saddle point. As just seen in Equation 4.73, the integrand also has a Gaussian behavior at and around the saddle point. Hence, using the approximation of Equation 4.73 and running the contour of integration through the saddle point so that $u = u_o + j\omega$ along the integration contour, the tail probability is approximated by

$$\begin{aligned}
\Pr(X \geq x_o) &\approx \frac{\exp(\lambda(u_o))}{2\pi j} \int_{u_o-j\infty}^{u_o+j\infty} \exp\left(\frac{1}{2}\lambda''(u_o)(u - u_o)^2\right) du \\
&= \frac{\exp(\lambda(u_o))}{2\pi} \int_{-\infty}^{\infty} \exp\left(-\frac{1}{2}\lambda''(u_o)\omega^2\right) d\omega \\
&= \frac{\exp(\lambda(u_o))}{\sqrt{2\pi\,\lambda''(u_o)}} \\
&= \frac{M_X(u_o)\exp(-u_ox_o)}{u_o\sqrt{2\pi\,\lambda''(u_o)}}. \qquad (4.74)
\end{aligned}$$

The third step is accomplished using the normalization integral for Gaussian PDFs.

The saddle point approximation is usually quite accurate provided that $x_o \gg E[X]$. That is, the farther we go out into the tail of the distribution, the better the approximation. If it is required to calculate $\Pr(X \geq x_o)$ for $x_o < E[X]$, it is usually better to calculate the left tail probability, in which case the saddle point approximation is

$$\Pr(X \geq x_o) \approx -\frac{M_X(u_o)\exp(-u_ox_o)}{u_o\sqrt{2\pi\,\lambda''(u_o)}}, \qquad (4.75)$$

where in this case the saddle point, u_o, must be negative. ∎

EXAMPLE 4.29: In this example, we will form the saddle point approximation to the Q-function that is the right tail probability for a standard normal random variable. The corresponding moment generating function is $M_X(u) = \exp(u^2/2)$. To find the saddle point, we note that

$$\lambda(u) = \frac{u^2}{2} - \ln(u) - ux_0.$$

We will need the first two derivatives of this function:

$$\lambda'(u) = u - \frac{1}{u} - x_0, \quad \lambda''(u) = 1 + \frac{1}{u^2}.$$

The saddle point is the solution to $\lambda'(u_0) = 0$ that results in a quadratic equation whose roots are

$$u_0 = \frac{x_0 \pm \sqrt{x_0^2 + 4}}{2}.$$

When calculating the right tail probability, the saddle point must be to the right of the imaginary axis, hence the positive root must be used:

$$u_0 = \frac{x_0 + \sqrt{x_0^2 + 4}}{2}.$$

The saddle point approximation then becomes

$$Q(x_0) \approx \frac{M_X(u_0)\exp(-u_0 x_0)}{u_0\sqrt{2\pi\lambda''(u_0)}} = \frac{\exp\left(-\dfrac{u_0^2}{2} - u_0 x_0\right)}{\sqrt{2\pi(1 + u_0^2)}}.$$

The exact value of the Q-function and the saddle point approximation are compared in Figure 4.6. As long as x_0 is not close to zero, this approximation is quite accurate.

4.11 Engineering Application: Scalar Quantization

In many applications, it is convenient to convert a signal that is analog in nature to a digital one. This is typically done in three steps. First, the signal is sampled,

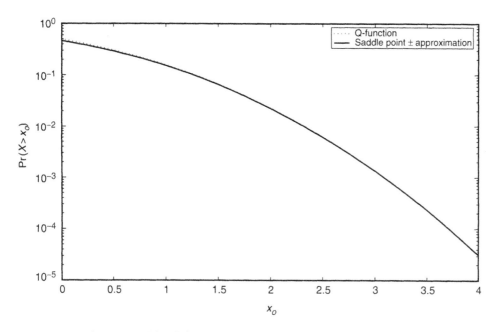

Figure 4.6 The Q-function and its saddle point approximation.

which converts the signal from continuous time to discrete time. Then samples of the signal are quantized. This second action converts the signal from one with a continuous amplitude to one whose amplitude can only take on discrete values. Once the signal is converted to discrete time/discrete amplitude, the signal can then easily be represented by a sequence of bits. In this third step, each discrete amplitude level is represented by a binary code word. While the first step (sampling) and the third step (encoding) are invertible, the second step (quantization) is not. That is, we can perfectly recover the analog signal from its discrete time samples (provided the samples are taken at a rate above the Nyquist rate), and the discrete amplitude levels can easily be recovered from the code words that represent them (provided a lossless source code is used). However, the act of quantization causes distortion of the signal that cannot be undone. Given the discrete amplitude of a sample, it is not possible to determine the exact value of the original (continuous amplitude) sample. For this reason, careful attention is paid to the quantization process to minimize the amount of distortion.

In order to determine efficient ways to quantize signals, we must first quantify this concept of signal distortion. Suppose a signal is sampled and we focus attention on one of those samples. Let the random variable X represent the value of that sample, which in general will draw from a continuous sample space. Now suppose that sample is quantized (using some quantization function $q(x)$) to form a new (discrete) random variable $Y = q(X)$. The difference between the original sample value and its quantized value, $X - q(X)$, is the error caused by the quantizer, or the quantizer noise. It is common to measure signal distortion as the mean squared quantizer error,

$$d = E[(X - q(X))^2] = \int_{-\infty}^{\infty} (x - q(x))^2 f_X(x)\,dx. \tag{4.76}$$

We will see Chapter 10 in the text, that the mean square value of a signal has the physical interpretation of the signal's power. Thus, the quantity d can be interpreted as the quantization noise power. Often the fidelity of a quantized signal is measured in terms of the ratio of the original signal power, $E[X^2]$, to the quantization noise power. This is referred to as the signal–to–quantization noise power ratio (SQNR):

$$\text{SQNR} = \frac{E[X^2]}{E[(X - q(X))^2]}. \tag{4.77}$$

The goal of the quantizer design is to choose a quantization function that minimizes the distortion, d. Normally, the quantizer maps the sample space of X into one of $M = 2^n$ levels. Then each quantization level can be represented by a unique n-bit code word. We refer to this as an n-bit quantizer. As indicated in Equation 4.76, the expected value is with respect to the PDF of X. Hence, the function $q(x)$ that minimizes the distortion will depend on the distribution of X.

To start with, consider a random variable X that is uniformly distributed over the interval $(-a/2, a/2)$. Since the sample X is equally likely to fall anywhere in the region, it would make sense for the quantizer to divide that region into M equally spaced subintervals of width $\Delta = a/M$. For each subinterval, the quantization level (i.e., the value of $q(x)$ for that subinterval) should be chosen as the midpoint of the subinterval. This is referred to as a uniform quantizer. A 3-bit uniform quantizer is illustrated in Figure 4.7. For example, if $X \in (0, a/8)$, then $q(X) = a/16$. To measure the distortion for this signal together with the uniform quantizer, assume that the signal falls within one of the quantization intervals, given that quantization interval, and then use the theorem of total probability:

$$d = E[(X - q(X))^2] = \sum_{k=1}^{8} E[(X - q(X))^2 | X \in X_k]\Pr(X \in X_k), \tag{4.78}$$

where X_k refers to the kth quantization interval. Consider, for example, $X_5 = (0, a/8)$, so that

$$E[(X - q(X))^2 | X \in X_5] = E[X - a/16)^2 | X \in (0, a/8)]. \tag{4.79}$$

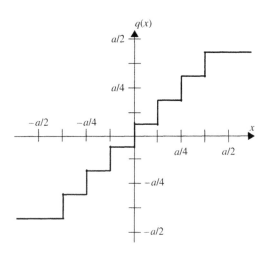

Figure 4.7 A 3-bit uniform quantizer on the interval $(-a/2, a/2)$

To calculate this conditional expected value requires the conditional PDF of X. From Equation 3.41 this is

$$f_X(x|X \in (0, a/8)) = \frac{f_X(x)}{\Pr(X \in (0, a/8))} = \frac{1/a}{1/8} = \frac{8}{a}, \quad x \in (0, a/8). \tag{4.80}$$

Not surprisingly, conditioned on $X \in (0, a/8)$, X is uniformly distributed over $(0, a/8)$. The conditional expected value is then

$$E[(X - a/16)^2 | X \in (0, a/8)] = \frac{8}{a} \int_0^{a/8} (x - a/16)^2 dx = \frac{a^2}{768}. \tag{4.81}$$

Due to the symmetry of the uniform distribution, this conditional distortion is the same regardless of what quantization interval the signal falls in. Hence, Equation 4.81 is also the unconditional distortion. Note that the power of the original signal is

$$E[X^2] = \frac{1}{a} \int_{-a/2}^{a/2} x^2 dx = \frac{a^2}{12}. \tag{4.82}$$

The resulting SQNR is then

$$\text{SQNR} = \frac{E[X^2]}{d} = \frac{a^2/12}{a^2/768} = 64 = 18.06 \text{ dB}. \tag{4.83}$$

The preceding result for the 3-bit uniform quantizer can be generalized to any uniform quantizer. In general, for an n-bit uniform quantizer, there will be $M = 2^n$ quantization intervals of width $\Delta = a/M$. Consider the quantization interval $(0, \Delta)$ and suppose that quantization level for that interval is chosen to be the midpoint, $\Delta/2$. Then the distortion for that interval (and hence the distortion for the quantizer) is

$$E[(X - q(X))^2|X \in (0, \Delta)] = \frac{1}{\Delta} \int_0^\Delta (x - \Delta/2)^2 dx = \Delta^2/12. \qquad (4.84)$$

The SQNR for an n-bit uniform quantizer with a uniformly distributed input is

$$\text{SQNR} = \frac{E[X^2]}{d} = \frac{a^2/12}{\Delta^2/12} = M^2 = 2^{2n} \text{ or } \text{SQNR(dB)} = 2n\log_{10}(2) = 6.02n \text{ dB}.$$

$$(4.85)$$

This is the so-called 6 dB rule whereby the SQNR is increased by approximately 6 dB for each extra bit added to the quantizer. For example, in wireline digital telephony, 8-bit quantization is used that would result in an SQNR of approximately 48 dB.

The previous results assumed that the input to the quantizer followed a uniform probability distribution. This is rarely the case. Speech signals, for example, are commonly modeled using a Laplace (two-sided exponential) distribution. For such signals, small sample values are much more frequent than larger values. In such a situation, it would make sense to use finer resolution (i.e., narrower quantization intervals) for the more frequent smaller amplitudes in order to keep the distortion minimized in those regions, at the cost of more distortion in the less frequent larger amplitude regions.

Given that an n-bit quantizer is to be used, the design of an optimum quantizer involves two separate problems. First, the ranges for each quantization interval must be specified; then the quantization level for each interval must be chosen. The following theorem specifies how each of these two tasks should be accomplished.

THEOREM 4.9: A random variable X with PDF $f_X(x)$ is to be quantized with an M-level quantizer that produces a discrete random variable Y according to

$$y = q(x) = y_i, \quad \text{for} \quad x_{i-1} < x < x_i, \quad i = 1, 2, \ldots, M, \qquad (4.86)$$

where it is assumed that the lower limit of the first quantization interval is $x_0 = -\infty$ and the upper limit of the last quantization interval is $x_M = \infty$. The (mean-squared) distortion is minimized by choosing the quantization intervals (i.e., the x_i) and the quantization levels (i.e., the y_i) according to

(i) $y_i = E[X|x_{i-1} < X < x_i], \quad i = 1, 2, \ldots, M$, (the conditional mean criterion),

$$(4.87)$$

(ii) $x_i = \dfrac{y_i + y_{i+1}}{2}, \quad i = 1, 2, \ldots, M - 1$ (the midpoint criterion). $\qquad (4.88)$

These two criteria provide a system of $2M - 1$ equations with which to solve for the $2M - 1$ quantities $(x_1, x_2, \ldots, x_{M-1}, y_1, y_2, \ldots, y_M)$ that specify the optimum M-level quantizer.

PROOF: The distortion is given by

$$d = \sum_{i=1}^{M} \int_{x_{i-1}}^{x_i} (x - y_i)^2 f_X(x)\, dx. \tag{4.89}$$

To minimize d with respect to y_j,

$$\frac{\partial d}{\partial y_j} = -2 \int_{x_{i-1}}^{x_i} (x - y_i) f_X(x)\, dx = 0. \tag{4.90}$$

Solving for y_j in this equation establishes the conditional mean criterion. Similarly, differentiating with respect to x_j gives

$$\frac{\partial d}{\partial x_j} = (x_j - y_j)^2 f_X(x_j) - (x_j - y_{j+1})^2 f_X(x_j) = 0. \tag{4.91}$$

Solving for x_j produces the midpoint criterion. ∎

EXAMPLE 4.30: Using the criteria set forth in Theorem 4.9, an ideal nonuniform 2-bit quantizer will be designed for a signal whose samples have a Laplace distribution, $f_X(x) = (1/2) \exp(-|x|)$. A 2-bit quantizer will have four quantization levels, $\{y_1, y_2, y_3, y_4\}$, and four corresponding quantization intervals that can be specified by three boundary points, $\{x_1, x_2, x_3\}$. The generic form of the 2-bit quantizer is illustrated in Figure 4.8. Due to the symmetry of the Laplace distribution, it seems reasonable to expect that the quantizer should have a negative symmetry about the y-axis. That is, $x_1 = -x_3$, $x_2 = 0$, $y_1 = -y_4$, and $y_2 = -y_3$. Hence, it is sufficient to determine just three unknowns, such as $\{x_3, y_3, y_4\}$. The rest can be inferred from the symmetry. Application of the conditional mean criterion and the midpoint criterion leads to the following set of three equations:

$$y_3 = E[X|0 < X < x_3] = \frac{\int_0^{x_3} \dfrac{x}{2} \exp(-x)\, dx}{\int_0^{x_3} \dfrac{1}{2} \exp(-x)\, dx} = 1 - \frac{x_3}{e^{x_3} - 1},$$

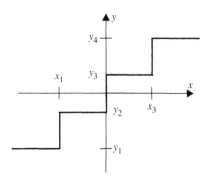

Figure 4.8   A 2-bit quantizer.

$$y_4 = E[X|X > x_3] = \dfrac{\displaystyle\int_{x_3}^{\infty} \dfrac{x}{2} \exp(-x)\,dx}{\displaystyle\int_{x_3}^{\infty} \dfrac{1}{2} \exp(-x)\,dx} = x_3 + 1,$$

$$x_3 = \frac{y_3 + y_4}{2}.$$

Plugging the expressions for y_3 and y_4 into the last equation results in a single equation to solve for the variable x_3. Unfortunately, the equation is transcendental and must be solved numerically. Doing so results in the solution $\{x_1, y_3, y_4\} = \{1.5940, 0.594, 2.594\}$. The (mean-square) distortion of this 2-bit quantizer is given by

$$d = 2\int_0^{x_3}(x - y_3)^2 f_X(x)\,dx + 2\int_{x_3}^{\infty}(x - y_4)^2 f_X(x)\,dx = 0.3524.$$

Note that the power in the original (unquantized) signal is $E[X^2] = 2$ so that the SQNR of this quantizer is

$$\text{SQNR} = \frac{E[X^2]}{d} = \frac{2}{0.3524} = 5.675 = 7.54 \text{ dB}.$$

EXAMPLE 4.31: In this example, we generalize the results of the last example for an arbitrary number of quantization levels. When the number of quantization levels gets large, the number of equations to solve becomes too much to do by hand, so we use MATLAB to help us with this task. Again, we assume that the random variable X

follows a Laplace distribution given by $f_X(x) = (1/2)\exp(-|x|)$. Because of the symmetry of this distribution, we again take advantage of the fact that the optimum quantizer will be symmetric. We design a quantizer with M levels for positive X (and hence M levels for negative X as well, for a total of $2M$ levels). The M quantization levels are at $y_1, y_2, \ldots y_M$, and the quantization bin edges are at $x_0 = 0, x_1, x_2, \ldots, x_{N-1}, x_N = \infty$. We compute the optimum quantizer in an iterative fashion. We start by arbitrarily setting the quantization bins in a uniform fashion. We choose to start with $x_i = 2i/M$. We then iterate between computing new quantization levels according to Equation 4.87 and new quantization bin edges according to Equation 4.88. After going back and forth between these two equations many times, the results eventually converge toward a final optimum quantizer. For the Laplace distribution, the conditional mean criterion results in

$$y_i = \frac{(x_{i-1}+1)\exp(-x_{i-1}) - (x_i+1)\exp(-x_i)}{\exp(-x_{i-1}) - \exp(-x_i)}.$$

At each iteration stage (after computing new quantization levels), we also compute the SQNR. By observing the SQNR at each stage, we can verify that this iterative process is in fact improving the quantizer design at each iteration. For this example, the SQNR is computed according to

$$\text{SQNR} = \frac{1}{1 - \dfrac{1}{2}\displaystyle\sum_{i=1}^{M} p_i y_i^2}, \quad p_i = \exp(-x_{i-1}) - \exp(-x_i).$$

The MATLAB code we used to implement this process (see Exercise 4.51) follows. Figure 4.9 shows the results of running this code for the case of $M = 8$ (16-level, 4-bit quantizer).

```
M=8;
x=[0 2*[1:M-1]/M];   % Initialize quantization bin edges.
iterations=50;
for k=1:iterations
    % Update quantization levels.
    x1=x(1:M-1);
    x2=x(2:M);
    y=(x1+1).*exp(-x1)-(x2+1).*exp(-x2);
    y=y./(exp(-x1)-exp(-x2));
    y=[y x(length(x))+1];
    % Calculate SQNR.
    p=exp(-x1)-exp(-x2);
```

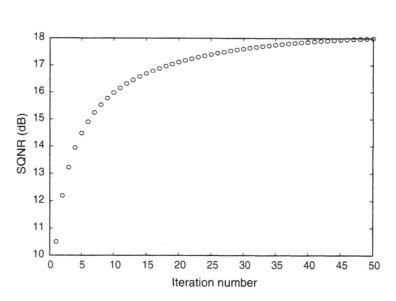

Figure 4.9   SQNR measurements for the iterative quantizer design in Example 4.31.

```
        p=[p exp(-x(length(x)))];
        SQNR(k)=1/(1-(y.^2)*p'/2);
        % Update quantization bin edges.
        y1=y(1:M-1);
        y2=y(2:M);
        x=[0 (y1+y2)/2];
    end
    plot(10*log10(SQNR), 'o')
```

4.12 Engineering Application: Entropy and Source Coding

The concept of information is something we hear about frequently. After all, we supposedly live in the "information age" with the Internet often referred to as the "information superhighway." But, what is information? In this section, we will give a quantitative definition of information and show how this concept is used in the world of digital communications. To motivate the forthcoming definition, imagine a situation in which we have a genie who can tell us about certain future events.

Suppose that this genie tells us that on July 15 of next year, the high temperature in the state of Texas will be above 90°F. Anyone familiar with the weather trends in Texas would know that our genie has actually given us very little information. Since we know that the temperature in July in Texas is above 90°F with probability approaching 1, the statement made by the genie does not tell us anything new. Next, suppose that the genie tells us that on July 15 of next year, the high temperature in the state of Texas will be below 80°F. Since this event is improbable, in this case the genie will be giving us a great deal of information.

To define a numerical quantity that we will call information, we note from the previous discussion that

- Information should be a function of various events. Observing (or being told) that some event A occurs (or will occur) provides a certain amount of information, $I(A)$.
- The amount of information associated with an event should be inversely related to the probability of the event. Observing highly probable events provides very little information, while observing very unlikely events gives a large amount of information.

At this point, there are many definitions which could satisfy these two bullet items. We include one more observation that will limit the possibilities:

- If it is observed that two events, A and B, have occurred and if these two events are independent, then we expect that the information $I(A \cap B) = I(A) + I(B)$.

Since we observed that information should be a function of the probability of an event, the last bullet item requires us to define information as a function of probability that satisfies

$$I(p_A p_B) = I(p_A) + I(p_B). \tag{4.92}$$

Since a logarithmic function satisfies this property, we obtain the following definition.

DEFINITION 4.11: If some event A occurs with probability p_A, then observing the event A provides an amount of *information* given by

$$I(A) = -\log(p_A). \tag{4.93}$$

The units associated with this measure of information depend on the base of the logarithm used. If base 2 logs are used, then the unit of information is a "bit"; if natural logs are used, the unit of information is the "nat."

Note that with this definition, an event that is sure to happen ($p_A = 1$) provides $I(A) = 0$ bits of information. This makes sense since if we know the event must happen, observing that it does happen provides us with no information.

Next, suppose we conduct some experiment that has a finite number of outcomes. The random variable X will be used to map these outcomes into the set of integers, $0, 1, 2, \ldots, n-1$. How much information do we obtain when we observe the outcome of the experiment? Since information is a function of the probability of each outcome, the amount of information is random and depends on which outcome occurs. We can, however, talk about the average information associated with the observation of the experiment.

DEFINITION 4.12: Suppose a discrete random variable X takes on the values $0, 1, 2, \ldots, n-1$ with probabilities $p_0, p_1, \ldots, p_{n-1}$. The *average information* or *(Shannon) entropy* associated with observing a realization of X is

$$H(X) = \sum_{k=1}^{n-1} \Pr(X = k) I(X = k) = \sum_{k=1}^{n-1} p_k \log\left(\frac{1}{p_k}\right).$$
(4.94)

Entropy provides a numerical measure of how much randomness or uncertainty there is in a random variable. In the context of a digital communication system, the random variable might represent the output of a data source. For example, suppose a binary source outputs the letters $X = 0$ and $X = 1$ with probabilities p and $1 - p$, respectively. The entropy associated with each letter the source outputs is

$$H(X) = p \log\left(\frac{1}{p}\right) + (1 - p) \log\left(\frac{1}{1-p}\right) = H(p).$$
(4.95)

The function $H(p)$ is known as the binary entropy function and is plotted in Figure 4.10. Note that this function has a maximum value of 1 bit when $p = 1/2$. Consequently, to maximize the information content of a binary source, the source symbols should be equally likely.

Next, suppose a digital source described by a discrete random variable X periodically outputs symbols. Then the information rate of the source is given by $H(X)$ bits/source symbol. Furthermore, suppose we wish to represent the source symbols with binary code words such that the resulting binary representation of the source outputs uses r bits/symbol. A fundamental result of source coding, which we will not attempt to prove here, is that if we desire the source code to be lossless (that is, we can always recover the original source symbols from their binary representation), then the source code rate must satisfy $r \geq H(X)$. In other words, the entropy of a source provides a lower bound on the average number of bits needed to represent each source output.

EXAMPLE 4.32: Consider a source that outputs symbols from a four-letter alphabet. That is, suppose $X \in \{a, b, c, d\}$. Let $p_a = 1/2, p_b = 1/4,$

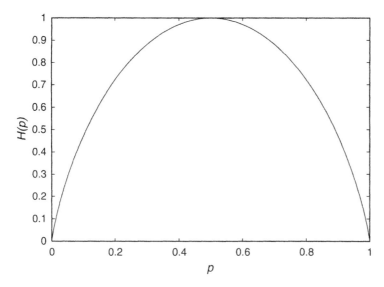

Figure 4.10   The binary entropy function.

and $p_c = p_d = 1/8$ be the probability of each of the four source symbols. Given this source distribution, the source has an entropy of

$$H(X) = \tfrac{1}{2}\log(2) + \tfrac{1}{4}\log(4) + 2 * \tfrac{1}{8}\log(8) = 1.75 \text{ bits/source symbol.}$$

Table 4.2 shows several different possible binary representations of this source. This first code is the simplest and most obvious representation. Since there are four letters, we can always assign a unique 2-bit code word to represent each source letter. This results in a code rate of $r_1 = 2$ bits/symbol, which is indeed greater than the entropy of the source. The second code uses variable length code words. The average code word length is

$$r_2 = \tfrac{1}{2} * 1 + \tfrac{1}{4} * 2 + \tfrac{1}{4} * 3 = 1.75 \text{ bits/symbol.}$$

Table 4.2   Three Possible Codes for a Four-Letter Source

	Source letters			
	a	b	c	d
Code 1	00	01	10	11
Code 2	0	10	110	111
Code 3	0	1	10	01

Hence, this code produces the most efficient representation of any loss-less source coded since the code rate is equal to the source entropy. Note that Code 3 from Table 4.2 produces a code rate of

$$r_3 = \tfrac{3}{4} * 1 + \tfrac{1}{4} * 2 = 1.25 \text{ bits/symbol,}$$

which is lower than the entropy, but this code is not lossless. This can easily be seen by noting that the source sequences "d" and "a, b" both lead to the same encoded sequence "01."

Exercises

4.1 Calculate the mean value, second moment, and variance of each of the following random variables:

   (a) binomial, $P_X(k) = \dbinom{n}{k} p^k (1 - p)^{n-k}$, $k = 0, 1, 2, \ldots, n$;

   (b) Poisson, $P_X(k) = \dfrac{\alpha^k}{k!} e^{-\alpha}$, $k = 0, 1, 2, \ldots$;

   (c) Laplace, $f_X(x) = \dfrac{1}{2b} \exp\left(-\dfrac{|x|}{b}\right)$;

   (d) gamma, $f_X(x) = \dfrac{(x/b)^{c-1} \exp\left(-\dfrac{x}{b}\right)}{b\Gamma(c)} u(x)$.

4.2 Imagine that you are trapped in a circular room with three doors symmetrically placed around the perimeter. You are told by a mysterious voice that one door leads to the outside after a two-hour trip through a maze. However, the other two doors lead to mazes that terminate back in the room after a two-hour trip, at which time you are unable to tell through which door you exited or entered. What is the average time for escape to the outside? Can you guess the answer ahead of time? If not, can you provide a physical explanation for the answer you calculate?

4.3 A communication system sends data in the form of packets of fixed length. Noise in the communication channel may cause a packet to be received incorrectly. If this happens, then the packet is retransmitted. Let the probability that a packet is received incorrectly be q. Determine the average number of transmissions that are necessary before a packet is received correctly.

4.4 In Exercise 4.3 let the transmission time be T_t seconds for a packet. If the packet was received incorrectly, then a message is sent back to the transmitter that states that the message was received incorrectly. Let the time for sending such a message be T_j. Assume that if the packet is received correctly, we do not send an acknowledgment. What is the average time for a successful transmission?

4.5 For a Gaussian random variable, derive expressions for the coefficient of skewness and the coefficient of kurtosis in terms of the mean and variance, μ and σ^2.

4.6 Prove that all odd central moments of a Gaussian random variable are equal to zero. Furthermore, develop an expression for all even central moments of a Gaussian random variable.

4.7 Show that the variance of a Cauchy random variable is undefined (infinite).

4.8 Let c_n be the nth central moment of a random variable and μ_n be its nth moment. Find a relationship between c_n and $\mu_k, k = 0, 1, 2, \ldots, n$.

4.9 Let X be a random variable with $E[X] = 1$ and $\text{var}(X) = 4$. Find the following:

(a) $E[2X - 4]$,

(b) $E[X]^2$,

(c) $E[(2X - 4)^2]$.

4.10 Suppose X is a Gaussian random variable with a mean of μ and a variance of σ^2 (i.e., $X \sim N(\mu, \sigma^2)$). Find an expression for $E[|X|]$.

4.11 A random variable X has a uniform distribution over the interval $(-a/2, a/2)$ for some positive constant a.

(a) Find the coefficient of skewness for X.

(b) Find the coefficient of kurtosis for X.

(c) Compare the results of (a) and (b) with the same quantities for a standard normal random variable.

4.12 Suppose a random variable X has a PDF which is nonzero on only the interval $[0, \infty)$. That is, the random variable cannot take on negative values.

Prove that

$$E[X] = \int_0^\infty [1 - F_X(x)]\,dx.$$

4.13 Show that the concept of total probability can be extended to expected values. That is, if $\{A_i\}$, $i = 1, 2, 3, \ldots, n$ is a set of mutually exclusive and exhaustive events, then

$$E[X] = \sum_{k=1}^n E[X|A_k]\Pr(A_k).$$

4.14 Prove *Jensen's Inequality*, which states that for any convex function $g(x)$ and any random variable X,

$$E[g(X)] \geq g(E[X]).$$

4.15 Suppose Θ is a random variable uniformly distributed over the interval $[0, 2\pi)$.

(a) Find the PDF of $Y = \sin(\Theta)$.
(b) Find the PDF of $Z = \cos(\Theta)$.
(c) Find the PDF of $W = \tan(\Theta)$.

4.16 Suppose X is uniformly distributed over $(-a, a)$, where a is some positive constant. Find the PDF of $Y = X^2$.

4.17 Suppose X is a random variable with an exponential PDF of the form $f_X(x) = 2e^{-2x}u(x)$. A new random variable is created according to the transformation $Y = 1 - X$.

(a) Find the range for X and Y.
(b) Find $f_Y(y)$.

4.18 Let X be a standard normal random variable (i.e., $X \sim N(0, 1)$). Find the PDF of $Y = |X|$.

4.19 Repeat Exercise 4.18 if the transformation is

$$Y = \begin{cases} X & X > 0, \\ 0 & X \leq 0. \end{cases}$$

4.20 Suppose a random variable, X, has a Gaussian PDF with zero mean and variance σ_X^2. The random variable is transformed by the device whose input/output relationship is shown in the accompanying figure. Find and sketch the PDF of the transformed random variable, Y.

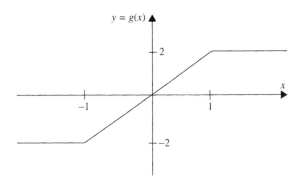

4.21 Let X be a Gaussian random variable with zero mean and arbitrary variance, σ^2. Given the transformation $Y = X^3$, find $f_Y(y)$.

4.22 A real number between 0 and 100 is randomly selected according to a uniform distribution and rounded off to the nearest integer. For example, 36.5001 is rounded off to 37; $\sqrt{3}$ is rounded off to 2; and 69.49 is rounded off to 69. Define a random variable to be $X = $ (number selected) $-$ (nearest integer).

(a) What is the range of this random variable?
(b) Determine the PDF for X.
(c) Determine the mean square value of X.

4.23 A Gaussian random variable with zero mean and variance σ_X^2 is applied to a device that has only two possible outputs, 0 or 1. The output 0 occurs when the input is negative, and the output 1 occurs when the input is 0 or positive.

(a) What is the probability mass function of the output?
(b) Rework the problem when $\mu_X = 1/2$ and $\sigma_X^2 = 1$.

4.24 Let X be a Cauchy random variable whose PDF is given by

$$f_X(x) = \frac{b/\pi}{b^2 + x^2}.$$

Find the PDF of $Y = 1/X$.

4.25 Let X be a chi-square random variable with a PDF given by

$$f_X(x) = \frac{x^{c-1} \exp(-x/2)u(x)}{2^c \Gamma(c)},$$

where $c = n/2$ for any positive integer n. Find the PDF of $Y = \sqrt{X}$.

4.26 Suppose a random variable has some PDF given by $f_X(x)$. Find a function $g(x)$ such that $Y = g(X)$ is a uniform random variable over the interval $(0, 1)$. Next, suppose that X is a uniform random variable. Find a function $g(x)$ such that $Y = g(X)$ has some specified PDF, $f_Y(y)$.

4.27 Suppose X is uniformly distributed over $(0, 1)$. Using the results of the previous problem, find transformations $Y = g(X)$ to produce random variables with the following distributions:

(a) exponential,

(b) Rayleigh,

(c) Cauchy,

(d) geometric,

(e) Poisson.

4.28 A random variable X has a characteristic function, $\phi_X(\omega)$. Write the characteristic function of $Y = aX + b$ in terms of $\phi_X(\omega)$ and the constants a and b.

4.29 Prove that the characteristic function of any random variable must satisfy the following properties.

(a) $\phi_X^*(\omega) = \phi_X(-\omega)$.

(b) $\phi_X(0) = 1$.

(c) For real ω, $|\phi_X(\omega)| \leq 1$.

(d) If the PDF is symmetric about the origin (i.e, an even function), then $\phi_X(\omega)$ is real.

(e) $\phi_X(\omega)$ cannot be purely imaginary.

4.30 Suppose X is an integer-valued random variable. Show that in this case, $\phi_X(2\pi n) = 1$ for any integer, n. Likewise, prove the reverse is also true. That is, show that if $\phi_X(2\pi n) = 1$ for any integer, n, the random variable X must be integer-valued.

4.31 For a Laplace random variable whose PDF is given by $f_X(x) = \frac{1}{2b}\exp\left(-\frac{|x|}{b}\right)$, find the following:

(a) the characteristic function, $\phi_X(\omega)$,

(b) the Taylor series expansion of $\phi_X(\omega)$,

(c) a general expression for the kth moment of X.

4.32 Derive a formula expressing the variance of a random variable in terms of its factorial moments.

4.33 Derive a relationship between the kth factorial moment for a nonnegative, integer-valued random variable and the coefficients of the Taylor series expansion of its probability generating function, $H_X(z)$, about the point $z = 1$.

4.34 Use the characteristic function (or the moment generating function or the probability generating function) to show that a Poisson PMF is the limit of a binomial PMF with n approaching infinity and p approaching zero in such a way that $np = \mu = $ constant.

4.35 For a Poisson random variable whose PMF is given by $P_X(k) = \frac{\alpha^k}{k!}e^{-\alpha}$, $k = 0, 1, 2, \ldots$, find the following:

(a) the probability generating function, $H_X(z)$,

(b) the Taylor series expansion of $H_X(z)$ about the point $z = 1$,

(c) a general expression for the kth factorial moment.

4.36 A certain random variable has a probability generating function given by

$$H_X(z) = \frac{1}{n}\frac{1 - z^n}{1 - z}.$$

Find the PMF for this random variable.

4.37 Derive an expression for the moment generating function of a Rayleigh random variable whose PDF is

$$f_X(x) = x\exp\left(-\frac{x^2}{2}\right)u(x).$$

4.38 Suppose X is a Rician random variable with a PDF given by

$$f_X(x) = x \exp\left(-\frac{x^2 + a^2}{2}\right) I_0(ax)u(x).$$

Derive an expression for $E\left[e^{uX^2}\right]$. Note that this is not quite the moment generating function, but it can be used in a similar way.

4.39 Prove that for a random variable X with mean μ_X,

$$\Pr(|X - \mu_X| > \varepsilon) < \frac{E[|X - \mu_X|^n]}{\varepsilon^n}.$$

where n is any positive integer.

4.40 Suppose we are interested in finding the left tail probability for a random variable, X. That is, we want to find $\Pr(X \leq x_0)$. Re-derive an expression for the Chernoff bound for the left tail probability.

4.41 Suppose X is a Poisson random variable with PMF, $P_X(k) = \dfrac{\alpha^k}{k!}\exp(-\alpha)$, $k = 0, 1, 2, \ldots$. Find the Chernoff bound for the tail probability, $\Pr(X \geq n_0)$.

4.42 Suppose X is a gamma random variable with PDF, $f_X(x) = \dfrac{(x/b)^{c-1}\exp(-x/b)}{b\Gamma(c)}u(x)$. Find the Chernoff bound for the tail probability, $\Pr(X > x_0)$.

4.43 Let X be a Erlang random variable with PDF, $f_X(x) = \dfrac{x^{n-1}e^{-x}u(x)}{(n-1)!}$. Derive a saddle point approximation for the left tail probability, $\Pr(X < x_0)$. Compare your result with the exact value for $0 \leq x_0 < E[X]$.

4.44 In Exercise 4.38, an expression was derived for $E[e^{uX^2}]$ for a Rician random variable. Use this function to obtain a saddle point approximation for the tail probability of a Rician random variable, $\Pr(X \geq x_0)$.
Hint: For one-sided random variables, $\Pr(X \geq x_0) = \Pr(X^2 \geq x_0^2)$.

4.45 Suppose a source sends symbols from a three-letter alphabet with $X \in \{a, b, c\}$, and $p_a = 1/2$, $p_b = 1/4$, $p_c = 1/4$ are the source symbol probabilities.

(a) Determine the entropy of this source.
(b) Give a source code that has an average code word length that matches the entropy.

MATLAB Exercises

4.46 Let X be a random variable that is uniformly distributed over the interval (0,100). Form a new random variable Y by rounding X to the nearest integer. In MATLAB code, this could be represented by Y=round(X). Finally, form the random roundoff error according to $Z = X - Y$.

    (a) Using analytical methods, find the PDF of Z as well as the mean squared value, $E[Z^2]$.

    (b) Using MATLAB, create a histogram for the probability densities for the random variable Z. Compare with the PDF found analytically in part (a).

4.47 Suppose you have a random variable X with PDF, $f_X(x) = 2x(u(x)-u(x-1))$ and that this random variable is transformed as $Y = 2 - X$. Calculate $f_Y(y)$. Repeat this problem using MATLAB. Compare the estimate of the PDF from MATLAB with the analytically determined PDF. Note that for this problem there is no function in MATLAB that provides a sequence of data samples that has the PDF specified in this problem for X. Thus, you must find an appropriate way to transform a uniform random variable to produce the desired X. The results of Exercise 4.26 will be helpful here.

4.48 Use MATLAB to generate a large number of samples from a Gaussian distribution with mean $\mu = 20$ and variance $\sigma = 4$. *Hint:* the MATLAB command sigma*randn(1,N)+mu will create N such numbers with mean mu and standard deviation sigma. Let $x_1, x_2, \ldots, x_N$ represent the samples you generated. Compute each of the following "mean" values:

    (a) sample mean, $\hat{\mu}_{\text{s.m.}} = \dfrac{1}{N}\sum_{k=1}^{N} x_k;$

    (b) geometric mean, $\hat{\mu}_{\text{g.m.}} = \left(\prod_{k=1}^{N} x_k\right)^{1/N};$

    (c) harmonic mean, $\hat{\mu}_{\text{h.m.}} = \left(\dfrac{1}{N}\sum_{k=1}^{N}\dfrac{1}{x_k}\right)^{-1};$

    (d) quadratic mean (root mean square), $\hat{\mu}_{\text{s.m.}} = \sqrt{\dfrac{1}{N}\sum_{k=1}^{N} x_k^2}.$

Which of these "estimates" give a decent estimate of the true mean?

4.49 Write a MATLAB program to simulate the problem described in Exercise 4.2. Estimate the average time until escape. Do your MATLAB results agree with your analytically determined results in Exercise 4.2?

4.50 Copy a segment of text into MATLAB as a string (you choose the source of the text). Then write a MATLAB program to count the relative frequency of each character (ignore all characters that do not correspond to one of the 26 letters and do not make a distinction between upper and lower case). Using the results of your program, calculate the entropy of a source that outputs the 26 English characters with the probabilities you calculated.

4.51 Suppose a random variable has a PDF given by

$$f_X(x) = \begin{cases} 1+x & -1 < x < 0 \\ 1-x & 0 \le x < 1 \\ 0 & |x| \ge 1 \end{cases}.$$

Following the procedure laid out in Example 4.31, write a MATLAB program to design an optimum 4-bit (16-level) quantizer for this random variable. Compute the SQNR in decibels of the quantizer you designed. How does this SQNR compare with that obtained in Example 4.31. Can you explain any differences?

Pairs of Random Variables

The previous two chapters dealt with the theory of single random variables. However, many problems of practical interest require the modeling of random phenomena using two or maybe even more random variables. This chapter extends the theory of Chapters 3 and 4 to consider pairs of random variables. Chapter 6 then generalizes these results to include an arbitrary number of random variables. A common example that involves two random variables is the study of a system with a random input. Due to the randomness of the input, the output will naturally be random as well. Quite often it is necessary to characterize the relationship between the input and the output. A pair of random variables can be used to characterize this relationship; one for the input and another for the output.

Another class of examples involving random variables is one involving spatial coordinates in two dimensions. A pair of random variables can be used to probabilistically describe the position of an object that is subject to various random forces. There are endless examples of situations in which we are interested in two random quantities that may or may not be related to one another, for example, the height and weight of a student, the grade point average and GRE scores of a student, or the temperature and relative humidity at a certain place and time.

To start with, consider an experiment E whose outcomes lie in a sample space, S. A two-dimensional random variable is a mapping of the points in the sample space to ordered pairs $\{x, y\}$. Usually, when dealing with a pair of random variables, the sample space naturally partitions itself so that it can be viewed as a combination of two simpler sample spaces. For example, suppose the experiment was to observe the height and weight of a typical student. The range of student heights could fall within some set, which we'll call sample space S_1, while the range of student weights could fall within the space S_2. The overall sample space of the experiment could then be viewed as $S = S_1 \times S_2$. For any outcome $s \in S$ of this experiment, the pair of random variables (X, Y) is merely a mapping of the outcome s to a pair of numerical values $(x(s), y(s))$. In the case of our height/weight experiment, it would be natural to choose $x(s)$ to be the height of the student (in inches perhaps) while $y(s)$ is the weight of the student (in pounds). Note that it is probably not sufficient to consider two separate experiments, one in which the student's height is measured and assigned to the random variable X and another in which a student's weight is measured and assigned to the random variable Y.

While the density functions $f_X(x)$ and $f_Y(y)$ do partially characterize the experiment, they do not completely describe the situation. It would be natural to expect that the height and weight are somehow related to each other. While it may not be rare to have a student 74 inches tall, nor unusual to have a student weighing nearly 120 pounds, it is probably rare indeed to have a student who is both 74 inches tall and weighs 120 pounds. A careful reading of the wording in the previous sentence makes it clear that in order to characterize the relationship between a pair of random variables, it is necessary to look at the joint probabilities of events relating to both random variables. We accomplish this through the joint cumulative distribution function and the joint probability density function in the next two sections.

5.1 Joint Cumulative Distribution Functions

When introducing the idea of random variables in Chapter 3, we started with the notion of a cumulative distribution function. In the same way, to probabilistically describe a pair of random variables, $\{X, Y\}$, we start with the notion of a joint cumulative distribution function.

DEFINITION 5.1: The *joint cumulative distribution function* of a pair of random variables, $\{X, Y\}$, is $F_{X,Y}(x, y) = \Pr(X \leq x, Y \leq y)$. That is, the joint CDF is the joint probability of the two events $\{X \leq x\}$ and $\{Y \leq y\}$.

As with the CDF of a single random variable, not any function can be a joint CDF. The joint CDF of a pair of random variables will satisfy properties similar to those satisfied by the CDFs of single random variables. First of all, since the joint CDF is a probability, it must take on a value between 0 and 1. Also, since the random variables X and Y are real-valued, it is impossible for either to take on a value less than $-\infty$ and both must be less than ∞. Hence, $F_{X,Y}(x, y)$ evaluated at either $x = -\infty$ or $y = -\infty$ (or both) must be zero and $F_{X,Y}(\infty, \infty)$ must be one. Next, for $x_1 \leq x_2$ and $y_1 \leq y_2$, $\{X \leq x_1\} \cap \{Y \leq y_1\}$ is a subset of $\{X \leq x_2\} \cap \{Y \leq y_2\}$ so that $F_{X,Y}(x_1, y_1) \leq F_{X,Y}(x_2, y_2)$. That is, the CDF is a monotonic, nondecreasing function of both x and y. Note that since the event $X \leq \infty$ must happen, then $\{X \leq \infty\} \cap \{Y \leq y\} = \{Y \leq y\}$ so that $F_{X,Y}(\infty, y) = F_Y(y)$. Likewise, $F_{X,Y}(x, \infty) = F_X(x)$. In the context of joint CDFs, $F_X(x)$ and $F_Y(y)$ are referred to as the *marginal* CDFs of X and Y, respectively.

Finally, consider using a joint CDF to evaluate the probability that the pair of random variables (X, Y) falls into a rectangular region bounded by the points (x_1, y_1), (x_2, y_1), (x_1, y_2), and (x_2, y_2). This calculation is illustrated in Figure 5.1; the desired rectangular region is the lightly shaded area. Evaluating $F_{X,Y}(x_2, y_2)$ gives the probability that the random variable falls anywhere below or to the left of the point (x_2, y_2); this includes all of the area in the desired rectangle, but it also includes everything below and to the left of the desired rectangle. The probability of the random variable falling to the left of the rectangle can be subtracted off using $F_{X,Y}(x_1, y_2)$. Similarly, the region below the rectangle can be subtracted off using $F_{X,Y}(x_2, y_1)$; these are the two medium-shaded regions in Figure 5.1. In subtracting off these two quantities, we have subtracted twice the probability of the pair falling both below and to the left of the desired rectangle (the dark-shaded region). Hence, we must add back this probability using $F_{X,Y}(x_1, y_1)$. All of these properties of joint CDFs are summarized as follows:

(1) $F_{X,Y}(-\infty, -\infty) = F_{X,Y}(-\infty, y) = F_{X,Y}(x, -\infty) = 0;$ \hfill (5.1a)

(2) $F_{X,Y}(\infty, \infty) = 1;$ \hfill (5.1b)

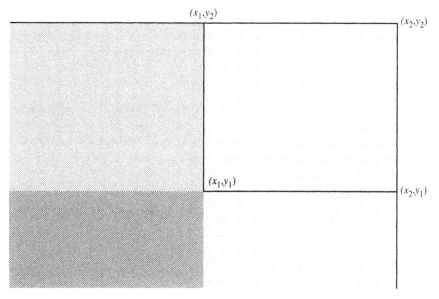

Figure 5.1 Illustrating the evaluation of the probability of a pair of random variables falling in a rectangular region.

(3) $0 \le F_{X,Y}(x,y) \le 1$;   (5.1c)

(4) $F_{X,Y}(x,\infty) = F_X(x), F_{X,Y}(\infty,y) = F_Y(y)$;   (5.1d)

(5) $\Pr(x_1 < X_1 \le x_2, y_1 < Y_1 \le y_2)$

$\qquad = F_{X,Y}(x_2,y_2) - F_{X,Y}(x_1,y_2) - F_{X,Y}(x_2,y_1) + F_{X,Y}(x_1,y_1) \ge 0.$ (5.1e)

With the exception of property (4), all of these properties are analogous to the ones listed in Equation 3.3 for CDFs of single random variables.

Property (5) tells us how to calculate the probability of the pair of random variables falling in a rectangular region. Often, we are interested in also calculating the probability of the pair of random variables falling in a region that is not rectangular (e.g., a circle or triangle). This can be done by forming the required region using many infinitesimal rectangles and then repeatedly applying property (5). In practice, however, this task is somewhat overwhelming, and hence we do not go into the details here.

EXAMPLE 5.1: One of the simplest examples (conceptually) of a pair of random variables is one that is uniformly distributed over the unit square (i.e., $0 < x < 1, 0 < y < 1$). The CDF of such a random variable is

$$F_{X,Y}(x,y) = \begin{cases} 0 & x < 0 \text{ or } y < 0 \\ x & 0 \le x \le 1, \, y > 1 \\ y & x > 1, \, 0 \le y \le 1 \\ xy & 0 \le x \le 1, \, 0 \le y \le 1 \\ 1 & x > 1, \, y > 1 \end{cases}$$

Even this very simple example leads to a rather cumbersome function. Nevertheless, it is straightforward to verify that this function does indeed satisfy all the properties of a joint CDF. From this joint CDF, the marginal CDF of X can be found to be

$$F_X(x) = F_{X,Y}(x,\infty) = \begin{cases} 0 & x < 0 \\ x & 0 \le x \le 1 \\ 1 & x > 1 \end{cases}$$

Hence, the marginal CDF of X is also a uniform distribution. The same statement holds for Y as well.

5.2 Joint Probability Density Functions

As seen in Example 5.1, even the simplest joint random variables can lead to CDFs that are quite unwieldy. As a result, working with joint CDFs can be difficult. In order to avoid extensive use of joint CDFs, attention is now turned to the two-dimensional equivalent of the PDF.

DEFINITION 5.2: The *joint probability density function* of a pair of random variables (X, Y) evaluated at the point (x, y) is

$$f_{X,Y}(x, y) = \lim_{\varepsilon_x \to 0, \varepsilon_y \to 0} \frac{\Pr(x \leq X < x + \varepsilon_x, y \leq Y < y + \varepsilon_y)}{\varepsilon_x \varepsilon_y}. \tag{5.2}$$

Similar to the one-dimensional case, the joint PDF is the probability that the pair of random variables (X, Y) lies in an infinitesimal region defined by the point (x, y) normalized by the area of the region.

For a single random variable, the PDF was the derivative of the CDF. By applying Equation 5.1e to the definition of the joint PDF, a similar relationship is obtained.

THEOREM 5.1: The joint PDF $f_{X,Y}(x, y)$ can be obtained from the joint CDF $F_{X,Y}(x, y)$ by taking a partial derivative with respect to each variable. That is

$$f_{X,Y}(x, y) = \frac{\partial^2}{\partial x \partial y} F_{X,Y}(x, y). \tag{5.3}$$

PROOF: Using Equation 5.1e,

$$\Pr(x \leq X < x + \varepsilon_x, y \leq Y < y + \varepsilon_y)$$

$$= F_{X,Y}(x + \varepsilon_x, y + \varepsilon_y) - F_{X,Y}(x, y + \varepsilon_y) - F_{X,Y}(x + \varepsilon_x, y) + F_{X,Y}(x, y)$$

$$= [F_{X,Y}(x + \varepsilon_x, y + \varepsilon_y) - F_{X,Y}(x, y + \varepsilon_y)] - [F_{X,Y}(x + \varepsilon_x, y) - F_{X,Y}(x, y)]. \tag{5.4}$$

Dividing by ε_x and taking the limit as $\varepsilon_x \to 0$ results in

$$\lim_{\varepsilon_x \to 0} \frac{\Pr(x \leq X < x + \varepsilon_x, y \leq Y < y + \varepsilon_y)}{\varepsilon_x}$$

$$= \lim_{\varepsilon_x \to 0} \frac{F_{X,Y}(x + \varepsilon_x, y + \varepsilon_y) - F_{X,Y}(x, y + \varepsilon_y)}{\varepsilon_x} - \lim_{\varepsilon_x \to 0} \frac{F_{X,Y}(x + \varepsilon_x, y) + F_{X,Y}(x, y)}{\varepsilon_x}$$

$$= \frac{\partial}{\partial x} F_{X,Y}(x, y + \varepsilon_y) - \frac{\partial}{\partial x} F_{X,Y}(x, y). \tag{5.5}$$

Then dividing by ε_y and taking the limit as $\varepsilon_y \to 0$ gives the desired result:

$$f_{X,Y}(x,y) = \lim_{\varepsilon_x \to 0, \varepsilon_y \to 0} \frac{\Pr(x \le X < x + \varepsilon_x, y \le Y < y + \varepsilon_y)}{\varepsilon_x \varepsilon_y}$$

$$= \lim_{\varepsilon_y \to 0} \frac{\frac{\partial}{\partial x}F_{X,Y}(x, y + \varepsilon_y) - \frac{\partial}{\partial x}F_{X,Y}(x,y)}{\varepsilon_y} = \frac{\partial^2}{\partial x \partial y}F_{X,Y}(x,y). \qquad (5.6)$$

∎

This theorem shows that we can obtain a joint PDF from a joint CDF by differentiating with respect to each variable. The converse of this statement would be that we could obtain a joint CDF from a joint PDF by integrating with respect to each variable. Specifically,

$$F_{X,Y}(x,y) = \int_{-\infty}^{y} \int_{-\infty}^{x} f_{X,Y}(u,v)\,du\,dv. \qquad (5.7)$$

EXAMPLE 5.2: From the joint CDF given in Example 5.1, it is easily found (by differentiating the joint CDF with respect to both x and y) that the joint PDF for a pair of random variables uniformly distributed over the unit square is

$$f_{X,Y}(x,y) = \begin{cases} 1 & 0 < x < 1, 0 < y < 1 \\ 0 & \text{otherwise} \end{cases}.$$

Note how much simpler the joint PDF is to specify than is the joint CDF.

From the definition of the joint PDF in Equation 5.2 as well as the relationships specified in Equation 5.3 and 5.7, several properties of joint PDFs can be inferred. These properties are summarized as follows:

(1) $f_{X,Y}(x,y) \ge 0;$ (5.8a)

(2) $\displaystyle\int_{-\infty}^{\infty} \int_{-\infty}^{\infty} f_{X,Y}(x,y)\,dx\,dy = 1;$ (5.8b)

(3) $\displaystyle F_{X,Y}(x,y) = \int_{-\infty}^{y} \int_{-\infty}^{x} f_{X,Y}(u,v)\,du\,dv;$ (5.8c)

(4) $\displaystyle f_{X,Y}(x,y) = \frac{\partial^2}{\partial x \partial y}F_{X,Y}(x,y);$ (5.8d)x

(5) $f_X(x) = \int_{-\infty}^{\infty} f_{X,Y}(x,y)\,dy, \quad f_Y(y) = \int_{-\infty}^{\infty} f_{X,Y}(x,y)\,dx;$ (5.8e)

(6) $\Pr(x_1 < X_1 \le x_2, y_1 < Y_1 \le y_2) = \int_{y_1}^{y_2} \int_{x_1}^{x_2} f_{X,Y}(x,y)\,dx dy.$ (5.8f)

Property (1) follows directly from the definition of the joint PDF in Equation 5.2 since both the numerator and denominator there are nonnegative. Property (2) results from the relationship in Equation 5.7 together with the fact that $F_{X,Y}(\infty, \infty) = 1$. This is the normalization integral for joint PDFs. These first two properties form a set of sufficient conditions for a function of two variables to be a valid joint PDF. Properties (3) and (4) have already been developed. Property (5) is obtained by first noting that the marginal CDF of X is $F_X(x) = F_{X,Y}(x, \infty)$. Using Equation 5.7 then results in $F_X(x) = \int_{-\infty}^{\infty} \int_{-\infty}^{x} f_{X,Y}(u,y)\,du dy$. Differentiating this expression with respect to x produces the expression in property (5) for the marginal PDF of x. A similar derivation produces the marginal PDF of y. Hence, the marginal PDFs are obtained by integrating out the unwanted variable in the joint PDF. The last property is obtained by combining Equations 5.1e and 5.7.

EXAMPLE 5.3: Suppose a pair of random variables is jointly uniformly distributed over the unit circle. That is, the joint PDF $f_{X,Y}(x,y)$ is constant anywhere such that $x^2 + y^2 < 1$:

$$f_{X,Y}(x,y) = \begin{cases} c & x^2 + y^2 < 1 \\ 0 & \text{otherwise} \end{cases}.$$

The constant c can be determined using the normalization integral for joint PDFs:

$$\iint_{x^2+y^2<1} c\,dx dy = 1 \Rightarrow c = \frac{1}{\pi}.$$

The marginal PDF of X is found by integrating y out of the joint PDF:

$$f_X(x) = \int_{-\infty}^{\infty} f_{X,Y}(x,y)\,dy = \int_{-\sqrt{1-x^2}}^{\sqrt{1-x^2}} \frac{1}{\pi}\,dy = \frac{2}{\pi}\sqrt{1-x^2}, \quad \text{for } -1 \le x \le 1.$$

By symmetry, the marginal PDF of Y would have the same functional form:

$$f_Y(y) = \frac{2}{\pi}\sqrt{1-y^2}, \quad \text{for } -1 \le x \le 1.$$

Although X and Y were jointly uniformly distributed, the marginal distributions are not uniform. Stated another way, suppose we are given just the marginal PDFs of X and Y as just specified. This information alone is not enough to determine the joint PDF. One may be able to form many joint PDFs that produce the same marginal PDFs. For example, suppose we form

$$f_{X,Y}(x,y) = \begin{cases} \dfrac{4}{\pi^2}\sqrt{(1-x^2)(1-y^2)} & -1 \leq x \leq 1,\ -1 \leq y \leq 1 \\ 0 & \text{otherwise} \end{cases}.$$

It is easy to verify that this is a valid joint PDF and leads to the same marginal PDFs. Yet, this is clearly a completely different joint PDF than the uniform distribution with which we started. This reemphasizes the need to specify the joint distributions of random variables and not just their marginal distributions.

Property (6) of joint PDFs given in Equation 5.8f specifies how to compute the probability that a pair of random variables takes on a value in a rectangular region. Often we are interested in computing the probability that the pair of random variables falls in a region that is not rectangularly shaped. In general, suppose we wish to compute $\Pr((X, Y) \in A)$, where A is the region illustrated in Figure 5.2. This general region can be approximated as a union of many nonoverlapping rectangular regions as shown in the figure. In fact, as we make the rectangles ever smaller, the approximation improves to the point where the representation becomes exact in the limit as the rectangles get infinitely small. That is, any region can be represented as an infinite number of infinitesimal rectangular regions so that $A = \cup R_i$, where R_i represents the ith rectangular region. The probability that the random pair falls in A is then computed as

$$\Pr((X, Y) \in A) = \sum_i \Pr((X, Y) \in R_i) = \sum_i \iint_{R_i} f_{X,Y}(x,y)\,dxdy. \qquad (5.9)$$

The sum of the integrals over the rectangular regions can be replaced by an integral over the original region A:

$$\Pr((X, Y) \in A) = \iint_A f_{X,Y}(x,y)\,dxdy. \qquad (5.10)$$

This important result shows that the probability of a pair of random variables falling in some two-dimensional region A is found by integrating the joint PDF of the two random variables over the region A.

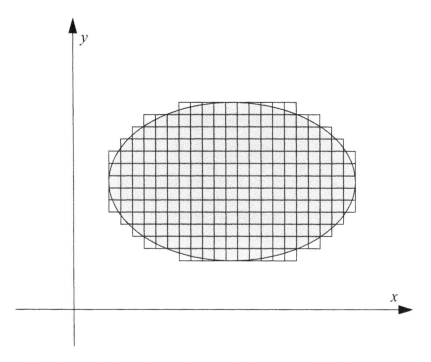

Figure 5.2 Approximation of an arbitrary region by a series of infinitesimal rectangles.

EXAMPLE 5.4: Suppose a pair of random variables has the joint PDF given by

$$f_{X,Y}(x,y) = \frac{1}{2\pi} \exp\left(-\frac{(x^2+y^2)}{2}\right)$$

The probability that the point (X, Y) falls inside the unit circle is given by

$$\Pr(X^2 + Y^2 < 1) = \iint\limits_{x^2+y^2<1} \frac{1}{2\pi} \exp\left(-\frac{(x^2+y^2)}{2}\right) dx\, dy.$$

Converting this integral to polar coordinates results in

$$\Pr(X^2 + Y^2 < 1) = \int_0^{2\pi}\int_0^1 \frac{r}{2\pi} \exp\left(-\frac{r^2}{2}\right) dr\, d\theta = \int_0^1 r \exp\left(-\frac{r^2}{2}\right) dr$$

$$= -\exp\left(-\frac{r^2}{2}\right)\Big|_0^1 = 1 - \exp\left(-\frac{1}{2}\right).$$

EXAMPLE 5.5: Now suppose that a pair of random variables has the joint PDF given by

$$f_{X,Y}(x,y) = c \exp\left(-x - \frac{y}{2}\right) u(x)u(y).$$

First, the constant c is found using the normalization integral

$$\int_0^\infty \int_0^\infty c \exp\left(-x - \frac{y}{2}\right) dxdy = 1 \Rightarrow c = \frac{1}{2}.$$

Next, suppose we wish to determine the probability of the event $\{X > Y\}$. This can be viewed as finding the probability of the pair (X, Y) falling in the region A that is now defined as $A = \{(x,y) : x > y\}$. This probability is calculated as

$$\Pr(X > Y) = \iint\limits_{x>y} f_{X,Y}(x,y)\, dxdy = \int_0^\infty \int_y^\infty \frac{1}{2}\exp\left(-x - \frac{y}{2}\right) dxdy$$

$$= \int_0^\infty \frac{1}{2}\exp\left(-\frac{3y}{2}\right) dy = \frac{1}{3}.$$

EXAMPLE 5.6: In many cases, evaluating the probability of a pair of random variables falling in some region may be quite difficult to calculate analytically. For example, suppose we modify Example 5.4 so that the joint PDF is now of the form

$$f_{X,Y}(x,y) = \frac{1}{2\pi}\exp\left(-\frac{((x-2)^2 + (y-3)^2)}{2}\right).$$

Again, we would like to evaluate the probability that the pair (X, Y) falls in the unit circle. To do this analytically we must evaluate

$$\iint\limits_{x^2+y^2<1} \frac{1}{2\pi}\exp\left(-\frac{((x-2)^2 + (y-3)^2)}{2}\right) dxdy.$$

Converting to polar coordinates, the integral becomes

$$\int_0^1 \int_0^{2\pi} \frac{r}{2\pi}\exp\left(-\frac{((r\cos(\theta) - 2)^2 + (r\sin(\theta) - 3)^2)}{2}\right) d\theta dr.$$

Either way, the double integral looks formidable. We can enlist MATLAB to help in one of two ways. First, we could randomly generate many samples of the pair of random variables according to the specified distribution and count the relative frequency of the number that fall within the unit circle. Alternatively, we could get MATLAB to calculate one of the preceding double integrals numerically. We will take the latter approach here and evaluate the double integral in polar coordinates. First, we must define a MATLAB function to evaluate the integrand:

```
function out=dblintegrand(q,r)
out=r.*exp(-((r*cos(q)-2).^2+(r*sin(q)-3).^2)/2);
```

MATLAB will then evaluate the integral by executing the command

```
dblquad('dblintegrand',0,2*pi,0,1)/(2*pi).
```

By executing these MATLAB commands, we find the value of the integral to be 0.002072.

5.3 Joint Probability Mass Functions

When the random variables are discrete rather than continuous, it is often more convenient to work with probability mass functions rather than PDFs or CDFs. It is straightforward to extend the concept of the probability mass function to a pair of random variables.

DEFINITION 5.3: The *joint probability mass function* for a pair of discrete random variables X and Y is given by $P_{X,Y}(x,y) = \Pr(\{X = x\} \cap \{Y = y\})$.

In particular, suppose the random variable X takes on values from the set $\{x_1, x_2, \ldots, x_M\}$ and the random variable Y takes on values from the set $\{y_1, y_2, \ldots, y_N\}$. Here, either M and/or N could be potentially infinite, or both could be finite. Several properties of the joint probability mass function analogous to those developed for joint PDFs should be apparent.

(1) $0 \leq P_{X,Y}(x_m, y_n) \leq 1;$ (5.11a)

(2) $\displaystyle\sum_{m=1}^{M} \sum_{n=1}^{N} P_{X,Y}(x_m, y_n) = 1;$ (5.11b)

$$(3) \sum_{n=1}^{N} P_{X,Y}(x_m, y_n) = P_X(x_m), \quad \sum_{m=1}^{M} P_{X,Y}(x_m, y_n) = P_Y(y_n); \tag{5.11c}$$

$$(4) \Pr((X, Y) \in A) = \sum_{(x,y) \in A} \sum P_{X,Y}(x, y). \tag{5.11d}$$

Furthermore, the joint PDF or the joint CDF of a pair of discrete random variables can be related to the joint PMF through the use of delta functions or step functions by

$$f_{X,Y}(x, y) = \sum_{m=1}^{M} \sum_{n=1}^{N} P_{X,Y}(x_m, y_n)\delta(x - x_m)\delta(y - y_m), \tag{5.12}$$

$$F_{X,Y}(x, y) = \sum_{m=1}^{M} \sum_{n=1}^{N} P_{X,Y}(x_m, y_n)u(x - x_m)u(y - y_m). \tag{5.13}$$

Usually, it is most convenient to work with PMFs when the random variables are discrete. However, if the random variables are mixed (i.e., one is discrete and one is continuous), then it becomes necessary to work with PDFs or CDFs since the PMF will not be meaningful for the continuous random variable.

EXAMPLE 5.7: A pair of discrete random variables N and M have a joint PMF given by

$$P_{N,M}(n, m) = \frac{(n+m)!}{n!m!} \frac{a^n b^m}{(a+b+1)^{n+m+1}}, \quad m = 0, 1, 2, 3, \ldots, \quad n = 0, 1, 2, 3, \ldots.$$

The marginal PMF of N can be found by summing over m in the joint PMF:

$$P_N(n) = \sum_{m=0}^{\infty} P_{N,M}(n, m) = \sum_{m=0}^{\infty} \frac{(n+m)!}{n!m!} \frac{a^n b^m}{(a+b+1)^{n+m+1}}.$$

To evaluate this series, the following identity is used:

$$\sum_{m=0}^{\infty} \frac{(n+m)!}{n!m!} x^m = \left(\frac{1}{1-x}\right)^{n+1}.$$

The marginal PMF then reduces to

$$P_N(n) = \frac{a^n}{(a+b+1)^{n+1}} \sum_{m=0}^{\infty} \frac{(n+m)!}{n!m!} \frac{b^m}{(a+b+1)^m}$$

$$= \frac{a^n}{(a+b+1)^{n+1}} \left(\frac{1}{1 - \dfrac{b}{a+b+1}} \right)^{n+1} = \frac{a^n}{(1+a)^{n+1}}.$$

Likewise, by symmetry, the marginal PMF of M is

$$P_M(m) = \frac{b^m}{(1+b)^{m+1}}.$$

Hence, the random variables M and N both follow a geometric distribution.

5.4 Conditional Distribution, Density, and Mass Functions

The notion of conditional distribution functions and conditional density functions was first introduced in Chapter 3. In this section, those ideas are extended to the case where the conditioning event is related to another random variable. For example, we might want to know the distribution of a random variable representing the score a student achieves on a test given the value of another random variable representing the number of hours the student studied for the test. Or, perhaps we want to know the probability density function of the outside temperature given that the humidity is known to be below 50 percent.

To start with, consider a pair of discrete random variables X and Y with a PMF, $P_{X,Y}(x,y)$. Suppose we would like to know the PMF of the random variable X given that the value of Y has been observed. Then, according to the definition of conditional probability:

$$\Pr(X = x | Y = y) = \frac{\Pr(X = x, Y = y)}{\Pr(Y = y)} = \frac{P_{X,Y}(x,y)}{P_Y(y)}. \tag{5.14}$$

We refer to this as the conditional PMF of X given Y. By way of notation we write $P_{X|Y}(x|y) = P_{X,Y}(x,y)/P_Y(y)$.

EXAMPLE 5.8: Using the joint PMF given in Example 5.7, along with the marginal PMF found in that example, it is found that

$$P_{N|M}(n|m) = \frac{P_{M,N}(m,n)}{P_M(m)} = \frac{(n+m)!}{n!m!}\frac{a^n b^m}{(a+b+1)^{n+m+1}}\frac{(1+b)^{m+1}}{b^m}$$

$$= \frac{(n+m)!}{n!m!}\frac{a^n(1+b)^{m+1}}{(a+b+1)^{n+m+1}}.$$

Note that the conditional PMF of N given M is quite different than the marginal PMF of N. That is, knowing M changes the distribution of N.

The simple result developed in Equation 5.14 can be extended to the case of continuous random variables and PDFs. The following theorem shows that the PMFs in Equation 5.14 can simply be replaced by PDFs.

THEOREM 5.2: The conditional PDF of a random variable X given that $Y = y$ is

$$f_{X|Y}(x|y) = \frac{f_{X,Y}(x,y)}{f_Y(y)}. \tag{5.15}$$

PROOF: Consider the conditioning event $A = y \leq Y < y + dy$. Then

$$f_{X|A}(x)\,dx = \Pr(x \leq X < x + dx | y \leq Y < y + dy)$$

$$= \frac{\Pr(x \leq X < x + dx,\ y \leq Y < y + dy)}{\Pr(y \leq Y < y + dy)}$$

$$= \frac{f_{X,Y}(x,y)\,dx\,dy}{f_Y(y)\,dy} = \frac{f_{X,Y}(x,y)\,dx}{f_Y(y)}.$$

Passing to the limit as $dy \to 0$, the event A becomes the event $\{Y = y\}$, producing the desired result. ∎

Integrating both sides of this equation with respect to x produces the appropriate result for CDFs:

$$F_{X|Y}(x|y) = \frac{\int_{-\infty}^{x} f_{X,Y}(x',y)\,dx'}{f_Y(y)}. \tag{5.16}$$

Usually, the conditional PDF is much easier to work with, so the conditional CDF will not be discussed further.

EXAMPLE 5.9: A certain pair of random variables has a joint PDF given by

$$f_{X,Y}(x,y) = \frac{2abc}{(ax + by + c)^3} u(x)u(y)$$

for some positive constants a, b, and c. The marginal PDFs are easily found to be

$$f_X(x) = \int_0^\infty f_{X,Y}(x,y)\, dy = \frac{ac}{(ax+c)^2} u(x)$$

and

$$f_Y(y) = \int_0^\infty f_{X,Y}(x,y)\, dx = \frac{bc}{(by+c)^2} u(y).$$

The conditional PDF of X given Y then works out to be

$$f_{X|Y}(x|y) = \frac{f_{X,Y}(x,y)}{f_Y(y)} = \frac{2a(by+c)^2}{(ax+by+c)^3} u(x).$$

The conditional PDF of Y given X could also be determined in a similar way:

$$f_{Y|X}(y|x) = \frac{f_{X,Y}(x,y)}{f_X(x)} = \frac{2b(ax+c)^2}{(ax+by+c)^3} u(y).$$

EXAMPLE 5.10: This example involves two Gaussian random variables. Suppose X and Y have a joint PDF given by

$$f_{X,Y}(x,y) = \frac{1}{\pi\sqrt{3}} \exp\left(-\frac{2}{3}(x^2 - xy + y^2)\right).$$

The marginal PDF is found as follows:

$$f_X(x) = \int_{-\infty}^\infty f_{X,Y}(x,y)\, dy = \frac{1}{\pi\sqrt{3}}\exp\left(-\frac{2}{3}x^2\right)\int_{-\infty}^\infty \exp\left(-\frac{2}{3}(y^2 - xy)\right) dy.$$

In order to evaluate the integral, complete the square in the exponent:

$$f_X(x) = \frac{1}{\pi\sqrt{3}}\exp\left(-\frac{2}{3}x^2\right)\exp\left(\frac{x^2}{6}\right)\int_{-\infty}^\infty \exp\left(-\frac{2}{3}\left(y^2 - xy + \frac{x^2}{4}\right)\right) dy$$

$$= \frac{1}{\pi\sqrt{3}}\exp\left(-\frac{x^2}{2}\right)\int_{-\infty}^\infty \exp\left(-\frac{2}{3}\left(y - \frac{x}{2}\right)^2\right) dy.$$

Now the integrand is a Gaussian-looking function. If the appropriate constant is added to the integrand, the integrand will be a valid PDF and, hence, must integrate out to one. In this case, the constant we need to add to the integrand to make the integral unity is $\sqrt{2/(3\pi)}$. Stated another way, the integral as just written must evaluate to $\sqrt{3\pi/2}$. Hence, the marginal PDF of X is

$$f_X(x) = \frac{1}{\sqrt{2\pi}} \exp\left(-\frac{x^2}{2}\right),$$

and we see that X is a zero-mean, unit-variance, Gaussian (i.e., standard normal) random variable. By symmetry, the marginal PDF of Y must also be of the same form. The conditional PDF of X given Y is

$$f_{X|Y}(x|y) = \frac{f_{X,Y}(x,y)}{f_Y(y)} = \frac{\frac{1}{\pi\sqrt{3}}\exp\left(-\frac{2}{3}(x^2 - xy + y^2)\right)}{\frac{1}{\sqrt{2\pi}}\exp\left(-\frac{y^2}{2}\right)}$$

$$= \sqrt{\frac{2}{3\pi}}\exp\left(-\frac{2}{3}\left(x^2 - xy + \frac{y^2}{4}\right)\right) = \sqrt{\frac{2}{3\pi}}\exp\left(-\frac{2}{3}\left(x - \frac{y}{2}\right)^2\right).$$

So, the conditional PDF of X given Y is also Gaussian. But, given that it is known that $Y = y$, the mean of X is now $y/2$ (instead of zero), and the variance of X is $3/4$ (instead of one). In this example, knowledge of Y has shifted the mean and reduced the variance of X.

In addition to conditioning on a random variable taking on a point value such as $Y = y$, the conditioning can also occur on an interval of the form $y_1 \leq Y \leq y_2$. To simplify notation, let the conditioning event A be $A = \{y_1 \leq Y \leq y_2\}$. The relevant conditional PMF, PDF, and CDF are then given respectively by:

$$P_{X|A}(x) = \frac{\sum_{y=y_1}^{y_2} P_{X,Y}(x,y)}{\sum_{y=y_1}^{y_2} P_Y(y)};$$ (5.17)

$$f_{X|A}(x) = \frac{\int_{y_1}^{y_2} f_{X,Y}(x,y)\,dy}{\int_{y_1}^{y_2} f_Y(y)\,dy};$$ (5.18)

$$F_{X|A}(x) = \frac{F_{X,Y}(x,y_2) - F_{X,Y}(x,y_1)}{F_Y(y_2) - F_Y(y_1)}.$$ (5.19)

It is left as an exercise for the reader to derive these expressions.

EXAMPLE 5.11: Using the joint PDF of Example 5.10, suppose we want to determine the conditional PDF of X given that $Y > y_o$. The numerator in Equation 5.18 is calculated according to

$$\int_{y_o}^{\infty} f_{X|Y}(x|y)\, dy = \int_{y_o}^{\infty} \frac{1}{\pi\sqrt{3}} \exp\left(-\frac{2}{3}(x^2 - xy + y^2)\right) dy$$

$$= \frac{1}{\sqrt{2\pi}} \exp\left(-\frac{x^2}{2}\right) \int_{y_o}^{\infty} \sqrt{\frac{2}{3\pi}} \exp\left(-\frac{2}{3}\left(y - \frac{x}{2}\right)^2\right) dy$$

$$= \frac{1}{\sqrt{2\pi}} \exp\left(-\frac{x^2}{2}\right) Q\left(\frac{2y_o - x}{\sqrt{3}}\right)$$

Since the marginal PDF of Y is a zero-mean, unit-variance Gaussian PDF, the denominator of Equation 5.18 becomes

$$\int_{y_o}^{\infty} f_Y(y)\, dy = \int_{y_o}^{\infty} \frac{1}{\sqrt{2\pi}} \exp\left(-\frac{y^2}{2}\right) dy = Q(y_o).$$

Therefore, the PDF of X conditioned on $Y > y_o$ is

$$f_{X|Y>y_o}(x) = \frac{1}{\sqrt{2\pi}} \exp\left(-\frac{x^2}{2}\right) \frac{Q\left(\dfrac{2y_o - x}{\sqrt{3}}\right)}{Q(y_o)}.$$

Note that when the conditioning event was a point condition on Y, the conditional PDF of X was Gaussian; yet, when the conditioning event is an interval condition on Y, the resulting conditional PDF of X is not Gaussian at all.

5.5 Expected Values Involving Pairs of Random Variables

The notion of expected value is easily generalized to pairs of random variables. To begin, we define the expected value of an arbitrary function of two random variables.

DEFINITION 5.4: Let $g(x, y)$ be an arbitrary two-dimensional function. The expected value of $g(X, Y)$, where X and Y are random variables, is

$$E[g(X, Y)] = \iint\limits_{-\infty}^{\infty} g(x, y) f_{X,Y}(x, y)\, dx\, dy. \qquad (5.20)$$

For discrete random variables, the equivalent expression in terms of the joint PMF is

$$E[g(X, Y)] = \sum_m \sum_n g(x_m, y_n) P_{X,Y}(x_m, y_n). \qquad (5.21)$$

If the function $g(x, y)$ is actually a function of only a single variable, say x, then this definition reduces to the definition of expected values for functions of a single random variable as given in Definition 4.2.

$$E[g(X)] = \iint\limits_{-\infty}^{\infty} g(x) f_{X,Y}(x, y)\, dx\, dy = \int_{-\infty}^{\infty} g(x) \left[\int_{-\infty}^{\infty} f_{X,Y}(x, y)\, dy \right] dx = \int_{-\infty}^{\infty} g(x) f_X(x)\, dx.$$

$$(5.22)$$

To start with, consider an arbitrary linear function of the two variables $g(x, y) = ax + by$, where a and b are constants. Then

$$E[aX + bY] = \iint\limits_{-\infty}^{\infty} [ax + by] f_{X,Y}(x, y)\, dx\, dy$$

$$= a \iint\limits_{-\infty}^{\infty} x f_{X,Y}(x, y)\, dx\, dy + b \iint\limits_{-\infty}^{\infty} y f_{X,Y}(x, y)\, dx\, dy$$

$$= aE[X] + bE[Y]. \qquad (5.23)$$

This result merely states that expectation is a linear operation.

In addition to the functions considered in Chapter 4 that led to such statistics as means, variances, and the like, functions involving both variables x and y will be considered here. These new functions will lead to statistics that will partially characterize the relationships between the two random variables.

DEFINITION 5.5: The *correlation* between two random variables is defined as

$$R_{X,Y} = E[XY] = \iint\limits_{-\infty}^{\infty} xy f_{X,Y}(x, y)\, dx\, dy. \qquad (5.24)$$

Furthermore, two random variables that have a correlation of zero are said to be *orthogonal*.

One instance in which the correlation appears is in calculating the second moment of a sum of two random variables. That is, consider finding the expected value of $g(X, Y) = (X + Y)^2$:

$$E[(X + Y)^2] = E[X^2 + 2XY + Y^2] = E[X^2] + E[Y^2] + 2E[XY]. \tag{5.25}$$

Hence, the second moment of the sum is the sum of the second moment plus twice the correlation.

DEFINITION 5.6: The *covariance* between two random variables is

$$\mathrm{Cov}(X, Y) = E[(X - \mu_X)(Y - \mu_Y)] = \int\!\!\!\int_{-\infty}^{\infty} (x - \mu_X)(y - \mu_Y)f_{X,Y}(x, y)\, dxdy. \tag{5.26}$$

If two random variables have a covariance of zero, they are said to be *uncorrelated*.

The correlation and covariance are strongly related to one another as shown by the following theorem.

THEOREM 5.3: $$\mathrm{Cov}(X, Y) = R_{X,Y} - \mu_X\mu_Y. \tag{5.27}$$

PROOF: $\mathrm{Cov}(X, Y) = E[(X - \mu_X)(Y - \mu_y)] = E[XY - \mu_X Y - \mu_Y X + \mu_X\mu_Y]$
$$= E[XY] - \mu_X E[Y] - \mu_Y E[X] + \mu_X\mu_Y = E[XY] - \mu_X\mu_Y.$$

∎

As a result, if either X or Y (or both) has a mean of zero, correlation and covariance are equivalent. The covariance function occurs when calculating the variance of a sum of two random variables:

$$\mathrm{Var}(X + Y) = \mathrm{Var}(X) + \mathrm{Var}(Y) + 2\mathrm{Cov}(X, Y). \tag{5.28}$$

This result can be obtained from Equation 5.25 by replacing X with $X - \mu_X$ and Y with $Y - \mu_Y$.

Another statistical parameter related to a pair of random variables is the correlation coefficient, which is nothing more than a normalized version of the covariance.

DEFINITION 5.7: The *correlation coefficient* of two random variables X and Y, ρ_{XY}, is defined as

$$\rho_{XY} = \frac{\mathrm{Cov}(X, Y)}{\sqrt{\mathrm{Var}(X)\mathrm{Var}(Y)}} = \frac{E[(X - \mu_X)(Y - \mu_Y)]}{\sigma_X\sigma_Y}. \tag{5.29}$$

The next theorem quantifies the nature of the normalization. In particular, it shows that a correlation coefficient can never be more than 1 in absolute value.

THEOREM 5.4: The correlation coefficient is less than 1 in magnitude.

PROOF: Consider taking the second moment of $X + aY$, where a is a real constant:

$$E[(X + aY)^2] = E[X^2] + 2aE[XY] + a^2E[Y^2] \geq 0.$$

Since this is true for any a, we can tighten the bound by choosing the value of a that minimizes the left-hand side. This value of a turns out to be

$$a = \frac{-E[XY]}{E[Y^2]}.$$

Plugging in this value gives

$$E[X^2] + \frac{(E[XY])^2}{E[Y^2]} - \frac{2(E[XY])^2}{E[Y^2]} \geq 0$$

$$\Rightarrow (E[XY])^2 \leq E[X^2]E[Y^2].$$

If we replace X with $X - \mu_X$ and Y with $Y - \mu_Y$, the result is

$$(\text{Cov}(X, Y))^2 \leq \text{Var}(X)\text{Var}(Y).$$

Rearranging terms then gives the desired result:

$$|\rho_{XY}| = \left| \frac{\text{Cov}(X, Y)}{\sqrt{\text{Var}(X)\text{Var}(Y)}} \right| \leq 1. \tag{5.30}$$

∎

Note that we can also infer from the proof that equality holds if Y is a constant times X. That is, a correlation coefficient of 1 (or -1) implies that X and Y are completely correlated (knowing Y determines X). Furthermore, uncorrelated random variables will have a correlation coefficient of zero. Hence, as its name implies, the correlation coefficient is a quantitative measure of the correlation between two random variables. It should be emphasized at this point that zero correlation is not to be confused with independence. These two concepts are not the same (more on this later).

The significance of the correlation, covariance, and correlation coefficient will be discussed further in the next two sections. For now, we present an example showing how to compute these parameters.

EXAMPLE 5.12: Consider once again the joint PDF of Example 5.10. The correlation for these random variables is

$$E[XY] = \int_{-\infty}^{\infty} \int_{-\infty}^{\infty} \frac{xy}{\pi\sqrt{3}} \exp\left(-\frac{2}{3}(x^2 - xy + y^2) \right) dy\,dx.$$

In order to evaluate this integral, the joint PDF is rewritten as $f_{X,Y}(x, y) = f_{Y|X}(y|x)f_X(x)$ and those terms involving only x are pulled outside the inner integral over y:

$$E[XY] = \int_{-\infty}^{\infty} \frac{x}{\sqrt{2\pi}} \exp\left(-\frac{x^2}{2}\right) \left[\int_{-\infty}^{\infty} y\sqrt{\frac{2}{3\pi}} \exp\left(-\frac{2}{3}\left(y - \frac{x}{2}\right)^2\right) dy\right] dx.$$

The inner integral (in square brackets) is the expected value of a Gaussian random variable with a mean of $x/2$ and a variance of $3/4$ which thus evaluates to $x/2$. Hence,

$$E[XY] = \frac{1}{2} \int_{-\infty}^{\infty} \frac{x^2}{\sqrt{2\pi}} \exp\left(-\frac{x^2}{2}\right) dx.$$

The remaining integral is the second moment of a Gaussian random variable with zero-mean and unit variance, which thus integrates to 1. The correlation of these two random variables is therefore $E[XY] = 1/2$. Since both X and Y have zero means, $\text{Cov}(X, Y)$ is also equal to $1/2$. Finally, the correlation coefficient is also $\rho_{XY} = 1/2$ due to the fact that both X and Y have unit variance.

The concepts of correlation and covariance can be generalized to higher order moments as given in the following definition.

DEFINITION 5.8: The (m, n)th joint moment of two random variables X and Y is

$$E[X^m Y^n] = \int_{-\infty}^{\infty}\!\!\int x^m y^n f_{X,Y}(x, y)\, dx\, dy. \tag{5.31}$$

The (m, n)th joint central moment is similarly defined as

$$E[(X - \mu_X)^m (Y - \mu_Y)^n] = \int_{-\infty}^{\infty}\!\!\int (x - \mu_X)^m (y - \mu_Y)^n f_{X,Y}(x, y)\, dx\, dy \tag{5.32}$$

These higher order joint moments are not frequently used and therefore are not considered further here.

As with single random variables, a conditional expected value can also be defined for which the expectation is carried out with respect to the appropriate conditional density function.

DEFINITION 5.9: The conditional expected value of a function $g(X)$ of a random variable X given that $Y = y$ is

$$E[g(X)|Y] = \int_{-\infty}^{\infty} g(x) f_{X|Y}(x|y)\, dx. \tag{5.33}$$

Conditional expected values can be particularly useful in calculating expected values of functions of two random variables that can be factored into the product of two one-dimensional functions. That is, consider a function of the form $g(x,y) = g_1(x)g_2(y)$. Then

$$E[g_1(X)g_2(Y)] = \int\int_{-\infty}^{\infty} g_1(x)g_2(y)f_{X,Y}(x,y)\,dxdy. \tag{5.34}$$

From Equation 5.15, the joint PDF is rewritten as $f_{X,Y}(x,y) = f_{Y|X}(y|x)f_X(x)$, resulting in

$$E[g_1(X)g_2(Y)] = \int_{-\infty}^{\infty} g_1(x)f_X(x)\left[\int_{-\infty}^{\infty} g_2(y)f_{Y|X}(y|x)\,dy\right]dx$$

$$= \int_{-\infty}^{\infty} g_1(x)f_X(x)E_Y[g_2(Y)|X]\,dx = E_X[g_1(X)E_Y[g_2(Y)|X]]. \tag{5.35}$$

Here the subscripts on the expectation operator have been included for clarity to emphasize that the outer expectation is with respect to the random variable X, while the inner expectation is with respect to the random variable Y (conditioned on X). This result allows us to break a two-dimensional expectation into two one-dimensional expectations. This technique was used in Example 5.12, where the correlation between two variables was essentially written as

$$R_{X,Y} = E_X[XE_Y[Y|X]]. \tag{5.36}$$

In that example, the conditional PDF of Y given X was Gaussian, and hence finding the conditional mean was trivial. The outer expectation then required finding the second moment of a Gaussian random variable, which is also straightforward.

5.6 Independent Random Variables

The concept of independent events was introduced in Chapter 2. In this section, we extend this concept to the realm of random variables. To make that extension, consider the events $A = \{X \leq x\}$ and $B = \{Y \leq y\}$ related to the random variables X and Y. The two events A and B are statistically independent if $\Pr(A,B) = \Pr(A)\Pr(B)$. Restated in terms of the random variables, this condition becomes

$$\Pr(X \leq x, Y \leq y) = \Pr(X \leq x)\Pr(Y \leq y) \Rightarrow F_{X,Y}(x,y) = F_X(x)F_Y(y). \tag{5.37}$$

Hence, two random variables are statistically independent if their joint CDF factors into a product of the marginal CDFs. Differentiating both sides of this equation with respect to both x and y reveals that the same statement applies to the PDF as well. That is, for statistically independent random variables, the joint PDF factors into a product of the marginal PDFs:

$$f_{X,Y}(x,y) = f_X(x)f_Y(y). \tag{5.38}$$

It is not difficult to show that the same statement applies to PMFs as well. The preceding condition can also be restated in terms of conditional PDFs. Dividing both sides of Equation 5.38 by $f_X(x)$ results in

$$f_{Y|X}(y|x) = f_Y(y). \tag{5.39}$$

A similar result involving the conditional PDF of X given Y could have been obtained by dividing both sides by the PDF of Y. In other words, if X and Y are independent, knowing the value of the random variable X should not change the distribution of Y and vice versa.

EXAMPLE 5.13: Returning once again to the joint PDF of Example 5.10, we saw in that example that the marginal PDF of X is

$$f_X(x) = \frac{1}{\sqrt{2\pi}} \exp\left(-\frac{x^2}{2}\right),$$

while the conditional PDF of X given Y is

$$f_{X|Y}(x|y) = \sqrt{\frac{2}{3\pi}} \exp\left(-\frac{2}{3}\left(x - \frac{y}{2}\right)^2\right).$$

Clearly, these two random variables are not independent.

EXAMPLE 5.14: Suppose the random variables X and Y are uniformly distributed on the square defined by $0 \le x, y \le 1$. That is

$$f_{X,Y}(x,y) = \begin{cases} 1 & 0 \le x, \ y \le 1 \\ 0 & \text{otherwise} \end{cases}.$$

The marginal PDFs of X and Y work out to be

$$f_X(x) = \begin{cases} 1 & 0 \le x \le 1 \\ 0 & \text{otherwise} \end{cases}, \qquad f_Y(y) = \begin{cases} 1 & 0 \le y \le 1 \\ 0 & \text{otherwise} \end{cases}.$$

These random variables are statistically independent since $f_{X,Y}(x,y) = f_X(x)f_Y(y)$.

THEOREM 5.5: Let X and Y be two independent random variables and consider forming two new random variables $U = g_1(X)$ and $V = g_2(Y)$. These new random variables U and V are also independent.

PROOF: To show that U and V are independent, consider the events $A = \{U \leq u\}$ and $B = \{V \leq v\}$. Next, define the region R_u to be the set of all points x such that $g_1(x) \leq u$. Similarly, define R_v to be the set of all points y such that $g_2(y) \leq v$. Then

$$\Pr(U \leq u, V \leq v) = \Pr(X \in R_u, Y \in R_v) = \int_{R_v} \int_{R_u} f_{X,Y}(x,y)\,dx\,dy.$$

Since X and Y are independent, their joint PDF can be factored into a product of marginal PDFs resulting in

$$\Pr(U \leq u, V \leq v) = \int_{R_u} f_X(x)\,dx \int_{R_v} f_Y(y)\,dy$$

$$= \Pr(X \in R_u)\Pr(Y \in R_v) = \Pr(U \leq u)\Pr(V \leq v).$$

Since we have shown that $F_{U,V}(u,v) = F_U(u)F_V(v)$, the random variables U and V must be independent.   ∎

Another important result deals with the correlation, covariance, and correlation coefficients of independent random variables.

THEOREM 5.6: If X and Y are independent random variables, then $E[XY] = \mu_X \mu_Y$, $\mathrm{Cov}(X,Y) = 0$, and $\rho_{X,Y} = 0$.

PROOF:
$$E[XY] = \int\!\!\!\int_{-\infty}^{\infty} xy f_{X,Y}(x,y)\,dx\,dy = \int_{-\infty}^{\infty} x f_X(x)\,dx \int_{-\infty}^{\infty} y f_Y(y)\,dy = \mu_X \mu_Y.$$

The conditions involving covariance and correlation coefficient follow directly from this result.   ∎

Therefore, independent random variables are necessarily uncorrelated, but the converse is not always true. Uncorrelated random variables do not have to be independent as demonstrated by the next example.

EXAMPLE 5.15: Consider a pair of random variables X and Y that are uniformly distributed over the unit circle so that

$$f_{X,Y}(x,y) = \begin{cases} \dfrac{1}{\pi} & x^2 + y^2 \leq 1 \\ 0 & \text{otherwise} \end{cases}.$$

The marginal PDF of X can be found as follows:

$$f_X(x) = \int_{-\infty}^{\infty} f_{X,Y}(x,y)\,dy = \int_{-\sqrt{1-x^2}}^{\sqrt{1-x^2}} \frac{1}{\pi}\,dy = \frac{2}{\pi}\sqrt{1-x^2}, \quad -1 \le x \le 1.$$

By symmetry, the marginal PDF of Y must take on the same functional form. Hence, the product of the marginal PDFs is

$$f_X(x)f_Y(y) = \frac{4}{\pi^2}\sqrt{(1-x^2)(1-y^2)}, \quad -1 \le x, y \le 1.$$

Clearly, this is not equal to the joint PDF, and hence the two random variables are dependent. This conclusion could have been determined in a simpler manner. Note that if we are told that $X = 1$, then necessarily $Y = 0$, whereas if we know that $X = 0$, then Y can range anywhere from -1 to 1. Therefore, conditioning on different values of X leads to different distributions for Y.

Next, the correlation between X and Y is calculated.

$$E[XY] = \iint_{x^2+y^2 \le 1} \frac{xy}{\pi}\,dx\,dy = \frac{1}{\pi}\int_{-1}^{1} x\left[\int_{-\sqrt{1-x^2}}^{\sqrt{1-x^2}} y\,dy\right]dx$$

Since the inner integrand is an odd function (of y) and the limits of integration are symmetric about zero, the integral is zero. Hence, $E[XY] = 0$. Note from the marginal PDFs just found that both X and Y are zero-mean. So, it is seen for this example that while the two random variables are uncorrelated, they are not independent.

EXAMPLE 5.16: Suppose we wish to use MATLAB to generate samples of a pair of random variables (X, Y) that are uniformly distributed over the unit circle. That is, the joint PDF is

$$f_{X,Y}(x,y) = \begin{cases} \dfrac{1}{\pi} & x^2 + y^2 < 1 \\ 0 & \text{otherwise} \end{cases}.$$

If we generated two random variables independently according to the MATLAB code X=rand(1); Y=rand(1); this would produce a pair of random variables uniformly distributed over the square $0 < x < 1, 0 < y < 1$. One way to achieve the desired result is to generate random variables uniformly over some region that includes the unit circle, and then keep only those pairs of samples which fall inside the unit circle. In this case, it is straight-forward to generate random variables that are uniformly distributed over the square $-1 < x < 1, -1 < y < 1$, which circumscribes the unit circle. Then we keep only those samples drawn from within this square that also fall within the unit circle. The code that follows illustrates this technique. We also show how to generate a three-dimensional plot of an estimate of the joint PDF from the random data generated. To get a decent estimate of the joint PDF, we need to generate a rather large number of samples (we found that 100,000 worked pretty well). This requires that we create and perform several operations on some very large vectors. Doing so tends to make the program run very slowly. In order to speed up the operation of the program, we choose to create shorter vectors of random variables (1,000 in this case) and then repeat the procedure several times (100 in this case). Although this makes the code a little longer and probably a little harder to follow, by avoiding the creation of very long vectors, it substantially speeds up the program. The results of this program are shown in Figure 5.3.

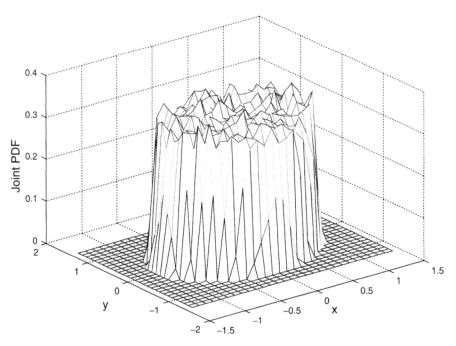

Figure 5.3   Estimate of the joint PDF of a pair of random variables uniformly distributed over the unit circle from the data generated in Example 5.16.

```
clear
N=1000;                              % Number of samples per
                                       iteration.
bw=0.1;                              % Bin widths for histogram.
xbins=[-1.4:bw:1.4];
ybins=[-1.4:bw:1.4];                 % Histogram bins.
iterations=100;                      % Number of iterations.
M=length(xbins);
Nsamples=zeros(M);                   % Initialize matrix for
                                       storing data.
count=0;                             % Initialize counter.

for ii=1:iterations
x=2*rand(1,N)-1; y=2*rand(1,N)-1;    % Generate variables over
                                       square.

% Keep only those within the unit circle.
X=[]; Y=[];
for k=1:N
   if x(k)^2+y(k)^2<1
      X=[X x(k)];
      Y=[Y y(k)];
   end                               % End if statement.
end                                  % End k loop.
count=count+length(X);               % Count random samples
                                       generated.

% Compute number of samples that fall within each bin.
for m=1:length(xbins)
   for n=1:length(ybins)
      temp1=(abs(X-xbins(m))<bw/2);
      temp2=(abs(Y-ybins(n))<bw/2);
      Nsamples(m,n)=Nsamples(m,n)+sum(temp1.*temp2);
   end                               % End n loop.
end                                  % End m loop.
end                                  % End iterations.

PDFest=Nsamples/(count*bw^2);        % Convert to probability
                                       densities.
mesh(xbins,ybins,PDFest)             % Plot estimate of joint
                                       PDF.
xlabel('x'); ylabel('y');            % Label plot axes.
zlabel('Joint PDF');
```

5.7 Jointly Gaussian Random Variables

As with single random variables, the most common and important example of a two-dimensional probability distribution is that of a joint Gaussian distribution. We begin by defining what is meant by a joint Gaussian distribution.

DEFINITION 5.10: A pair of random variables X and Y are said to be *jointly Gaussian* if their joint PDF is of the general form

$$f_{X,Y}(x,y) = \frac{1}{2\pi\sigma_X\sigma_Y\sqrt{1-\rho_{XY}^2}}$$

$$\times \exp\left(-\left[\frac{\left(\frac{x-\mu_X}{\sigma_X}\right)^2 - 2\rho_{XY}\left(\frac{x-\mu_X}{\sigma_X}\right)\left(\frac{y-\mu_Y}{\sigma_Y}\right) + \left(\frac{y-\mu_Y}{\sigma_Y}\right)^2}{2(1-\rho_{XY}^2)}\right]\right)$$

(5.40)

where μ_X and μ_Y are the means of X and Y, respectively; σ_X and σ_Y are the standard deviations of X and Y, respectively; and ρ_{XY} is the correlation coefficient of X and Y.

It is left as an exercise for the reader to verify that this joint PDF results in marginal PDFs that are Gaussian. That is

$$f_X(x) = \int_{-\infty}^{\infty} f_{X,Y}(x,y)\,dy = \frac{1}{\sqrt{2\pi\sigma_X^2}}\exp\left(-\frac{(x-\mu_X)^2}{2\sigma_X^2}\right),$$

$$f_Y(y) = \int_{-\infty}^{\infty} f_{X,Y}(x,y)\,dx = \frac{1}{\sqrt{2\pi\sigma_Y^2}}\exp\left(-\frac{(y-\mu_Y)^2}{2\sigma_Y^2}\right).$$

(5.41)

It is also left as an exercise for the reader (see Exercise 5.15) to demonstrate that if X and Y are jointly Gaussian, then the conditional PDF of X given $Y = y$ is also Gaussian, with a mean of $\mu_X + \rho_{XY}(\sigma_X/\sigma_Y)(y - \mu_Y)$ and a variance of $\sigma_X^2(1 - \rho_{XY}^2)$. An example of this was shown in Example 5.10 and the general case can be proven following the same steps shown in that example.

Figure 5.4 shows the joint Gaussian PDF for three different values of the correlation coefficient. In Figure 5.4(a), the correlation coefficient is $\rho_{XY} = 0$, and hence the two random variables are uncorrelated (and as we will see shortly, independent). Figure 5.4(b) shows the joint PDF when the correlation coefficient is large and positive, $\rho_{XY} = 0.9$. Note how the surface has become taller and thinner and

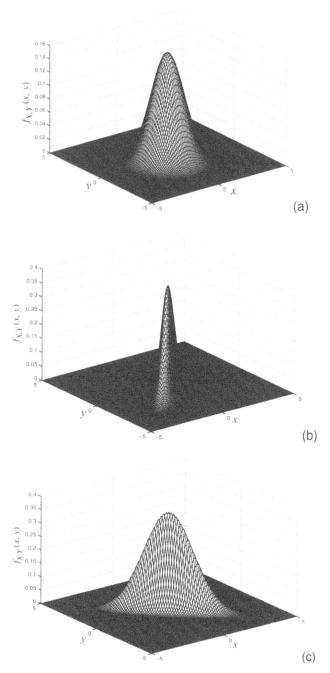

Figure 5.4 (a) The joint Gaussian PDF: $\mu_X = \mu_Y = 0$, $\sigma_X = \sigma_Y = 1$, $\rho_{XY} = 0$; (b) the joint Gaussian PDF: $\mu_X = \mu_Y = 0$, $\sigma_X = \sigma_Y = 1$, $\rho_{XY} = 0.9$; (c) the joint Gaussian PDF: $\mu_X = \mu_Y = 0$, $\sigma_X = \sigma_Y = 1$, $\rho_{XY} = -0.9$.

largely lies above the line $y = x$. In Figure 5.4(c), the correlation is now large and negative, $\rho_{XY} = -0.9$. Note that this is the same image as in Figure 5.4(b), except that it has been rotated by 90°. Now the surface lies largely above the line $y = -x$. In all three figures, the means of both X and Y are zero and the variances of both X and Y are 1. Changing the means would simply translate the surface but would not change the shape. Changing the variances would expand or contract the surface along either the X or Y axis, depending on which variance was changed.

EXAMPLE 5.17: The joint Gaussian PDF is given by

$$f_{X,Y}(x,y) = \frac{1}{2\pi\sigma_X\sigma_Y\sqrt{1-\rho_{XY}^2}}\exp\left(-\frac{1}{2(1-\rho_{XY}^2)}\left[\left(\frac{x-\mu_X}{\sigma_X}\right)^2\right.\right.$$

$$\left.\left.-2\rho_{XY}\left(\frac{x-\mu_X}{\sigma_X}\right)\left(\frac{y-\mu_Y}{\sigma_Y}\right)+\left(\frac{y-\mu_Y}{\sigma_Y}\right)^2\right]\right).$$

Suppose we equate the portion of this equation that is within the square brackets to a constant. That is

$$\left(\frac{x-\mu_X}{\sigma_X}\right)^2 - 2\rho_{XY}\left(\frac{x-\mu_X}{\sigma_X}\right)\left(\frac{y-\mu_Y}{\sigma_Y}\right) + \left(\frac{y-\mu_Y}{\sigma_Y}\right)^2 = c^2.$$

This is the equation for an ellipse. Plotting these ellipses for different values of c results in what is known as a contour plot. Figure 5.5 shows such plots for the two-dimensional joint Gaussian PDF. The following code can be used to generate such plots. The reader is encouraged to try creating similar plots for different values of the parameters in the Gaussian distribution.

```
clear
[X,Y]=meshgrid(-8:0.1:8);    % Generate x and y array to
                             % be used for contour plot.
mux=0; muy=0;                % Set means.
stdx=3; stdy=3;              % Set standard deviations.
varx=stdx^2; vary=stdy^2;    % Compute variances.
rho=0.5;                     % Set correlation coefficient.
% Compute exponent of two-dimensional Gaussian PDF.
X1=(X-mux)/stdx;
Y1=(Y-muy)/stdy;
Z=X1.^2-2*rho*X1.*Y1+Y1.^2;
```

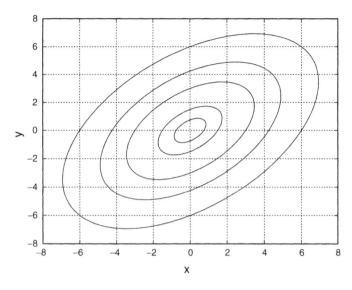

Figure 5.5   Contour plots for Example 5.17.

```
c=[1/16 1/4 1 2 4];              % Set contour levels.
contour(X,Y,Z,c)                 % Produce contour plot grid.
                                 % Turn on grid lines.
xlabel('x'); ylabel('y')         % Label axes.
```

THEOREM 5.7: Uncorrelated Gaussian random variables are independent.

PROOF: Uncorrelated Gaussian random variables have a correlation coefficient of zero. Plugging $\rho_{XY} = 0$ into the general joint Gaussian PDF results in

$$f_{X,Y}(x,y) = \frac{1}{2\pi\sigma_X\sigma_Y}\exp\left(-\left[\frac{\left(\frac{x-\mu_X}{\sigma_X}\right)^2 + \left(\frac{y-\mu_Y}{\sigma_Y}\right)^2}{2}\right]\right).$$

This clearly factors into the product of the marginal Gaussian PDFs.

$$f_{X,Y}(x,y) = \frac{1}{\sqrt{2\pi\sigma_x^2}}\exp\left(-\frac{(x-m_x)^2}{2\sigma_x^2}\right)\frac{1}{\sqrt{2\pi\sigma_y^2}}\exp\left(-\frac{(y-m_y)^2}{2\sigma_y^2}\right) = f_X(x)f_Y(y)$$

■

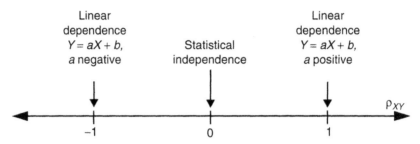

Figure 5.6 Interpretation of the correlation coefficient for jointly Gaussian random variables.

While Example 5.15 demonstrates that this property does not hold for all random variables, it is true for Gaussian random variables. This allows us to give a stronger interpretation to the correlation coefficient when dealing with Gaussian random variables. Previously it was stated that the correlation coefficient is a quantitative measure of the amount of correlation between two variables. While this is true, it is a rather vague statement. After all, what does "correlation" mean? In general, we cannot equate correlation and statistical dependence. Now, however, we see that in the case of Gaussian random variables, we can make the connection between correlation and statistical dependence. Hence, for jointly Gaussian random variables, the correlation coefficient can indeed be viewed as a quantitative measure of statistical dependence. This relationship is illustrated in Figure 5.6.

5.8 Joint Characteristic and Related Functions

When computing the joint moments of random variables, it is often convenient to use characteristic functions, moment generating functions, or probability generating functions. Since a pair of random variables is involved, the "frequency domain" function must now be two-dimensional. We start with a description of the joint characteristic function, which is similar to a two-dimensional Fourier transform of the joint PDF.

DEFINITION 5.11: Given a pair of random variables X and Y with a joint PDF, $f_{XY}(x, y)$, the *joint characteristic function* is

$$\Phi_{X,Y}(\omega_1, \omega_2) = E\big[e^{j(\omega_1 X + \omega_2 Y)}\big] = \iint\limits_{-\infty}^{\infty} e^{j(\omega_1 x + \omega_2 y)} f_{X,Y}(x, y)\, dxdy. \qquad (5.42)$$

The various joint moments can be evaluated from the joint characteristic function using techniques similar to those used for single random variables. It is left as an exercise for the reader to establish the following relationship:

$$E[X^m Y^n] = (-j)^{m+n} \frac{\partial^m}{\partial \omega_1^m} \frac{\partial^n}{\partial \omega_2^n} \Phi_{X,Y}(\omega_1, \omega_2)\Big|_{\omega_1 = \omega_2 = 0}. \qquad (5.43)$$

EXAMPLE 5.18: Consider a pair of zero-mean, unit-variance, jointly Gaussian random variables whose joint PDF is

$$f_{X,Y}(x,y) = \frac{1}{2\pi\sqrt{1-\rho^2}} \exp\left(-\frac{(x^2 - 2\rho xy + y^2)}{2(1 - \rho^2)}\right).$$

One way to calculate the joint characteristic function is to break the problem into two one-dimensional problems:

$$\Phi_{X,Y}(\omega_1, \omega_2) = E\left[e^{j(\omega_1 X + \omega_2 Y)}\right] = E_Y\left[e^{j\omega_2 Y} E_X\left[e^{j\omega_1 X}|Y\right]\right].$$

Conditioned on Y, X is a Gaussian random variable with a mean of ρY and a variance of $1 - \rho^2$. The general form of the characteristic function (one-dimensional) of a Gaussian random variable with mean μ_X and variance σ_X^2 (see Example 4.20) is

$$\Phi_X(\omega) = \exp\left(j\mu_X\omega - \frac{\omega^2 \sigma_X^2}{2}\right).$$

Hence, the inner expectation evaluates to

$$E_X\left[e^{j\omega_1 X}|Y\right] = \exp\left(j\rho Y\omega_1 - \frac{\omega_1^2(1 - \rho^2)}{2}\right).$$

The joint characteristic function is then

$$\Phi_{X,Y}(\omega_1, \omega_2) = E_Y\left[\exp\left(j\omega_2 Y + j\rho Y\omega_1 - \frac{\omega_1^2(1 - \rho^2)}{2}\right)\right]$$

$$= \exp\left(-\frac{\omega_1^2(1 - \rho^2)}{2}\right) E_Y\left[e^{j(\rho\omega_1 + \omega_2)Y}\right].$$

The remaining expectation is the characteristic function of a zero-mean, unit-variance Gaussian random variable evaluated at $\omega = \rho\omega_1 + \omega_2$.

The resulting joint characteristic function is then found to be

$$\Phi_{X,Y}(\omega_1, \omega_2) = \exp\left(-\frac{\omega_1^2(1 - \rho^2)}{2}\right) \exp\left(-\frac{(\rho\omega_1 + \omega_2)^2}{2}\right)$$

$$= \exp\left(-\frac{\omega_1^2 + 2\rho\omega_1\omega_2 + \omega_2^2}{2}\right).$$

From this expression, various joint moments can be found. For example, the correlation is

$$E[XY] = -\frac{\partial}{\partial\omega_1}\frac{\partial}{\partial\omega_2}\exp\left(-\frac{\omega_1^2 + 2\rho\omega_1\omega_2 + \omega_2^2}{2}\right)\Big|_{\omega_1=\omega_2=0}$$

$$= \frac{\partial}{\partial\omega_2}\rho\omega_2\exp\left(-\frac{\omega_2^2}{2}\right)\Big|_{\omega_2=0} = \rho.$$

Since the two random variables were zero-mean, $\text{Cov}(X, Y) = \rho$. Furthermore, since the two random variables were unit variance, ρ is also the correlation coefficient. We have therefore proved that the parameter ρ that shows up in the joint Gaussian PDF is indeed the correlation coefficient.

We could easily compute higher order moments as well. For example, suppose we need to compute $E[X^2Y^2]$. It can be computed in a similar manner to the preceding:

$$E[X^2Y^2] = \frac{\partial^2}{\partial\omega_1^2}\frac{\partial^2}{\partial\omega_2^2}\exp\left(-\frac{\omega_1^2 + 2\rho\omega_1\omega_2 + \omega_2^2}{2}\right)\Big|_{\omega_1=\omega_2=0}$$

$$= -\frac{\partial^2}{\partial\omega_2^2}[1 - (\rho\omega_2)^2]\exp\left(-\frac{\omega_2^2}{2}\right)\Big|_{\omega_2=0} = 1 + 2\rho^2.$$

DEFINITION 5.12: For a pair of discrete random variables defined on a two-dimensional lattice of nonnegative integers, one can define a *joint probability generating function* as

$$H_{X,Y}(z_1, z_2) = E\left[z_1^X z_2^Y\right] = \sum_{m=0}^{\infty}\sum_{n=0}^{\infty}P_{X,Y}(m, n)z_1^m z_2^n. \tag{5.44}$$

The reader should be able to show that the joint partial derivatives of the joint probability generating function evaluated at zero are related to the terms in the joint probability mass function, whereas those same derivatives evaluated at 1 lead to joint factorial moments. Specifically,

$$P_{X,Y}(k,l) = \frac{1}{k!l!} \frac{\partial^k}{\partial z_1^k} \frac{\partial^l}{\partial z_2^l} H_{X,Y}(z_1, z_2)\Big|_{z_1=z_2=0}, \tag{5.45}$$

$$E[X(X-1)\cdots(X-k+1)Y(Y-1)\cdots(Y-l+1)] = \frac{\partial^k}{\partial z_1^k} \frac{\partial^l}{\partial z_2^l} H_{X,Y}(z_1, z_2)\Big|_{z_1=z_2=1}. \tag{5.46}$$

EXAMPLE 5.19: Consider the joint PMF given in Example 5.7:

$$P_{N,M}(n,m) = \frac{(n+m)!}{n!m!} \frac{a^n b^m}{(a+b+1)^{n+m+1}}.$$

It is not too difficult to work out the joint probability generating function for this pair of discrete random variables.

$$H_{N,M}(z_1 z_2) = \sum_{n=0}^{\infty} \sum_{m=0}^{\infty} \frac{(n+m)!}{n!m!} \frac{(az_1)^n (bz_2)^m}{(a+b+1)^{n+m+1}}$$

$$= \sum_{n=0}^{\infty} \frac{(az_1)^n}{(a+b+1)^{n+1}} \sum_{m=0}^{\infty} \frac{(n+m)!}{n!m!} \left(\frac{bz_2}{a+b+1}\right)^m$$

$$= \sum_{n=0}^{\infty} \frac{(az_1)^n}{(a+b+1)^{n+1}} \left(\frac{1}{1-\dfrac{bz_2}{a+b+1}}\right)^{n+1} = \sum_{n=0}^{\infty} \frac{(az_1)^n}{(a+b(1-z_2)+1)^{n+1}}$$

$$= \frac{1}{a+b(1-z_2)+1} \sum_{n=0}^{\infty} \left(\frac{az_1}{a+b(1-z_2)+1}\right)^n = \frac{1}{1+a(1-z_1)+b(1-z_2)}$$

It should be noted that the closed form expression used for the various preceding series limits the range in the (z_1, z_2) plane for which these expressions are valid; thus, care must be taken when evaluating this function and its derivatives at various points. However, for this example, the expression is valid in and around the points of interest (i.e., $(z_1, z_2) = (0, 0)$ and $(z_1, z_2) = (1, 1)$).

Now that the joint probability generating function has been found, joint moments are fairly easy to compute. For example,

$$E[NM] = \frac{\partial}{\partial z_1} \frac{\partial}{\partial z_2} \frac{1}{1+a(1-z_1)+b(1-z_2)}\Big|_{z_1=z_2=1}$$

$$= \frac{\partial}{\partial z_1} \frac{b}{[1+a(1-z_1)]^2}\Big|_{z_1=1} = 2ab,$$

$$E[N(N-1)M] = \frac{\partial^2}{\partial z_1^2}\frac{\partial}{\partial z_2}\frac{1}{1+a(1-z_1)+b(1-z_2)}\Big|_{z_1=z_2=1}$$

$$= \frac{\partial^2}{\partial z_1^2}\frac{b}{[1+a(1-z_1)]^2}\Big|_{z_1=1} = 6a^2b.$$

Putting these two results together, it is found that

$$E[N^2M] = E[N(N-1)M] + E[NM] = 6a^2b + 2ab.$$

By symmetry, we can also conclude that $E[NM(M-1)] = 6ab^2$ and $E[NM^2] = 6ab^2 + 2ab$. As one last example, we note that

$$E[N(N-1)M(M-1)] = \frac{\partial^2}{\partial z_1^2}\frac{\partial^2}{\partial z_2^2}\frac{1}{1+a(1-z_1)+b(1-z_2)}\Big|_{z_1=z_2=1}$$

$$= \frac{\partial^2}{\partial z_1^2}\frac{2b^2}{[1+a(1-z_1)]^3}\Big|_{z_1=1} = 24a^2b^2.$$

From this and the previous results, we can find $E[N^2M^2]$ as follows:

$$E[N^2M^2] = E[N(N-1)M(M-1)] + E[NM^2] + E[N^2M] - E[NM]$$

$$= 24a^2b^2 + 6ab^2 + 6a^2b + 2ab.$$

The moment generating function can also be generalized in a manner virtually identical to what was done for the characteristic function. We leave the details of this extension to the reader.

5.9 Transformations of Pairs of Random Variables

In this section, we consider forming a new random variable as a function of a pair of random variables. When a pair of random variables is involved, there are two classes of such transformations. The first class of problems deals with the case where a single new variable is created as a function of two random variables. The second class of problems involves creating two new random variables as two functions of two random variables. These two distinct but related problems are treated in this section.

Consider first a single function of two random variables, $Z = g(X, Y)$. If the joint PDF of X and Y is known, can the PDF of the new random variable Z be found? Of course, the answer is yes, and there are a variety of techniques to solve these types of problems depending on the nature of the function $g(\cdot)$. The first technique to be developed is an extension of the approach we used in Chapter 4 for functions of a single random variable.

The CDF of Z can be expressed in terms of the variables X and Y as

$$F_Z(z) = \Pr(Z \leq z) = \Pr(g(X, Y) \leq z) = \iint_{g(x,y) \leq z} f_{X,Y}(x, y)\,dx\,dy. \tag{5.47}$$

The inequality $g(x, y) \leq z$ defines a region in the (x, y) plane. By integrating the joint PDF of X and Y over that region, the CDF of Z is found. The PDF can then be found by differentiating with respect to z. In principle, one can use this technique with any transformation; however, the integral to be computed may or may not be analytically tractable, depending on the specific joint PDF and the transformation.

To illustrate, consider a simple yet very important example where the transformation is just the sum of the random variables, $Z = X + Y$. Then

$$F_Z(z) = \iint_{x+y \leq z} f_{X,Y}(x, y)\,dx\,dy = \int_{-\infty}^{\infty} \int_{-\infty}^{z-y} f_{X,Y}(x, y)\,dx\,dy. \tag{5.48}$$

Differentiating to form the PDF results in

$$f_Z(z) = \frac{d}{dz} \int_{-\infty}^{\infty} \int_{-\infty}^{z-y} f_{X,Y}(x, y)\,dx\,dy = \int_{-\infty}^{\infty} f_{X,Y}(z - y, y)\,dy. \tag{5.49}$$

The last step in Equation 5.49 is completed using Liebnitz's rule[1]. An important special case results when X and Y are independent. In this case, the joint PDF factors into the product of the marginals, producing

$$f_Z(z) = \int_{-\infty}^{\infty} f_X(z - y)f_Y(y)\,dy. \tag{5.50}$$

Note that this integral is a convolution. Thus, the following important result has been proved:

THEOREM 5.8: If X and Y are statistically independent random variables, then the PDF of $Z = X + Y$ is given by the convolution of the PDFs of X and Y, $f_Z(z) = f_X(z) * f_Y(z)$.

---

[1]Liebnitz's rule states that

$$\frac{\partial}{\partial x} \int_{a(x)}^{b(x)} f(x, y)\,dy = \frac{\partial b}{\partial x} f(x, b(x)) - \frac{\partial a}{\partial x} f(x, a(x)) + \int_{a(x)}^{b(x)} \frac{\partial}{\partial x} f(x, y)\,dy.$$

EXAMPLE 5.20: Suppose X and Y are independent and both have exponential distributions,

$$f_X(x) = a\exp(-ax)u(x), \qquad f_Y(y) = b\exp(-by)u(y).$$

The PDF of $Z = X + Y$ is then found by performing the necessary convolution:

$$f_Z(z) = \int_{-\infty}^{\infty} f_X(z-y)f_Y(y)\,dy = ab\int_{-\infty}^{\infty} \exp(-a(z-y))\exp(-by)u(z-y)u(y)\,dy$$

$$= abe^{-az}\int_0^z \exp((a-b)y)\,dy\,u(z)$$

$$= \frac{ab}{a-b}\left[e^{-az}e^{(a-b)y}\Big|_{y=0}^{y=z}\right]u(z) = \frac{ab}{a-b}\left[e^{-by} - e^{-az}\right]u(z).$$

This result is valid assuming that $a \neq b$. If $a = b$, then the convolution works out to be

$$f_Z(z) = a^2 z e^{-az}u(z).$$

Students familiar with the study of signals and systems should recall that the convolution integral appears in the context of passing signals through linear time-invariant systems. In that context, most students develop a healthy respect for the convolution and will realize that quite often the convolution can be a cumbersome operation. To avoid difficult convolutions, these problems can often be solved using a frequency domain approach in which a Fourier or Laplace transform is invoked to replace the convolution with a much simpler multiplication. In the context of probability, the characteristic function or the moment generating function can fulfill the same role. Instead of finding the PDF of $Z = X + Y$ directly via convolution, suppose we first find the characteristic function of Z:

$$\Phi_Z(\omega) = E\left[e^{j\omega Z}\right] = E\left[e^{j\omega(X+Y)}\right] = E\left[e^{j\omega X}e^{j\omega Y}\right]. \tag{5.51}$$

If X and Y are independent, then the expected value of the product of a function of X times a function of Y factors into the product of expected values:

$$\Phi_Z(\omega) = E\left[e^{j\omega X}\right]E\left[e^{j\omega Y}\right] = \Phi_X(\omega)\Phi_Y(\omega). \tag{5.52}$$

Once the characteristic function of Z is found, the PDF can be found using an inverse Fourier transform.

Again, the characteristic function can be used to simplify the amount of computation involved in calculating PDFs of sums of independent random variables. Furthermore, we have also developed a new approach to find the PDFs of a general function of two random variables. Returning to a general transformation of the form $Z = g(X, Y)$, one can first find the characteristic function of Z according to

$$\Phi_Z(\omega) = E[e^{j\omega g(X,Y)}] = \int\int_{-\infty}^{\infty} e^{j\omega g(x,y)} f_{X,Y}(x,y)\, dx\, dy. \qquad (5.53)$$

An inverse transform of this characteristic function will then produce the desired PDF. In some cases, this method will provide a simpler approach to the problem, while in other cases the direct method may be easier.

EXAMPLE 5.21: Suppose X and Y are independent, zero-mean, unit variance Gaussian random variables. The PDF of $Z = X^2 + Y^2$ can be found using either of the methods described thus far. Using characteristic functions,

$$\Phi_Z(\omega) = E[e^{j\omega(X^2+Y^2)}] = E[e^{j\omega X^2}]E[e^{j\omega Y^2}].$$

The expected values are evaluated as follows:

$$E[e^{j\omega X^2}] = \int_{-\infty}^{\infty} \frac{1}{\sqrt{2\pi}} e^{j\omega x^2} e^{-x^2/2}\, dx$$

$$= \frac{1}{\sqrt{1-2j\omega}} \int_{-\infty}^{\infty} \sqrt{\frac{1-2j\omega}{2\pi}} e^{-(1-2j\omega)x^2/2}\, dx = \frac{1}{\sqrt{1-2j\omega}}.$$

The last step is accomplished using the normalization integral for Gaussian functions. The other expected value is identical to the first since X and Y have identical distributions. Hence,

$$\Phi_Z(\omega) = \left(\frac{1}{\sqrt{1-2j\omega}}\right)^2 = \frac{1}{1-2j\omega}.$$

The PDF is found from the inverse Fourier transform to be

$$f_Z(z) = \frac{1}{2}\exp\left(-\frac{x}{2}\right)u(x).$$

The other approach is to find the CDF as follows:

$$F_Z(z) = \Pr(X^2 + Y^2 \le z) = \int\int_{x^2+y^2 \le z} \frac{1}{2\pi}\exp\left(-\frac{x^2+y^2}{2}\right) dx\, dy.$$

Converting to polar coordinates,

$$F_Z(z) = \int_0^{\sqrt{z}} \int_0^{2\pi} \frac{r}{2\pi} \exp\left(-\frac{r^2}{2}\right) d\theta \, dr u(z)$$

$$= \int_0^{\sqrt{z}} r \exp\left(-\frac{r^2}{2}\right) dr u(z) = \left[1 - exp\left(-\frac{z}{2}\right)\right] u(z).$$

Finally, differentiating with respect to z results in

$$f_Z(z) = \frac{d}{dz}\left[1 - \exp\left(-\frac{z}{2}\right)\right] u(z) = \frac{1}{2}\exp\left(-\frac{z}{2}\right) u(z).$$

Another approach to solving these types of problems uses conditional distributions. Consider a general transformation, $Z = g(X, Y)$. Next, suppose we condition on one of the two variables, say $X = x$. Conditioned on $X = x$, $Z = g(x, Y)$ is now a single variable transformation. Hence, the conditional PDF of Z given X can be found using the general techniques presented in Chapter 4. Once $f_{Z|X}(z|x)$ is known, the desired (unconditional) PDF of Z can be found according to

$$f_Z(z) = \int_{-\infty}^{\infty} f_{Z,X}(z, x) \, dx = \int_{-\infty}^{\infty} f_{Z|X}(z|x) f_X(x) \, dx. \tag{5.54}$$

EXAMPLE 5.22: Suppose X and Y are independent zero-mean, unit-variance Gaussian random variables and we want to find the PDF of $Z = Y/X$. Conditioned on $X = x$, the transformation $Z = Y/x$ is a simple linear transformation and

$$f_{Z|X}(z|x) = |x| f_Y(xz) = \frac{|x|}{\sqrt{2\pi}} \exp\left(-\frac{x^2 z^2}{2}\right).$$

Multiplying the conditional PDF by the marginal PDF of X and integrating out x gives the desired marginal PDF of Z.

$$f_Z(z) = \int_{-\infty}^{\infty} f_{Z|X}(z|x) f_X(x) \, dx = \int_{-\infty}^{\infty} \frac{|x|}{\sqrt{2\pi}} \exp\left(-\frac{x^2 z^2}{2}\right) \frac{1}{\sqrt{2\pi}} \exp\left(-\frac{x^2}{2}\right) dx$$

$$= \frac{1}{2\pi} \int_{-\infty}^{\infty} |x| \exp\left(-\frac{(1+x^2)z^2}{2}\right) dx$$

$$= \frac{1}{\pi} \int_0^{\infty} x \exp\left(-\frac{(1+z^2)x^2}{2}\right) dx = \frac{1}{\pi} \frac{1}{1+z^2}$$

Evaluating the integral in the last step can be accomplished by making the substitution $u = (1 + z^2)x^2/2$. Hence, the quotient of two independent Gaussian random variables follows a Cauchy distribution.

Up to this point, three methods have been developed for finding the PDF of $Z = g(X, Y)$ given the joint PDF of X and Y; they can be summarized as follows.

5.9.1 Method 1, CDF Approach

Define a set $R(z) = \{(x, y): g(x, y) \le z\}$. The CDF of Z is the integral of the joint PDF of X and Y over the region $R(z)$. The PDF is then found by differentiating the expression for the CDF:

$$f_Z(z) = \frac{d}{dz} \iint\limits_{R(z)} f_{X,Y}(x, y)\, dx\, dy. \tag{5.55}$$

5.9.2 Method 2, Characteristic Function Approach

First, find the characteristic function of Z according to

$$\Phi_Z(\omega) = E\left[e^{j\omega g(X,Y)}\right]. \tag{5.56}$$

Then compute the inverse transform to get the PDF of Z.

5.9.3 Method 3, Conditional PDF Approach

Fix either $X = x$ or $Y = y$ (whichever is more convenient). The conditional PDF of Z can then be found using the techniques developed for single random variables in Chapter 4. Once the conditional PDF of Z is found, the unconditional PDF is given by

$$f_Z(z) = \int_{-\infty}^{\infty} f_{Z|Y}(z|y) f_Y(y)\, dy \quad \text{or} \quad f_Z(z) = \int_{-\infty}^{\infty} f_{Z|X}(z|x) f_X(x)\, dx. \tag{5.57}$$

Next, our attention moves to solving a slightly more general class of problems. Given two random variables X and Y, suppose we now create two new random variables W and Z according to some 2×2 transformation of the general form

$$Z = g_1(X, Y),$$

$$W = g_2(X, Y). \tag{5.58}$$

The most common example of this type of problem involves changing coordinate systems. Suppose, for example, that the variables X and Y represent the random position of some object in Cartesian coordinates. In some problems, it may be easier to view the object in a polar coordinate system, in which case two new variables R and Θ could be created to describe the location of the object in polar coordinates. Given the joint PDF of X and Y, how can we find the joint PDF of R and Θ?

The procedure for finding the joint PDF of Z and W for a general transformation of the form given in Equation 5.58 is an extension of the technique used for a 1×1 transformation. First, recall the definition of the joint PDF given in Equation 5.2, which states that for an infinitesimal region $A_{x,y} = (x, x + \varepsilon_x) \times (y, y + \varepsilon_y)$, the joint PDF, $f_{X,Y}(x, y)$, has the interpretation

$$\Pr((X, Y) \in A_{x,y}) = f_{X,Y}(x, y)\varepsilon_x\varepsilon_y = f_{X,Y}(x, y)(\text{Area of } A_{x,y}). \tag{5.59}$$

Assume for now that the transformation is invertible. In that case, the transformation maps the region $A_{x,y}$ into a corresponding region $A_{z,w}$ in the (z, w)-plane. Furthermore,

$$\Pr((X, Y) \in A_{x,y}) = \Pr((Z, W) \in A_{z,w}) = f_{Z,W}(z, w)(\text{Area of } A_{z,w}). \tag{5.60}$$

Putting the two previous equations together results in

$$f_{Z,W}(z, w) = f_{X,Y}(x, y)\frac{\text{Area of } A_{x,y}}{\text{Area of } A_{z,w}}. \tag{5.61}$$

A fundamental result of multivariable calculus states that if a transformation of the form given in Equation 5.58 maps an infinitesimal region $A_{x,y}$, to a region $A_{z,w}$, then the ratio of the areas of these regions is given by the absolute value of the Jacobian of the transformation:

$$\frac{\text{Area of } A_{x,y}}{\text{Area of } A_{z,w}} = \left| J\begin{pmatrix} x & y \\ z & w \end{pmatrix} \right| = \left| \det \begin{bmatrix} \dfrac{\partial x}{\partial z} & \dfrac{\partial y}{\partial z} \\ \dfrac{\partial x}{\partial w} & \dfrac{\partial y}{\partial w} \end{bmatrix} \right|. \tag{5.62}$$

The PDF of Z and W is then given by

$$f_{Z,W}(z, w) = f_{X,Y}(x, y)\left| J\begin{pmatrix} x & y \\ z & w \end{pmatrix} \right|. \tag{5.63}$$

If it is more convenient to take derivatives of z and w with respect to x and y rather than vice versa, we can alternatively use

$$\frac{\text{Area of } A_{z,w}}{\text{Area of } A_{x,y}} = \left| J\begin{pmatrix} z & w \\ x & y \end{pmatrix} \right| = \left| \det \begin{bmatrix} \frac{\partial z}{\partial x} & \frac{\partial z}{\partial y} \\ \frac{\partial w}{\partial x} & \frac{\partial w}{\partial y} \end{bmatrix} \right|, \tag{5.64}$$

$$f_{Z,W}(z,w) = \frac{f_{X,Y}(x,y)}{\left| J\begin{pmatrix} z & w \\ x & y \end{pmatrix} \right|}. \tag{5.65}$$

Whether Equation 5.63 or 5.65 is used, any expressions involving x or y must be replaced with the corresponding functions of z and w. Let the inverse transformation of Equation 5.58 be written as

$$X = h_1(Z, W),$$
$$Y = h_2(Z, W). \tag{5.66}$$

Then these results can be summarized as

$$f_{Z,W}(z,w) = \left. \frac{f_{X,Y}(x,y)}{\left| J\begin{pmatrix} z & w \\ x & y \end{pmatrix} \right|} \right|_{\substack{x = h_1(z,w) \\ y = h_2(z,w)}} = \left. f_{X,Y}(x,y) \left| J\begin{pmatrix} x & y \\ z & w \end{pmatrix} \right| \right|_{\substack{x = h_1(z,w) \\ y = h_2(z,w)}} \tag{5.67}$$

If the original transformation is not invertible, then the inverse transformation may have multiple roots. In this case, as with transformations involving single random variables, the expression in Equation 5.67 must be evaluated at each root of the inverse transformation and the results summed together. This general procedure for transforming pairs of random variables is demonstrated next through a few examples.

EXAMPLE 5.23: A classical example of this type of problem involves the transformation of two independent Gaussian random variables from Cartesian to polar coordinates. Suppose

$$f_{X,Y}(x,y) = \frac{1}{2\pi\sigma^2} \exp\left(-\frac{x^2 + y^2}{2\sigma^2}\right).$$

We seek the PDF of the polar magnitude and phase given by

$$R = \sqrt{X^2 + Y^2},$$
$$\Theta = \tan^{-1}(Y/X).$$

The inverse transformation is

$$X = R\cos(\Theta),$$

$$Y = R\sin(\Theta).$$

In this case, the inverse transformation takes on a simpler functional form and so we elect to use this form to compute the Jacobian.

$$J\begin{pmatrix} x & y \\ r & \theta \end{pmatrix} = \det \begin{bmatrix} \dfrac{\partial x}{\partial r} & \dfrac{\partial x}{\partial \theta} \\ \dfrac{\partial y}{\partial r} & \dfrac{\partial y}{\partial \theta} \end{bmatrix} = \det \begin{bmatrix} \cos(\theta) & -r\sin(\theta) \\ \sin(\theta) & r\cos(\theta) \end{bmatrix}$$

$$= r\cos^2(\theta) + r\sin^2(\theta) = r$$

The joint PDF of R and Θ is then

$$f_{R,\Theta}(r,\theta) = f_{X,Y}(x,y)\left| J\begin{pmatrix} x & y \\ r & \theta \end{pmatrix} \right| \Bigg|_{\substack{x = h_1(r,\theta) \\ y = h_2(r,\theta)}}$$

$$= \frac{r}{2\pi\sigma^2}\exp\left(-\frac{x^2+y^2}{2\sigma^2}\right)\Bigg|_{\substack{x = r\cos(\theta) \\ y = r\sin(\theta)}}$$

$$= \frac{r}{2\pi\sigma^2}\exp\left(-\frac{r^2}{2\sigma^2}\right), \quad \begin{array}{l} r \geq 0 \\ 0 \leq \theta < 2\pi \end{array}.$$

Note that in these calculations, we do not have to worry about taking the absolute value of the Jacobian since for this problem the Jacobian $(= r)$ is always nonnegative. If we were interested, we could also find the marginal distributions of R and Θ to be

$$f_R(r) = \frac{r}{\sigma^2}\exp\left(-\frac{r^2}{2\sigma^2}\right)u(r) \quad \text{and} \quad f_\Theta(\theta) = \frac{1}{2\pi}, \quad 0 \leq \theta < 2\pi.$$

The magnitude follows a Rayleigh distribution while the phase is uniformly distributed over $(0, 2\pi)$.

EXAMPLE 5.24: Suppose X and Y are independent and both uniformly distributed over $(0, 1)$, so that

$$f_{X,Y}(x,y) = \begin{cases} 1 & 0 \leq x,\, y < 1 \\ 0 & \text{otherwise} \end{cases}.$$

Consider forming the two new random variables

$$Z = \sqrt{-2\ln(X)}\cos(2\pi Y),$$

$$W = \sqrt{-2\ln(X)}\sin(2\pi Y).$$

The inverse transformation in this case is found to be

$$X = \exp\left(\frac{Z^2 + W^2}{2}\right),$$

$$Y = \frac{1}{2\pi}\tan^{-1}\left(\frac{W}{Z}\right).$$

In this example, we will compute the Jacobian by taking derivatives of z and w with respect to x and y to produce

$$J\begin{pmatrix} z & w \\ x & y \end{pmatrix} = \det \begin{bmatrix} \dfrac{\partial z}{\partial x} & \dfrac{\partial z}{\partial y} \\[2mm] \dfrac{\partial w}{\partial x} & \dfrac{\partial w}{\partial y} \end{bmatrix}$$

$$= \det \begin{bmatrix} -\dfrac{1}{x}\dfrac{\cos(2\pi y)}{\sqrt{-2\ln(x)}} & -2\pi\sqrt{-2\ln(X)}\sin(2\pi Y) \\[4mm] -\dfrac{1}{x}\dfrac{\sin(2\pi y)}{\sqrt{-2\ln(x)}} & 2\pi\sqrt{-2\ln(X)}\cos(2\pi Y) \end{bmatrix}$$

$$= -\frac{2\pi}{x}\left[\cos^2(2\pi y) + \sin^2(2\pi y)\right] = -\frac{2\pi}{x}.$$

Note that since x is always nonnegative, the absolute value of the Jacobian will just be $2\pi/x$. The joint PDF of Z and W is then found to be

$$f_{Z,W}(z,w) = \frac{f_{X,Y}(x,y)}{\left|J\begin{pmatrix} z & w \\ x & y \end{pmatrix}\right|}\Bigg|_{\substack{x=h_1(z,w) \\ y=h_2(z,w)}} = \frac{x}{2\pi}\Bigg|_{\substack{x=\exp\left(-\frac{z^2+w^2}{2}\right) \\ y=\frac{1}{2\pi}\tan^{-1}\left(\frac{w}{z}\right)}}$$

$$= \frac{1}{2\pi}\exp\left(-\frac{z^2 + w^2}{2}\right).$$

This transformation is known as the Box-Muller transformation. It transforms a pair of independent uniform random variables into a pair of independent Gaussian random variables. This transformation has application in the world of computer simulations. Techniques for generating uniform random variables are well known. This transformation then allows us to generate Gaussian random variables as well. More material on this subject is given in Chapter 12, Simulation Techniques.

EXAMPLE 5.25: Suppose X and Y are independent Gaussian random variables, both with zero mean and unit variance. Two new random variables, Z and W, are formed through a linear transformation of the form

$$Z = aX + bY,$$

$$W = cX + dY.$$

The inverse transformation is given by

$$X = \frac{d}{ad - bc} Z - \frac{b}{ad - bc} W,$$

$$Y = -\frac{c}{ad - bc} Z + \frac{a}{ad - bc} W.$$

With this general linear transformation, the various partial derivatives are trivial to compute and the resulting Jacobian is

$$J\begin{pmatrix} z & w \\ x & y \end{pmatrix} = \det \begin{bmatrix} a & b \\ c & d \end{bmatrix} = ad - bc.$$

Plugging these results into the general formula results in

$$f_{Z,W}(z,w) = \frac{f_{X,Y}(x,y)}{\left| J\begin{pmatrix} z & w \\ x & y \end{pmatrix} \right|} \Bigg|_{\substack{x = h_1(z,w) \\ y = h_2(z,w)}} = \frac{\frac{1}{2\pi} \exp\left(-\frac{x^2 + y^2}{2} \right)}{|ad - bc|} \Bigg|_{\substack{x = \frac{d}{ad-bc}z - \frac{b}{ad-bc}w \\ y = \frac{c}{ad-bc}z + \frac{a}{ad-bc}w.}}$$

$$= \frac{1}{2\pi \sqrt{(ad - bc)^2}} \exp\left(-\frac{(c^2 + d^2)z^2 - 2(bd + ac)zw + ((a^2 + b^2)w)^2}{2(ad - bc)^2} \right)$$

With a little algebraic manipulation, it can be shown that this joint PDF fits the general form of a joint Gaussian PDF. In particular,

$$f_{Z,W}(z,w) = \frac{1}{2\pi \sigma_Z \sigma_W \sqrt{1 - \rho_{ZW}^2}}$$

$$\times \exp\left(-\frac{(z/\sigma_Z)^2 - 2\rho_{ZW}(z/\sigma_Z)(w/\sigma_W) + (w/\sigma_W)^2}{2(1 - \rho_{ZW}^2)} \right),$$

where $\sigma_Z^2 = a^2 + b^2$, $\sigma_W^2 = c^2 + d^2$, and $\rho_{ZW}^2 = (ac + bd)^2(a^2 + b^2)^{-1}(c^2 + d^2)^{-1}$.

A few remarks about the significance of the result of Example 5.25 are appropriate. First, we have performed an arbitrary linear transformation on a pair of independent Gaussian random variables and produced a new pair of Gaussian random variables (which are no longer independent). In the next chapter, it will be shown that a linear transformation of any number of jointly Gaussian random variables always produces jointly Gaussian random variables. Second, if we look at this problem in reverse, two correlated Gaussian random variables Z and W can be transformed into a pair of uncorrelated Gaussian random variables X and Y using an appropriate linear transformation. More information will be given on this topic in the next chapter as well.

5.10 Complex Random Variables

In engineering practice, it is common to work with quantities that are complex. Usually, a complex quantity is just a convenient shorthand notation for working with two real quantities. For example, a sinusoidal signal with amplitude, A, frequency, ω, and phase, θ, can be written as

$$s(t) = A\cos(\omega t + \theta) = Re\left[Ae^{j\theta}e^{j\omega t}\right], \tag{5.68}$$

where $j = \sqrt{-1}$. The complex number $Z = Ae^{j\theta}$ is known as a phasor representation of the sinusoidal signal. It is a complex number with real part of $X = Re[Z] = A\cos(\theta)$ and imaginary part of $Y = Im[Z] = A\sin(\theta)$. The phasor Z can be constructed from two real quantities (either A and θ or X and Y).

Suppose a complex quantity that we are studying is composed of two real quantities that happen to be random. For example, the preceding sinusoidal signal might have a random amplitude and/or a random phase. In either case, the complex number Z will also be random. Unfortunately, our formulation of random variables does not allow for complex quantities. When we began to describe a random variable via its CDF in the beginning of Chapter 3, the CDF was defined as $F_Z(z) = \Pr(Z \leq z)$. This definition makes no sense if Z is a complex number: what does it mean for a complex number to be less than another number? Nevertheless, the engineering literature is filled with complex random variables and their distributions.

The concept of a complex random variable can often be the source of great confusion to many students, but it doesn't have to be as long as we realize that a complex random variable is nothing more than a shorthand representation of two real random variables. To motivate the concept of a complex random variable, we use the most common example of a pair of independent, equal variance, jointly Gaussian random variables X and Y. The joint PDF is of the form

$$f_{X,Y}(x,y) = \frac{1}{2\pi\sigma^2} \exp\left(-\frac{(x-\mu_X)^2 + (y-\mu_Y)^2}{2\sigma^2}\right). \tag{5.69}$$

This joint PDF (of two real random variables) naturally lends itself to be written in terms of some complex variables. Define $Z = X + jY$, $z = x + jy$, and $\mu_Z = \mu_X + j\mu_Y$. Then

$$f_{X,Y}(x,y) = f_Z(z) = \frac{1}{2\pi\sigma^2} \exp\left(-\frac{|z-\mu_Z|^2}{2\sigma^2}\right). \tag{5.70}$$

We reemphasize at this point that this is not to be interpreted as the PDF of a complex random variable (since such an interpretation would make no sense); rather, this is just a compact representation of the joint PDF of two real random variables. This density is known as the *circular Gaussian density function* (since the contours of $f_Z(z) = $ constant form circles in the complex z-plane).

Note that the PDF in Equation 5.70 has two parameters, μ_Z and σ. The parameter μ_Z is interpreted as the mean of the complex quantity, $Z = X + jY$:

$$\mu_Z = E[Z] = E[X + jY] = \int_{-\infty}^{\infty} (x+jy)f_{X,Y}(x,y)\,dxdy = \mu_X + j\mu_Y. \tag{5.71}$$

But what about σ^2? We would like to be able to interpret it as the variance of $Z = X + jY$. To do so, we need to redefine what we mean by variance of a complex quantity. If we used the definition we are used to (for real quantities), we would find

$$E[(Z-\mu_Z)^2] = E\{[(X-\mu_X) + j(Y-\mu_Y)]^2\} = \text{Var}(X) - \text{Var}(Y) + 2j\,\text{Cov}(X,Y). \tag{5.72}$$

In the case of our independent Gaussian random variables, since $\text{Cov}(X,Y) = 0$ and $\text{Var}(X) = \text{Var}(Y)$, this would lead to $E[(Z-\mu_Z)^2] = 0$. To overcome this inconsistency, we redefine the variance for a complex quantity as follows.

DEFINITION 5.13: For a complex random quantity, $Z = X + jY$, the variance is defined as

$$\text{Var}(Z) = \frac{1}{2}E\left[|Z-\mu_Z|^2\right] = \frac{1}{2}\text{Var}(X) + \frac{1}{2}\text{Var}(Y). \tag{5.73}$$

We emphasize at this point that this definition is somewhat arbitrary and was chosen so that the parameter σ^2 that shows up in Equation 5.70 can be interpreted as the variance of Z. Many textbooks do not include the factor of $1/2$ in the definition, while many others (besides this one) do include the $1/2$. Hence, there seems to be no way to avoid a little bit of confusion here. The student just needs to be aware that there are two inconsistent definitions prevalent in the literature.

DEFINITION 5.14: For two complex random variables $Z_1 = X_1 + jY_1$ and $Z_2 = X_2 + jY_2$, the correlation and covariance are defined as

$$R_{1,2} = \frac{1}{2}E[Z_1 Z_2^*] = \frac{1}{2}\{E[X_1 X_2] + E[Y_1 Y_2] - jE[X_1 Y_2] + jE[X_2 Y_1]\}, \qquad (5.74)$$

$$C_{1,2} = \frac{1}{2}E[(Z_1 - \mu_{Z_1})(Z_2 - \mu_{Z_2})^*]. \qquad (5.75)$$

As with real random variables, complex quantities are said to be orthogonal if their correlation is zero, whereas they are uncorrelated if their covariance is zero.

5.11 Engineering Application: Mutual Information, Channel Capacity, and Channel Coding

In Section 4.12, we introduced the idea of the entropy of a random variable that is a quantitative measure of how much randomness there is in a specific random variable. If the random variable represents the output of a source, the entropy tells us how much mathematical information there is in each source symbol. We can also construct similar quantities to describe the relationships between random variables. Consider two random variables X and Y that are statistically dependent upon one another. Each random variable has a certain entropy associated with it, $H(X)$ and $H(Y)$, respectively. Suppose it is observed that $Y = y$. Since X and Y are related, knowing Y will tell us something about X, and hence the amount of randomness in X will be changed. This could be quantified using the concept of conditional entropy.

DEFINITION 5.15: The *conditional entropy* of a discrete random variable X given knowledge of a particular realization of a related random variable $Y = y$ is

$$H(X|Y = y) = \sum_x \Pr(X = x|Y = y) \log\left(\frac{1}{\Pr(X = x|Y = y)}\right). \qquad (5.76)$$

Averaging over all possible conditioning events produces

$$H(X|Y) = \sum_x \sum_y \Pr(Y = y)\Pr(X = x|Y = y) \log \left(\frac{1}{\Pr(X = x|Y = y)} \right)$$

$$= \sum_x \sum_y \Pr(X = x, Y = y) \log \left(\frac{1}{\Pr(X = x|Y = y)} \right). \qquad (5.77)$$

The conditional entropy tells how much uncertainty remains in the random variable X after we observe the random variable Y. The amount of information provided about X by observing Y can be determined by forming the difference between the entropy in X before and after observing Y.

DEFINITION 5.16: The *mutual information* between two discrete random variables X and Y is

$$I(X;Y) = H(X) - H(X|Y) = \sum_x \sum_y \Pr(X = x, Y = y) \log \left(\frac{\Pr(X = x|Y = y)}{\Pr(X = x)} \right).$$

$$(5.78)$$

We leave it as an exercise for the reader to prove the following properties of mutual information:

- *Nonnegative*: $I(X;Y) \geq 0$.
- *Independence*: $I(X;Y) = 0$ if and only if X and Y are independent.
- *Symmetry*: $I(X;Y) = I(Y;X)$.

Now we apply the concept of mutual information to a digital communication system. Suppose we have some digital communication system that takes digital symbols from some source (or from the output of a source encoder) and transmits them via some modulation format over some communications medium. At the receiver, a signal is received and processed and ultimately a decision is made as to what symbol(s) was most likely sent. We will not concern ourselves with the details of how the system operates, but rather we will model the entire process in a probabilistic sense. Let X represent the symbol to be sent, which is randomly drawn from some n-letter alphabet according to some distribution $p = (p_0, p_1, \ldots, p_{n-1})$. Furthermore, let Y represent the decision made by the receiver, with Y taken to be a random variable on an m-letter alphabet. It is not unusual to have $m \neq n$, but in order to keep this discussion as simple as possible, we will consider only the case where $m = n$ so that the input and output of our communication system are taken from the same alphabet. Also, we assume the system to be memoryless so that decisions made on one symbol are not affected by previous decisions, nor do they affect future decisions. In this case, we can describe the operation of the digital

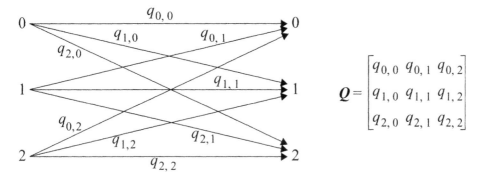

Figure 5.7   A transition diagram for a ternary (three-letter) communication channel.

communication system using a transition diagram as illustrated in Figure 5.7 for a three-letter alphabet. Mathematically, the operation of this communication system can be described by a matrix Q whose elements are $q_{i,j} = \Pr(Y = i|X = j)$.

We can now ask ourselves how much information the communication system can carry. Or, in other words, if we observe the output of the system, how much information does this give us about what was really sent? The mutual information answers this question. In terms of the channel (as described by Q) and the input (as described by p), the mutual information is

$$I(X, Y) = \sum_i \sum_j p_j q_{i,j} \log\left(\frac{q_{i,j}}{p_j}\right). \tag{5.79}$$

Note that the amount of information carried by the system is a function not only of the channel but also of the source. As an extreme example, suppose the input distribution were $p = (1, 0, \ldots, 0)$. In that case it is easy to show that $I(X, Y) = 0$; that is, the communication system carries no information. This is not because the communication system is incapable of carrying information, but because what we are feeding into the system contains no information. To describe the information carrying capability of a communication channel, we need a quantity that is a function of the channel only and not of the input to the channel.

DEFINITION 5.17: Given a discrete communications channel described by a transition probability matrix Q, the channel capacity is given by

$$C = \max_p I(X; Y) = \max_p \sum_i \sum_j p_j q_{i,j} \log\left(\frac{q_{i,j}}{p_j}\right). \tag{5.80}$$

The maximization of the mutual information is with respect to any valid probability distribution p.

EXAMPLE 5.26: As a simple example, consider the so-called binary symmetric channel (BSC) described by the transition probability matrix

$$Q = \begin{bmatrix} 1 - q & q \\ q & 1 - q \end{bmatrix}.$$

The BSC is described by a single parameter q, which has the interpretation of the probability of bit error of the binary communications system. That is, q is the probability of the receiver deciding a 0 was sent when a 1 was actually sent, and it is also the probability of the receiver deciding a 1 was sent when a 0 was actually sent. Since the input to this channel is binary, its distribution can also be described by a single parameter. That is, $p = (p, 1 - p)$. The mutual information for the BSC is

$$I(X;Y) = p(1 - q) \log \left(\frac{1 - q}{p} \right) + pq \log \left(\frac{q}{p} \right) + (1 - p)(1 - q) \log \left(\frac{1 - q}{1 - p} \right)$$

$$+ (1 - p)q \log \left(\frac{q}{1 - p} \right).$$

Some straightforward algebraic manipulations reveal that this expression can be simplified to $I(X;Y) = H(p) - H(q)$, where $H(\cdot)$ is the binary entropy function. Maximization with respect to p is now straightforward. The mutual information is maximized when the input distribution is $p = (0.5, 0.5)$ and the resulting capacity is

$$C = 1 - H(q).$$

This function is illustrated in Figure 5.8.

The channel capacity provides a fundamental limitation on the amount of information that can reliably be sent over a channel. For example, suppose we wanted to transmit information across the binary symmetric channel of Example 5.26. Furthermore, suppose the error probability of the channel was $q = 0.1$. Then the capacity is $C = 1 - H(0.1) = 0.53$ bit. That is, every physical bit that is transmitted across the channel must contain less than 0.53 bit of mathematical information. This is achieved through the use of redundancy via channel coding. Consider the block diagram of the digital communication system in Figure 5.9. The binary source produces independent but identically distributed (i.i.d., or IID) bits that are equally likely to be 0 or 1. (We will discuss IID random variables further in Chapter 7.) This source has an entropy of 1 bit/source symbol. Since the channel has a capacity of 0.53 bit, the information content of the source must be reduced before these

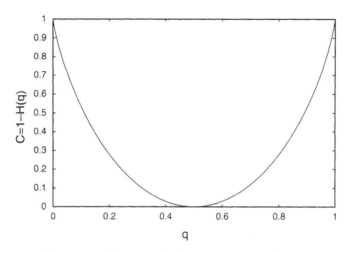

Figure 5.8   Capacity of a binary symmetric channel.

Figure 5.9   A functional block diagram of a digital communications system.

symbols are sent across the channel. This is achieved by the channel coder, which takes blocks of k information bits and maps them to n-bit code words where $n > k$. Each code word contains k bits of information and so each coded bit contains k/n bits of mathematical information. By choosing the code rate, k/n, to be less than the channel capacity, C, we can assure that the information content of the symbols being input to the channel is no greater than the information-carrying capability of the channel.

Viewed from a little more concrete perspective, the channel used to transmit physical bits has an error rate of 10 percent. The purpose of the channel code is to add redundancy to the data stream to provide the ability to correct the occasional errors caused by the channel. A fundamental result of information theory known as the *channel coding theorem* states that as k and n go to infinity in such a way that $k/n < C$, it is possible to construct a channel code (along with the appropriate decoder) that will provide error-free communication. That is, the original information bits will be provided to the destination with an arbitrarily small probability of error. The channel coding theorem does not tell us how to construct such a code, but significant progress has been made in recent years toward finding practical techniques to achieve what information theory promises is possible.

Exercises

5.1 For positive constants a and b, a pair of random variables has a joint PDF specified by

$$f_{X,Y}(x,y) = abe^{-(ax+by)} u(x) u(y).$$

(a) Find the joint CDF, $F_{X,Y}(x,y)$.

(b) Find the marginal PDFs, $f_X(x)$ and $f_Y(y)$.

(c) Find $\Pr(X > Y)$.

(d) Find $\Pr(X > Y^2)$.

5.2 For positive constants a, b, c, and positive integer n, a pair of random variables has a joint PDF specified by

$$f_{X,Y}(x,y) = \frac{d}{(ax + by + c)^n} u(x) u(y).$$

(a) Find the constant d in terms of a, b, c, and n.

(b) Find the marginal PDFs, $f_X(x)$ and $f_Y(y)$.

(c) Find $\Pr(X > Y)$.

5.3 A pair of random variables has a joint PDF specified by

$$f_{X,Y}(x,y) = d \exp(-(ax^2 + bxy + cy^2)).$$

(a) Find the constant d in terms of a, b, and c. Also find any restrictions needed for a, b, and c themselves for this to be a valid PDF.

(b) Find the marginal PDFs, $f_X(x)$ and $f_Y(y)$.

(c) Find $\Pr(X > Y)$.

5.4 A pair of random variables has a joint PDF specified by

$$f_{X,Y}(x,y) = \begin{cases} c\sqrt{1 - x^2 - y^2} & x^2 + y^2 \le 1 \\ 0 & \text{otherwise} \end{cases}.$$

(a) Find the constant c.

(b) Find $\Pr(X^2 + Y^2 > 1/4)$.

(c) Find $\Pr(X > Y)$.

5.5 A pair of random variables has a joint PDF specified by

$$f_{X,Y}(x,y) = \frac{1}{8\pi} \exp\left(-\frac{(x-1)^2 + (y+1)^2}{8}\right).$$

(a) Find $\Pr(X > 2, Y < 0)$.
(b) Find $\Pr(0 < X < 2, |Y+1| > 2)$.
(c) Find $\Pr(Y > X)$. *Hint:* Set up the appropriate double integral and then use the change of variables: $u = x - y, v = x + y$.

5.6 For some integer L and constant c, two discrete random variables have a joint PMF given by

$$P_{M,N}(m,n) = \begin{cases} c & m \geq 0, n \geq 0, \ m + n < L \\ 0 & \text{otherwise} \end{cases}.$$

(a) Find the value of the constant c in terms of L.
(b) Find the marginal PMFs, $P_M(m)$ and $P_N(n)$.
(c) Find $\Pr(M + N < L/2)$.

5.7 A pair of random variables has a joint PDF specified by

$$f_{X,Y}(x,y) = \frac{1}{2\pi\sqrt{3}} \exp\left(-\frac{x^2 + 2xy + 4y^2}{6}\right).$$

(a) Find the marginal PDFs, $f_X(x)$ and $f_Y(y)$.
(b) Based on the results of part (a), find $E[X]$, $E[Y]$, $\text{Var}(X)$, and $\text{Var}(Y)$.
(c) Find the conditional PDF, $f_{X|Y}(x|y)$.
(d) Based on the results of part (c), find $E[XY]$, $\text{Cov}(X,Y)$, and $\rho_{X,Y}$.

5.8 A pair of random variables is uniformly distributed over the ellipse defined by $x^2 + 4y^2 \leq 1$.

(a) Find the marginal PDFs, $f_X(x)$ and $f_Y(y)$.
(b) Based on the results of part (a), find $E[X]$, $E[Y]$, $\text{Var}(X)$, and $\text{Var}(Y)$.
(c) Find the conditional PDFs, $f_{X|Y}(x|y)$ and $f_{Y|X}(y|x)$.
(d) Based on the results of part (c), find $E[XY]$, $\text{Cov}(X,Y)$, and $\rho_{X,Y}$.

5.9 Prove that if two random variables are linearly related (i.e., $Y = aX + b$ for constants $a \neq 0$ and b), then

$$\rho_{X,Y} = \text{sgn}(a) = \begin{cases} 1 & \text{if } a > 0 \\ -1 & \text{if } a < 0 \end{cases}.$$

Also, prove that if two random variables have $|\rho_{X,Y}| = 1$, then they are linearly related.

5.10 Prove the triangle inequality that states that

$$\sqrt{E[(X + Y)^2]} \leq \sqrt{E[X^2]} + \sqrt{E[Y^2]}.$$

5.11 Two random variables X and Y have, $\mu_X = 2$, $\mu_Y = -1$, $\sigma_X = 1$, $\sigma_Y = 4$, and $\rho_{X,Y} = 1/4$. Let $U = X + 2Y$ and $V = 2X - Y$. Find the following quantities:

(a) $E[U]$ and $E[V]$;
(b) $E[U^2]$, $E[V^2]$, Var(U), and Var(V);
(c) $E[UV]$, Cov(U, V), and $\rho_{U,V}$.

5.12 Suppose two random variables are related by $Y = aX^2$ and assume that $f_X(x)$ is symmetric about the origin. Show that $\rho_{X,Y} = 0$.

5.13 Find an example (other than the one given in Example 5.15) of two random variables that are uncorrelated but not independent.

5.14 Starting from the general form of the joint Gaussian PDF given in Equation 5.40, show that the resulting marginal PDFs are both Gaussian.

5.15 Starting from the general form of the joint Gaussian PDF given in Equation 5.40 and using the results of Exercise 5.14, show that conditioned on $Y = y$, X is Gaussian with a mean of $\mu_X + \rho_{XY}(\sigma_X/\sigma_Y)(y - \mu_Y)$ and a variance of $\sigma_X^2(1 - \rho_{XY}^2)$.

5.16 Let X and Y be random variables with means μ_X and μ_Y, variances σ_X^2 and σ_Y^2, and correlation coefficient $\rho_{X,Y}$.

(a) Find the value of the constant a which minimizes $[E(Y - aX)^2]$.
(b) Find the value of $[E(Y - aX)^2]$ when a is given as determined in part (a).

5.17 Let X and Y be zero-mean random variables with a correlation coefficient of ρ and unequal variances of σ_X^2 and σ_Y^2.

(a) Find the joint characteristic function, $\Phi_{X,Y}(\omega_1, \omega_2)$.
(b) Find the correlation, covariance, and correlation coefficient.
(c) Find $E[X^2Y^2]$.

5.18 Find the general form of the joint characteristic function of two jointly Gaussian random variables.

5.19 A quarterback throws a football at a target marked out on the ground 40 yards from his position. Assume that the PDF for the football's hitting the target is Gaussian within the plane of the target. Let the coordinates of the plane of the target be denoted by the x and y axes. Thus, the joint PDF of (X, Y) is a two-dimensional Gaussian PDF. The average location of the hits is at the origin of the target, and the standard deviation in each direction is the same and is denoted as σ. Assuming X and Y are independent, find the probability that the hits will be located within an annular ring of width dr located a distance r from the origin; that is, find the probability density function for hits as a function of the radius from the origin.

5.20 Let X and Y be independent and both exponentially distributed with

$$f_X(v) = f_Y(v) = be^{-bv} u(v).$$

Find the PDF of $Z = X - Y$.

5.21 Let X and Y be jointly Gaussian random variables. Show that $Z = aX + bY$ is also a Gaussian random variable. Hence, any linear transformation of two Gaussian random variables produces a Gaussian random variable.

5.22 Let X and Y be jointly Gaussian random variables with $E[X] = 1$, $E[Y] = -2$, $\text{Var}(X) = 4$, $\text{Var}(Y) = 9$, and $\rho_{X,Y} = 1/3$. Find the PDF of $Z = 2X - 3Y - 5$. *Hint:* To simplify this problem, use the result of Exercise 5.21.

5.23 Let X and Y be independent Rayleigh random variables such that

$$f_X(v) = f_Y(v) = v \exp\left(-\frac{v^2}{2}\right) u(v).$$

(a) Find the PDF of $Z = \max(X, Y)$.
(b) Find the PDF of $W = \min(X, Y)$.

5.24 Suppose X and Y are independent and exponentially distributed both with unit-mean. Consider the roots of the quadratic equation $z^2 + Xz + Y = 0$.

(a) Find the probability that the roots are real.
(b) Find the probability that the roots are complex.
(c) Find the probability that the roots are equal.

5.25 Suppose X is a Rayleigh random variable and Y is an arcsine random variable, so that

$$f_X(x) = \frac{x}{\sigma^2} \exp\left(-\frac{x^2}{2\sigma^2}\right) u(x) \quad \text{and} \quad f_Y(y) = \frac{1}{\pi\sqrt{1-y^2}} |y| < 1.$$

Furthermore, assume X and Y are independent. Find the PDF of $Z = XY$.

5.26 Let X and Y be independent and both uniformly distributed over $(0, 2\pi)$. Find the PDF of $Z = (X + Y) \bmod 2\pi$.

5.27 Let X be a Gaussian random variable, and let Y be a Bernoulli random variable with $\Pr(Y = 1) = p$ and $\Pr(Y = -1) = 1 - p$. If X and Y are independent, find the PDF of $Z = XY$. Under what conditions is Z a Gaussian random variable?

5.28 Let X and Y be independent zero-mean, unit-variance Gaussian random variables. Consider forming the new random variable U, V according to

$$U = X \cos(\theta) - Y \sin(\theta),$$

$$V = X \sin(\theta) + Y \cos(\theta).$$

Note that this transformation produces a coordinate rotation through an angle of (θ). Find the joint PDF of U and V. *Hint:* The result of Example 5.25 will be helpful here.

5.29 Let X and Y be zero-mean, unit-variance Gaussian random variables with correlation coefficient, ρ. Suppose we form two new random variables using a linear transformation:

$$U = aX + bY,$$

$$V = cX + dY.$$

Find constraints on the constants a, b, c, and d such that U and V are independent.

5.30 Suppose X and Y are independent and Gaussian with means of μ_X and μ_Y, respectively, and equal variances of σ^2. The polar variables are formed according to $R = \sqrt{X^2 + Y^2}$ and $\Theta = \tan^{-1}(Y/X)$.

(a) Find the joint PDF of R and Θ.

(b) Show that the marginal PDF of R follows a Rician distribution.

5.31 Suppose X and Y are independent, zero-mean Gaussian random variables with variances of σ_X^2 and σ_Y^2, respectively. Find the joint PDF of

$$Z = X^2 + Y^2 \quad \text{and} \quad W = Z^2 - Y^2.$$

5.32 Suppose X and Y are independent, Cauchy random variables. Find the joint PDF of

$$Z = X^2 + Y^2 \quad \text{and} \quad W = XY.$$

5.33 A complex random variable is defined by $Z = Ae^{j\Theta}$, where A and Θ are independent and Θ is uniformly distributed over $(0, 2\pi)$.

(a) Find $E[Z]$.
(b) Find $\text{Var}(Z)$. For this part, leave your answer in terms of the moments of A.

5.34 Suppose $Q = \begin{bmatrix} 0.8 & 0.1 & 0.1 \\ 0.1 & 0.8 & 0.1 \\ 0.1 & 0.1 & 0.8 \end{bmatrix}$ in Figure 5.7 and $p_i = 1/3$, $i = 1, 2, 3$.

Determine the mutual information for this channel.

5.35 Repeat the previous problem if $Q = \begin{bmatrix} 0.9 & 0.1 & 0 \\ 0 & 0.9 & 0.1 \\ 0 & 0.1 & 0.9 \end{bmatrix}$.

5.36 Repeat Exercise 5.35 if $Q = \begin{bmatrix} 1/3 & 1/3 & 1/3 \\ 1/3 & 1/3 & 1/3 \\ 1/3 & 1/3 & 1/3 \end{bmatrix}$. Can you give an interpretation for your result?

MATLAB Exercises

5.37 Provide contour plots for the ellipses discussed in Example 5.17. Consider the following cases:

(a) $\sigma_X = \sigma_Y$ and $\rho_{XY} = 0$;
(b) $\sigma_X < \sigma_Y$ and $\rho_{XY} = 0$;
(c) $\sigma_X > \sigma_Y$ and $\rho_{XY} = 0$;
(d) $\sigma_X = \sigma_Y$ and $\rho_{XY} \neq 0$.

Let c^2 be the same for each case. Discuss the effect that σ_X, σ_Y, and ρ_{XY}, have on the shape of the contour. Now select one of the cases and let c^2 increase and decrease. What is the significance of c^2?

5.38 Let X and Y have a joint PDF given by

$$f_{X,Y}(x,y) = \frac{1}{2\pi} \exp\left(-\frac{((x-2)^2 + (y-3)^2)}{2}\right)$$

as in Example 5.6. Write a MATLAB program to generate many samples of this pair of random variables. Note that X and Y are independent, Gaussian random variables with unit variances and means of 2 and 3, respectively. After a large number of sample pairs have been generated, compute the relative frequency of the number of pairs that fall within the unit circle, $X^2 + Y^2 < 1$. Compare your answer with that obtained in Example 5.6. How many random samples must you generate in order to get a decent estimate of the probability?

5.39 Let X and Y have a joint PDF given by

$$f_{X,Y}(x,y) = \frac{1}{2\pi} \exp\left(-\frac{((x-1)^2 + y^2)}{2}\right).$$

Write a MATLAB program to evaluate $\Pr((X, Y \in \Re))$, where $\Re$ is the shaded region bounded by the lines $y = x$ and $y = -x$, as shown in the accompanying figure. You should set up the appropriate double integral and use MATLAB to evaluate the integral numerically. Note in this case that one of the limits of integration is infinite. How will you deal with this?

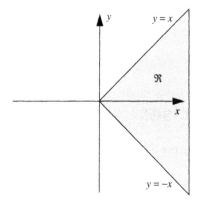

5.40 Write a MATLAB program to generate pairs of random variables that are uniformly distributed over the ellipse $x^2 + 4y^2 < 1$. Use the technique employed in Example 5.16. Also, create a three-dimensional plot of an estimate of the PDF obtained from the random data you generated.

Multiple Random Variables

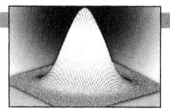

In many applications, it is necessary to deal with a large number of random variables. Often, the number of variables can be arbitrary. In this chapter, the concepts developed previously for single random variables and pairs of random variables are extended to allow for an arbitrary number of random variables. Much of the focus of this chapter is on multidimensional Gaussian random variables, since most non-Gaussian random variables are difficult to deal with in many dimensions. One of our main goals here is to develop a vector/matrix notation that will allow us to represent potentially large sequences of random variables with a compact notation. Many of the concepts developed in Chapter 5 can be extended to multiple dimensions in a very straightforward manner; thus we will devote minimal time to those concepts in our current discussion. Rather, attention is focused on those ideas that require more than a trivial extension to the work done in previous chapters.

6.1 Joint and Conditional PMFs, CDFs, and PDFs

The concepts of probability mass function, conditional distribution function, and probability density function are easily extended to an arbitrary number of random variables. Their definitions follow.

DEFINITION 6.1: For a set of N random variables $X_1, X_2, \ldots, X_N$, the joint PMF, CDF, and PDF are given respectively by

$$P_{X_1, X_2, \ldots, X_N}(x_{k_1}, x_{k_2}, \ldots, x_{k_N}) = \Pr(X_1 = x_{k_1}, X_2 = x_{k_2}, \ldots, X_N = x_{k_N}); \qquad (6.1)$$

$$F_{X_1, X_2, \ldots, X_N}(x_1, x_2, \ldots, x_N) = \Pr(X_1 \le x_1, X_2 \le x_2, \ldots, X_N \le x_N); \quad (6.2)$$

$$f_{X_1, X_2, \ldots, X_N}(x_1, x_2, \ldots, x_N) = \frac{\partial^N}{\partial x_1 \partial x_2 \ldots \partial x_N} F_{X_1, X_2, \ldots, X_N}(x_1, x_2, \ldots, x_N). \quad (6.3)$$

When large numbers of random variables are involved, this notation can get cumbersome, so it is convenient to introduce a vector notation to write these quantities in a more compact fashion. Let $X = [X_1, X_2, \ldots, X_N]^T$ be a column vector[1] consisting of the N random variables and similarly define $x = [x_1, x_2, \ldots, x_N]^T$. Then the preceding functions can be expressed respectively as $P_X(x)$, $F_X(x)$, and $f_X(x)$.

Marginal CDFs can be found for a subset of the variables by evaluating the joint CDF at infinity for the unwanted variables. For example,

$$F_{X_1, X_2, \ldots, X_M}(x_1, x_2, \ldots, x_M) = F_{X_1, X_2, \ldots, X_N}(x_1, x_2, \ldots, x_M, \infty, \infty, \ldots, \infty). \quad (6.4)$$

Marginal PDFs are found from the joint PDF by integrating out the unwanted variables. Similarly, marginal PMFs are obtained from the joint PMF by summing out the unwanted variables:

$$f_{X_1, X_2, \ldots, X_M}(x_1, x_2, \ldots, x_M) = \int\!\!\!\int_{-\infty}^{\infty} \cdots \int f_{X_1, X_2, \ldots, X_N}(x_1, x_2, \ldots, x_N) dx_{M+1} dx_{M+2} \ldots dx_N,$$

$$(6.5)$$

$$P_{X_1, X_2, \ldots, X_M}(x_{k_1}, x_{k_2}, \ldots, x_{k_M}) = \sum_{k_{M+1}} \sum_{k_{M+2}} \cdots \sum_{k_N} P_{X_1, X_2, \ldots, X_N}(x_{k_1}, x_{k_2}, \ldots, x_{k_N}). \quad (6.6)$$

Similar to that done for pairs of random variables in Chapter 5, we can also establish conditional PMFs and PDFs.

DEFINITION 6.2: For a set of N random variables $X_1, X_2, \ldots, X_N$, the conditional PMF and PDF of $X_1, X_2, \ldots, X_M$ conditioned on $X_{M+1}, X_{M+2}, \ldots, X_N$ are given by

$$P_{X_1, \ldots, X_M | X_{M+1}, \ldots, X_N}(x_{k_1}, \ldots, x_{k_M} \mid x_{k_{M+1}}, \ldots, x_{k_N}) = \frac{\Pr(X_1 = x_{k_1}, \ldots, X_N = x_{k_N})}{\Pr(X_{M+1} = x_{k_{M+1}}, \ldots, X_N = x_{k_N})}, \quad (6.7)$$

$$f_{X_1, \ldots, X_M | X_{M+1}, \ldots, X_N}(x_1, \ldots, x_M | x_{M+1}, \ldots, x_N) = \frac{f_{X_1, \ldots, X_N}(x_1, \ldots, x_N)}{f_{X_{M+1}, \ldots, X_N}(x_{M+1}, \ldots, x_N)}. \quad (6.8)$$

---

[1] We use T to represent the matrix transpose operation so that if v is a row vector, then v^T is a column vector. Also, to avoid confusion throughout the text, we use boldface variables to represent vector and matrix quantities and regular face variables for scalar quantities.

Using conditional PDFs, many interesting factorization results can be established for joint PDFs involving multiple random variables. For example, consider four random variables, X_1, X_2, X_3, X_4:

$$f_{X_1, X_2, X_3, X_4}(x_1, x_2, x_3, x_4) = f_{X_1 | X_2, X_3, X_4}(x_1 \mid x_2, x_3, x_4) f_{X_2, X_3, X_4}(x_2, x_3, x_4)$$

$$= f_{X_1 | X_2, X_3, X_4}(x_1 \mid x_2, x_3, x_4) f_{X_2 | X_3, X_4}(x_2 \mid x_3, x_4) f_{X_3, X_4}(x_3, x_4)$$

$$= f_{X_1 | X_2, X_3, X_4}(x_1 \mid x_2, x_3, x_4) f_{X_2 | X_3, X_4}(x_2 \mid x_3, x_4) f_{X_3 | X_4}(x_3 \mid x_4) f_{X_4}(x_4). \quad (6.9)$$

Almost endless other possibilities exist as well.

DEFINITION 6.3: A set of N random variables is statistically independent if any subset of the random variables are independent of any other disjoint subset. In particular, any joint PDF of $M \leq N$ variables should factor into a product of the corresponding marginal PDFs.

As an example, consider three random variables, X, Y, Z. For these three random variables to be independent, we must have each pair independent. This implies that

$$f_{X,Y}(x,y) = f_X(x) f_Y(y), \ f_{X,Z}(x,z) = f_X(x) f_Z(z), \ f_{Y,Z}(y,z) = f_Y(y) f_Z(z). \quad (6.10)$$

In addition, the joint PDF of all three must also factor into a product of the marginals,

$$f_{X,Y,Z}(x, y, z) = f_X(x) f_Y(y) f_Z(z). \quad (6.11)$$

Note that all three conditions in Equation 6.10 follow directly from the single condition in Equation 6.11. Hence, Equation 6.11 is a necessary and sufficient condition for three variables to be statistically independent. Naturally, this result can be extended to any number of variables. That is, the elements of a random vector $X = [X_1, X_2, \dots, X_N]^T$ are independent if

$$f_X(x) = \prod_{n=1}^{N} f_{X_n}(x_n). \quad (6.12)$$

6.2 Expectations Involving Multiple Random Variables

For a vector of random variables $X = [X_1, X_2, \dots, X_N]^T$, we can construct a corresponding mean vector that is a column vector of the same dimension and whose components are the means of the elements of X. Mathematically, we say $\mu = E[X] = [E[X_1], E[X_2], \ldots, E[X_N]]^T$. Two other important quantities associated with the random vector are the correlation and covariance matrices. (To be consistent, the vector μ should be bold; however, we have elected not to do so.)

DEFINITION 6.4: For a random vector $X = [X_1, X_2, \ldots, X_N]^T$, the correlation matrix is defined as $R_{XX} = E[XX^T]$. That is, the (i, j)th element of the $N \times N$ matrix R_{XX} is $E[X_i X_j]$. Similarly, the covariance matrix is defined as $C_{XX} = E[(X - \mu)(X - \mu)^T]$ so that the (i, j)th element of C_{XX} is $\text{Cov}(X_i, X_j)$.

THEOREM 6.1: Correlation matrices and covariance matrices are symmetric and positive definite.

PROOF: Recall that a square matrix, R_{XX}, is symmetric if $R_{XX} = R_{XX}^T$. Equivalently, the (i, j)th element must be the same as the (j, i)th element. This is clearly the case here since $E[X_i X_j] = E[X_j X_i]$. Recall that the matrix R_{XX} is positive definite if $z^T R_{XX} z > 0$ for any vector z such that $||z|| > 0$.

$$z^T R_{XX} z = z^T E[XX^T] z = E[z^T XX^T z] = E[(z^T X)^2]. \qquad (6.13)$$

Note that $z^T X$ is a scalar random variable (a linear combination of the components of X). Since the second moment of any random variable is positive (except for the pathological case of a random variable that is identically equal to zero), then the correlation matrix is positive definite. As an aside, this also implies that the eigenvalues of the correlation matrix are all positive. Identical steps can be followed to prove the same properties hold for the covariance matrix. ∎

Next, consider a linear transformation of a vector random variable. That is, create a new set of M random variables, $Y = [Y_1, Y_2, \ldots, Y_M]^T$, according to

$$Y_1 = a_{1,1} X_1 + a_{1,2} X_2 + \cdots + a_{1,N} X_N + b_1,$$

$$Y_2 = a_{2,1} X_1 + a_{2,2} X_2 + \cdots + a_{2,N} X_N + b_2, \qquad (6.14)$$

$$\cdots$$

$$Y_M = a_{M,1} X_1 + a_{M,2} X_2 + \cdots + a_{M,N} X_N + b_M.$$

The number of new variables, M, does not have to be the same as the number of original variables, N. To write this type of linear transformation in a compact fashion, define a matrix A whose (i, j)th element is the coefficient $a_{i,j}$ and a column vector, $b = [b_1, b_2, \ldots, b_M]^T$. Then the linear transformation of Equation 6.14 is written in vector/matrix form as $Y = AX + b$. The next theorem describes the relationship between the means of X and Y and the correlation matrices of X and Y.

THEOREM 6.2: For a linear transformation of vector random variables of the form $Y = AX + b$, the means of X and Y are related by

$$\mu_Y = A\mu_X + b. \tag{6.15}$$

Also, the correlation matrices of X and Y are related by

$$R_{YY} = AR_{XX}A^T + A\mu_X b^T + b\mu_X^T A^T + bb^T, \tag{6.16}$$

and the covariance matrices of X and Y are related by

$$C_{YY} = AC_{XX}A^T. \tag{6.17}$$

PROOF: For the mean vector,

$$\mu_Y = E[Y] = E[AX + b] = AE[X] + b = A\mu_X + b. \tag{6.18}$$

Similarly, for the correlation matrix,

$$\begin{aligned}
R_{YY} &= E[YY^T] = E[(AX + b)(AX + b)^T] \\
&= E[AXX^T A^T] + E[bX^T A^T] + E[AXb^T] + E[bb^T] \\
&= AE[XX^T]A^T + bE[X^T]A^T + AE[X]b^T + bb^T \\
&= AR_{XX}A^T + A\mu_X b^T + b\mu_X^T A^T + bb^T. \tag{6.19}
\end{aligned}$$

To prove the result for the covariance matrix, write $Y - \mu_Y$ as

$$Y - \mu_Y = (AX + b) - (A\mu_X + b) = A(X - \mu_X). \tag{6.20}$$

Then,

$$\begin{aligned}
C_{YY} &= E[(Y - \mu_Y)(Y - \mu_Y)^T] = E[\{A(X - \mu_X)\}\{A(X - \mu_X)\}^T] \\
&= E[A(X - \mu_X)(X - \mu_X)^T A^T] = AE[(X - \mu_X)(X - \mu_X)^T]A^T = AC_{XX}A^T. \tag{6.21}
\end{aligned}$$

∎

6.3 Gaussian Random Variables in Multiple Dimensions

Recall from the study of two-dimensional random variables in the previous chapter that the functional form of the joint Gaussian PDF was fairly complicated.

It would seem that the prospects of forming a joint Gaussian PDF for an arbitrary number of dimensions are grim. However, the vector/matrix notation developed in the previous sections make this task manageable and, in fact, the resulting joint Gaussian PDF is quite simple.

DEFINITION 6.5: The joint Gaussian PDF for a vector of N random variables, X, with mean vector μ_X and covariance matrix C_{XX} is given by[2]

$$f_X(x) = \frac{1}{\sqrt{(2\pi)^N \det(C_{XX})}} \exp\left(-\frac{1}{2}(x - \mu_X)^T C_{XX}^{-1}(x - \mu_X)\right). \qquad (6.22)$$

EXAMPLE 6.1: To demonstrate the use of this matrix notation, suppose X is a two-element vector and the mean vector and covariance matrix are given by their general forms

$$\mu_X = \begin{bmatrix} \mu_1 \\ \mu_2 \end{bmatrix} \text{ and } C_{XX} = \begin{bmatrix} \sigma_1^2 & \rho\sigma_1\sigma_2 \\ \rho\sigma_1\sigma_2 & \sigma_2^2 \end{bmatrix}.$$

The determinant of the covariance matrix is

$$\det(C_{XX}) = \sigma_1^2\sigma_2^2 - (\rho\sigma_1\sigma_2)^2 = \sigma_1^2\sigma_2^2(1 - \rho^2),$$

while the inverse is

$$C_{XX}^{-1} = \frac{\begin{bmatrix} \sigma_2^2 & -\rho\sigma_1\sigma_2 \\ -\rho\sigma_1\sigma_2 & \sigma_1^2 \end{bmatrix}}{\sigma_1^2\sigma_2^2(1 - \rho^2)} = \frac{\begin{bmatrix} \sigma_1^{-2} & -\rho\sigma_1^{-1}\sigma_2^{-1} \\ -\rho\sigma_1^{-1}\sigma_2^{-1} & \sigma_2^{-2} \end{bmatrix}}{(1 - \rho^2)}.$$

The quadratic form in the exponent then works out to be

$$(x - \mu_X)^T C_{XX}^{-1}(x - \mu_X)$$

$$= [x_1 - \mu_1 \quad x_2 - \mu_2] \frac{\begin{bmatrix} \sigma_1^{-2} & -\rho\sigma_1^{-1}\sigma_2^{-1} \\ -\rho\sigma_1^{-1}\sigma_2^{-1} & \sigma_2^{-2} \end{bmatrix}}{(1 - \rho^2)} \begin{bmatrix} x_1 - \mu_1 \\ x_2 - \mu_2 \end{bmatrix}$$

$$= \frac{\left(\dfrac{x_1 - \mu_1}{\sigma_1}\right)^2 - 2\rho\left(\dfrac{x_1 - \mu_1}{\sigma_1}\right)\left(\dfrac{x_2 - \mu_2}{\sigma_2}\right) + \left(\dfrac{x_2 - \mu_2}{\sigma_2}\right)^2}{(1 - \rho^2)}.$$

---

[2]The notation $\det(A)$ refers to the determinant of the matrix A, while A^{-1} is the inverse of A.

Plugging all these results into the general form for the joint Gaussian PDF gives

$$f_{X_1, X_2}(x_1, x_2) = \frac{1}{\sqrt{(2\pi)^2 \sigma_1^2 \sigma_2^2 (1 - \rho^2)}}$$

$$\times \exp\left(-\frac{\left(\dfrac{x_1 - \mu_1}{\sigma_1}\right)^2 - 2\rho\left(\dfrac{x_1 - \mu_1}{\sigma_1}\right)\left(\dfrac{x_2 - \mu_2}{\sigma_2}\right) + \left(\dfrac{x_2 - \mu_2}{\sigma_2}\right)^2}{2(1 - \rho^2)}\right).$$

This is exactly the form of the two-dimensional joint Gaussian PDF given in the previous chapter.

EXAMPLE 6.2: As a special case, suppose a vector of N jointly Gaussian random variables are all mutually uncorrelated. This means that $\text{Cov}(X_i, X_j) = 0$ for all $i \neq j$. A direct result of this is that all of the off-diagonal elements of the covariance matrix of X are zero. In other words, C_{XX} is a diagonal matrix of the general form

$$C_{XX} = \begin{bmatrix} \sigma_1^2 & 0 & \cdots & 0 \\ 0 & \sigma_2^2 & \cdots & 0 \\ \cdots & \cdots & & \cdots \\ 0 & 0 & \cdots & \sigma_N^2 \end{bmatrix}.$$

The determinant of a diagonal matrix is the product of the diagonal entries so that in this case $\det(C_{XX}) = \sigma_1^2 \sigma_2^2 \ldots \sigma_N^2$. The inverse is also trivial to compute and takes on the form

$$C_{XX}^{-1} = \begin{bmatrix} \sigma_1^{-2} & 0 & \cdots & 0 \\ 0 & \sigma_2^{-2} & \cdots & 0 \\ \cdots & \cdots & & \cdots \\ 0 & 0 & \cdots & \sigma_N^{-2} \end{bmatrix}.$$

The quadratic form that appears in the exponent of the Gaussian PDF becomes

$$(x - \mu_X)^T C_{XX}^{-1} (x - \mu_X)$$

$$= \begin{bmatrix} x_1 - \mu_1 x_2 - \mu_2 \dots x_N - \mu_N \end{bmatrix} \begin{bmatrix} \sigma_1^{-2} & 0 & \dots & 0 \\ 0 & \sigma_2^{-2} & \dots & 0 \\ \dots & \dots & & \dots \\ 0 & 0 & \dots & \sigma_N^{-2} \end{bmatrix} \begin{bmatrix} x_1 - \mu_1 \\ x_2 - \mu_2 \\ \dots \\ x_N - \mu_N \end{bmatrix}$$

$$= \sum_{n=1}^{N} \left(\frac{x_n - \mu_n}{\sigma_n} \right)^2 .$$

The joint Gaussian PDF for a vector of uncorrelated random variables is then

$$f_X(x) = \frac{1}{\sqrt{(2\pi)^n \sigma_1^2 \sigma_2^2 \dots \sigma_N^2}} \exp \left(-\frac{1}{2} \sum_{n=1}^{N} \left(\frac{x_n - \mu_n}{\sigma_n} \right)^2 \right)$$

$$= \prod_{n=1}^{N} \frac{1}{\sqrt{2\pi \sigma_n^2}} \exp \left(-\frac{(x_n - \mu_n)^2}{2\sigma_n^2} \right) .$$

This shows that for any number of uncorrelated Gaussian random variables, the joint PDF factors into the product of marginal PDFs, and hence uncorrelated Gaussian random variables are independent. This is a generalization of the same result that was proved in Chapter 5 for two Gaussian random variables.

EXAMPLE 6.3: In this example, we use MATLAB's symbolic capabilities to compute the form of a three-dimensional Gaussian PDF. Suppose we have three jointly Gaussian random variables $[X, Y, Z]^T$ with a mean vector $\mu = [1, 2, 3]^T$ and covariance matrix

$$C = \begin{bmatrix} 9 & 4 & 1 \\ 4 & 9 & 4 \\ 1 & 4 & 9 \end{bmatrix} .$$

The three-dimensional joint PDF, $f_{X,Y,Z}(x,y,z)$, can be found with the following MATLAB code:

```
x=sym('x','real');          % Define x, y, and z as symbolic.
y=sym('y','real');
z=sym('z','real');
pi=sym('pi');               % Disable numeric definition of pi.
C=[9 4 1; 4 9 4; 1 4 9];    % Covariance matrix.
mu=[1; 2; 3];               % Mean vector.
% Compute PDF symbolically.
v=[x; y; z]-mu;
f=exp(-v'*(inv(C))*v/2)/sqrt((2*pi)^3*det(C));
simplify(f)
```

Executing this program, MATLAB finds the joint PDF to be

$$f_{X,Y,Z}(x,y,z) = \frac{1}{464\sqrt{58\pi^3}}\exp\left(-\frac{65}{928}x^2 + \frac{11}{232}x - \frac{125}{232} + \frac{2}{29}xy\right.$$

$$\left. +\frac{2}{29}y - \frac{7}{464}xz + \frac{69}{232}z - \frac{5}{58}y^2 + \frac{2}{29}yz - \frac{65}{928}z^2\right).$$

The reader is encouraged to try different mean vectors and covariance matrices in the preceding program.

6.4 Transformations Involving Multiple Random Variables

In this section, we discuss various transformations of vector random variables. To exhaustively cover this topic would require much more space than we can devote to it here. Instead, we chose to cover some of the more common transformations encountered in engineering practice. To start with, we extend the formula for 2×2 transformations developed in the previous chapter to the case of $N \times N$ transforms. Let $Y = g(X)$ be a vector transformation,

$$Y_1 = g_1(X_1, X_2, \ldots, X_N),$$

$$Y_2 = g_2(X_1, X_2, \ldots, X_N),$$

$$\ldots \tag{6.23}$$

$$Y_N = g_N(X_1, X_2, \ldots, X_N),$$

and let $X = h(Y)$ be the inverse transformation. Given the PDF of X, the PDF of Y is found by

$$
f_Y(y) = \frac{f_X(x)}{\left| \det \left[J \begin{pmatrix} y_1 & y_2 & \cdots & y_N \\ x_1 & x_2 & \cdots & x_N \end{pmatrix} \right] \right|} \Bigg|_{x = h(y)}
$$

$$
= f_X(x) \left| \det \left[J \begin{pmatrix} x_1 & x_2 & \cdots & x_N \\ y_1 & y_2 & \cdots & y_N \end{pmatrix} \right] \right| \Bigg|_{x = h(y)}. \tag{6.24}
$$

As in Chapter 5, it needs to be understood that if the transformation is not one-to-one, the preceding expression must be evaluated at each root and summed together. This result can be proved using the same sort of derivation that was used for the case of 2×2 transformations.

6.4.1 Linear Transformations

Perhaps the single most important class of transformations is that involving linear transformations of Gaussian random variables. Consider a linear transformation of the general form $Y = AX + b$ when the random vector X has a joint Gaussian PDF as given in Equation 6.22. To begin, consider the case where the dimensionality of X and Y are the same (i.e., both are N-element vectors). In that case, the matrix A is a square matrix. Furthermore, it is assumed that the matrix A is invertible $(\det(A) \neq 0)$. In this case, the Jacobian of the linear transformation is

$$
J \begin{pmatrix} y_1 & y_2 & \cdots & y_N \\ x_1 & x_2 & \cdots & x_N \end{pmatrix} = \begin{bmatrix} \dfrac{\partial y_1}{\partial x_1} & \dfrac{\partial y_1}{\partial x_2} & \cdots & \dfrac{\partial y_1}{\partial x_N} \\[2mm] \dfrac{\partial y_2}{\partial x_1} & \dfrac{\partial y_2}{\partial x_2} & \cdots & \dfrac{\partial y_2}{\partial x_N} \\[2mm] \cdots & \cdots & & \cdots \\[2mm] \dfrac{\partial y_N}{\partial x_1} & \dfrac{\partial y_N}{\partial x_2} & \cdots & \dfrac{\partial y_N}{\partial x_N} \end{bmatrix} = \begin{bmatrix} a_{1,1} & a_{1,2} & \cdots & a_{1,N} \\ a_{2,1} & a_{2,2} & \cdots & a_{2,N} \\ \cdots & \cdots & & \cdots \\ a_{N,1} & a_{N,2} & \cdots & a_{N,N} \end{bmatrix} = A.
$$

$$\tag{6.25}$$

Also, the inverse transformation is linear and can be written as $X = A^{-1}(Y - b)$. The joint PDF for the vector Y is then

$$
f_Y(y) = \frac{f_X(x)}{|\det(A)|} \Bigg|_{x = A^{-1}(y-b)}. \tag{6.26}
$$

Plugging in the form of the Gaussian PDF for $f_X(x)$ results in

$$f_Y(y) = \frac{1}{|\det(A)|\sqrt{(2\pi)^N \det(C_{XX})}} \exp\left(-\frac{1}{2}(x - \mu_X)^T C_{XX}^{-1}(x - \mu_X)\right)\Bigg|_{x = A^{-1}(y-b)}.$$

(6.27)

To simplify this result, write

$$x - \mu_X|_{x=A^{-1}(y-b)} = A^{-1}(y - b) - \mu_X = A^{-1}(y - (b + A\mu_X)) = A^{-1}(y - \mu_Y).$$

(6.28)

The quadratic form in the exponent is then

$$(x - \mu_X)^T C_{XX}^{-1}(x - \mu_X)\Bigg|_{x=A^{-1}(y-b)} = [A^{-1}(y - \mu_Y)]^T C_{XX}^{-1}[A^{-1}(y - \mu_Y)]$$

$$= (y - \mu_Y)^T (A^{-1})^T C_{XX}^{-1} A^{-1}(y - \mu_Y). \quad (6.29)$$

In addition, we can write

$$|\det(A)|\sqrt{\det(C_{XX})} = \sqrt{[\det(A)]^2 \det(C_{XX})}$$

$$= \sqrt{\det(A)\det(C_{XX})\det(A^T)} = \sqrt{\det(AC_{XX}A^T)}. \quad (6.30)$$

These steps are carried out using the fact that for a square matrix, $\det(A) = \det(A^T)$ and also that the determinant of a product of matrices is equal to the product of the determinants. At this point we have established that

$$f_Y(y) = \frac{1}{\sqrt{(2\pi)^N \det(AC_{XX}A^T)}} \exp\left(-\frac{1}{2}(y - \mu_Y)^T (A^{-1})^T C_{XX}^{-1} A^{-1}(y - \mu_Y)\right). \quad (6.31)$$

Finally, recall that for a linear transformation, $C_{YY} = AC_{XX}A^T$. Furthermore, from this relationship, we can also determine that $C_{YY}^{-1} = (AC_{XX}A^T)^{-1} = (A^T)^{-1} C_{XX}^{-1} A^{-1} = (A^{-1})^T C_{XX}^{-1} A^{-1}$. Hence, the PDF for Y can be written as

$$f_Y(y) = \frac{1}{\sqrt{(2\pi)^N \det(C_{YY})}} \exp\left(-\frac{1}{2}(y - \mu_Y)^T C_{YY}^{-1}(y - \mu_Y)\right). \quad (6.32)$$

This is the general form of a joint Gaussian PDF. Thus, we have shown that any linear transformation of any number of jointly Gaussian random variables produces more jointly Gaussian random variables. Note that this statement applies to more than just $N \times N$ linear transformations. Suppose we wanted to transform N jointly Gaussian random variables to $M(M < N)$ new random variables through a linear transformation. We could always form an $N \times N$ transformation producing N new jointly Gaussian random variables. Any subset of M out of N of these random variables will also be jointly Gaussian. In summary, we have proved the following theorem.

THEOREM 6.3: Given a vector X of N jointly Gaussian random variables, any linear transformation to a set of $M(M \leq N)$ new variables, Y, will produce jointly Gaussian random variables.

Next, suppose we want to create a set of N jointly Gaussian random variables, Y, with a specified covariance matrix, C. We could start with a set of uncorrelated Gaussian random variables (as might be generated by a typical Gaussian random number generator) and then perform a linear transformation to produce a new set of Gaussian random variables with the desired covariance matrix. But, how should the transformation be selected to produce the desired covariance matrix? To answer this question, recall that any covariance matrix is symmetric and any symmetric matrix can be decomposed into

$$C = Q \Lambda Q^T, \tag{6.33}$$

where Λ is a diagonal matrix of the eigenvalues of C, and Q is an orthogonal matrix whose columns are the corresponding eigenvectors of C. Note also that C is positive definite and hence its eigenvalues are all positive. Thus, the matrix Λ is not only diagonal, but its diagonal elements are all positive and as a result, the matrix Λ is a valid covariance matrix. That is, suppose we create a set of N uncorrelated Gaussian random variables, X, with a covariance matrix $C_{XX} = \Lambda$. Then, the matrix Q will transform this set of uncorrelated Gaussian random variables to a new set of Gaussian random variables with the desired covariance matrix. If we form $Y = QX$, then according to Theorem 6.2, the covariance matrix of Y will be of the form

$$C_{YY} = Q C_{XX} Q^T = Q \Lambda Q^T = C. \tag{6.34}$$

At this point, the problem has been reduced from creating a set of random variables with an arbitrary covariance matrix to creating a set of random variables with a diagonal covariance matrix. Typical Gaussian random number generators create random variables with a unit variance. To create random variables with unequal variances, simply scale each component by the appropriate value. In particular, suppose[3] $\Lambda = \mathrm{diag}(\lambda_1, \lambda_2, \ldots, \lambda_N)$. Given a set of unit variance uncorrelated

---

[3]The notation $A = \mathrm{diag}(a_1, a_2, \ldots, a_N)$ means that A is a diagonal matrix with diagonal elements $a_1, a_2, \ldots, a_N$.

Gaussian random variables $Z = [Z_1, Z_2, \ldots, Z_N]^T$, one could form X with the desired variance according to $X_i = \sqrt{\lambda_i} Z_i$, $i = 1, 2, \ldots, N$. In matrix notation we write

$$X = \sqrt{\Lambda} Z, \tag{6.35}$$

where $\sqrt{\Lambda}$ is understood to mean $\sqrt{\Lambda} = \mathrm{diag}(\sqrt{\lambda_1}, \sqrt{\lambda_2}, \ldots, \sqrt{\lambda_N})$.

In summary, we have a two-step linear transformation. Given a vector of uncorrelated, unit variance Gaussian random variables, we form $X = \sqrt{\Lambda} Z$ and then $Y = QX$ to produce the vector of Gaussian random variables with the desired covariance matrix. Naturally, these two consecutive linear transformations can be combined into a single transformation,

$$Y = QX = Q\sqrt{\Lambda} Z. \tag{6.36}$$

It is common to write the matrix $A = Q\sqrt{\Lambda}$ as $\sqrt{C}$ since $AA^T = Q\sqrt{\Lambda}\sqrt{\Lambda}^T Q^T = Q\Lambda Q^T = C$.

Finally, note that if Z is zero-mean, then Y will be zero-mean as well. If it is desired to create Y with a nonzero mean, then a constant term can be added to the transformation to shift the mean to the specified value. This will not alter the covariance matrix of Y. In summary, we have the result shown in Theorem 6.4.

THEOREM 6.4: Given a vector Z of zero-mean, unit variance, uncorrelated random variables, then a new set of random variables, Y, with arbitrary mean vector, μ, and covariance, matrix C can be formed using the linear transformation

$$Y = \sqrt{C} Z + \mu. \tag{6.37}$$

Furthermore, if Z is a Gaussian random vector, then Y will be as well.

If a Gaussian random number generator is not available[4], one can always use a uniform random number generator together with the Box-Muller transformation described in Example 5.24 to produce Gaussian random variables.

Sometimes it is desirable to transform a set of correlated random variables into a new set of uncorrelated random variables. Later in the text, when studying noise, this process will be referred to as "whitening." For now, it is seen that this process is the opposite of the problem just solved. That is, given a random vector Y with mean, μ, and covariance matrix, C, then a vector of zero-mean, unit-variance, uncorrelated random variables can be formed according to

$$Z = (\sqrt{C})^{-1} (Y - \mu). \tag{6.38}$$

---

[4]Many high-level programming languages come with a built-in uniform random number generator, but not a Gaussian random number generator. See Chapter 12 for more details on random number generators.

Here, the expression $(\sqrt{C})^{-1}$ is interpreted as[5]

$$(\sqrt{C})^{-1} = (Q\sqrt{\Lambda})^{-1} = (\sqrt{\Lambda})^{-1}Q^{-1} = \Lambda^{-1/2}Q^T. \tag{6.39}$$

EXAMPLE 6.4: Let's suppose we desire to create a vector of four random variables with a mean vector of $\mu = [1, 0, 3, -2]^T$ and covariance matrix of

$$C = \begin{bmatrix} 30 & -10 & -20 & 4 \\ -10 & 30 & 4 & -20 \\ -20 & 4 & 30 & -10 \\ 4 & -20 & -10 & 30 \end{bmatrix}.$$

The eigenvalue matrix and eigenvector matrix are calculated (in this instance using MATLAB) to be

$$\Lambda = \begin{bmatrix} 4 & 0 & 0 & 0 \\ 0 & 16 & 0 & 0 \\ 0 & 0 & 36 & 0 \\ 0 & 0 & 0 & 64 \end{bmatrix} \quad \text{and} \quad Q = \frac{1}{2}\begin{bmatrix} 1 & -1 & 1 & 1 \\ 1 & 1 & 1 & -1 \\ 1 & -1 & -1 & -1 \\ 1 & 1 & -1 & 1 \end{bmatrix}.$$

Hence, the appropriate transformation matrix is

$$A = Q\sqrt{\Lambda} = \frac{1}{2}\begin{bmatrix} 1 & -1 & 1 & 1 \\ 1 & 1 & 1 & -1 \\ 1 & -1 & -1 & -1 \\ 1 & 1 & -1 & 1 \end{bmatrix}\begin{bmatrix} 2 & 0 & 0 & 0 \\ 0 & 4 & 0 & 0 \\ 0 & 0 & 6 & 0 \\ 0 & 0 & 0 & 8 \end{bmatrix} = \begin{bmatrix} 1 & -2 & 3 & 4 \\ 1 & 2 & 3 & -4 \\ 1 & -2 & -3 & -4 \\ 1 & 2 & -3 & 4 \end{bmatrix}.$$

Thus, given a vector of zero-mean, unit-variance, and uncorrelated random variables, Z, the required transformation is

$$Y = \begin{bmatrix} 1 & -2 & 3 & 4 \\ 1 & 2 & 3 & -4 \\ 1 & -2 & -3 & -4 \\ 1 & 2 & -3 & 4 \end{bmatrix}Z + \begin{bmatrix} 1 \\ 0 \\ 3 \\ -2 \end{bmatrix}.$$

---

[5]Since Q is an orthogonal matrix, $Q^{-1} = Q^T$. Also, $\Lambda^{-1/2} = \text{diag}(\lambda_1^{-1/2}, \lambda_2^{-1/2}, \ldots, \lambda_N^{-1/2})$.

The MATLAB code to perform the necessary eigendecomposition of our example is very straightforward and is as follows:

```
C = [30 -10 -20 4; -10 30 4 -20; -20 4 30 -10; 4 -20 -10 30]
[Q,lambda]=eig(C)
A=Q*sqrt(lambda)
```

6.4.2 Quadratic Transformations of Gaussian Random Vectors

In this section, we show how to calculate the PDFs of various quadratic forms of Gaussian random vectors. In particular, given a vector of N zero-mean Gaussian random variables, X, with an arbitrary covariance matrix, C_{XX}, we form a scalar quadratic function of the vector X of the general form

$$Z = X^T B X, \tag{6.40}$$

where B is an arbitrary $N \times N$ matrix. We would then like to find the PDF of the random variable, Z. These types of problems occur frequently in the study of noncoherent communication systems.

One approach to this problem would be first to form the CDF, $F_Z(z) = \Pr(X^T B X \leq z)$. This could be accomplished by computing

$$F_Z(z) = \int_{A(z)} f_X(x) dx, \tag{6.41}$$

where $A(z)$ is the region defined by $x^T B x \leq z$. While conceptually straightforward, defining the regions and performing the required integration can get quite involved. Instead, we elect to calculate the PDF of Z by first finding its characteristic function. Once the characteristic function is found, the PDF can be found through an inverse transformation.

For the case of Gaussian random vectors, finding the characteristic function of a quadratic form turns out to be surprisingly straightforward:

$$\Phi_Z(\omega) = E[e^{j\omega X^T B X}] = \int_{-\infty}^{\infty} \frac{1}{\sqrt{(2\pi)^N \det(C_{XX})}} \exp\left(-\frac{1}{2}(x^T[C_{XX}^{-1} - 2j\omega B]x)\right) dx. \tag{6.42}$$

This integral is understood to be over the entire N-dimensional x-plane. To evaluate this integral, we simply manipulate the integrand into the standard form of an N-dimensional Gaussian distribution and then use the normalization integral for Gaussian PDFs. Toward that end, define the matrix F according to $F^{-1} = C_{XX}^{-1} - 2j\omega B$. Then

$$\Phi_Z(\omega) = \int_{-\infty}^{\infty} \frac{1}{\sqrt{(2\pi)^N \det(C_{XX})}} \exp\left(-\frac{1}{2}(x^T F^{-1} x)\right) dx$$

$$= \sqrt{\frac{\det(F)}{\det(C_{XX})}} \int_{-\infty}^{\infty} \frac{1}{\sqrt{(2\pi)^N \det(F)}} \exp\left(-\frac{1}{2}(x^T F^{-1} x)\right) dx = \sqrt{\frac{\det(F)}{\det(C_{XX})}} \tag{6.43}$$

where the integral in unity because the integral is a Gaussian distribution.

Using the fact that $\det(F^{-1}) = (\det(F))^{-1}$, this can be rewritten in the more convenient form

$$\Phi_Z(\omega) = \sqrt{\frac{\det(F)}{\det(C_{XX})}} = \frac{1}{\sqrt{\det(F^{-1})\det(C_{XX})}}$$

$$= \frac{1}{\sqrt{\det(F^{-1}C_{XX})}} = \frac{1}{\sqrt{\det(I - 2j\omega BC_{XX})}}. \tag{6.44}$$

To get a feel for the functional form of the characteristic function, note that the determinant of a matrix can be written as the product of its eigenvalues. Furthermore, for a matrix of the form $A = I + cD$ for a constant c, the eigenvalues of A, $\{\lambda_A\}$, can be written in terms of the eigenvalues of the matrix D, $\{\lambda_D\}$, according to $\lambda_A = 1 + c\lambda_D$. Hence,

$$\Phi_Z(\omega) = \prod_{n=1}^{N} \frac{1}{\sqrt{1 - 2j\omega\lambda_n}}, \tag{6.45}$$

where the λ_ns are the eigenvalues of the matrix BC_{XX}. The particular functional form of the resulting PDF depends on the specific eigenvalues. Two special cases are considered as examples next.

6.4.2.1 Special Case #1: $B = I \Rightarrow Z = \sum_{n=1}^{N} X_n^2$

In this case, let's assume further that the X_n are uncorrelated and equal variance so that $C_{XX} = \sigma^2 I$. Then the matrix BC_{XX} has N repeated eigenvalues all equal to σ^2. The resulting characteristic function is

$$\Phi_Z(\omega) = (1 - 2j\omega\sigma^2)^{-N/2}. \tag{6.46}$$

This is the characteristic function of a chi-square random variable with N degrees of freedom. The corresponding PDF is

$$f_Z(z) = \frac{z^{(N/2)-1}}{(2\sigma^2)^{N/2}\Gamma(N/2)} \exp\left(-\frac{z}{2\sigma^2}\right) u(z). \tag{6.47}$$

6.4.2.2 Special Case #2: $N=4$, $B=\dfrac{1}{2}\begin{bmatrix} 0 & 1 & 0 & 0 \\ 1 & 0 & 0 & 0 \\ 0 & 0 & 0 & 1 \\ 0 & 0 & 1 & 0 \end{bmatrix} \Rightarrow Z=X_1X_2+X_3X_4$

Again, we take the X_i to be uncorrelated with equal variance so that $C_{XX}=\sigma^2 I$. In this case, the product matrix BC_{XX} has two pairs of repeated eigenvalues of values $\pm\sigma^2/2$. The resulting characteristic function is

$$\Phi_Z(\omega)=\frac{1}{1+j\omega\sigma^2}\frac{1}{1-j\omega\sigma^2}=\frac{1}{1+(\omega\sigma^2)^2}. \qquad (6.48)$$

This is the characteristic function of a two-sided exponential (Laplace) random variable,

$$f_Z(z)=\frac{1}{2\sigma^2}\exp\left(-\frac{|z|}{\sigma^2}\right). \qquad (6.49)$$

6.4.3 Order Statistics

Suppose a vector of random variables has elements that are independent and identically distributed. In many applications, we need to find the PDF of the largest element in the vector. Or, as a more general problem, we might be interested in the PDF of the mth largest element. Let $X_1,X_2,\ldots,X_N$ be a sequence of random variables. We create a new set of random variables $Y_1,Y_2,\ldots,Y_N$ such that Y_1 is the smallest of the X_ns, Y_2 is the second smallest, and so on. The sequence of Y_ns are referred to as *order statistics* of the sequence of X_ns. Given that each of the X_ns follows a common PDF, $f_X(x)$, we seek to find the PDF of the Y_ns.

First, we find the PDF of Y_1 by finding its CDF.

$$F_{Y_1}(y)=\Pr(Y_1\leq y)=1-\Pr(Y_1\geq y)=1-\Pr(X_1\geq y,X_2\geq y,\ldots,X_N\geq y). \qquad (6.50)$$

This expression follows from the observation that if the smallest of a sequence is larger than some threshold, then all elements in the sequence must be above that threshold. Next, using the fact that the X_n are independent and all have the same distribution, our expression simplifies to

$$F_{Y_1}(y)=1-\Pr(X_1\geq y)\Pr(X_2\geq y)\ldots\Pr(X_N\geq y)=1-(1-F_X(y))^N. \qquad (6.51)$$

Differentiating with respect to y then produces the desired PDF,

$$f_{Y_1}(y)=Nf_X(y)(1-F_X(y))^{N-1}. \qquad (6.52)$$

A similar procedure can be followed to determine the PDF of the mth smallest, Y_m. First, we work out an expression for the CDF.

$$F_{Y_m}(y) = \Pr(Y_m \le y) = \Pr(m \text{ or more of the } X_n s \text{ are less than } y)$$

$$= \sum_{k=m}^{N} \Pr(k \text{ of the } X_n s \text{ are less than } y) \qquad (6.53)$$

To evaluate the probability of the event $\{k$ of the $X_n s$ are less than $y\}$, it is noted that one way for this event to occur is if $X_1 \le y,\ X_2 \le y, \ldots,\ X_k \le y,\ X_{k+1} > y, \ldots, X_N > y$. The probability of this event is

$$\Pr(X_1 \le y,\ X_2 \le y, \ldots,\ X_k \le y,\ X_{k+1} > y, \ldots, X_n > y) = (F_X(y))^k (1 - F_X(y))^{N-k}. \quad (6.54)$$

Of course, we don't have to have the first k elements of the sequence smaller than y. We are looking for the probability that *any* k of the N elements are below y. Hence, we need to count the number of combinations of k out of N variables. This is merely the binomial coefficient. Hence,

$$\Pr(k \text{ of the } X_n s \text{ are less than } y) = \binom{N}{k} (F_X(y))^k (1 - F_X(y))^{N-k}. \qquad (6.55)$$

Summing over k gives the desired CDF:

$$F_{Y_m}(y) = \sum_{k=m}^{N} \binom{N}{k} (F_X(y))^k (1 - F_X(y))^{N-k}. \qquad (6.56)$$

Differentiating with respect to y then gives the expression

$$f_{Y_m}(y) = f_X(y) \sum_{k=m}^{N} \binom{N}{k} k (F_X(y))^{k-1} (1 - F_X(y))^{N-k}$$

$$- f_X(y) \sum_{k=m}^{N} \binom{N}{k} (N-k)(F_X(y))^k (1 - F_X(y))^{N-k-1}. \qquad (6.57)$$

It is left as an exercise for the reader (see Exercise 6.14) to show that this expression reduces to the form

$$f_{Y_m}(y) = \frac{N!}{(m-1)!(N-m)!} f_X(y)(F_X(y))^{m-1}(1 - F_X(y))^{N-m}. \qquad (6.58)$$

An alternative approach to deriving this expression is outlined in Exercise 6.15. The next example illustrates one possible use of order statistics.

EXAMPLE 6.5: Suppose we observe a sequence of $N=2k-1$ independent and identically distributed (i.i.d or IID) random variables, $X_1,\ldots,X_{2k-1}$, and we wish to estimate the mean of the common distribution. One method to do this would be to use the median (middle) element in the sequence as an estimate of the mean. In terms of order statistics, the median is simply Y_k:

$$f_{Y_k}(y) = \frac{(2k-1)!}{[(k-1)!]^2} f_X(y)[F_X(y)]^{k-1}[1-F_X(y)]^{k-1}.$$

For example, if the X_n are all uniformly distributed over $(0, 1)$, then

$$f_{Y_k}(y) = \frac{(2k+1)!}{[(k-1)!]^2}[y(1-y)]^{k-1}, \quad 0 \le y \le 1.$$

Some straightforward calculations reveal that this distribution has a mean of $E[Y_k]=1/2$ and a variance of $\sigma_{Y_k}^2 = 1/(4(2k+1))$. Note that, "on the average," the median is equal to the true mean of the distribution. We say that this estimator is *unbiased*. Furthermore, we also see that as we observe more samples (k gets larger), the variance of the median gets smaller. In other words, as the sample size increases, the median becomes increasingly more precise as an estimator of the mean. In fact, in the limit as $k \to \infty$, the variance goes to zero, which means that the median becomes equal to the mean. We will discuss this problem of estimating means of sequences of IID random variables in the next chapter.

6.4.4 Coordinate Systems in Three Dimensions

Coordinate system transformations in three dimensions follow the same procedure as was derived for the two-dimensional problems. Given a random vector X and a corresponding joint PDF $f_X(x)$, the joint PDF of $Y=(g(X))$ is given by the general formula expressed in Equation 6.24. An example is included next to illustrate the procedure.

EXAMPLE 6.6 (Cartesian-to-Spherical Coordinates): Let the random variables X, Y, and Z in Cartesian coordinates be transformed to spherical coordinates according to

$$R = \sqrt{X^2+Y^2+Z^2}$$

$$\Theta = \cos^{-1}\left(\frac{Z}{\sqrt{X^2+Y^2+Z^2}}\right).$$

$$\Phi = \tan^{-1}\left(\frac{Y}{X}\right)$$

The reverse transformation is probably more familiar to most readers and is given by

$$X = R\sin(\Theta)\cos(\Phi)$$

$$Y = R\sin(\Theta)\sin(\Phi).$$

$$Z = R\cos(\Theta)$$

The Jacobian of this transformation is

$$J\begin{pmatrix} x & y & z \\ r & \theta & \phi \end{pmatrix} = \begin{bmatrix} \sin(\theta)\cos(\phi) & \sin(\theta)\sin(\phi) & \cos(\theta) \\ r\cos(\theta)\cos(\phi) & r\cos(\theta)\sin(\phi) & -r\sin(\theta) \\ -r\sin(\theta)\sin(\phi) & r\sin(\theta)\cos(\phi) & 0 \end{bmatrix},$$

and the determinant of this matrix works out to be

$$\det\left[J\begin{pmatrix} x & y & z \\ r & \theta & \phi \end{pmatrix}\right] = r^2\sin(\theta).$$

Suppose, X, Y, and Z are jointly Gaussian with a joint PDF given by

$$f_{X,Y,Z}(x,y,z) = \frac{1}{(2\pi\sigma^2)^{3/2}}\exp\left(-\frac{x^2+y^2+z^2}{2\sigma^2}\right).$$

Then, the joint PDF of R, Θ, and Φ is found to be

$$f_{R,\Theta,\Phi}(r,\theta,\phi) = f_{X,Y,Z}(x,y,z)\left|\det\left[J\begin{pmatrix} x & y & z \\ r & \theta & \phi \end{pmatrix}\right]\right|_{\substack{x=r\sin(\theta)\cos(\phi) \\ y=r\sin(\theta)\sin(\phi) \\ z=r\cos(\theta)}}$$

$$= \frac{r^2\sin(\theta)}{(2\pi\sigma^2)^{3/2}}\exp\left(-\frac{r^2}{2\sigma^2}\right), \quad \begin{matrix} r\geq 0 \\ 0\leq\theta\leq\pi \\ 0\leq\phi\leq 2\pi \end{matrix}.$$

The marginal PDFs are found by integrating the unwanted variables out of this joint PDF. In this case, the required integrations are fairly

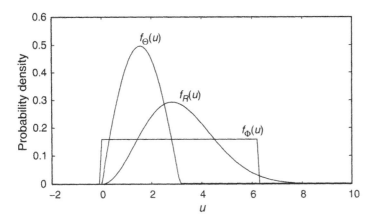

Figure 6.1   PDFs of spherical coordinate variables for Example 6.6.

straightforward, resulting in

$$f_R(r) = \sqrt{\frac{2}{\pi}} \frac{r^2}{\sigma^3} \exp\left(-\frac{r^2}{2\sigma^2}\right) u(r),$$

$$f_\Theta(\theta) = \frac{1}{2}\sin(\theta), \quad 0 \le \theta \le \pi,$$

$$f_\Phi(\phi) = \frac{1}{2\pi}, \quad 0 \le \phi \le 2\pi.$$

Note also that for this example, $f_{R,\Theta,\Phi}(r,\theta,\phi) = f_R(r)f_\Theta(\theta)f_\Phi(\phi)$ so that R, Θ, and Φ are all independent. These PDFs are plotted in Figure 6.1.

6.5 Engineering Application: Linear Prediction of Speech

In many applications, we are interested in predicting future values of a waveform given current and past samples. This is used extensively in speech coders where the signal-to-quantization noise associated with a quantizer can be greatly increased if only the prediction error is quantized. A fairly simple speech coder that utilizes this idea is illustrated in Figure 6.2. In Section 4.11, we introduced the idea of scalar quantization. The process of sampling (at or above the Nyquist rate), quantizing,

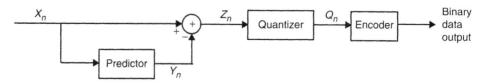

Figure 6.2   Block diagram of a simple speech coder using differential pulse code modulation.

and then encoding each quantization level with some binary code word is known as *pulse code modulation (PCM)*. In Figure 6.2, we consider a slight modification to the basic PCM technique known as *differential PCM (or DPCM)*. The basic idea here is that if we can reduce the range of the signal that is being quantized, then we can either reduce the number of quantization levels needed (and hence reduce the bit rate of the speech coder) or reduce the amount of quantization noise and hence increase the SQNR.

A typical speech signal has a frequency content in the range from about 300–3500 Hz. In order to be able to recover the signal from its samples, a typical sampling rate of 8 KHz is used, which is slightly higher than the Nyquist rate. However, most of the energy content of a speech signal is below 1 kHz; hence when sampled at 8 kHz, a great deal of the speech signal does not change substantially from one sample to the next. Stated another way, when the speech signal is sampled at 8 kHz, we should be able to predict future sample values from current and past samples with pretty good accuracy. The DPCM encoder does exactly that and then only quantizes and encodes the portion of the signal that it is not able to predict.

In Figure 6.2, the X_n represent samples of a speech waveform. These samples are input to the predictor whose job is to make its best estimate of X_n given $X_{n-1}, X_{n-2}, X_{n-3}, \ldots$ as inputs. It is common to use linear prediction, in which case the predictor output is a linear combination of the inputs. That is, assuming the predictor uses the last m samples to form its estimate, the predictor output is of the form

$$Y_n = \sum_{i=1}^{m} a_i X_{n-i},\tag{6.59}$$

where the a_i are constants that we select to optimize the performance of the predictor. The quantity $Z_n = X_n - Y_n$ is the predictor error, which we want to make as small as possible. This error is quantized with a scalar quantizer that uses 2^b levels, and each level is encoded with a b bit code word. The overall bit rate of the speech coder is $b * f_s$ bits/second, where f_s is the rate (in Hz) at which the speech is sampled. For example, if a 16-level quantizer were used with a speech sampling rate of 8 kHz, the DPCM speech coder would have a bit rate of 32 kbits/second.

An important question is, Can the original samples be recovered from the binary representation of the signal? Given the encoded bit stream, we can construct the sequence of quantizer outputs, Q_n. As with any quantization scheme, we can never recover the exact quantizer input from the quantizer output, but if we use enough levels in the quantizer, the quantization noise can be kept fairly small. The speech samples are reconstructed according to $X_n = Y_n + Z_n$. Since we don't have Z_n, we use Q_n in its place and form

$$\widehat{X}_n = Y_n + Q_n = X_n + \varepsilon_n, \tag{6.60}$$

where $\varepsilon = Q_n - Z_n$ is the quantization noise in the nth sample. To complete the process of recovering the sample values, the decoder must also form the Y_n. It can do this by employing an identical predictor as used at the encoder. Unfortunately, the predictor at the decoder does not have access to the same input as the predictor at the encoder. That is, at the decoder, we cannot use the true values of the past speech samples, but rather must use the quantized (noisy) versions. This can be problematic since the predictor at the decoder will now form

$$\widehat{Y} = \sum_{i=1}^{m} a_i \widehat{X}_{n-1}. \tag{6.61}$$

If the $\widehat{X}_n$ are noisy versions of the X_n, then the $\widehat{Y}_n$ will also be noisy. Now, not only do we have quantization noise, but that noise propagates from one sample to the next through the predictor. This leads to the possibility of a snowballing effect, where the noise in our recovered samples gets progressively larger from one sample to the next.

This problem is circumvented using the modified DPCM encoder shown in Figure 6.3; the corresponding decoder is shown in the figure as well. The difference between this DPCM system and the one in Figure 6.2 is that now the predictor used in the encoder bases its predictions on the quantized samples rather than on the true samples. By doing this, the predicted value may be slightly degraded (but not much if the number of quantization levels is sufficient), but there will be no propagation of errors in the decoder, since the predictor at the decoder now uses the same inputs as the predictor at the encoder.

Now that we have the design of the speech encoder and decoder squared away, we shift our attention to the problem of designing the predictor. Assuming a linear predictor, the problem is essentially to choose the coefficients a_i in Equation 6.59 to minimize the prediction error:

$$Z_n = X_n - Y_n = X_n - \sum_{i=1}^{m} a_i X_{n-i}. \tag{6.62}$$

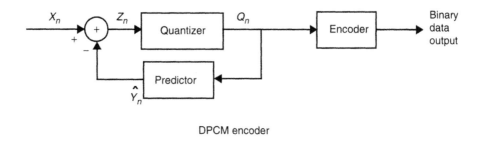

DPCM encoder

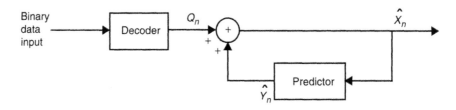

DPCM decoder

Figure 6.3   Block diagram of a modified speech coder using differential pulse code modulation.

In this example, we choose to minimize the second moment of the error:

$$E[Z_n^2] = E\left[\left(X_n - \sum_{i=1}^{m} a_i X_{n-i}\right)^2\right].$$

(6.63)

Differentiating the preceding expression with respect to a_j and setting it equal to zero provides the following set of equations that the predictor coefficients should satisfy:

$$\sum_{i=1}^{m} a_i E[X_{n-i}X_{n-j}] = E[X_n X_{n-j}], \quad j=1,2,\ldots,m.$$

(6.64)

Define the correlation parameter r_k to be the correlation between two samples spaced by k sampling intervals. Then our system of equations can be expressed in matrix form as

$$\begin{bmatrix} r_0 & r_1 & r_2 & \cdots & r_{m-1} \\ r_1 & r_0 & r_1 & \cdots & r_{m-2} \\ r_2 & r_1 & r_0 & \cdots & r_{m-3} \\ \cdots & \cdots & \cdots & & \cdots \\ r_{m-1} & r_{m-2} & r_{m-3} & \cdots & r_0 \end{bmatrix} = \begin{bmatrix} a_1 \\ a_2 \\ a_3 \\ \cdots \\ a_m \end{bmatrix} = \begin{bmatrix} r_1 \\ r_2 \\ r_3 \\ \cdots \\ r_m \end{bmatrix}.$$

(6.65)

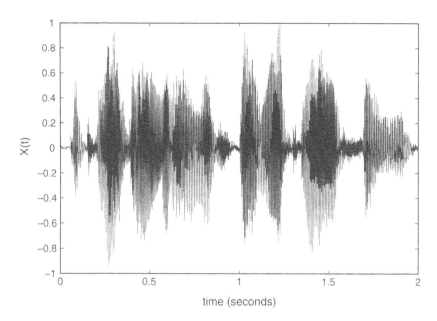

Figure 6.4 Speech segment used in Example 6.7.

EXAMPLE 6.7: Figure 6.4 shows a segment of speech that has a duration of about 2 seconds, which was sampled at a rate of 8kHz. From this data, (using MATLAB) we estimated the correlation parameters $r_k = E[X_n X_{n+k}]$; found the linear prediction coefficients, a_i, $i = 1, 2, \dots, m$; and then calculated the mean squared estimation error, $\text{MSE} = E[X_n - Y_n)^2]$. The results are shown in Table 6.1. We should note a couple of observations. First, even with a simple one-tap predictor, the size of the error signal is much smaller than the original signal (compare the values of MSE with r_0 in the table). Second, we note that (for this example) there does not seem to be much benefit gained from using more than two previous samples to form the predictor.

Finally, we compare the quality of the encoded speech as measured by the SQNR for PCM and the DPCM scheme of Figure 6.3 using the two-tap predictor specified in Table 6.1. For an equal number of bits per sample, the DPCM scheme improves the SQNR by more than 20 dB. Alternatively, the DPCM scheme can use 3 bits/sample fewer than the PCM scheme and still provide better SQNR.

Note that the SQNR is an objective measure of speech quality and is poorly correlated with speech quality. For example, one can introduce a time delay to the estimated signal and obtain a very low SQNR value, yet the speech quality will be good.

Table 6.1   Results of Linear Prediction of Speech Segment from Figure 6.4

	$r_0 = 0.0591$	$r_1 = 0.0568$	$r_2 = 0.0514$	$r_3 = 0.0442$	$r_4 = 0.0360$	
$m=1$	$a_1 = 0.9615$					MSE $= 0.004473$
$m=2$	$a_1 = 1.6564$	$a_2 = -0.7228$				MSE $= 0.002144$
$m=3$	$a_1 = 1.7166$	$a_2 = -0.8492$	$a_3 = 0.0763$			MSE $= 0.002132$
$m=4$	$a_1 = 1.7272$	$a_2 = -1.0235$	$a_3 = 0.4276$	$a_4 = -0.2052$		MSE $= 0.002044$

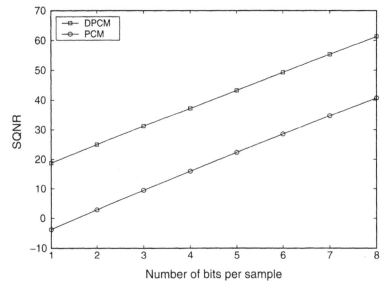

Figure 6.5   SQNR comparison of PCM and DPCM speech coders for the speech segment in Figure 6.4.

Exercises

6.1 Suppose we flip a coin three times, thereby forming a sequence of heads and tails. Form a random vector by mapping each outcome in the sequence to 0 if a head occurs or to 1 if a tail occurs.

(a) How many realizations of the vector may be generated? List them.

(b) Are the realizations independent of one another?

6.2 Let $X = [X_1, X_2, X_3]^T$ represent a three-dimensional vector of random variables that is uniformly distributed over the unit sphere. That is,

$$f_X(x) = \begin{cases} c & ||x|| \leq 1 \\ 0 & ||x|| > 1 \end{cases}.$$

(a) Find the constant c.

(b) Find the marginal PDF for a subset of two of the three random variables. For example, find $f_{X_1,X_2}(x_1,x_2)$.

(c) Find the marginal PDF for one of the three random variables. That is, find $f_{X_1}(x_1)$.

(d) Find the conditional PDFs $f_{X_1|X_2,X_3}(x_1|x_2,x_3)$ and $f_{X_1,X_2|X_3}(x_1,x_2|x_3)$. *Extra:* Can you extend this problem to N-dimensions?

6.3 Let $X=[X_1,X_2,\ldots X_N]^T$ represent an N-dimensional vector of random variables that is uniformly distributed over the region $x_1+x_2+\cdots+x_N \le 1$, $x_i \ge 0$, $i=1,2,\ldots,N$. That is,

$$f_X(x) = \begin{cases} c & \sum_{i=1}^{N} x_i \le 1, x_i \ge 0 \\ 0 & \text{otherwise} \end{cases}.$$

(a) Find the constant c.

(b) Find the marginal PDF for a subset of M of the N random variables.

(c) Are the X_i independent? Are the X_i identically distributed?

6.4 Consider a vector of N random variables, $X=[X_1,X_2,\ldots,X_N]^T$. Suppose we form a new random variable Z by performing a weighted average of the components of X. That is,

$$Z=\sum_{i=1}^{N} b_i X_i, \text{ where } b_i \ge 0 \quad \text{and} \quad \sum_{i=1}^{N} b_i = 1.$$

Find the values of the constants b_i such that the variance of Z is minimized.

6.5 Show that all correlation matrices are nonnegative definite.

6.6 A random vector is generated by rolling a die and observing the outcome. The components of the random vector are determined by successive rolls of the die. If the die is rolled two times:

(a) list the possible realizations of the random vector;

(b) determine the probability of each realization;

(c) determine the mean vector;

(d) determine the covariance matrix.

6.7 Repeat parts (c) and (d) of Exercise 6.6 if a three-element vector is formed from three rolls of a die.

6.8 Let X be a two-element, zero-mean random vector. Suppose we construct a new random vector Y according to a linear transformation, $Y = TX$. Find the transformation matrix, T, such that Y has a covariance matrix of

$$C = \begin{bmatrix} 5 & 1 \\ 1 & 2 \end{bmatrix}.$$

6.9 A three-dimensional vector random variable, X, has a covariance matrix of

$$C = \begin{bmatrix} 3 & 1 & -1 \\ 1 & 5 & -1 \\ -1 & -1 & 3 \end{bmatrix}.$$

Find a transformation matrix A, such that the new random variables $Y = AX$ will be uncorrelated.

6.10 Let X_1, X_2, and X_3 be a set of three zero-mean Gaussian random variables with a covariance matrix of the form

$$C = \sigma^2 \begin{bmatrix} 1 & \rho & \rho \\ \rho & 1 & \rho \\ \rho & \rho & 1 \end{bmatrix}.$$

Find the following expected values:

(a) $E[X_1 | X_2 = x_2, X_3 = x_3]$,

(b) $E[X_1 X_2 | X_3 = x_3]$,

(c) $E[X_1 X_2 X_3]$.

6.11 Define the N-dimensional characteristic function for a random vector $X = [X_1, X_2, \ldots, X_N]^T$, according to $\Phi_X(\Omega) = E[e^{j\Omega^T X}]$ where $\Omega = [\omega_1, \omega_2, \ldots, \omega_N]^T$. Show that the N-dimensional characteristic function for a zero-mean Gaussian random vector is given by

$$\Phi_X(\Omega) = \exp\left(-\frac{\Omega^T C_{XX} \Omega}{2}\right).$$

6.12 For any four zero-mean Gaussian random variables X_1, X_2, X_3, and X_4, show that

$$E[X_1 X_2 X_3 X_4] = E[X_1 X_2]E[X_3 X_4] + E[X_1 X_3]E[X_2 X_4] + E[X_1 X_4]E[X_2 X_3].$$

Hint: You might want to use the result of Exercise 6.11.

6.13 Suppose X_m, $m = 1, 2, \ldots, n$ are a sequence of independent and exponentially distributed random variables with

$$f_{X_m}(x) = \frac{1}{\mu} \exp(-x/\mu) u(x).$$

Assume that n is an odd number ($n = 2k - 1$ for some integer k).

(a) Find the PDF of the median of the sequence.

(b) Find the expected value of the median of the sequence. Is the median an unbiased estimate of the mean of the underlying exponential distribution?

(c) Find the variance of the median of the sequence.

6.14 Show that the derivative of

$$F_{Y_m}(y) = \sum_{k=m}^{N} \binom{N}{k} (F_X(y))^k (1 - F_X(y))^{N-k}$$

reduces to the form

$$f_{Y_m}(y) = \frac{N!}{(m-1)!(N-m)!} f_X(y)(F_X(y))^{m-1}(1 - F_X(y))^{N-m}.$$

6.15 In this problem we formulate an alternative derivation of Equation 6.58 that gives the PDF of the order statistic, Y_m, which is the mth largest of a sequence of N random variables, $X_1, X_2, \ldots, X_N$. Start by writing $f_{Y_m}(y) dy = \Pr(y < Y_m < y + dy)$. Then note that

$\Pr(y < Y_m < y + dy) = \Pr(\{m-1 \text{ of the } X\text{s are less than } y\} \cap$

$\{1\ X \text{ is between } y \text{ and } y + dy\} \cap \{n - m \text{ of the } X\text{s are greater than } y\}).$

Find the probability of this event and by doing so, prove that the PDF of Y_m is as given by Equation 6.58.

6.16 The traffic managers of toll roads and toll bridges need specific information to properly staff the toll booths so that the queues are minimized (i.e., the waiting time is minimized).

(a) Assume that there is one toll booth on a busy interstate highway and that the number of cars per minute approaching the toll booth follows a Poisson PMF with $\alpha = 10$. The traffic manager wants you to determine the probability that exactly 11 cars will approach this toll booth in the minute from noon to one minute past noon.

(b) Now assume there are N toll booths at a toll plaza and that the number of cars per minute approaching the plaza follows a Poisson PMF with $\alpha = 30$. The traffic manager wants you to calculate the minimum number of toll booths that need to be staffed if the probability is at least 0.05 that not more than five cars approach each toll booth in one minute. For this part, assume the traffic approaching the plaza divides itself among the N booths such that the traffic approaching each booth is independent and follows a Poisson PMF with $\alpha = 30/N$.

6.17 A sequence of zero mean unit variance independent random variables, X_n, $n = 0,1,2,\ldots,N-1$ are input to a filter that produces an output sequence according to $Y_n = (X_n + X_{n-1})/2$ for $n = 0,1,2,\ldots,N-1$. For initialization purposes, X_{-1} is taken to be zero.

(a) Find the covariance (correlation) matrix of the Y_n.
(b) Now let the variance of the X_n be σ_X^2. Find the covariance (correlation) matrix of the Y_n.

6.18 Repeat Exercise 6.17 with the filter changed to $Y_n = X_n - X_{n-1}$.

6.19 Suppose a zero-mean random sequence X_n has correlation parameters given by $r_k = E[X_n X_{n+k}] = c^{|k|}$. An estimate of a future value of X_n is $\widehat{X}_n = a_1 X_{n-1} + a_2 X_{n-2}$, which is a special case of Equation 6.59.

(a) Use Equation 6.65 to find the a_j.
(b) What is the mean squared error, $E[(X_n - \widehat{X}_n)^2]$?

MATLAB Exercises

6.20 Three jointly Gaussian random variables $[X,Y,Z]^T$ have a mean vector $\mu = [1,0,-1]^T$ and covariance matrix

$$C = \begin{bmatrix} 4 & 2 & -1 \\ 2 & 4 & 2 \\ -1 & 2 & 4 \end{bmatrix}.$$

Use MATLAB to help you find the form of the three-dimensional joint PDF, $f_{X,Y,Z}(x,y,z)$.

6.21 For each of the following matrices, determine if the matrix is a valid correlation matrix. In some cases, you may want to use MATLAB to check if the matrix is positive definite.

(a) $C_a = \begin{bmatrix} 3 & -2 & 1 \\ 2 & 6 & 0 \\ -1 & 0 & 2 \end{bmatrix}$,

(b) $C_b = \begin{bmatrix} 3 & -2 & 3 \\ -2 & 6 & 0 \\ 3 & 0 & 2 \end{bmatrix}$,

(c) $C_c = \begin{bmatrix} 3 & -2 & 1 \\ -2 & 6 & 0 \\ 1 & 0 & 2 \end{bmatrix}$,

(d) $C_d = \begin{bmatrix} 1 & -\dfrac{1}{2} & \dfrac{1}{4} & -\dfrac{1}{8} \\ -\dfrac{1}{2} & 1 & -\dfrac{1}{2} & \dfrac{1}{4} \\ \dfrac{1}{4} & -\dfrac{1}{2} & 1 & -\dfrac{1}{2} \\ -\dfrac{1}{8} & \dfrac{1}{4} & -\dfrac{1}{2} & 1 \end{bmatrix}$,

(e) $C_e = \begin{bmatrix} 11 & -3 & 7 & 5 \\ -3 & 11 & 5 & 7 \\ 7 & 5 & 11 & -3 \\ 5 & 7 & -3 & 11 \end{bmatrix}$,

(f) $C_f = \begin{bmatrix} 5 & 1 & 3 & -1 \\ 1 & 5 & -1 & 3 \\ 3 & -1 & 5 & 1 \\ -1 & 3 & 1 & 5 \end{bmatrix}$.

6.22 For each matrix in the previous problem that is a valid correlation matrix, find a transformation matrix that will transform a set of independent unit variance random variables into a set of random variables with the specified correlation matrix.

6.23 Given a random sequence $X = [X_1, X_2, X_3, X_4]$ with a covariance matrix

$$C_X = \begin{bmatrix} 1 & 0.3 & 0.09 & 0.027 \\ 0.3 & 1 & 0.3 & 0.027 \\ 0.09 & 0.3 & 1 & 0.3 \\ 0.0027 & 0.09 & 0.3 & 1 \end{bmatrix},$$

find a linear transformation that will produce a random sequence $Y = [Y_1, Y_2, Y_3, Y_4]$ with a covariance matrix

$$C_Y = \begin{bmatrix} 1 & 0.1 & 0.2 & 0.3 \\ 0.1 & 1 & 0.1 & 0.2 \\ 0.2 & 0.1 & 1 & 0.1 \\ 0.3 & 0.2 & 0.1 & 1 \end{bmatrix}.$$

Random Sequences and Series

7

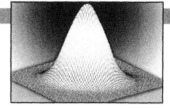

This chapter forms a bridge between the study of random variables in the previous chapters and the study of random processes to follow. A random process is simply a random function of time. If time is discrete, then such a random function could be viewed as a sequence of random variables. Even when time is continuous, we often choose to sample waveforms (whether they are deterministic or random) in order to work with discrete time sequences rather than continuous time waveforms. Hence, sequences of random variables will naturally occur in the study of random processes. In this chapter, we'll develop some basic results regarding both finite and infinite sequences of random variables and random series.

7.1 Independent and Identically Distributed Random Variables

In many applications, we are able to observe an experiment repeatedly. Each new observation can occur with an independent realization of whatever random phenomena control the experiment. This sort of situation gives rise to *independent and identically distributed* (*i.i.d.* or *IID*) random variables.

DEFINITION 7.1: A sequence of random variables $X_1, X_2, \ldots, X_n$ is IID if

$$F_{X_i}(x) = F_X(x) \quad \forall i = 1, 2, \ldots, n \text{ (identically distributed)}, \tag{7.1}$$

and

$$F_{X_1, X_2, \ldots, X_n}(x_1, x_2, \ldots, x_n) = \prod_{i=1}^{n} F_{X_i}(x_i) \text{ (independent)}. \tag{7.2}$$

For continuous random variables, the CDFs can be replaced with PDFs in Equations 7.1 and 7.2, while for discrete random variables the CDFs can be replaced by PMFs.

Suppose, for example, we wish to measure the voltage produced by a certain sensor. The sensor might be measuring the relative humidity outside. Our sensor converts the humidity to a voltage level that we can then easily measure. However, as with any measuring equipment, the voltage we measure is random due to noise generated in the sensor as well as in the measuring equipment. Suppose the voltage we measure is represented by a random variable X given by $X = v(h) + N$, where $v(h)$ is the true voltage that should be presented by the sensor when the humidity is h, and N is the noise in the measurement. Assuming that the noise is zero-mean, then $E[X] = v(h)$. That is, on the average, the measurement will be equal to the true voltage $v(h)$. Furthermore, if the variance of the noise is sufficiently small, then the measurement will tend to be close to the true value we are trying to measure. But what if the variance is not small? Then the noise will tend to distort our measurement making our system unreliable. In such a case, we might be able to improve our measurement system by taking several measurements. This will allow us to "average out" the effects of the noise.

Suppose we have the ability to make several measurements and observe a sequence of measurements $X_1, X_2, \ldots, X_n$. It might be reasonable to expect that the noise that corrupts a given measurement has the same distribution each time (and hence the X_i are identically distributed) and is independent of the noise in any other measurement (so that the X_i are independent). Then the n measurements form a sequence of IID random variables. A fundamental question is then How do we process an IID sequence to extract the desired information from it? In the preceding case above, the parameter of interest, $v(h)$, happens to be the mean of the distribution of the X_i. This turns out to be a fairly common problem, so we start by examining in some detail the problem of estimating the mean from a sequence of IID random variables.

7.1.1 Estimating the Mean of IID Random Variables

Suppose the X_i have some common PDF, $f_X(x)$, which has some mean value, μ_X. Given a set of IID observations, we wish to form some function

$$\hat{\mu} = g(X_1, X_2, \ldots, X_n), \tag{7.3}$$

which will serve as an estimate of the mean. But what function should we choose? Even more fundamentally, what criterion should we use to select a function?

There are many criteria that are commonly used. To start with, we would like the average value of the estimate of the mean to be equal to the true mean. That is, we want $E[\hat{\mu}] = \mu_X$. If this criterion is met, we say that $\hat{\mu}$ is an *unbiased* estimate of μ_X. Given that the estimate is unbiased, we would also like the error in the estimate to be as small as possible. Define the estimation error to be $\varepsilon = \hat{\mu} - \mu_X$. A common criterion is to choose the estimator that minimizes the second moment of the error (mean-square error), $E\left[\varepsilon^2\right] = E\left[(\hat{\mu} - \mu_X)^2\right]$. If this criterion is met, we say that $\hat{\mu}$ is an *efficient* estimator of μ_X. To start with a relatively simple approach, suppose we desire to find a linear estimator. That is, we will limit ourselves to estimators of the form

$$\hat{\mu} = a_1 X_1 + a_2 X_2 + \cdots + a_n X_n = \sum_{i=1}^{n} a_i X_i. \tag{7.4}$$

Then, we seek to find the constants $a_1, a_2, \ldots, a_n$ such that the estimator (1) is unbiased and (2) minimizes the mean-square error. Such an estimator is referred to as the *best linear unbiased estimator (BLUE)*.

To simplify notation in this problem, we write $X = [X_1, X_2, \ldots, X_n]^T$ and $a = [a_1, a_2, \ldots, a_n]^T$. The linear estimator $\hat{\mu}$ can then be written as $\hat{\mu} = a^T X$. First, for the estimator to be unbiased, we need

$$\mu_X = E[\hat{\mu}] = E\left[a^T X\right] = a^T E[X]. \tag{7.5}$$

Since the X_i are all IID, they all have means equal to μ_X. Hence, the mean vector for X is just $\mu_X \mathbf{1}_n$, where $\mathbf{1}_n$ is an n-element column vector of all 1s. The linear estimator will then be unbiased if

$$\sum_{i=1}^{n} a_i = a^T \mathbf{1}_n = 1. \tag{7.6}$$

The mean square error is given by

$$E\left[\varepsilon^2\right] = E\left[(a^T X - \mu_X)^2\right] = a^T E\left[XX^T\right]a - 2\mu_X a^T E[X] + \mu_X^2$$
$$= a^T Ra - 2\mu_X^2 a^T \mathbf{1}_n + \mu_X^2. \tag{7.7}$$

In this expression, $R = E[XX^T]$ is the correlation matrix for the vector X. Using the constraint of Equation 7.6, the mean square error simplifies to

$$E[\varepsilon^2] = a^T Ra - \mu_X^2. \tag{7.8}$$

The problem then reduces to minimizing the function $a^T Ra$ subject to the constraint $a^T \mathbf{1}_n = 1$.

To solve this multidimensional optimization problem, we use standard Lagrange multiplier techniques. Form the auxiliary function

$$h(\lambda) = a^T R a + \lambda a^T 1_n. \tag{7.9}$$

Then solve the equation $\nabla h = 0$. It is not difficult to show that the gradient of the function h works out to be $\nabla h = 2Ra + \lambda 1_n$. Hence, the optimum vector a will satisfy

$$Ra = \left(-\frac{\lambda}{2}\right) 1_n. \tag{7.10}$$

Solving for a in this equation and then applying the constraint $a^T 1_n = 1$ results in the solution

$$a = \frac{R^{-1} 1_n}{1_n^T R^{-1} 1_n}. \tag{7.11}$$

Due to the fact that the X_i are IID, the form of the correlation matrix can easily be shown to be

$$R = \mu_X^2 1_{nxn} + \sigma_X^2 I, \tag{7.12}$$

where 1_{nxn} is an nxn matrix consisting of all 1s and I is an identity matrix. Also, σ_X^2 is the variance of the IID random variables. It can be shown using the matrix inversion lemma[1] that the inverse of this correlation matrix is

$$R^{-1} = \sigma_X^{-2} \left[I - \frac{\mu_X^2/\sigma_X^2}{1 + n\mu_X^2/\sigma_X^2} 1_{nxn} \right]. \tag{7.13}$$

From here, it is easy to demonstrate that $R^{-1} 1_n$ is proportional to 1_n, and hence the resulting vector of optimum coefficients is

$$a = \frac{1}{n} 1_n. \tag{7.14}$$

In terms of the estimator $\hat{\mu}$, the best linear unbiased estimator of the mean of an IID sequence is

$$\hat{\mu} = \frac{1}{n} 1_n^T X = \frac{1}{n} \sum_{i=1}^{n} X_i. \tag{7.15}$$

---

[1] The matrix inversion lemma gives a formula to find the inverse of a rank one update of another matrix whose inverse is known. In particular, suppose $A = B + xx^T$, where x is a column vector and the inverse of B is known. Then

$$A^{-1} = B^{-1} - \frac{B^{-1} x x^T B^{-1}}{1 + x^T B^{-1} x}.$$

This estimator is commonly referred to as the *sample mean*. The preceding derivation proves Theorem 7.1, which follows.

THEOREM 7.1: Given a sequence of IID random variables $X_1, X_2, \ldots, X_n$, the sample mean is BLUE.

Another common approach to estimate various parameters of a distribution is the *maximum likelihood (ML) approach*. In the ML approach, the distribution parameters are chosen to maximize the probability of the observed sample values. Suppose, as in the preceding discussion, we are interested in estimating the mean of a distribution. Given a set of observations, $X_1 = x_1, X_2 = x_2, \ldots, X_n = x_n$, the ML estimate of μ_X would be the value of μ_X that maximizes $f_X(x)$. A few examples will clarify this concept.

EXAMPLE 7.1: Suppose the X_i are jointly Gaussian so that

$$f_X(x) = \frac{1}{(2\pi\sigma^2)^{n/2}} \exp\left(-\frac{1}{2\sigma^2}\sum_{i=1}^{n}(x_i - \mu)^2\right).$$

The value of μ that maximizes this expression will minimize

$$\sum_{i=1}^{n}(x_i - \mu)^2.$$

Differentiating and setting equal to zero gives the equation

$$-2\sum_{i=1}^{n}(x_i - \mu) = 0.$$

The solution to this equation works out to be

$$\hat{\mu}_{ML} = \frac{1}{n}\sum_{i=1}^{n}x_i.$$

Hence, the sample mean is also the ML estimate of the mean when the random variables follow a Gaussian distribution.

EXAMPLE 7.2: Now suppose the random variables have an exponential distribution,

$$f_X(x) = \prod_{i=1}^{n}\frac{1}{\mu}\exp\left(-\frac{x_i}{\mu}\right)u(x_i) = \frac{1}{\mu^n}\exp\left(-\frac{1}{\mu}\sum_{i=1}^{n}x_i\right)\prod_{i=1}^{n}u(x_i).$$

Differentiating with respect to μ and setting equal to zero results in

$$-\frac{n}{\mu^{n+1}} \exp\left(-\frac{1}{\mu} \sum_{i=1}^{n} x_i\right) + \frac{1}{\mu^2} \left(\sum_{i=1}^{n} x_i\right) \frac{1}{\mu^n} \exp\left(-\frac{1}{\mu} \sum_{i=1}^{n} x_i\right) = 0.$$

Solving for μ results in

$$\hat{\mu}_{ML} = \frac{1}{n} \sum_{i=1}^{n} x_i.$$

Once again, the sample mean is the maximum likelihood estimate of the mean of the distribution.

Since the sample mean occurs so frequently, it is beneficial to study this estimator in a little more detail. First, we note that the sample mean is itself a random variable since it is a function of the n IID random variables. We have already seen that the sample mean is an unbiased estimate of the true mean; that is, $E[\hat{\mu}] = \mu_X$. It is instructive also to look at the variance of this random variable.

$$\text{Var}(\hat{\mu}) = E\left[\left(\frac{1}{n} \sum_{i=1}^{n} X_i - \mu_X\right)^2\right] = E\left[\left(\frac{1}{n} \sum_{i=1}^{n} (X_i - \mu_X)\right)^2\right]$$

$$= \frac{1}{n^2} \sum_{i=1}^{n} \sum_{j=1}^{n} E[(X_i - \mu_X)(X_j - \mu_X)]$$

$$= \frac{1}{n^2} \sum_{i=1}^{n} \sum_{j=1}^{n} E[(X_i - \mu_X)(X_j - \mu_X)] = \frac{1}{n^2} \sum_{i=1}^{n} \sum_{j=1}^{n} \text{Cov}(X_i, X_j) \qquad (7.16)$$

All terms in the double series in the previous equation are zero except for the ones where $i = j$ since X_i and X_j are uncorrelated for all $i \neq j$. Hence, the variance of the sample mean is

$$\text{Var}(\hat{\mu}) = \frac{1}{n^2} \sum_{i=1}^{n} \text{Var}(X_i) = \frac{1}{n^2} \sum_{i=1}^{n} \sigma_X^2 = \frac{\sigma_X^2}{n}. \qquad (7.17)$$

This means that if we use n samples to estimate the mean, the variance of the resulting estimate is reduced by a factor of n relative to what the variance would be if we used only one sample.

Consider what happens in the limit as $n \to \infty$. As long as the variance of each of the samples is finite, the variance of the sample mean approaches zero. Of course, we never have an infinite number of samples in practice, but this does mean that the sample mean can achieve any level of precision (i.e., arbitrarily small variance) if a sufficient number of samples is taken. We will study this limiting behavior in more detail in section 7.3. For now, we turn our attention to estimating other parameters of a distribution.

7.1.2 Estimating the Variance of IID Random Variables

Now that we have a handle on how to estimate the mean of IID random variables, suppose we would like also to estimate the variance (or equivalently, the standard deviation). Since the variance is not a linear function of the random variables, it would not make much sense to try to form a linear estimator. That is, to talk about an estimator of the variance being BLUE does not make much sense. Hence, we take the ML approach here. As with the problem of estimating the mean, we seek the value of the variance that maximizes the joint PDF of the IID random variables evaluated at their observed values.

EXAMPLE 7.3: Suppose that the random variables are jointly Gaussian so that

$$f_X(x) = \frac{1}{(2\pi\sigma^2)^{n/2}} \exp\left(-\frac{1}{2\sigma^2}\sum_{i=1}^{n}(x_i-\mu)^2\right).$$

Differentiating the joint PDF with respect to σ results in

$$\frac{d}{d\sigma}f_X(x) = \left(-\frac{n}{\sigma}+\frac{1}{\sigma^3}\sum_{i=1}^{n}(x_i-\mu)^2\right)\frac{1}{(2\pi\sigma^2)^{n/2}}\exp\left(-\frac{1}{2\sigma^2}\sum_{i=1}^{n}(x_i-\mu)^2\right).$$

Setting this expression equal to zero and solving results in

$$\hat{\sigma}_{ML}^2 = \frac{1}{n}\sum_{i=1}^{n}(x_i-\mu)^2.$$

The result of Example 7.3 seems to make sense. The only problem with this estimate is that it requires knowledge of the mean in order to form the estimate. What if we don't know the mean? One obvious approach would be to replace the true mean in the previous result with the sample mean. That is, one could estimate

the variance of an IID sequence using

$$\hat{s}^2 = \frac{1}{n}\sum_{i=1}^{n}(x_i - \hat{\mu})^2, \quad \hat{\mu} = \frac{1}{n}\sum_{i=1}^{n}x_i. \tag{7.18}$$

This approach, however, can lead to problems. It is left as an exercise for the reader to show that this estimator is biased; that is, in this case, $E[\hat{\sigma}^2] \neq \sigma^2$. To overcome this problem, it is common to adjust the previous form. The following estimator turns out to be an unbiased estimate of the variance:

$$\hat{s}^2 = \frac{1}{n-1}\sum_{i=1}^{n}(x_i - \hat{\mu})^2. \tag{7.19}$$

This is known as the sample variance and is the most commonly used estimate for the variance of IID random variables. In the previous expression, $\hat{\mu}$ is the usual sample mean.

In summary, given a set of IID random variables, the variance of the distribution is estimated according to

$$\hat{s}^2 = \begin{cases} \dfrac{1}{n}\sum_{i=1}^{n}(x_i - \mu)^2 & \text{if } \mu \text{ is known} \\[2ex] \dfrac{1}{n-1}\sum_{i=1}^{n}(x_i - \hat{\mu})^2 & \text{if } \mu \text{ is unknown} \end{cases} \tag{7.20}$$

 EXAMPLE 7.4: Suppose we form a random variable Z according to $Z = \sqrt{X^2 + Y^2}$ where X and Y are independent Gaussian random variables with means of μ and variances of σ^2. In this example, we will estimate the mean and variance of Z using the sample mean and sample variance of a large number of MATLAB-generated realizations of the random variable Z. The MATLAB code to accomplish this follows. Upon running this code, we obtain a sample mean of $\hat{\mu} = 5.6336$ and a sample variance of $\hat{s}^2 = 8.5029$. Note that the true mean and variance of the Rician random variable Z can be found analytically (with some effort). For this example, the PDF of the random variable Z is found to take on a Rician form

$$f_Z(z) = \frac{z}{\sigma^2}\exp\left(-\frac{z^2 + 2\mu^2}{2\sigma^2}\right)I_0\left(\frac{\sqrt{2}\mu z}{\sigma^2}\right)u(z).$$

Using the expressions given in Appendix D, Summary of Common Random Variables (see Equations D.52 and D.53), for the mean and variance of a Rician random variable, it is determined that the true mean and variance should be

$$\mu_Z = \sqrt{\frac{\pi\sigma^2}{2}} \exp\left(-\frac{\mu^2}{2\sigma^2}\right)\left[\left(1+\frac{\mu^2}{\sigma^2}\right)I_0\left(\frac{\mu^2}{2\sigma^2}\right)+\frac{\mu^2}{\sigma^2}I_1\left(\frac{\mu^2}{2\sigma^2}\right)\right],$$

$$\sigma_Z^2 = 2\sigma^2 + 2\mu^2 - \mu_Z^2.$$

For the values of $\mu = 2$ and $\sigma = 4$ used in the following program, the resulting mean and variance of the Rician random variable should be $\mu_Z = 5.6211$ and $\sigma_Z^2 = 8.4031$.

```
N=10000;
mu=2; sigma=4;                  % Set mean and standard
                                  deviation of X and Y.
X=sigma*randn(1,N)+mu;          % Generate samples of X.
Y=sigma*randn(1,N)+mu;          % Generate samples of Y.
Z=sqrt(X.^2+Y.^2);              % Create Z (Rician RVs).
mu_hat=sum(Z)/N                 % Sample mean.
s_hat2=sum((Z-mu_hat).^2)/(N-1) % Sample variance.
```

7.1.3 Estimating the CDF of IID Random Variables

Suppose instead of estimating the parameters of a distribution, we were interested in estimating the distribution itself. This can be done using some of the previous results. The CDF of the underlying distribution is $F_X(x) = \Pr(X \leq x)$. For any specific value of x, define a set of related variables $Y_1, Y_2, \ldots, Y_n$ such that

$$Y_i = \begin{cases} 1 & \text{if } X_i \leq x \\ 0 & \text{if } X_i > x \end{cases}. \tag{7.21}$$

It should be fairly evident that if the X_i are IID, then the Y_i must be IID as well. Note that for these Bernoulli random variables, the mean is $E[Y_i] = \Pr(X_i \leq x)$. Hence, estimating the CDF of the X_i is equivalent to estimating the mean of the Y_i, which is done using the sample mean:

$$\widehat{F}_X(x) = \frac{1}{n}\sum_{i=1}^{n} Y_i = \frac{1}{n}\sum_{i=1}^{n}[1 - u(X_i - x)]. \tag{7.22}$$

This estimator is nothing more than the relative frequency interpretation of probability. To estimate $F_X(x)$ from a sequence of n IID observations, we merely count the number of observations that satisfy $X_i \leq x$.

EXAMPLE 7.5: To illustrate this procedure of estimating the CDF of IID random variables, suppose the X_i are all uniformly distributed over $(0, 1)$. The plot in Figure 7.1 shows the results of one realization of estimating this CDF using n IID random variables for $n = 10$, $n = 100$, and $n = 1000$. Clearly, as n gets larger, the estimate gets better. The MATLAB code that follows can be used to generate a plot similar to the one in the figure. The reader is encouraged to try different types of random variables in this program as well.

```
N=100;                    % Set number of samples.
z=[-0.5:0.01:1.5];        % Define variable for horizontal axis.
x=rand(1,N);              % Generate uniform random samples.
F=zeros(1,length(z));     % Initialize CDF estimate.
for n=1:N                 % Estimate CDF.
    F=F+(x(n)<z);
end
```

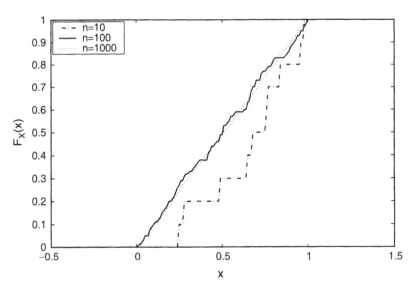

Figure 7.1 Estimate of the CDF of a uniform random variable obtained from n IID random variables, $n = 10, 100$, and 1000.

```
F=F/N;
plot(z,F)                    % Plot results.
xlabel('x'); ylabel('F_X(x)')
```

7.2 Convergence Modes of Random Sequences

In many engineering applications, it is common to use various iterative procedures. In such cases, it is often important to know under what circumstances an iterative algorithm converges to the desired result. The reader is no doubt familiar with many such applications in the deterministic world. For example, suppose we wish to solve for the root of some equation $g(x) = 0$. One could do this with a variety of iterative algorithms (e.g., Newton's method). The convergence of these algorithms is a quite important topic. That is, suppose x_i is the estimate of the root at the ith iteration of Newton's method. Does the sequence $x_1, x_2, x_3, \ldots$ converge to the true root of the equation? In this section, we study the topic of random sequences and in particular the issue of convergence of random sequences.

As an example of a random sequence, suppose we started with a set of IID random variables, $X_1, X_2, \ldots, X_n$, and then formed the sample mean according to

$$S_n = \frac{1}{n} \sum_{i=1}^{n} X_i. \tag{7.23}$$

The sequence $S_1, S_2, S_3, \ldots$ is a sequence of random variables. It is desirable that this sequence converge to the true mean of the underlying distribution. An estimator satisfying this condition is called *consistent*. But in what sense can we say that the sequence converges? If a sequence of deterministic numbers $s_1, s_2, s_3, \ldots$ were being considered, the sequence would be convergent to a fixed value s if

$$\lim_{i \to \infty} s_i = s. \tag{7.24}$$

More specifically, if for any $\varepsilon > 0$, there exists an i_ε such that $|s_i - s| < \varepsilon$ for all $i > i_\varepsilon$, then the sequence is said to converge to s.

Suppose an experiment, E, is run resulting in a realization, ζ. Each realization is mapped to a particular sequence of numbers. For example, the experiment might be to observe a sequence of IID random variables $X_1, X_2, X_3, \ldots$ and then map them into a sequence of sample means $S_1, S_2, S_3, \ldots$. Each realization, ζ, leads to a specific deterministic sequence, some of which might converge (in the previous sense) while others might not converge. Convergence for a sequence of random variables is not straightforward to define and can occur in a variety of manners.

7.2.1 Convergence Everywhere

The first and strictest form of convergence is what is referred to as convergence everywhere (a.k.a. sure convergence). A sequence is said to converge everywhere if every realization, ζ, leads to a sequence, $s_n(\zeta)$, that converges to $s(\zeta)$. Note that the limit may depend on the particular realization. That is, the limit of the random sequence may be a random variable.

7.2.2 Convergence Almost Everywhere

In many examples, it may be possible to find one or several realizations of the random sequence that do not converge, in which case the sequence (obviously) does not converge everywhere. However, it may be the case that such realizations are so rare that we might not want to concern ourselves with such cases. In particular, suppose that the only realizations that lead to a sequence that does not converge occur with probability zero. Then we say the random sequence converges almost everywhere (a.k.a. almost sure convergence or convergence with probability 1). Mathematically, let A be the set of all realizations that lead to a convergent sequence. Then the sequence converges almost everywhere if $\Pr(A) = 1$.

7.2.3 Convergence in Probability

A random sequence $S_1, S_2, S_3, \ldots$ converges in probability to a random variable S if for any $\varepsilon > 0$,

$$\lim_{n \to \infty} \Pr\left(|S_n - S| > \varepsilon\right) = 0. \tag{7.25}$$

7.2.4 Convergence in the Mean Square (MS) Sense

A random sequence $S_1, S_2, S_3, \ldots$ converges in the MS sense to a random variable S if

$$\lim_{n \to \infty} E\left[|S_n - S|^2\right] = 0. \tag{7.26}$$

7.2.5 Convergence in Distribution

Suppose the sequence of random variables $S_1, S_2, S_3, \ldots$ has CDFs given by $F_{S_n}(s)$, and the random variable S has a CDF, $F_S(s)$. Then, the sequence converges in distribution if

$$\lim_{n \to \infty} F_{S_n}(s) = F_S(s) \qquad (7.27)$$

for any s that is a point of continuity of $F_S(s)$.

7.3 The Law of Large Numbers

Having described the various ways in which a random sequence can converge, we return now to the study of sums of random variables. In particular, we look in more detail at the sample mean. The following very well known result is known as the *weak law of large numbers*.

THEOREM 7.2 (The Weak Law of Large Numbers): Let S_n be the sample mean computed from n IID random variables, $X_1, X_2, \ldots, X_n$. The sequence of sample means, S_n, converges in probability to the true mean of the underlying distribution, $F_X(x)$.

PROOF: Recall that if the distribution $F_X(x)$ has a mean of μ and variance σ^2, then the sample mean, S_n, has mean μ and variance σ^2/n. Applying Chebyshev's inequality,

$$\Pr(|S_n - \mu| > \varepsilon) \leq \frac{\text{Var}(S_n)}{\varepsilon^2} = \frac{\sigma^2}{n\varepsilon^2}. \qquad (7.28)$$

Hence, $\lim_{n \to \infty} \Pr(|S_n - \mu| > \varepsilon) = 0$ for any $\varepsilon > 0$. Thus, the sample mean converges in probability to the true mean. ∎

The implication of this result is that we can estimate the mean of a random variable with any amount of precision with arbitrary probability if we use a sufficiently large number of samples. A stronger result known as the *strong law of large numbers* shows that the convergence of the sample mean is not just in probability but also almost everywhere. We do not give a proof of this result in this text.

As was demonstrated in Section 7.2.3, the sample mean can be used to estimate more than just means. Suppose we are interested in calculating the probability that some event A results from a given experiment. Assuming that the experiment is repeatable and that each time the results of the experiment are independent of all other trials, then $\Pr(A)$ can easily be estimated. Simply define a random variable X_i

that is an indicator function for the event A on the ith trial. That is, if the event A occurs on the ith trial, then $X_i = 1$; otherwise, $X_i = 0$. Then

$$\Pr(A) = \Pr(X_i = 1) = E[X_i]. \tag{7.29}$$

The sample mean,

$$\hat{\mu}_n = \frac{1}{n} \sum_{i=1}^{n} X_i, \tag{7.30}$$

will give an unbiased estimate of the true probability, $\Pr(A)$. Furthermore, the law of large numbers tells us that as the sample size gets large, the estimate will converge to the true value. The weak law of large numbers tells us that the convergence is in probability while the strong law of large numbers tells us that the convergence is also almost everywhere.

The technique we've described for estimating the probability of events is known as the *Monte Carlo simulation*. It is commonly used, for example, to estimate the bit error probability of a digital communication system. A program is written to simulate transmission and detection of data bits. After a large number of data bits have been simulated, the number of errors is counted and divided by the total number of bits transmitted. This gives an estimate of the true probability of bit error of the system. If a sufficiently large number of bits are simulated, arbitrary precision of the estimate can be obtained.

 EXAMPLE 7.6: This example shows how the sample mean and sample variance converge to the true mean for a few different random variables. The results of running the MATLAB code that follows are shown in Figure 7.2. Plot (a) shows the results for a Gaussian distribution, whereas plot (b) shows the same results for an arcsine random variable. In each case, the parameters have been set so that the true mean is $\mu = 3$ and the variance of each sample is 1. Since the variance of the sample mean depends on only the variance and the number of the samples, crudely speaking the "speed" of convergence should be about the same in both cases.

```
N=100;

% Create Gaussian random variables.
mu1=3; sigma1=1;
X1=sigma1*randn(1,N)+mu1;
mu_hat1=cumsum(X1)./[1:N];   % sample means.

% Create Arcsine random variables.
mu2=3; b=sqrt(2); sigma2=b^2/2;
```

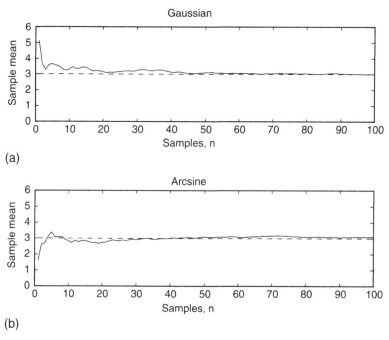

Figure 7.2   Convergence of the sample mean for Gaussian (a) and Arcsine (b) random variables.

```
X2=b*cos(2*pi*rand(1,N))+mu2;
mu_hat2=cumsum(X2)./[1:N];  % sample means.

subplot(2,1,1)
plot([1:N],mu_hat1,'-',[1:N], mu1,  '-')
xlabel('n'); ylabel('S_n'); title('Gaussian')
axis([0,N,0,2*mu1])
subplot(2,1,2)
plot([1:N],mu_hat2,'-',[1:N], mu2,  '-')
xlabel('n'); ylabel('S_n'); title('Arcsine')
axis([0,N,0,2*mu2])
```

7.4 The Central Limit Theorem

Probably the most important result dealing with sums of random variables is the central limit theorem, which states that under some mild conditions, these sums converge to a Gaussian random variable in distribution. This result provides the basis for many theoretical models of random phenomena. It also explains why the Gaussian random variable is of such great importance and why it occurs so frequently. In this section, we prove a simple version of the central limit theorem and then discuss some of the generalizations.

THEOREM 7.3 (The Central Limit Theorem): Let X_i be a sequence of IID random variables with mean μ_X and variance σ_X^2. Define a new random variable, Z, as a sum of the X_i:

$$Z = \frac{1}{\sqrt{n}} \sum_{i=1}^{n} \frac{X_i - \mu_X}{\sigma_x}. \tag{7.31}$$

Note that Z has been constructed such that $E[Z] = 0$ and $\text{Var}(Z) = 1$. In the limits as n approaches infinity, the random variable Z converges in distribution to a standard normal random variable.

PROOF: The most straightforward approach to prove this important theorem is using characteristic functions. Define the random variable $\widehat{X}_i$ as $\widehat{X}_i = (X_i - \mu_X)/\sigma_X$. The characteristic function of Z is computed as

$$\Phi_Z(\omega) = E[e^{j\omega Z}] = E\left[\exp\left(\frac{j\omega}{\sqrt{n}} \sum_{i=1}^{n} \widehat{X}_i\right)\right] = E\left[\prod_{i=1}^{n} \exp\left(\frac{j\omega \widehat{X}_i}{\sqrt{n}}\right)\right]$$

$$= \prod_{i=1}^{n} E\left[\exp\left(\frac{j\omega \widehat{X}_i}{\sqrt{n}}\right)\right] = \prod_{i=1}^{n} \phi_{\widehat{X}}\left(\frac{\omega}{\sqrt{n}}\right) = \left[\phi_{\widehat{X}}\left(\frac{\omega}{\sqrt{n}}\right)\right]^n. \tag{7.32}$$

Next, recall Taylor's theorem[2], which states that any function $g(x)$ can be expanded in a power series of the form

$$g(x) = g(x_o) + \frac{dg}{dx}\bigg|_{x=x_o}(x - x_o) + \cdots + \frac{1}{k!}\frac{d^k g}{dx^k}\bigg|_{x=x_o}(x - x_o)^k + r_k(x, x_o), \tag{7.33}$$

where the remainder $r_k(x, x_o)$ is small compared to $(x - x_o)^k$ as $x \to x_o$. Applying the Taylor series expansion about the point $\omega = 0$ to the characteristic function of $\widehat{X}$ results in

$$\phi_{\widehat{X}}(\omega) = \phi_{\widehat{X}}(0) + \phi_{\widehat{X}}'(0)\omega + \frac{1}{2}\phi_{\widehat{X}}''(0)\omega^2 + r_3(\omega), \tag{7.34}$$

---

[2]See, for example, Marsden, J. and A. Tromba, *Vector Calculus*, 5th ed., New York: W. H. Freeman, 2004.

where $r_3(\omega)$ is small compared to ω^2 as $\omega \to 0$. Furthermore, we note that $\phi_{\widehat{X}}(0) = 1$, $\phi'_{\widehat{X}}(0) = jE[\widehat{X}] = 0$, and $\phi''_{\widehat{X}}(0) = -E[\widehat{X}^2] = -1$. Hence Equation 7.34 reduces to

$$\phi_{\widehat{X}}(\omega) = 1 - \frac{\omega^2}{2} + r_3(\omega). \tag{7.35}$$

The characteristic function of Z is then

$$\Phi_Z(\omega) = \left(1 - \frac{\omega^2}{2n} + r_3\left(\frac{\omega}{\sqrt{n}}\right)\right)^n. \tag{7.36}$$

Note that as $n \to \infty$, the argument of $r_3()$ goes to zero for any finite ω. Hence, as $n \to \infty$, $r_3(\omega/\sqrt{n})$ becomes negligible compared to ω^2/n. Therefore, in the limit, the characteristic function of Z approaches[3]

$$\lim_{n \to \infty} \Phi_Z(\omega) = \lim_{n \to \infty} \left(1 - \frac{\omega^2}{2n}\right)^n = \exp\left(-\frac{\omega^2}{2}\right). \tag{7.37}$$

This is the characteristic function of a standard normal random variable. ∎

Several remarks about this theorem are in order at this point. First, no restrictions were put on the distribution of the X_i. The preceding proof applies to any infinite sum of IID random variables, regardless of the distribution. Also, the central limit theorem guarantees that the sum converges in *distribution* to Gaussian, but this does not necessarily imply convergence in *density*. As a counterexample, suppose that the X_i are discrete random variables. Then the sum must also be a discrete random variable. Strictly speaking, the density of Z would then not exist and it would not be meaningful to say that the density of Z is Gaussian. From a practical standpoint, the probability density of Z would be a series of impulses. While the envelope of these impulses would have a Gaussian shape to it, the density is clearly not Gaussian. If the X_i are continuous random variables, the convergence in density generally occurs as well.

The proof of the central limit theorem just given assumes that the X_i are IID. This assumption is not needed in many cases. The central limit theorem also applies to independent random variables that are not necessarily identically distributed.

---

[3]Here we have used the well-known fact that $\lim_{n \to \infty} (1 + (x/n))^n = e^x$. To establish this result, the interested reader is encouraged to expand both sides in a Taylor series and show that in the limit the two expansions become equivalent.

Loosely speaking[4], all that is required is that no single term (or small number of terms) dominate the sum and the resulting infinite sum of independent random variables will approach a Gaussian distribution in the limit as the number of terms in the sum goes to infinity. The central limit theorem also applies to some cases of dependent random variables, but we will not consider such cases here.

From a practical standpoint, the central limit theorem implies that for the sum of a sufficiently large (but finite) number of random variables, the sum is *approximately* Gaussian distributed. Of course, the goodness of this approximation depends on how many terms are in the sum and also on the distribution of the individual terms in the sum. The next examples show some illustrations to give the reader a feel for the Gaussian approximation.

EXAMPLE 7.7: Suppose the X_i are all independent and uniformly distributed over $(-1/2, 1/2)$. Consider the sum

$$Z = \sqrt{\frac{12}{n}} \sum_{i=1}^{n} X_i.$$

The sum has been normalized so that Z has zero-mean and unit variance. It was shown previously that the PDF of the sum of independent random variables is just the convolution of the individual PDFs. Hence, if we define Y to be $Y = X_1 + X_2 + \cdots + X_n$, then

$$f_Y(z) = f_X(z) * f_X(z) * \cdots * f_X(z), \text{ and } f_Z(z) = \sqrt{\frac{n}{12}} f_Y\left(z\sqrt{\frac{n}{12}}\right).$$

The results of performing this n-fold convolution are shown in Figure 7.3 for several values of n. Note that for as few as $n = 4$ or $n = 5$ terms in the series, the resulting PDF of the sum looks very much like the Gaussian PDF.

EXAMPLE 7.8: In this example, suppose the X_i are now discrete Bernoulli distributed random variables such that $\Pr(X_i = 1) = \Pr(X_i = 0) = 0.5$. In this case, the sum $Y = X_1 + X_2 + \cdots + X_n$ is

---

[4]Formal conditions can be found in Papoulis, A. and Pillai, S. Unnikrishna, *Probability, Random Variables, and Stochastic Processes*, 4th ed., New York: McGraw-Hill, 2001.

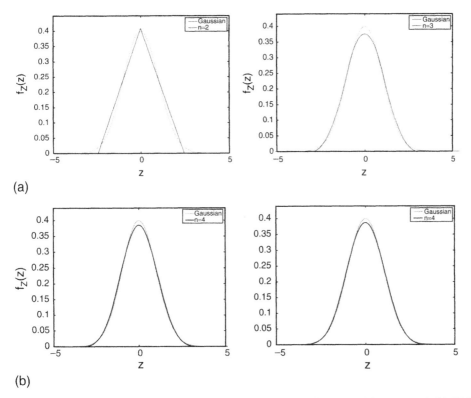

(a)

(b)

Figure 7.3 (a) PDF of the sum of independent uniform random variables, $n = 2, 3$; (b) PDF of the sum of independent uniform random variables, $n = 4, 5$.

a binomial random variable with PMF given by

$$\Pr(Y = k) = \binom{n}{k}\left(\frac{1}{2}\right)^n, \quad k = 0, 1, 2, \ldots, n.$$

The corresponding CDF is

$$F_Y(y) = \sum_{k=0}^{n} \binom{n}{k}\left(\frac{1}{2}\right)^n u(y - k).$$

The random variable Y has a mean of $E[Y] = n/2$ and a variance of $\text{Var}(Y) = n/4$. In Figure 7.4, this binomial distribution is compared to a Gaussian distribution with the same mean and variance. It is seen that for this discrete random variable, many more terms are needed in the sum before good convergence to a Gaussian distribution is achieved.

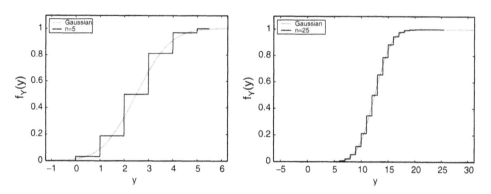

Figure 7.4   CDF of the sum of independent Bernoulli random variables, $n = 5, 25$.

7.5 Confidence Intervals

Consider once again the problem of estimating the mean of a distribution from n IID observations. When the sample mean $\hat{\mu}$ is formed, what have we actually learned? Loosely speaking, we might say that our best guess of the true mean is $\hat{\mu}$. However, in most cases, we know that the event $\{\hat{\mu} = \mu\}$ occurs with zero probability (since if $\hat{\mu}$ is a continuous random variable, the probability of it taking on any point value is zero). Alternatively, it could be said that (hopefully) the true mean is "close" to the sample mean. While this is a vague statement, with the help of the central limit theorem, we can make the statement mathematically precise.

If a sufficient number of samples is taken, the sample mean can be well approximated by a Gaussian random variable with a mean of $E[\hat{\mu}] = \mu_x$ and $\text{Var}(\hat{\mu}) = \sigma_X^2/n$. Using the Gaussian distribution, the probability of the sample mean being within some amount ε of the true mean can be easily calculated:

$$\Pr\left(\left|\hat{\mu} - \mu_X\right| < \varepsilon\right) = \Pr\left(\mu_X - \varepsilon < \hat{\mu} < \mu_X + \varepsilon\right) = 1 - 2Q\left(\varepsilon\sqrt{n}/\sigma_X\right). \qquad (7.38)$$

Stated another way, let ε_α be the value of ε such that the right-hand side of the preceding equation is $1 - \alpha$; that is,

$$\varepsilon_\alpha = \frac{\sigma_X}{\sqrt{n}}Q^{-1}\left(\frac{\alpha}{2}\right). \qquad (7.39)$$

where Q^{-1} is the inverse of Q.

Then, given n samples that lead to a sample mean $\hat{\mu}$, the true mean will fall in the interval $(\hat{\mu} - \varepsilon_\alpha, \hat{\mu} + \varepsilon_\alpha)$ with probability $1 - \alpha$. The interval $(\hat{\mu} - \varepsilon_\alpha, \hat{\mu} + \varepsilon_\alpha)$ is referred to as the *confidence interval* while the probability $1 - \alpha$ is the *confidence*

Table 7.1   Constants Used to Calculate Confidence Intervals

Confidence level $(1 - \alpha) * 100\%$	Level of significance $\alpha * 100\%$	$c_\alpha = Q^{-1}\left(\frac{\alpha}{2}\right)$
90%	10%	1.64
95%	5%	1.96
99%	1%	2.58
99.9%	0.1%	3.29
99.99%	0.01%	3.89

level or, alternatively, α is the *level of significance*. The confidence level and level of significance are usually expressed as percentages. The corresponding values of the quantity $c_\alpha = Q^{-1}(\alpha/2)$ are provided in Table 7.1 for several typical values of α. Other values not included in the table can be found from tables of the Q-function (such as those provided in Appendix E).

EXAMPLE 7.9: Suppose the IID random variables each have a variance of $\sigma_X^2 = 4$. A sample of $n = 100$ values is taken and the sample mean is found to be $\hat{\mu} = 10.2$. Determine the 95 percent confidence interval for the true mean μ_X. In this case, $\sigma_X / \sqrt{n} = 0.2$ and the appropriate value of c_α is $c_{0.05} = 1.96$ from Table 7.1. The 95 percent confidence interval is then

$$\left(\hat{\mu} - \frac{\sigma_X}{\sqrt{n}} c_{0.05}, \hat{\mu} + \frac{\sigma_X}{\sqrt{n}} c_{0.05}\right) = (9.808, 10.592).$$

EXAMPLE 7.10: Looking again at Example 7.9, suppose we want to be 99 percent confident that the true mean falls within ± 0.5 of the sample mean. How many samples need to be taken in forming the sample mean? To ensure this level of confidence, it is required that

$$\frac{\sigma_X}{\sqrt{n}} c_{0.01} = 0.5$$

and hence,

$$n = \left(\frac{c_{0.01}\sigma_X}{0.5}\right)^2 = \left(\frac{2.58 * 2}{0.5}\right)^2 = 106.5.$$

Since n must be an integer, it is concluded that at least 107 samples must be taken.

In summary, to achieve a level of significance specified by α, we note that by virtue of the central limit theorem, the sum

$$\widehat{Z}_n = \frac{\hat{\mu} - \mu_X}{\sigma_X / \sqrt{n}} \tag{7.40}$$

approximately follows a standard normal distribution. We can then easily specify a symmetric interval about zero in which a standard normal random variable will fall with probability $1 - \alpha$. As long as n is sufficiently large, the original distribution of the IID random variables does not matter.

Note that in order to form the confidence interval as specified, the standard deviation of the X_i must be known. While in some cases this may be a reasonable assumption, in many applications the standard deviation is also unknown. The most obvious thing to do in that case would be to replace the true standard deviation in Equation 7.40 with the sample standard deviation. That is, we form a statistic

$$\widehat{T}_n = \frac{\hat{\mu} - \mu_X}{\hat{s} / \sqrt{n}} \tag{7.41}$$

and then seek a symmetric interval about zero $(-t_\alpha, t_\alpha)$ such that the probability that $\widehat{T}_n$ falls in that interval is $1 - \alpha$. For very large n, the sample standard deviation will converge to the true standard deviation and thus $\widehat{T}_n$ will approach $\widehat{Z}_n$. Hence, in the limit as $n \to \infty$, $\widehat{T}_n$ can be treated as having a standard normal distribution and the confidence interval is found in the same manner we've described. That is, as $n \to \infty, t_\alpha \to c_\alpha$. For values of n that are not very large, the actual distribution of the statistic $\widehat{T}_n$ must be calculated in order to form the appropriate confidence interval.

Naturally, the distribution of $\widehat{T}_n$ will depend on the distribution of the X_i. One case where this distribution has been calculated for finite n is when the X_i are Gaussian random variables. In this case, the statistic $\widehat{T}_n$ follows the so-called Student's t-distribution[5] with $n - 1$ degrees of freedom:

$$f_{\widehat{T}_n}(t) = \frac{(1 + t^2/n)^{-(n+1)/2} \Gamma((n+1)/2)}{\sqrt{n\pi} \, \Gamma(n/2)}. \tag{7.42}$$

where Γ is the gamma function (See Chapter 3 (3.22) and Appendix E (E.39)).

From this PDF one can easily find the appropriate confidence interval for a given level of significance, α, and sample size, n. Tables of the appropriate confidence interval, t_α, can be found in any text on statistics. It is common to use the t-distribution to form confidence intervals even if the samples are

---

[5]The Student's t-distribution was developed by the English mathematician W. S. Gossett, who published under the pseudonym A. Student.

not Gaussian distributed. Hence, the *t*-distribution is very commonly used for statistical calculations.

Many other statistics associated with related parameter estimation problems are encountered and have been carefully expounded in the statistics literature. Indeed, many freshman- and sophomore-level statistics courses simply list all the cases and the corresponding statistical distributions without explaining the underlying probability theory. Left to memorize seemingly endless distributions and statistical tests, many students have been frightened away from the study of statistics before they ever have a chance to appreciate it. Rather than take that approach, we believe that with the probability theory developed to this point, the motivated student can now easily understand the motivation and justification for the variety of statistical tests that appear in the literature.

EXAMPLE 7.11: Suppose we wish to estimate the failure probability of some system. We might design a simulator for our system and count the number of times the system fails during a long sequence of operations of the system. Examples might include bit errors in a communications system, defective products in an assembly line, or the like. The failure probability can then be estimated as discussed at the end of Section 7.3. Suppose the true failure probability is p (which of course is unknown to us). We simulate operation of the system n times and count the number of errors observed, N_e. The estimate of the true failure probability is then just the relative frequency,

$$\hat{p} = \frac{N_e}{n}.$$

If errors occur independently, then the number of errors we observe in n trials is a binomial random variable with parameters n and p. That is,

$$P_{N_e}(k) = \binom{n}{k} p^k (1-p)^{n-k}, \quad k = 0, 1, 2, \ldots, n.$$

From this we infer that the mean and variance of the estimated failure probability is $E[\hat{p}] = p$ and $\text{Var}(\hat{p}) = n^{-1}p(1-p)$. From this we can develop confidence intervals for our failure probability estimates. The MATLAB code that follows creates estimates as just described and plots the results, along with error bars indicating the confidence intervals associated with each estimate. The plot resulting from running this code is shown in Figure 7.5.

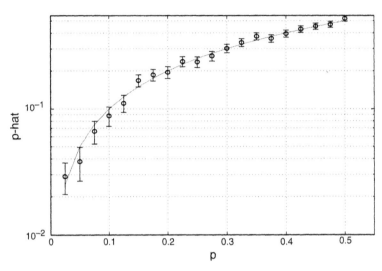

Figure 7.5 Estimates of failure probabilities along with confidence intervals. The solid line is the true probability while the circles represent the estimates.

```
N=1000;                          % Number of samples generated.
c=1.64;                          % For 90 percent confidence
                                 level.
points=[0.025:0.025:0.5];        % Values of p.

for k=1:length(points)
  p=points(k);
  X=rand(1,N)<p;                 % 1=error, 0=no error.
  p_hat(k)=sum(X)/N;             % Relative frequency.
  sigma=sqrt(p*(1-p)/N);
  eps(k)=sigma*c;                % Compute confidence interval.
end

% Plot results.
semilogy(points,points,'-')      % True values.
axis([0 0.55 0.01 0.55])
grid on
xlabel('p')
ylabel('p-hat')
hold on
errorbar(points,p_hat,eps,'o')   % Estimated values with
hold off                         % confidence intervals.
```

7.6 Random Sums of Random Variables

The sums of random variables considered up to this point have always had a fixed number of terms. Occasionally, one also encounters sums of random variables where the number of terms in the sum is also random. For example, a node in a communication network may queue packets of variable length while they are waiting to be transmitted. The number of bytes in each packet, X_i, might be random as well as the number of packets in the queue at any given time, N. The total number of bytes stored in the queue would then be a random sum of the form

$$S = \sum_{i=1}^{N} X_i. \tag{7.43}$$

THEOREM 7.4: Given a set of IID random variables X_i with mean μ_X and variance σ_X^2 and an independent random integer N, the mean and variance of the random sum of the form given in Equation 7.43 are given by

$$E[S] = \mu_X E[N], \tag{7.44}$$

$$\text{Var}(S) = E[N]\sigma_X^2 + \text{Var}(N)\mu_X^2. \tag{7.45}$$

PROOF: To calculate the statistics of S, it is easiest to first condition on N and then average the resulting conditional statistics with respect to N. To start with, consider the mean:

$$E[S] = E_N[E[S|N]] = E_N\left[E\left[\sum_{i=1}^{N} X_i \middle| N\right]\right] = E_N[N\mu_X] = \mu_X E[N]. \tag{7.46}$$

The variance is found following a similar procedure. The second moment of S is found according to

$$E[S^2] = E_N[E[S^2|N]] = E_N\left[E\left[\sum_{i=1}^{N}\sum_{j=1}^{N} X_i X_j \middle| N\right]\right] = E_N\left[\sum_{i=1}^{N}\sum_{j=1}^{N} E[X_i X_j]\right]. \tag{7.47}$$

Note that X_i and X_j are uncorrelated unless $i = j$. Hence, this expected value works out to be

$$E[S^2] = E_N[(N^2 - N)\mu_X^2 + NE[X_i^2]] = E[N^2]\mu_X^2 + E[N]\sigma_X^2. \tag{7.48}$$

Finally, using $\text{Var}(S) = E[S^2] - (E[S])^2$ results in

$$\text{Var}(S) = E[N^2]\mu_X^2 + E[N]\sigma_X^2 - (E[N])^2\mu_X^2 = \text{Var}(N)\mu_X^2 + E[N]\sigma_X^2. \qquad (7.49)$$

One could also derive formulas for higher order moments in a similar manner. ∎

THEOREM 7.5: Given a set of IID random variables X_i with a characteristic function $\Phi_X(\omega)$ and an independent random integer N with a probability generating function $H_N(z)$, the characteristic function of the random sum of the form given in Equation 7.43 is given by

$$\Phi_S(\omega) = H_N(\Phi_X(\omega)). \qquad (7.50)$$

PROOF: Following a derivation similar to the last theorem:

$$\Phi_S(\omega) = E[e^{j\omega S}] = E_N[E[e^{j\omega S}|N]] = E_N\left[E\left[\exp\left(j\omega \sum_{i=1}^{N} X_i\right)\bigg| N\right]\right]$$

$$= E_N\left[E\left[\prod_{i=1}^{N} \exp(j\omega X_i)\bigg| N\right]\right] = E_N[(\Phi_X(\omega))^N]$$

$$= \sum_k \Pr(N = k)(\Phi_X(\omega))^k = H_N(\Phi_X(\omega)). \qquad (7.51)$$

∎

EXAMPLE 7.12: Suppose the X_i are Gaussian random variables with zero mean and unit variance and N is a binomial random variable with a PMF,

$$\Pr(N = k) = \binom{n}{k} p^k (1 - p)^{n-k}.$$

The mean and variance of this discrete distribution are $E[N] = np$ and $\text{Var}(N) = np(1 - p)$, respectively. From the results of Theorem 7.4 it is found that

$$E[S] = \mu_X E[N] = 0 \text{ and } \text{Var}(S) = \text{Var}(N)\mu_X^2 + E[N]\sigma_X^2 = np.$$

The corresponding characteristic function of X_i and probability generating function of N are given by

$$\Phi_X(\omega) = \exp\left(-\frac{\omega^2}{2}\right) \quad \text{and} \quad H_N(z) = (1 - p + pz)^n.$$

The characteristic function of the random sum is then

$$\Phi_S(\omega) = \left(1 - p + p \exp\left(-\frac{\omega^2}{2}\right)\right)^n.$$

It is interesting to note that the sum of any (fixed) number of Gaussian random variables produces a Gaussian random variable. Yet, the preceding characteristic function is clearly not that of a Gaussian random variable, and hence a random sum of Gaussian random variables is not Gaussian.

All of the results presented thus far in this section have made the assumption that the IID variables, X_i, and the number of terms in the series, N, are statistically independent. Quite often these two quantities are dependent. For example, one might be interested in accumulating terms in the sum until the sum exhibits a specified characteristic (e.g., until the sample standard deviation falls below some threshold). Then the number of terms in the sum would clearly be dependent on the values of the terms themselves. In such a case, the preceding results would not apply and similar results for dependent variables would have to be developed. The following application section considers such a situation.

7.7 Engineering Application: A Radar System

In this section, we consider a simple radar system like that depicted in Figure 1.3. At known instants of time, the system transmits a known pulse and then waits for a reflection. Suppose the system is looking for a target at a known range and so the system can determine exactly when the reflection should appear at the radar receiver. To make this discussion as simple as possible, suppose that the system has the ability to "sample" the received signal at the appropriate time instant and, further, that each sample is a random variable X_j that is modeled as a Gaussian random variable with variance σ^2. Let A_1 be the event that there is indeed a target present, in which case X_j is taken to have a mean of μ; whereas, A_0 is the event that there is no target present and the resulting mean is zero. That is, our received sample consists of a signal part (if it is present) that is some fixed voltage, μ, plus a noise part that we model as Gaussian and zero-mean. As with many radar systems, we assume that the reflection is fairly weak (μ is not large compared to σ), and hence if we try to decide whether or not a target is present based on a single observation, we will likely end up with a very unreliable decision. As a result, our system is designed to transmit several pulses (at nonoverlapping time instants) and observe several returns, $X_j, j = 1, 2, \ldots, n$, that we take to be IID. The problem is to determine how to process these returns in order to make the best decision and also how many returns to collect in order to have our decisions attain a prescribed reliability.

We consider two possible approaches. In the first approach, we decide ahead of time how many returns to collect and call that fixed number, n. We then process that random vector $X = (X_1, X_2, \ldots, X_n)^T$ and form a decision. While there are many ways to process the returns, we will use what is known as a probability ratio test. That is, given $X = x$, we want to determine if the ratio $\Pr(A_1|x)/\Pr(A_0|x)$ is greater or less than 1. Recall that

$$\Pr(A_i|x) = \frac{f_X(x|A_i)\Pr(A_i)}{f_X(x)}, \quad i = 0, 1. \tag{7.52}$$

Hence, the probability ratio test makes the following comparison:

$$\frac{\Pr(A_1|x)}{\Pr(A_0|x)} = \frac{f_X(x|A_1)\Pr(A_1)}{f_X(x|A_0)\Pr(A_0)} \mathrel{\overset{?}{\underset{?}{\gtrless}}} 1. \tag{7.53}$$

This can be written in terms of an equivalent likelihood ratio test:

$$\Lambda(x) \mathrel{\overset{?}{\underset{?}{\gtrless}}} \Lambda, \tag{7.54}$$

where $\Lambda(x) = f_X(x|A_1)/f_X(x|A_0)$ is the likelihood ratio and the threshold $\Lambda = \Pr(A_0)/\Pr(A_1)$ depends on the a priori probabilities. In practice, we may have no idea about the a priori probabilities of whether or not a target is present. However, we can still proceed by choosing the threshold for the likelihood ratio test to provide some prescribed level of performance.

Let the false alarm probability be defined as $P_{fa} = \Pr(\Lambda(X) > \Lambda|A_0)$. This is the probability that the system declares a target is present when in fact there is none. Similarly, define the correct detection probability as $P_d = \Pr(\Lambda(X) > \Lambda|A_1)$. This is the probability that the system correctly identifies a target as being present. These two quantities, P_{fa} and P_d, will specify the performance of our radar system.

Given that the X_j are IID Gaussian as described, the likelihood ratio works out to be

$$\Lambda(x) = \frac{f_X(x|A_1)}{f_X(x|A_0)} = \frac{(2\pi\sigma^2)^{-n/2}\exp\left(-\frac{1}{2\sigma^2}\sum_{j=1}^{n}(x_j - \mu)^2\right)}{(2\pi\sigma^2)^{-n/2}\exp\left(-\frac{1}{2\sigma^2}\sum_{j=1}^{n}x_j^2\right)}$$

$$= \exp\left(\frac{n\mu}{\sigma^2}\left[\left(\frac{1}{n}\sum_{j=1}^{n}x_j\right) - \frac{\mu}{2}\right]\right). \tag{7.55}$$

Clearly, comparing this with a threshold is equivalent to comparing the sample mean with a threshold. That is, for IID Gaussian returns, the likelihood ratio test simplifies to

$$\hat{\mu} = \frac{1}{n}\sum_{j=1}^{n}x_j \underset{?}{\overset{?}{\underset{<}{\gtrless}}} \mu_0, \tag{7.56}$$

where the threshold μ_0 is set to produce the desired system performance. Since the X_j are Gaussian, the sample mean is also Gaussian. Hence, when there is no target present $\hat{\mu} \sim N(0, \sigma^2/n)$ and when there is a target present $\hat{\mu} \sim N(\mu, \sigma^2/n)$. With these distributions, the false alarm and detection probabilities work out to be

$$P_{fa} = Q\left(\frac{\sqrt{n}\mu_0}{\sigma}\right) \quad \text{and} \quad 1 - P_d = Q\left(\frac{\sqrt{n}(\mu - \mu_0)}{\sigma}\right). \tag{7.57}$$

By adjusting the threshold, we can trade off false alarms for missed detections. Since the two probabilities are related, it is common to write the detection probability in terms of the false alarm probability as

$$1 - P_d = Q\left(\frac{\sqrt{n}\mu}{\sigma} - Q^{-1}(P_{fa})\right). \tag{7.58}$$

From this equation, we can determine how many returns must be collected in order to attain a prescribed system performance specified by (P_{fa}, P_d). In particular,

$$n = \frac{\left[Q^{-1}(1 - P_d) + Q^{-1}(P_{fa})\right]^2}{\mu^2/\sigma^2}. \tag{7.59}$$

The quantity μ^2/σ^2 has the physical interpretation of the strength of the signal (when it is present) divided by the strength of the noise, or simply the signal-to-noise ratio.

Since the radar system must search at many different ranges and many different angles of azimuth, we would like to minimize the amount of time it has to spend collecting returns at each point. Presumably, the amount of time we spend observing each point in space depends on the number of returns we need to collect. We can often reduce the number of returns needed by noting that the number of returns required to attain a prescribed reliability as specified by (P_{fa}, P_d) will depend on the particular realization of returns encountered. For example, if the first few returns come back such that the sample mean is very large, we may be very certain that a target is present and hence there is no real need to collect more returns. In other instances, the first few returns may produce a sample mean near $\mu/2$. This inconclusive data would lead us to wait and collect more data before making a decision. Using a variable number of returns whose number depends on the data themselves is known as *sequential detection*.

The second approach we consider will use a sequential detection procedure whereby after collecting n returns, we compare the likelihood ratio with two thresholds, Λ_0 and Λ_1, and decide according to

$$\Lambda(x) \begin{cases} \geq \Lambda_1 & \text{decide a target is present} \\ \in (\Lambda_0, \Lambda_1) & \text{collect another return} \\ \leq \Lambda_0 & \text{decide no target is present} \end{cases} \tag{7.60}$$

The performance of a sequential detection scheme can be determined as follows. Define the region $R_1^{(n)}$ to be the set of data points $x^{(n)} = (x_1, x_2, \ldots, x_n)$ that lead to a decision in favor of A_1 after collecting exactly n data points. That is, $\Lambda(x^{(n)}) > \Lambda_1$ and $\Lambda_0 < \Lambda(x^{(j)}) < \Lambda_1$ for $j = 1, 2, \ldots, n-1$. Similarly, define the region $R_0^{(n)}$ to be the set of data points $x^{(n)}$ that lead to a decision in favor of A_0 after collecting exactly n data points. Let $P_{fa}^{(n)}$ be the probability of a false alarm occurring after collecting exactly n returns and $P_d^{(n)}$ the probability of making a correct detection after collecting exactly n returns. The overall false alarm and detection probabilities are then

$$P_{fa} = \sum_{n=1}^{\infty} P_{fa}^{(n)} \quad \text{and} \quad P_d = \sum_{n=1}^{\infty} P_d^{(n)}. \tag{7.61}$$

We are now in a position to establish the fundamental result, shown in Theorem 7.6, which will instruct us in how to set the decision thresholds in order to obtain the desired performance.

THEOREM 7.6 (Wald's Inequalities): For a sequential detection strategy, the false alarm and detection strategies satisfy

$$P_d \geq \Lambda_1 P_{fa}, \tag{7.62}$$

$$(1 - P_d) \leq \Lambda_0 (1 - P_{fa}). \tag{7.63}$$

PROOF: First note that

$$P_{fa}^{(n)} = \Pr(x^{(n)} \in R_1^{(n)} | A_0) = \int_{R_1^{(n)}} f_{X^{(n)}}(x^{(n)} | A_0) \, dx^{(n)}, \tag{7.64}$$

and similarly,

$$P_d^{(n)} = \Pr(x^{(n)} \in R_1^{(n)} | A_1) = \int_{R_1^{(n)}} f_{X^{(n)}}(x^{(n)} | A_1) \, dx^{(n)}. \tag{7.65}$$

For all $x^{(n)} \in R_1^{(n)}$, $f_{X^{(n)}}\left(x^{(n)} | A_1\right) \geq \Lambda_1 f_{X^{(n)}}\left(x^{(n)} | A_0\right)$ and hence,

$$P_d^{(n)} \geq \Lambda_1 \int_{R_1^{(n)}} f_{X^{(n)}}(x^{(n)} | A_0) \, dx^{(n)} = \Lambda_1 P_{fa}^{(n)}. \tag{7.66}$$

Summing over all n then produces Equation 7.62a. Equation 7.63b is derived in a similar manner. ∎

Since the likelihood ratio is often exponential in form, it is common to work with the log of the likelihood ratio, $\lambda(x) = \ln(\Lambda(x))$. For the case of Gaussian IID data, we get

$$\lambda(x^{(n)}) = \frac{n\mu}{\sigma^2} \left[\left(\frac{1}{n} \sum_{j=1}^{n} x_j \right) - \frac{\mu}{2} \right] = \lambda\left(x^{(n-1)}\right) + \frac{\mu}{\sigma^2} x_n - \frac{\mu^2}{2\sigma^2}. \tag{7.67}$$

In terms of log-likelihood ratios, the sequential decision mechanism is

$$\lambda(x) \begin{cases} \geq \lambda_1 & \text{decide a target is present} \\ \in (\lambda_0, \lambda_1) & \text{collect another return} \\ \leq \lambda_0 & \text{decide no target is present} \end{cases}, \tag{7.68}$$

where $\lambda_j = \ln(\Lambda_j)$, $j = 0, 1$. The corresponding versions of Wald's inequalities are then

$$\ln(P_d) \geq \lambda_1 + \ln(P_{fa}), \tag{7.69}$$

$$\ln(1 - P_d) \leq \lambda_0 + \ln(1 - P_{fa}). \tag{7.70}$$

For the case when the signal-to-noise ratio is small, each new datum collected adds a small amount to the sum in Equation 7.67, and it will typically take a large number of terms before the sum will cross one of the thresholds, λ_0 or λ_1. As a result, when the requisite number of data are collected so that the log-likelihood ratio crosses a threshold, it will usually be only incrementally above the threshold. Hence, Wald's inequalities will be approximate equalities and the decision thresholds that lead to (approximately) the desired performance can be found according to

$$\lambda_1 = \ln\left(\frac{P_d}{P_{fa}}\right) \quad \text{and} \quad \lambda_0 = \ln\left(\frac{1 - P_d}{1 - P_{fa}}\right). \tag{7.71}$$

Now that the sequential detection strategy can be designed to give any desired performance, we are interested in determining how much effort we save relative to the fixed sample size test. Let N be the instant at which the test terminates and to simplify notation, define

$$S_N = \lambda(\mathbf{X}^{(N)}) = \sum_{i=1}^{N} Z_i, \tag{7.72}$$

where from Equation 7.67, $Z_i = \mu(X_i - \mu/2)/\sigma^2$. Note that S_N is a random sum of IID random variables as studied in Section 7.6, except that now the random variable N is not independent of the Z_i. Even with this dependence, for this example it is still true that

$$E[S_N] = E[N]E[Z_i]. \tag{7.73}$$

The reader is led through a proof of this in Exercise 7.17. Note that when the test terminates:

$$S_N \cong \begin{cases} \lambda_1 & \text{if the test terminates in } A_1 \\ \lambda_0 & \text{if the test terminates in } A_0 \end{cases}, \tag{7.74}$$

and hence,

$$E[S_N] \cong \lambda_1 \Pr(\text{test terminates in } A_1) + \lambda_0 \Pr(\text{test terminates in } A_0). \tag{7.75}$$

Suppose that A_0 is true. Then the event that the test terminates in A_1 is simply a false alarm. Combining Equations 7.73 and 7.75 results in

$$E[N|A_0] \cong \frac{E[S_N|A_0]}{E[Z_i|A_0]} = \frac{\lambda_1 P_{fa} + \lambda_0(1 - P_{fa})}{\mu^2/2\sigma^2}. \tag{7.76}$$

Similarly, when A_1 is true:

$$E[N|A_1] \cong \frac{E[S_N|A_1]}{E[Z_i|A_1]} = \frac{\lambda_1 P_d + \lambda_0(1 - P_d)}{\mu^2/2\sigma^2}. \tag{7.77}$$

It is noted that not only is the number of returns collected a random variable, but the statistics of this random variable depend on whether or not a target is present.

This may work to our advantage in that the average number of returns we need to observe might be significantly smaller in the more common case when there is no target present.

EXAMPLE 7.13: Suppose we want to achieve a system performance specified by $P_{fa} = 10^{-6}$ and $P_d = 0.99$. Furthermore, suppose the signal-to-noise ratio for each return is $\mu^2/\sigma^2 = 0.1 = -10$dB. Then the fixed sample size test will use a number of returns given by

$$n = \frac{[Q^{-1}(1 - P_d) + Q^{-1}(P_{fa})]^2}{\mu^2/\sigma^2} = \frac{2.326 + 4.755]^2}{0.1} = 501.4.$$

Since n must be an integer, 502 samples must be taken to attain the desired performance. For the sequential test, the two thresholds for the log-likelihood test are set according to Wald's inequalities,

$$\lambda_0 \ln \left(\frac{1 - P_d}{1 - P_{fa}} \right) = -4.6 \quad \text{and} \quad \lambda_1 = \ln \left(\frac{P_d}{P_{fa}} \right) = 13.8.$$

With these thresholds set, the average number of samples needed for the test to terminate is

$$E[N|A_0] = -\frac{\lambda_1 P_{fa} + \lambda_0(1 - P_{fa})}{\mu^2/2\sigma^2} = 92.1$$

when there is no target present, and

$$E[N|A_1] = \frac{\lambda_1 P_d + \lambda_0(1 - P_d)}{\mu^2/2\sigma^2} = 272.4$$

when a target is present. Clearly, for this example, the sequential test saves us significantly in terms of the amount of data that needs to be collected to make a reliable decision.

Exercises

7.1 A random variable, X, has a normal PDF with mean 5 and unit variance. We measure 10 independent samples of the random variable.

(a) Determine the expected value of the sample mean.

(b) Determine the variance of the sample mean.

(c) Determine the expected value of the unbiased sample variance.

7.2 Two independent samples of a random variable X are taken. Determine the expected value and variance of the estimate of μ_X if the PDF is exponential, (i.e., $f_X(x) = \exp(-x)u(x)$).

7.3 The noise level in a room is measured N times. The error E for each measurement is independent of the others and is normally distributed with zero-mean and standard deviation $\sigma_E = 0.1$. In terms of the true mean, μ, determine the PDF of the sample mean, $\hat{\mu}$ for $N = 100$.

7.4 Suppose X is a vector of N IID random variables where each element has some PDF, $f_X(x)$. Find an example PDF such that the median is a better estimate of the mean than the sample mean.

7.5 Suppose the variance of an IID sequence of random variables is formed according to

$$\hat{\sigma}^2 = \frac{1}{n}\sum_{m=1}^{n}(X_m - \hat{\mu})^2,$$

where $\hat{\mu}$ is the sample mean. Find the expected value of this estimate and show that it is biased.

7.6 Find the variance of the sample standard deviation,

$$\hat{s}^2 = \frac{1}{n-1}\sum_{m=1}^{n}(X_m - \hat{\mu})^2,$$

assuming that the X_i are IID Gaussian random variables with mean μ and variance σ^2.

7.7 Show that if X_n, $n = 1, 2, 3, \ldots$ is a sequence of IID Gaussian random variables, the sample mean and sample variance are statistically independent.

7.8 A sequence of random variables, X_n, is to be approximated by a straight line using the estimate $\widehat{X}_n = a + bn$. Determine the least squares (i.e., minimum mean squared error) estimates for a and b if N samples of the sequence are observed.

7.9 (a) Prove that any sequence that converges in the mean square sense must also converge in probability. *Hint:* Use Markov's inequality.

   (b) Prove by counterexample that convergence in probability does not necessarily imply convergence in the mean square sense.

7.10 Suppose $X_1, X_2, \ldots, X_n$ is a sequence of IID positive random variables. Define

$$Y_n = \prod_{i=1}^{n} X_i.$$

Show that as $n \to \infty$, Y_n converges in distribution, and find the distribution to which it converges.

7.11 Suppose we wish to estimate the probability, p_A, of some event, A. We do so by repeating an experiment n times and observing whether or not the event A occurs during each experiment. In particular, let

$$X_i = \begin{cases} 1 & A \text{ occurred during } i\text{th experiment} \\ 0 & \text{otherwise} \end{cases}.$$

We then estimate p_A using the sample mean of the X_i,

$$\hat{p}_A = \frac{1}{n} \sum_{i=1}^{n} X_i.$$

   (a) Assuming n is large enough so that the central limit theorem applies, find an expression for $\Pr(|\hat{p}_A - p_A| < \varepsilon)$.

   (b) Suppose we want to be 95 percent certain that our estimate is within ± 10 percent of the true value. That is, we want $\Pr(|\hat{p}_A - p_A| < 0.1p_A) = 0.95$. How large does n need to be? In other words, how many times do we need to run the experiment?

   (c) Let Y_n be the number of times that we observe the event A during our n repetitions of the experiment. That is, let $Y_n = X_1 + X_2 + \cdots + X_n$. Assuming that n is chosen according to the results of part (b), find an expression for the average number of times the event A is observed, $E[Y_n]$. Show that for rare events (i.e., $p_A \ll 1$), $E[Y_n]$ is essential independent of p_A. Thus, even if we have no idea about the true value of p_A, we can run the experiment until we observe the event A for a predetermined number of times and be assured of a certain degree of accuracy in our estimate of p_A.

7.12 Suppose we wish to estimate the probability, p_A, of some event A as outlined in the previous exercise. As motivated by the result of part (c) of Exercise 7.11, suppose we repeat our experiment for a random number of trials, N. In particular, we run the experiment until we observe the event A exactly m times and then form the estimate of p_A according to

$$\hat{p}_A = \frac{m-1}{N-1}.$$

Here, the random variable N represents the number of trials until the mth occurrence of A.

(a) Find $E[\hat{p}_A]$. Is this estimate unbiased?

(b) Would it be better to use $\hat{p}_A = \dfrac{m}{N}$ as an estimate?

7.13 Independent samples are taken of a random variable X. If the PDF of X is uniform with amplitude $\sqrt{3}$ over the interval $[-1/\sqrt{12}, 1/\sqrt{12})$ and zero elsewhere, then approximate the density of the sample mean with a normal density, assuming the number of samples is large. Write the approximation as an equation.

7.14 A company manufactures 5-volt power supplies. However, since there are manufacturing tolerances, there are variations in the voltage design. The standard deviation in the design voltage is 5 percent. Using a 99 percent confidence level, determine whether or not the following samples fall within the confidence interval:

(a) 100 samples, the estimate of $\mu_X = 4.7$,

(b) 100 samples, the estimate of $\mu_X = 4.9$,

(c) 100 samples, the estimate of $\mu_X = 5.4$.

Hint: Refer to Equation 7.39.

7.15 You collect a sample size N_1 of data and find that a 90% confidence interval has width, w. What should the sample size N_2 be to increase the confidence level to 99.9% and yet maintain the same interval width, w?

7.16 Company A manufactures computer applications boards. They are concerned with the mean time between (or before) failures (MTBF), which they regularly measure. Denote the sample MTBF as $\hat{\mu}_M$ and the true MTBF as μ_M. Determine the number of failures that must be measured before $\hat{\mu}_M$ lies within 20 percent of the true μ_M with a 90 percent probability. Assume the PDF is exponential (i.e., $f_M(x) = (1/\mu_M) \exp(-x/\mu_M) u(x)$).

7.17 In this exercise, a proof of Equation 7.73 is constructed. Write the random sum as

$$S_N = \sum_{i=1}^{N} Z_i = \sum_{i=1}^{\infty} Y_i Z_i,$$

where Y_i is a Bernoulli random variable in which $Y_i = 1$ if $N \geq i$ and $Y_i = 0$ if $N < i$.

(a) Prove that Y_i and Z_i are independent and hence,

$$E[S_N] = \sum_{i=1}^{\infty} E[Y_i]E[Z_i].$$

(b) Prove that the equation of part (a) simplifies to

$$E[S_N] = E[Z_i]E[N].$$

MATLAB Exercises

7.18 Let X, Y, and Z be independent Gaussian random variables with equal means of $\mu = 3$ and variances of $\sigma^2 = 4$. Estimate the mean and variance of $W = \sqrt{X^2 + Y^2 + Z^2}$ by constructing a large number of realizations of this random variable in MATLAB and then computing the sample mean and sample variance. How many samples of the random variable were needed before the sample mean and sample variance seemed to converge to a fairly accurate estimate? (To answer this, you must define what you mean by "fairly accurate.")

7.19 For the random variable W described in Exercise 7.18, form an estimate of the CDF by following the procedure outlined in Example 7.5. Also, form an estimate of the PDF of this random variable. Explain the procedure you used to estimate the PDF.

7.20 A player engages in the following dice tossing game ("craps"). Two dice are rolled. If the player rolls the dice such that the sum is either 7 or 11, he immediately wins the game. If the sum is 2, 3, or 12, he immediately loses. If he rolls a 4, 5, 6, 8, 9, or 10, this number is called the "point" and the player continues to roll the dice. If he is able to roll the point again before he rolls a 7, he wins. If he rolls a 7 before he rolls the point again, he loses. Write a MATLAB program to simulate this dice game and estimate the probability of winning.

7.21 Let X_i, $i = 1, 2, \ldots, n$ be a sequence of IID random variables uniformly distributed over (0, 1). Suppose we form the sum $Z = \sum_{i=1}^{n} X_i$. First, find the mean and variance of Z. Then write a MATLAB program to compute the exact PDF of Z. Compare the exact PDF with a Gaussian PDF of the same mean and variance. Over what range of Z is the Gaussian approximation of the PDF within 1 percent of the true PDF? Repeat this problem for $n = 5, 10, 20, 50$, and 100.

7.22 Suppose you are given an observation of sample values of a sequence of random variables, x_n, $n = 1, 2, 3, \ldots, m$. Write a MATLAB program to plot these data points along with a least squares curve fit to the data (see the results of Exercise 7.8). Run your program using the following sequence:

$$(0, 1, 0, -1, 2, -3, 5, 0, -7, 8).$$

Random Processes

This chapter introduces the concept of a random process. Most of the treatment in this text views a random process as a random function of time. However, time need not be the independent variable. We can also talk about a random function of position, in which case there may be two or even three independent variables and the function is more commonly referred to as a random field. The concept of a random process allows us to study systems involving signals that are not entirely predictable. These random signals play a fundamental role in the fields of communications, signal processing, control systems, and many other engineering disciplines. This and the following chapters will extend the study of signal and system theory to include randomness. We introduce some basic concepts, terminologies, notations, and tools for studying random processes in this chapter and present several important examples of random processes as well.

8.1 Definition and Classification of Processes

In the study of deterministic signals, we often encounter four types or classes of signals:

(1) *Continuous time and continuous amplitude signals* are a function of a continuous independent variable, time. The range of the amplitude of the function is also continuous.

(2) *Continuous time and discrete amplitude signals* are a function of a continuous independent variable, time—but the amplitude is discrete.

(3) *Discrete time and continuous amplitude signals* are functions of a quantized or discrete independent time variable, while the range of amplitudes is continuous.

(4) *Discrete time and discrete amplitude signals* are functions where both the independent time variable and the amplitude are discrete.

In this text, we write a continuous function of time as $x(t)$, where t is the continuous time variable. For discrete time signals, the time variable is typically limited to regularly spaced discrete points in time, $t = nt_0$. In this case, we use the notation $x[n] = x(nt_0)$ to represent the discrete sequence of numbers. Most of the discussion that follows is presented in terms of continuous time signals, but the conversion to the discrete time case will be straightforward in most cases.

Recall from Chapter 3 that a random variable, X, is a function of the possible outcomes, ζ, of an experiment. Now, we would like to extend this concept so that a function of time $x(t)$ (or $x[n]$ in the discrete time case) is assigned to every outcome, ζ, of an experiment. The function, $x(t)$, may be real or complex and it can be discrete or continuous in amplitude. Strictly speaking, the function is really a function of two variables, $x(t, \zeta)$, but to keep the notation simple, we typically do not explicitly show the dependence on the outcome, just as we have not in the case of random variables. The function $x(t)$ may have the same general dependence on time for every outcome of the experiment or each outcome could produce a completely different waveform. In general, the function $x(t)$ is a member of an *ensemble* (family, set, collection) of functions. Just as we did for random variables, an ensemble of member functions, $X(t)$, is denoted with an upper case letter. Thus, $X(t)$ represents the random process, while $x(t)$ is one particular member or *realization* of the random process. In summary, we have the following definition of a random process:

DEFINITION 8.1: A random process is a function of the elements of a sample space, S, as well as another independent variable, t. Given an experiment, E, with sample space, S, the random process, $X(t)$, maps each possible outcome, $\zeta \in S$, to a function of t, $x(t, \zeta)$, as specified by some rule.

EXAMPLE 8.1: Suppose an experiment consists of flipping a coin. If the outcome is heads, $\zeta = H$, the random process takes on the functional form $x_H(t) = \sin(\omega_0 t)$; whereas, if the outcome is tails, $\zeta = T$, the realization $x_T(t) = \sin(2\omega_0 t)$ occurs, where ω_0 is some fixed frequency. The two realizations of this random process are illustrated in Figure 8.1.

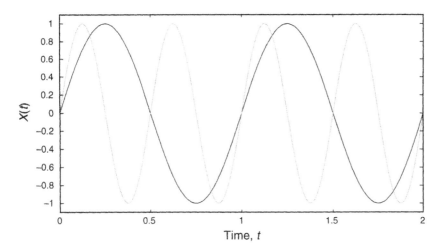

Figure 8.1 Member functions for the random process of Example 8.1.

The random process in Example 8.1 actually has very little randomness. There are only two possible realizations of the random process. Example 8.2 illustrates a property of sinusoids.

EXAMPLE 8.2: Now suppose that an experiment results in a random variable A that is uniformly distributed over $[0, 1)$. A random process is then constructed according to $X(t) = A \sin(\omega_0 t)$. Since the random variable is continuous, there are an uncountably infinite number of realizations of the random process. A few are shown in Figure 8.2. Given

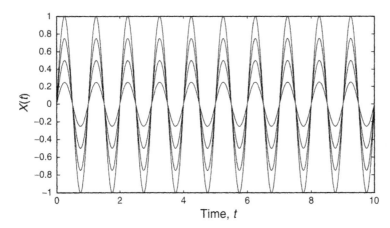

Figure 8.2 Some member functions for the random process of Example 8.2.

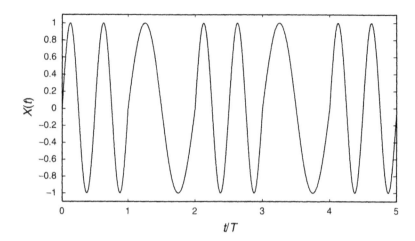

Figure 8.3 One possible realization for the random process of Example 8.3.

two observations of the realization of the random process at points in time, $X(t_1)$ and $X(t_2)$ one could determine the rest of the realization (as long as $\omega_0 t_1 \neq n\pi$ and $\omega_0 t_2 \neq n\pi$).

EXAMPLE 8.3: This example is a generalization of that given in Example 8.1. Suppose now that the experiment consists of flipping a coin repeatedly and observing the sequence of outcomes. The random process $X(t)$ is then constructed as $X(t) = \sin(\Omega_i t)$, $(i - 1)T \leq t < iT$, where $\Omega_i = \omega_o$ if the ith flip of the coin results in "heads" and $\Omega_i = 2\omega_o$ if the ith flip of the coin results in "tails." One possible realization of this random process is illustrated in Figure 8.3. This is the sort of signal that might be produced by a frequency shift keying (FSK) modem. In that application, the frequencies are not determined by coin tosses, but by random data bits instead.

EXAMPLE 8.4: As an example of a random process that is discrete in amplitude but continuous in time, we present the so-called random telegraph process. Let $T_1, T_2, T_3, \ldots$ be a sequence of IID random variables, each with an exponential distribution.

$$f_T(s) = \lambda e^{-\lambda s} u(s).$$

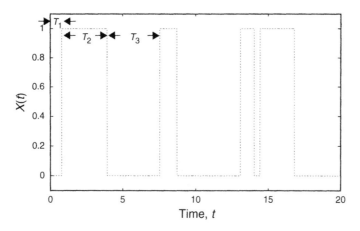

Figure 8.4   One possible realization for the random telegraph signal of Example 8.4.

At any time instant, the random telegraph signal, $X(t)$, takes on one of two possible states, $X(t) = 0$ or $X(t) = 1$. Suppose the process starts (at time $t = 0$) in the zero state. It then remains in that state for a time interval equal to T_1, at which point it switches to the state $X(t) = 1$. The process remains in that state for another interval of time equal in length to T_2 and then switches states again. The process then continues to switch after waiting for time intervals specified by the sequence of exponential random variables. One possible realization is shown in Figure 8.4. The MATLAB code for generating such a process follows.

```
N=10;                    % Number of switches in
                         realization.
Fs=100;                  % Sample rate (samples per second).
lambda=1/2;              % Switching rate (switches per
                         second).

X=[];
S=rand(1,N);             % Uniform random variables.
T=-log(S)/lambda;        % Transform to exponential RVs.
V=cumsum(T);             % Switching times.
state=0; Nsold=1;
for k=1:N
  Nsnew=ceil(V(k)*Fs);   % New switching time.
  Ns=Nsnew-Nsold;        % Number of samples in current.
                         % Switching interval.
  X=[X state*ones(1,Ns)];
  state=1-state;         % Switch state.
  Nsold=Nsnew;
end
```

```
t=[1:length(X)]/Fs;          % Time axis.
plot(t,X)                    % Plot results.
xlabel('time, t'); ylabel('X(t)')
axis([0 max(t) -0.1 1.1])    % Manual scale of axes.
```

EXAMPLE 8.5: As an example of a discrete time random process, suppose each outcome of an experiment produces a sequence of IID, zero-mean Gaussian random variables, $W_1, W_2, W_3, \ldots$. A discrete-time random process $X[n]$ could be constructed according to

$$X[n] = X[n-1] + W_n,$$

with the initial condition $X[0] = 0$. The value of the process at each point in time is equal to the value of the process at the previous point in time plus a random increase (decrease) which follows a Gaussian distribution. Note also that $X[n]$ is merely the sum of the first n terms in the sequence of W_i. A sample realization of this random process is shown in Figure 8.5. The MATLAB code for generating this process is provided here. The reader is encouraged to run this program several times to see several different realizations of the same random process.

```
N=25;                        % Number of time instants in process.
W=randn(1,N);                % Gaussian random variables.
```

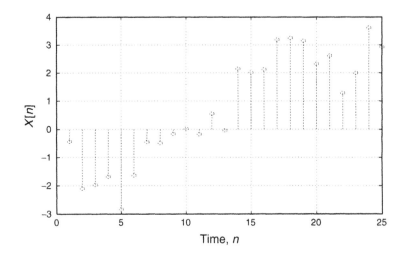

Figure 8.5   One possible realization for the discrete-time random process of Example 8.5.

```
X=[0 cumsum(W)];            % Samples of X[n].
stem([0:N],X,'o')           % Plot realization of X[n].
xlabel('time, n'); ylabel('X[n]');
```

8.2 Mathematical Tools for Studying Random Processes

As with random variables, we can mathematically describe a random process in terms of a cumulative distribution function, a probability density function, or a probability mass function. In fact, given a random process, $X(t)$, which is sampled at some specified point in time, $t = t_k$, the result is a random variable, $X_k = X(t_k)$. This random variable can then be described in terms of its PDF, $f_X(x_k; t_k)$. Note that an additional time variable has been added to the PDF. This is necessary due to the fact that the PDF of the sample of the random process may depend on when the process is sampled. If desired, the CDF or PMF can be used rather than the PDF to describe the sample of the random process.

EXAMPLE 8.6: Consider the random telegraph signal of Example 8.4. Since this process is binary-valued, any sample will be a Bernoulli random variable. The only question is, What is the probability that $X_k = X(t_k)$ is equal to 1 (or 0)? Suppose that there are exactly n switches in the time interval $[0, t_k)$. Then $X(t_k) = n \bmod 2$. Stated another way, define $S_n = T_1 + T_2 + \cdots + T_n$. There will be exactly n switches in the time interval $[0, t_k)$ provided that $S_n < t_k < S_{n+1}$. Hence,

$$\Pr(n \text{ switches in } [0, t_k)) = \Pr(S_n < t_k < S_{n+1})$$

$$= \int \Pr(S_n < t_k < S_{n+1}|S_n = s)f_{S_n}(s)\,ds$$

$$= \int_0^{t_k} \Pr(t_k < S_{n+1}|S_n = s)f_{S_n}(s)\,ds = \int_0^{t_k} \Pr(T_{n+1} > t_k - s)f_{S_n}(s)\,ds.$$

Since the T_i are IID and exponential, S_n will follow a Gamma distribution. Using the Gamma PDF for $f_{S_n}(s)$ and the fact that

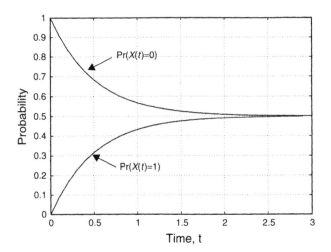

Figure 8.6 Time dependence of the PMF for the random telegraph signal.

$\Pr(T_{n+1} > t_k - s) = \exp(-\lambda(t_k - s))$ results in

$$\Pr(n \text{ switches in } [0, t_k)) = \int_0^{t_k} e^{-\lambda(t_k - s)} \frac{\lambda^n s^{n-1}}{(n-1)!} e^{-\lambda s} \, ds$$

$$= \frac{\lambda^n e^{-\lambda t_k}}{(n-1)!} \int_0^{t_K} s^{n-1} \, ds = \frac{(\lambda t_k)^n}{n!} e^{-\lambda t_k}.$$

So, it is seen that the number of switches in the interval $[0, t_k)$ follows a Poisson distribution. The sample of the random process will be equal to 0 if the number of switches is even. Hence,

$$\Pr(X(t_k) = 0) = \sum_{n \text{ even}} \Pr(n \text{ switches in } [0, t_k))$$

$$= \sum_{n \text{ even}} \frac{(\lambda t_k)^n}{n!} e^{-\lambda t_k} = e^{-\lambda t_k} \cosh(\lambda t_k) = \tfrac{1}{2} + \tfrac{1}{2} e^{-2\lambda t_k}.$$

Likewise,

$$\Pr(X(t_k) = 1) = \sum_{n \text{ odd}} \Pr(n \text{ switches in } [0, t_k))$$

$$= \sum_{n \text{ odd}} \frac{(\lambda t_k)^n}{n!} e^{-\lambda t_k} = e^{-\lambda t_k} \sinh(\lambda t_k) = \tfrac{1}{2} - \tfrac{1}{2} e^{-2\lambda t_k}.$$

The behavior of this distribution as it depends on time should make intuitive sense. For very small values of t_k, it is most likely that there are no switches in the interval $[0, t_k)$, in which case $\Pr(X(t_k) = 0)$ should be close to 1. On the other hand, for large t_k, many switches will likely occur and it should be almost equally likely that the process will take on the values of 0 or 1.

EXAMPLE 8.7: Now consider the PDF of a sample of the discrete-time process of Example 8.5. Note that since $X[n]$ is formed by summing n IID Gaussian random variables, $X[n]$ will itself be a Gaussian random variable with mean of $E[X[n]] = n\mu_W$ and variance $\text{Var}(X[n]) = n\sigma_W^2$. In this case, since the W_i were taken to be zero mean, the PDF of $X[n]$ is

$$f_X(x; n) = \frac{1}{\sqrt{2\pi n\sigma_W^2}} \exp\left(-\frac{x^2}{2n\sigma_W^2}\right).$$

Once again, we see that for this example, the form of the PDF does indeed depend on when the sample is taken.

The PDF (or CDF or PMF) of a sample of a random process taken at an arbitrary point in time goes a long way toward describing the random process, but it is not a complete description. To see this, consider two samples, $X_1 = X(t_1)$ and $X_2 = X(t_2)$, taken at two arbitrary points in time. The PDF, $f_X(x; t)$, describes both X_1 and X_2, but it does not describe the relationship between X_1 and X_2. For some random processes, it might be reasonable to expect that X_1 and X_2 would be highly correlated if t_1 is near t_2, while X_1 and X_2 might be virtually uncorrelated if t_1 and t_2 are far apart. To characterize relationships of this sort, a joint PDF of the two samples would be needed. Hence, it would be necessary to construct a joint PDF of the form $f_{X_1,X_2}(x_1, x_2; t_1, t_2)$. This is referred to as a *second order PDF* of the random process $X(t)$.

Continuing with this reasoning, in order to completely describe the random process, it is necessary to specify an nth order PDF for an arbitrary n. That is, suppose random process is sampled at time instants $t_1, t_2, \ldots, t_n$, producing the random variables $X_1 = X(t_1), X_2 = X(t_2), \ldots, X_n = X(t_n)$. The joint PDF of the n samples, $f_{X_1,X_2,\ldots,X_n}(x_1, x_2, \ldots, x_n; t_1, t_2, \ldots, t_n)$, for an arbitrary n and arbitrary sampling times will give a complete description of the random process. In order to make this notation more compact, the vectors $\mathbf{X} = (X_1, X_2, \ldots, X_n)^T$, $\mathbf{x} = (x_1, x_2, \ldots, x_n)^T$, and $\mathbf{t} = (t_1, t_2, \ldots, t_n)^T$ are introduced and the nth order joint PDF is written as $f_{\mathbf{X}}(\mathbf{x}; \mathbf{t})$.

Unfortunately, for many realistic random processes, the prospects of writing down an nth order PDF is rather daunting. One notable exception is the Gaussian random process, which will be described in detail in Section 8.5. However, for most other cases, specifying a joint PDF of n samples may be exceedingly difficult, and hence it is necessary to resort to a simpler but less complete description of the random process. The simplest is the mean function of the process.

DEFINITION 8.2: The *mean function* of a random process is simply the expected value of the process. For continuous time processes this is written as

$$\mu_X(t) = E[X(t)] = \int_{-\infty}^{\infty} x f_X(x; t)\, dx, \qquad (8.1)$$

while for discrete time processes, the following notation is used:

$$\mu_X[n] = E[X[n]] = \int_{-\infty}^{\infty} x f_X(x; n)\, dx. \qquad (8.2)$$

In general, the mean of a random process may change with time, but in many cases, this function is constant. Also, it is noted that only the first order PDF of the process is needed to compute the mean function.

EXAMPLE 8.8: Consider the random telegraph process of Example 8.4. It was shown in Example 8.6 that the first order PMF of this process was described by a Bernoulli distribution with

$$\Pr(X(t) = 1) = \tfrac{1}{2} - \tfrac{1}{2}\exp(-\lambda t).$$

The mean function then follows as

$$\mu_X(t) = E[X(t)] = 1^* \Pr(X(t) = 1) + 0^* \Pr(X(t) = 0) = \tfrac{1}{2} - \tfrac{1}{2}\exp(-\lambda t).$$

EXAMPLE 8.9: Next, consider the sinusoidal random process of Example 8.2 where $X(t) = A\sin(\omega_o t)$ and A was a uniform random variable over $[0, 1)$. In this case,

$$\mu_X(t) = E[X(t)] = E[A\sin(\omega_o t)] = E[A]\sin(\omega_o t) = \tfrac{1}{2}\sin(\omega_o t).$$

This example illustrates a very important concept in that quite often it is not necessary to explicitly evaluate the first order PDF of a random process in order to evaluate its mean function.

EXAMPLE 8.10: Now suppose the random process of the previous example is slightly modified. In particular, consider a sine-wave process where the random variable is the phase, Θ, which is uniformly distributed over $[0, 2\pi)$. That is, $X(t) = a \sin(\omega_o t + \Theta)$. For this example, the amplitude of the sine wave, a, is taken to be fixed (not random). The mean function is then

$$\mu_X(t) = E[X(t)] = E[a \sin(\omega_o t + \Theta)] = a \int f_\Theta(\theta) \sin(\omega_o t + \theta)\, d\theta$$

$$= \frac{a}{2\pi} \int_0^{2\pi} \sin(\omega_o t + \theta)\, d\theta = 0,$$

which is a constant. Why is the mean function of the previous example a function of time and this one is not? Consider the member functions of the respective ensembles for the two random processes.

EXAMPLE 8.11: Now consider a sinusoid with a random frequency $X(t) = \cos(2\pi F t)$, where F is a random variable uniformly distributed over some interval $(0, f_o)$. The mean function can be readily determined to be

$$\mu_X(t) = E[\cos(2\pi F t)] = \frac{1}{f_o} \int_0^{f_o} \cos(2\pi f t)\, df = \frac{\sin(2\pi f_o t)}{2\pi f_o t} = \text{sinc}\,(2f_o t).$$

We can also estimate the mean function through simulation. We provide some MATLAB code to produce many realizations of this random process. The mean function is then found by taking the sample mean of all the realizations created. The sample mean and the ensemble mean are shown in Figure 8.7. Naturally, more or less accuracy in the sample mean can be obtained by varying the number of realizations generated.

```
fo=2;                        % Maximum frequency.
N=1000;                      % Number of realizations.
t=[-4.995:0.01:4.995];       % Time axis.
F=fo*rand(N,1);              % Uniform frequencies.
x=cos(2*pi*F*t);             % Each row is a
                             % realization of process.
sample_mean=sum(x)/N;        % Compute sample mean.
```

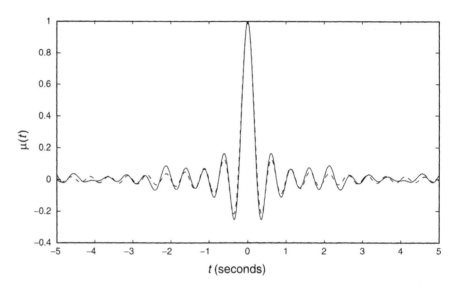

Figure 8.7   Comparison of the sample mean and ensemble mean for the sinusoid with random frequency of Example 8.11. The solid line is the sample mean, while the dashed line is the ensemble mean.

```
true_mean=sin(2*pi*fo*t)./(2*pi*fo*t);    % Compute ensemble
                                            mean.
plot(t,sample_mean,'-',t,true_mean,'--')  % Plot results.
xlabel('t (seconds)'); ylabel('mu(t)');
```

To partially describe the second order characteristics of a random process, the autocorrelation function is introduced.

DEFINITION 8.3: The *autocorrelation function*, $R_{XX}(t_1, t_2)$, of a continuous time random process, $X(t)$, is defined as the expected value of the product $X(t_1)X(t_2)$:

$$R_{XX}(t_1, t_2) = E[X(t_1)X(t_2)] = \int\!\!\int_{-\infty}^{\infty} x_1 x_2 f_{X_1, X_2}(x_1, x_2; t_1, t_2)\, dx_1\, dx_2. \qquad (8.3)$$

For discrete time processes, the autocorrelation function is

$$R_{XX}[n_1, n_2] = E[X[n_1]X[n_2]] = \int\!\!\int_{-\infty}^{\infty} x_1 x_2 f_{X_1, X_2}(x_1, x_2; n_1, n_2)\, dx_1\, dx_2. \qquad (8.4)$$

Naturally, the autocorrelation function describes the relationship (correlation) between two samples of a random process. This correlation will depend on when the samples are taken; thus, the autocorrelation function is, in general, a function of two time variables. Quite often we are interested in how the correlation between two samples depends on how far apart the samples are spaced. To explicitly draw out this relationship, define a time difference variable, $\tau = t_2 - t_1$, and the autocorrelation function can then be expressed as

$$R_{XX}(t, t + \tau) = E[X(t)X(t + \tau)], \tag{8.5}$$

where we have replaced t_1 with t to simplify the notation even further.

EXAMPLE 8.12: Consider the sine wave process with a uniformly distributed amplitude as described in Examples 8.2 and 8.9, where $X(t) = A\sin(\omega_o t)$. The autocorrelation function is found as

$$R_{XX}(t_1, t_2) = E[X(t_1)X(t_2)] = E[A^2 \sin(\omega_o t_1)\sin(\omega_o t_2)]$$

$$= \tfrac{1}{3}\sin(\omega_o t_1)\sin(\omega_o t_2),$$

or

$$R_{XX}(t, t + \tau) = \tfrac{1}{3}\sin(\omega_o t)\sin(\omega_o(t + \tau)).$$

EXAMPLE 8.13: Now consider the sine wave process with random phase of Example 8.10 where $X(t) = a\sin(\omega_o t + \Theta)$. Then

$$R_{XX}(t_1, t_2) = E[X(t_1)X(t_2)] = E[a^2 \sin(\omega_o t_1 + \theta)\sin(\omega_o t_2 + \theta)].$$

To aid in calculating this expected value, we use the trigonometric identity

$$\sin(x)\sin(y) = \tfrac{1}{2}\cos(x - y) - \tfrac{1}{2}\cos(x + y).$$

The autocorrelation then simplifies to

$$R_{XX}(t_1, t_2) = \frac{a^2}{2}E[\cos(\omega_o(t_2 - t_1))] + \frac{a^2}{2}E[\cos(\omega_o(t_1 + t_2 + 2\theta))]$$

$$= \frac{a^2}{2}\cos(\omega_o(t_2 - t_1)),$$

or

$$R_{XX}(t, t + \tau) = \frac{a^2}{2}\cos(\omega_o \tau).$$

Note that in this case, the autocorrelation function is only a function of the difference between the two sampling times. That is, it does not matter where the samples are taken, only how far apart they are.

EXAMPLE 8.14: Recall the random process of Example 8.5, where $X[n] = X[n-1] + W_n$, $X[0] = 0$ and the W_n were a sequence of IID zero mean Gaussian random variables. In this case, it is easier to calculate the autocorrelation function using the alternative expression,

$$X[n] = \sum_{i=1}^{n} W_i.$$

Then

$$R_{XX}[n_1, n_2] = E[X[n_1]X[n_2]] = E\left[\sum_{i=1}^{n_1} W_i \sum_{j=1}^{n_2} W_j\right] = \sum_{i=1}^{n_1}\sum_{j=1}^{n_2} E[W_i W_j].$$

Since the W_i are IID and zero-mean, $E[W_i W_j] = 0$ unless $i = j$. Hence,

$$R_{XX}[n_1, n_2] = \min(n_1, n_2)\sigma_W^2.$$

DEFINITION 8.4: The autocovariance function, $C_{XX}(t_1, t_2)$, of a continuous time random process, $X(t)$, is defined as the covariance of $X(t_1)$ and $X(t_2)$:

$$C_{XX}(t_1, t_2) = \text{Cov}(X(t_1), X(t_2)) = E[(X(t_1) - \mu_X(t_1))(X(t_2) - \mu_X(t_2))]. \qquad (8.6)$$

The definition is easily extended to discrete time random processes.

As with the covariance function for random variables, the autocovariance function can be written in terms of the autocorrelation function and the mean function:

$$C_{XX}(t_1, t_2) = R_{XX}(t_1, t_2) - \mu_X(t_1)\mu_X(t_2). \qquad (8.7)$$

Once the mean and autocorrelation functions of a random process have been computed, the autocovariance function is trivial to find.

The autocovariance function is helpful when studying random processes that can be represented as the sum of a deterministic signal, $s(t)$, plus a zero-mean noise process, $N(t)$. If $X(t) = s(t) + N(t)$, then the autocorrelation function of $X(t)$ is

$$R_{XX}(t_1, t_2) = E[(s(t_1) + N(t_1))(s(t_2) + N(t_2))] = s(t_1)s(t_2) + R_{NN}(t_1, t_2), \qquad (8.8)$$

using the fact that $\mu_N(t) = 0$. If the signal is strong compared to the noise, the deterministic part will dominate the autocorrelation function, and hence $R_{XX}(t_1, t_2)$ will not tell us much about the randomness in the process $X(t)$. On the other hand, the autocovariance function is

$$C_{XX}(t_1, t_2) = R_{XX}(t_1, t_2) - s(t_1)s(t_2) = R_{NN}(t_1, t_2) = C_{NN}(t_1, t_2). \qquad (8.9)$$

Hence, the autocovariance function allows us to isolate the noise that is the source of randomness in the process.

DEFINITION 8.5: For a pair of random processes $X(t)$ and $Y(t)$, the crosscorrelation function is defined as

$$R_{XY}(t_1, t_2) = E[X(t_1)Y(t_2)]. \qquad (8.10)$$

Likewise, the cross-covariance function is

$$C_{XY}(t_1, t_2) = E[(X(t_1) - \mu_X(t_1))(Y(t_2) - \mu_Y(t_2))]. \qquad (8.11)$$

EXAMPLE 8.15: Suppose $X(t)$ is a zero-mean random process with autocorrelation function $R_{XX}(t_1, t_2)$. A new process $Y(t)$ is formed by delaying $X(t)$ by some amount t_d. That is, $Y(t) = X(t - t_d)$. Then the crosscorrelation function is

$$R_{XY}(t_1, t_2) = E[X(t_1)Y(t_2)] = E[X(t_1)X(t_2 - t_d)] = R_{XX}(t_1, t_2 - t_d).$$

In a similar fashion, it is seen that $R_{YX}(t_1, t_2) = R_{XX}(t_1 - t_d, t_2)$ and $R_{YY}(t_1, t_2) = R_{XX}(t_1 - t_d, t_2 - t_d)$.

8.3 Stationary and Ergodic Random Processes

From the few simple examples given in the preceding section, we conclude that the mean function and the autocorrelation (or autocovariance) function can provide information about the temporal structure of a random process. We will delve into the properties of the autocorrelation function in more detail later in the chapter, but first the concepts of *stationarity* and *ergodicity* must be introduced.

DEFINITION 8.6: A continuous time random process $X(t)$ is *strict sense stationary* if the statistics of the process are invariant to a time shift. Specifically, for any time shift τ and any integer $n \geq 1$,

$$f_{X_1, X_2, \ldots, X_n}(x_1, x_2, \ldots, x_n; t_1, t_2, \ldots, t_n)$$

$$= f_{X_1, X_2, \ldots, X_n}(x_1, x_2, \ldots, x_n; t_1 + \tau, t_2 + \tau, \ldots, t_n + \tau). \qquad (8.12)$$

In general, it is quite difficult to show that a random process is strict sense stationary since to do so, one needs to be able to express the general nth order PDF. On the other hand, to show that a process is not strict sense stationary, one needs to show only that one PDF of any order is not invariant to a time shift. One example of a process that can be shown to be stationary in the strict sense is an IID process. That is, suppose $X(t)$ is a random process that has the property that $X(t)$ has an identical distribution for any t and that $X(t_1)$ and $X(t_2)$ are independent for any $t_1 \neq t_2$. In this case,

$$f_{X_1, X_2, \ldots, X_n}(x_1, x_2, \ldots, x_n; t_1, t_2, \ldots, t_n) = \prod_{i=1}^{n} f_{X_i}(x_i; t_i). \qquad (8.13)$$

Since the nth order PDF is the product of first order PDFs and the first order PDF is invariant to a time shift, then the nth order PDF must be invariant to a time shift.

EXAMPLE 8.16: Consider the sinusoidal process with random amplitude from Example 8.2, where $X(t) = A\sin(\omega_o t)$. This process is clearly not stationary since if we take any realization of the process, $x(t) = a\sin(\omega_o t)$, then a time shift $x(t + \tau) = a\sin(\omega_o(t + \tau))$ would not be a realization in the original ensemble. Now suppose the process has a random phase rather than a random amplitude as in Example 8.10, resulting in $X(t) = a\sin(\omega_o t + \Theta)$. It was already shown in Example 8.10 that $\mu_X(t) = 0$ for this process, and hence the mean function is invariant to a time shift. Furthermore, in Example 8.13 it was shown that $R_{XX}(t, t+\tau) = (a^2/2)\cos(\omega_o\tau)$, and hence the autocorrelation function is also invariant to a time shift. It is not difficult to show that the first order PDF follows an arcsine distribution:

$$f_X(x; t) = \frac{1}{\pi\sqrt{1 - x^2}}, \quad -1 < x < 1,$$

and it is also independent of time and thus invariant to a time shift. It seems that this process might be stationary in the strict sense, but it would be rather cumbersome to prove it because the nth order PDF is difficult to specify.

As was seen in Example 8.16, it may be possible in some examples to determine that some of the statistics of a random process are invariant to time shifts, but determining stationarity in the strict sense may be too big of a burden. In those cases, we often settle for a looser form of stationarity.

DEFINITION 8.7: A random process is *wide sense stationary* (WSS) if the mean function and autocorrelation function are invariant to a time shift. In particular, this implies that

$$\mu_X(t) = \mu_X = \text{constant}, \tag{8.14}$$

$$R_{XX}(t, t + \tau) = R_{XX}(\tau) \quad \text{(function only of } \tau\text{).} \tag{8.15}$$

All strict sense stationary random processes are also WSS, provided that the mean and autocorrelation function exist. The converse is not true. A WSS process does not necessarily need to be stationary in the strict sense. We refer to a process that is not WSS as *nonstationary*.

EXAMPLE 8.17: Suppose we form a random process $Y(t)$ by modulating a carrier with another random process, $X(t)$. That is, let $Y(t) = X(t) \cos(\omega_0 t + \Theta)$ where Θ is uniformly distributed over $[0, 2\pi)$ and independent of $X(t)$. Under what conditions is $Y(t)$ WSS? To answer this, we calculate the mean and autocorrelation function of $Y(t)$.

$$\mu_Y(t) = E[X(t) \cos(\omega_0 t + \Theta)] = E[X(t)]E[\cos(\omega_0 t + \Theta)] = 0;$$

$$R_{YY}(t, t+\tau) = E[X(t)X(t+\tau)\cos(\omega_0 t)\cos(\omega_0(t+\tau))]$$

$$= E[X(t)X(t+\tau)]\left\{ \tfrac{1}{2}\cos(\omega_0\tau) + \tfrac{1}{2}E[\cos(\omega_0(2t+\tau) + 2\Theta)] \right\}$$

$$= \tfrac{1}{2}R_{XX}(t, t+\tau)\cos(\omega_0\tau)$$

While the mean function is a constant, the autocorrelation is not necessarily only a function of τ. The process $Y(t)$ will be WSS provided that $R_{XX}(t, t+\tau) = R_{XX}(\tau)$. Certainly if $X(t)$ is WSS, then $Y(t)$ will be as well.

EXAMPLE 8.18: Let $X(t) = At + B$ where A and B are independent random variables, both uniformly distributed over the interval $(-1, 1)$. To determine whether this process is WSS, calculate the mean and autocorrelation functions:

$$\mu_X(t) = E[At + B] = E[A]t + E[B] = 0;$$

$$R_{XX}(t, t + \tau) = E[(At + B)(A(t + \tau) + B)]$$

$$= E[A^2]t(t + \tau) + E[B^2] + E[AB](2t + \tau) = \tfrac{1}{3}t(t + \tau) + \tfrac{1}{3}.$$

Clearly, this process is not WSS.

Many of the processes we deal with are WSS and hence have a constant mean function and an autocorrelation function that depends only on a single time variable. Hence, in the remainder of the text, when a process is known to be WSS or if we are assuming it to be WSS, then we will represent its autocorrelation function by $R_{XX}(\tau)$. If a process is nonstationary or if we do not know if the process is WSS, then we will explicitly write the autocorrelation function as a function of two variables, $R_{XX}(t, t + \tau)$. For example, if we say that a process has a mean function of $\mu_X = 1$ and an autocorrelation function, $R_{XX}(\tau) = \exp(-|\tau|)$, then the reader can infer that the process is WSS, even if it is not explicitly stated.

In order to calculate the mean or autocorrelation function of a random process, it is necessary to perform an ensemble average. In many cases, this may not be possible as we may not be able to observe all realizations (or a large number of realizations) of a random process. In fact, quite often we may be able to observe only a single realization. This would occur in situations where the conditions of an experiment cannot be duplicated and hence the experiment is not repeatable. Is it possible to calculate the mean and/or autocorrelation function from a single realization of a random process? The answer is, sometimes, depending on the nature of the process.

To start with, consider the mean. Suppose a WSS random process $X(t)$ has a mean μ_X. We are able to observe one realization of the random process, $x(t)$, and wish to try to determine μ_X from this realization. One obvious approach would be to calculate the time average[1] of the realization:

$$\langle x(t) \rangle = \lim_{t_o \to \infty} \frac{1}{2t_o} \int_{-t_o}^{t_o} x(t) \, dt. \tag{8.16}$$

---

[1]Throughout the text, angular brackets $\langle \, \rangle$ are used as a shorthand notation to represent the time average operator.

However, it is not obvious if the time average of one realization is necessarily equal to the ensemble average. If the two averages are the same, then we say that the random process is *ergodic in the mean*.

One could take the same approach for the autocorrelation function. Given a single realization, $x(t)$, form the time-average autocorrelation function:

$$\Re_{xx}(\tau) = \langle x(t)x(t+\tau) \rangle = \lim_{t_0 \to \infty} \frac{1}{2t_0} \int_{-t_0}^{t_0} x(t)x(t+\tau)\,dt. \tag{8.17}$$

If $\Re_{xx}(\tau) = R_{XX}(\tau)$ for any realization, $x(t)$, then the random process is said to be *ergodic in the autocorrelation*. In summary, we have the following definition of ergodicity:

DEFINITION 8.8: A WSS random process is *ergodic* if ensemble averages involving the process can be calculated using time averages of any realization of the process. Two limited forms of ergodicity are

(1) ergodic in the mean: $\langle x(t) \rangle = E[X(t)]$;
(2) ergodic in the autocorrelation: $\langle x(t+\tau) \rangle = E[X(t)X(t+\tau)]$.

EXAMPLE 8.19: As a simple example, suppose $X(t) = A$, where A is a random variable with some arbitrary PDF $f_A(a)$. Note that this process is stationary in the strict sense since for any realization, $x(t) = x(t+\tau)$. That is, not only are the statistics of the process invariant to time shifts, but every realization is also invariant to any time shift. If we take the time average of a single realization, $x(t) = a$, we get $\langle x(t) \rangle = a$. Hence, each different realization will lead to a different time average and will not necessarily give the ensemble mean, μ_A. Although this process is stationary in the strict sense, it is not ergodic in any sense.

EXAMPLE 8.20: Now consider the sinusoid with random phase $X(t) = a\sin(\omega_0 t + \Theta)$, where Θ is uniform over $[0, 2\pi)$. It was demonstrated in Example 8.13 that this process is WSS. But is it ergodic? Given any realization $x(t) = a\sin(\omega_0 t + \theta)$, the time average is $\langle x(t) \rangle = \langle a\sin(\omega_0 t + \theta) \rangle = 0$. That is, the average value of any sinusoid is zero. So this process is ergodic in the mean since the ensemble average of this process was also zero. Next, consider the sample autocorrelation function:

$$\langle x(t)x(t+\tau) \rangle = a^2 \langle \sin(\omega_0 t + \theta)\sin(\omega_0 t + \omega_0 \tau + \theta) \rangle$$

$$= \frac{a^2}{2}\langle \cos(\omega_0 \tau) \rangle - \frac{a^2}{2}\langle \cos(2\omega_0 t + \omega_0 \tau + 2\theta) \rangle = \frac{a^2}{2}\cos(\omega_0 \tau).$$

This also is exactly the same expression obtained for the ensemble averaged autocorrelation function. Hence, this process is also ergodic in the autocorrelation.

EXAMPLE 8.21: For a process that is known to be ergodic, the autocorrelation function can be estimated by taking a sufficiently long time average of the autocorrelation function of a single realization. We demonstrate this via MATLAB for a process that consists of a sum of sinusoids of fixed frequencies and random phases,

$$X(t) = \sum_{k=1}^{n} \cos(2\pi f_k t + \theta_k),$$

where the θ_k are IID and uniform over $(0, 2\pi)$. For an arbitrary signal $x(t)$, we note the similarity between the time averaged autocorrelation function and the convolution of $x(t)$ and $x(-t)$. If we are given a single realization, $x(t)$, which lasts only for the time interval, $(-t_o, t_o)$, then these two expressions are given by

$$\langle x(t)x(t+\tau)\rangle = \frac{1}{2t_o - \tau} \int_{-t_o}^{t_o-\tau} x(t)x(t+\tau)\, dt,$$

$$x(t) * x(-t) = \int_{-t_o}^{t_o-\tau} x(t)x(t+\tau)\, dt,$$

for $\tau > 0$. In general, we have the relationship

$$\langle x(t)x(t+\tau)\rangle = \frac{x(t) * x(-t)}{2t_o - |\tau|}.$$

By using the MATLAB convolution function, conv, the time averaged autocorrelation can easily be computed. This is demonstrated in the code that follows. Figure 8.8 shows a comparison between the ensemble averaged autocorrelation and the time averaged autocorrelation taken from a single realization. From the figure, it is noted that the agreement between the two is good for $\tau \ll t_o$ but not good when $\tau \sim t_o$. This is due to the fact that when τ approaches t_o, the time window over which the time average is computed gets too small to produce an accurate estimate.

```
N=4;                                    % Number of sinusoids.
to=5;                                   % Time duration.
```

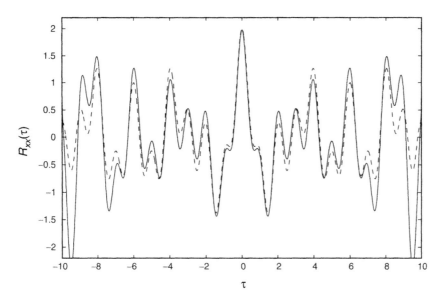

Figure 8.8 Comparison of the time-average autocorrelation and the ensemble-average autocorrelation for the sum of sinusoids process of Example 8.21. The solid line is the time-average autocorrelation, while the dashed line is the ensemble-average autocorrelation.

```
Ts=0.01;                                  % Sample interval.
t=[-to:Ts:to];                            % Time axis.
tau=[-2*to:Ts:2*to];                      % Tau axis.
theta=rand(1,N);                          % Random phases.
f=1./[1:N];                               % Frequencies (not
                                          random).

x=zeros(size(t));
True_Rxx=zeros(size(tau));
for k=1:N
  x=x+cos(2*pi*f(k)*t+2*pi*theta(k));     % Construct process.
  True_Rxx=True_Rxx+cos(2*pi*f(k)*tau)/2; % Compute Rxx(tau).
end
z=conv(x,fliplr(x));                      % x(t)*x(-t)
Rxx=Ts*z./(2*to-abs(tau));                % Time averaged Rxx.
plot(tau,Rxx,'-',tau,True_Rxx,'--')       % Plot results.
xlabel('tau'); ylabel('R_X_X(tau)')
axis([-2*to 2*to -1.1*N/2 1.1*N/2])
```

The previous examples show two different random processes, one that is ergodic and one that is not. What characteristics of a random process make it ergodic? To get some better insight toward answering this question, consider a discrete time process, $X[n]$, where each random variable in the sequence is IID and consider forming the time average:

$$\langle X[n] \rangle = \lim_{m \to \infty} \frac{1}{m} \sum_{n=1}^{m} X[n]. \tag{8.18}$$

The right-hand side of this equation is nothing more than the sample mean. By virtue of the law of large numbers, the limit will indeed converge to the ensemble mean of the random process, μ_X, and hence this process is ergodic in the mean. In this case, the time average converges to the ensemble average because each time sample of the random process gives an independent observation of the underlying randomness. If the samples were highly correlated, then taking more samples would not necessarily cause the sample mean to converge to the ensemble mean. So, it seems that the form of the autocorrelation function will play some role in determining if a random process is ergodic in the mean.

To formalize this concept, consider the time average of an arbitrary WSS random process:

$$\langle X(n) \rangle = \lim_{t_0 \to \infty} X_{t_0} \quad \text{where } X_{t_0} = \frac{1}{2t_0} \int_{-t_0}^{t_0} X(t) \, dt. \tag{8.19}$$

Note that X_{t_0} is a random variable with an expected value given by[2]

$$E[X_{t_0}] = E\left[\frac{1}{2t_0} \int_{-t_0}^{t_0} X(t) \, dt \right] = \frac{1}{2t_0} \int_{-t_0}^{t_0} E[X(t)] \, dt = \frac{1}{2t_0} \int_{-t_0}^{t_0} \mu_X \, dt = \mu_X. \tag{8.20}$$

Hence, X_{t_0} is an unbiased estimate of the true mean, but for the process to be ergodic in the mean, it is required that X_{t_0} converge to μ_X as $t_0 \to \infty$. This convergence will occur (in the mean square sense) if the variance of X_{t_0} goes to zero in the limit as $t_0 \to \infty$. To see under what conditions this occurs, we calculate the variance.

$$\text{Var}(X_{t_0}) = E[(X_{t_0} - \mu_X)^2] = E\left[\left(\frac{1}{2t_0} \int_{-t_0}^{t_0} (X(t) - \mu_X) \, dt \right)^2 \right]$$

$$= \frac{1}{4t_0^2} \int_{-t_0}^{t_0} \int_{-t_0}^{t_0} E[(X(t) - \mu_X)(X(s) - \mu_X)] \, dt \, ds = \frac{1}{4t_0^2} \int_{-t_0}^{t_0} \int_{-t_0}^{t_0} C_{XX}(t,s) \, dt \, ds. \tag{8.21}$$

---

[2]We exchange the order of expectation and integration since they are both linear operators.

Since the random process $X(t)$ is WSS, the autocovariance is only a function of $\tau = t - s$. As a result, the double integral can be converted to a single integral.[3] The result is

$$\text{Var}(X_{t_o}) = \frac{1}{2t_o} \int_{-2t_o}^{2t_o} \left(1 - \frac{|\tau|}{2t_o}\right) C_{XX}(\tau)\, d\tau = \frac{1}{t_o} \int_0^{2t_o} \left(1 - \frac{\tau}{2t_o}\right) C_{XX}(\tau)\, d\tau. \quad (8.22)$$

Thus, the random process will be ergodic in the mean if this expression goes to zero in the limit as $t_o \to \infty$. This proves the following theorem.

THEOREM 8.1: A continuous WSS random process $X(t)$ will be ergodic in the mean if

$$\lim_{t_o \to \infty} \frac{1}{t_o} \int_0^{2t_o} \left(1 - \frac{\tau}{2t_o}\right) C_{XX}(\tau)\, d\tau = 0. \quad (8.23)$$

One implication of Theorem 8.1 is that if $C_{XX}(\tau)$ tends toward a constant as $\tau \to \infty$, then that constant must be zero for the process to be ergodic. Stated in terms of the autocorrelation function, a sufficient condition for a process to be ergodic in the mean is that

$$\lim_{\tau \to \infty} R_{XX}(\tau) = \mu_X^2. \quad (8.24)$$

Similar relationships can be developed to determine when a process is ergodic in the autocorrelation, but that topic is beyond the intended scope of this text.

EXAMPLE 8.22: Consider the process $X(t) = A$ of Example 8.19. It is easily found that the autocovariance function of this process is $C_{XX}(\tau) = \sigma_A^2$ for all τ. Plugging this into the left-hand side of Equation 8.23 results in

$$\lim_{t_o \to \infty} \frac{1}{t_o} \int_0^{2t_o} \left(1 - \frac{\tau}{2t_o}\right) C_{XX}(\tau)\, d\tau = \lim_{t_o \to \infty} \frac{\sigma_A^2}{t_o} \int_0^{2t_o} \left(1 - \frac{\tau}{2t_o}\right) d\tau$$

$$= \lim_{t_o \to \infty} \frac{\sigma_A^2}{t_o} t_o = \sigma_A^2.$$

Since this limit is not equal to zero, the process clearly does not meet the condition for ergodicity. Next, consider the sinusoidal process of

---

[3]The procedure for doing this conversion will be described in detail in Chapter 10, where a similar integral will be encountered in the proof of the Wiener–Khintchine–Einstein theorem.

Example 8.20. In that case, the autocovariance function is $C_{XX}(\tau) = (a^2/2)\cos(\omega_0\tau)$ and the left-hand side of Equation 8.23 produces

$$\lim_{t_o\to\infty}\frac{1}{t_o}\int_0^{2t_o}\left(1-\frac{\tau}{2t_o}\right)C_{XX}(\tau)\,d\tau = \lim_{t_o\to\infty}\frac{a^2}{2t_o}\int_0^{2t_o}\left(1-\frac{\tau}{2t_o}\right)\cos(\omega_0\tau)\,d\tau$$

$$= \lim_{t_o\to\infty}a^2\left(\frac{1-\cos(2\omega_0 t_o)}{(2\omega_0 t_o)^2}\right) = 0.$$

So, even though this autocorrelation function does not approach zero in the limit at $\tau \to 0$ (it oscillates), it still meets the condition for ergodicity.

8.4 Properties of the Autocorrelation Function

Since the autocorrelation function, along with the mean, is considered to be a principal statistical descriptor of a WSS random process, we will now consider some properties of the autocorrelation function. It should quickly become apparent that not just any function of τ can be a valid autocorrelation function.

PROPERTY 8.4.1: The autocorrelation function evaluated at $\tau = 0$, $R_{XX}(0)$, is the *average normalized power* in the random process, $X(t)$.

To clarify this, note that $R_{XX}(0) = E[X^2(t)]$. Now suppose the random process $X(t)$ was a voltage measured at some point in a system. For a particular realization, $x(t)$, the instantaneous power would be $p(t) = x^2(t)/r$, where r is the impedance in Ohms (Ω). The average power (averaged over all realizations in the ensemble) would then be $P_{avg} = E[X^2(t)]/r = R_{XX}(0)/r$. If, on the other hand, $X(t)$ were a current rather than a voltage, then the average power would be $P_{avg} = R_{XX}(0)r$. From a systems level, it is often desirable not to concern ourselves with whether a signal is a voltage or a current. Accordingly, it is common to speak of a *normalized* power, which is the power measured using a 1-Ω impedance. With $r = 1$, the two expressions for average power are the same and equal to the autocorrelation function evaluated at zero.

PROPERTY 8.4.2: The autocorrelation function of a WSS random process is an even function; that is, $R_{XX}(\tau) = R_{XX}(-\tau)$.

This property can easily be established from the definition of autocorrelation. Note that $R_{XX}(-\tau) = E[X(t)X(t-\tau)]$. Since $X(t)$ is WSS, this expression is the same for any value of t. In particular, replace t in the previous expression with $t + \tau$ so that $R_{XX}(-\tau) = E[X(t+\tau)X(t)] = R_{XX}(\tau)$. As a result of this property, any function of τ which is not even cannot be a valid autocorrelation function.

PROPERTY 8.4.3: The autocorrelation function of a WSS random process is maximum at the origin; that is, $|R_{XX}(\tau)| \leq R_{XX}(0)$ for all τ.

This property is established using the fact that for any two random variables, X and Y,

$$(E[XY])^2 \leq E[X^2]E[Y^2]. \tag{8.25}$$

This fact was previously demonstrated in the proof of Theorem 5.4. Letting $X = X(t)$ and $Y = X(t + \tau)$ results in

$$R_{XX}^2(\tau) = \{E[X(t)X(t + \tau)]\}^2 \leq E[X^2(t)]E[X^2(t + \tau)] = R_{XX}^2(0). \tag{8.26}$$

Taking square roots of both sides results in Property 8.4.3.

PROPERTY 8.4.4: If $X(t)$ is ergodic and has no periodic components, then $\lim_{\tau \to \infty} R_{XX}(\tau) = \mu_X^2$.

PROPERTY 8.4.5: If $X(t)$ has a periodic component, then $R_{XX}(t)$ will have a periodic component with the same period.

From these properties it is seen that an autocorrelation function can oscillate, can decay slowly or rapidly, and can have a nonzero constant component. As the name implies, the auto-correlation function is intended to measure the extent of correlation of samples of a random process as a function of how far apart the samples are taken.

8.5 Gaussian Random Processes

One of the most important classes of random processes is the Gaussian random process, which is defined as follows.

DEFINITION 8.9: A random process, $X(t)$, for which any n samples, $X_1 = X(t_1), X_2 = X(t_2), \ldots, X_n = X(t_n)$, taken at arbitrary points in time $t_1, t_2, \ldots, t_n$, form a set of jointly Gaussian random variables for any $n = 1, 2, 3, \ldots$ is a Gaussian random process.

In vector notation, the vector of n samples, $X = [X_1, X_2, \ldots, X_n]^T$, will have a joint PDF given by

$$f_X(x) = \frac{1}{\sqrt{(2\pi)^n \det(C_{XX})}} \exp\left(-\tfrac{1}{2}(x - \mu_X)^T C_{XX}^{-1}(x - \mu_X)\right). \tag{8.27}$$

As with any joint Gaussian PDF, all that is needed to specify the PDF is the mean vector and the covariance matrix. When the vector of random variables consists of samples of a random process, to specify the mean vector, all that is needed is the mean function of the random process, $\mu_X(t)$, since that will give the mean for any sample time. Similarly, all that is needed to specify the elements of the covariance matrix, $C_{i,j} = \text{Cov}(X(t_i), X(t_j))$, would be the autocovariance function of the random process, $C_{XX}(t_1, t_2)$ or, equivalently, the autocorrelation function, $R_{XX}(t_1, t_2)$, together with the mean function. Hence, the mean and autocorrelation functions provide sufficient information to specify the joint PDF for any number of samples of a Gaussian random process. Note that since any nth order PDF is completely specified by $\mu_X(t)$ and $R_{XX}(t_1, t_2)$, if a Gaussian random process is WSS, then the mean and autocorrelation functions will be invariant to a time shift and hence any PDF will be invariant to a time shift. Therefore, any WSS Gaussian random process is also stationary in the strict sense.

EXAMPLE 8.23: Consider the random process $X(t) = A\cos(\omega_0 t) + B\sin(\omega_0 t)$, where A and B are independent, zero-mean Gaussian random variables with equal variances of σ^2. This random process is formed as a linear combination of two Gaussian random variables, and hence samples of this process are also Gaussian random variables. The mean and autocorrelation functions of this process are found as

$$\mu_X(t) = E[A\cos(\omega_0 t) + B\sin(\omega_0 t)] = E[A]\cos(\omega_0 t) + E[B]\sin(\omega_0 t) = 0,$$

$$R_{XXa}(t_1, t_2) = E[(A\cos(\omega_0 t_1) + B\sin(\omega_0 t_1))(A\cos(\omega_0 t_2) + B\sin(\omega_0 t_2))]$$

$$= E[A^2]\cos(\omega_0 t_1)\cos(\omega_0 t_2) + E[B^2]\sin(\omega_0 t_1)\sin(\omega_0 t_2)$$

$$+ E[AB]\{\cos(\omega_0 t_1)\sin(\omega_0 t_2) + \sin(\omega_0 t_1)\cos(\omega_0 t_2)\}$$

$$= \sigma^2\{\cos(\omega_0 t_1)\cos(\omega_0 t_2) + \sin(\omega_0 t_1)\sin(\omega_0 t_2)\}$$

$$= \sigma^2\cos(\omega_0(t_2 - t_1)).$$

Note that this process is WSS since the mean is constant and the autocorrelation function depends only on the time difference. Since the process is zero-mean, the first order PDF is that of a zero-mean Gaussian random variable:

$$f_X(x;t) = \frac{1}{\sqrt{2\pi\sigma^2}} \exp\left(-\frac{x^2}{2\sigma^2}\right).$$

This PDF is independent of time as would be expected for a stationary random process. Now consider the joint PDF of two samples, $X_1 = X(t)$ and $X_2 = X(t + \tau)$. Since the process is zero-mean, the mean vector is simply the all-zeros vector. The covariance matrix is then of the form

$$C_{XX} = \begin{bmatrix} R_{XX}(0) & R_{XX}(\tau) \\ R_{XX}(\tau) & R_{XX}(0) \end{bmatrix} = \sigma^2 \begin{bmatrix} 1 & \cos(\omega_o\tau) \\ \cos(\omega_o\tau) & 1 \end{bmatrix}.$$

The joint PDF of the two samples would then be

$$f_{X_1,X_2}(x_1,x_2;t,t+\tau) = \frac{1}{2\pi\sigma^2|\sin(\omega_o\tau)|} \exp\left(-\frac{x_1^2 - 2x_1x_2\cos(\omega_o\tau) + x_2^2}{2\sigma^2\sin^2(\omega_o\tau)}\right).$$

Note once again that this joint PDF is dependent only on time difference, τ, and not on absolute time t. Higher order joint PDFs could be worked out in a similar manner.

8.6 Poisson Processes

Consider a process $X(t)$ that counts the number of occurrences of some event in the time interval $[0, t)$. The event might be the telephone calls arriving at a certain switch in a public telephone network, customers entering a certain store, or the birth of a certain species of animal under study. Since the random process is discrete (in amplitude), we will describe it in terms of a probability mass function, $P_X(i;t) = \Pr(X(t) = i)$. Each occurrence of the event being counted is referred to as an *arrival*, or a *point*. These types of processes are referred to as *counting processes*, or *birth processes*. Suppose this random process has the following general properties:

- *Independent Increments* — The number of arrivals in two nonoverlapping intervals are independent. That is, for two intervals $[t_1, t_2)$ and $[t_3, t_4)$ such that $t_1 \leq t_2 \leq t_3 \leq t_4$, the number of arrivals in $[t_1, t_2)$ is statistically independent of the number of arrivals in $[t_3, t_4)$.

- *Stationary Increments* — The number of arrivals in an interval $[t, t + \tau)$ depends only on the length of the interval τ and not on where the interval occurs, t.
- *Distribution of Infinitesimal Increments* — For an interval of infinitesimal length, $[t, t + \Delta t)$, the probability of a single arrival is proportional to Δt, and the probability of having more than one arrival in the interval is negligible compared to Δt. Mathematically, we say that for some arbitrary constant λ:[4]

$$\Pr(\text{no arrivals in } [t, t + \Delta t)) = 1 - \lambda \Delta t + o(\Delta t), \tag{8.28}$$

$$\Pr(\text{one arrival in } [t, t + \Delta t)) = \lambda \Delta t + o(\Delta t), \tag{8.29}$$

$$\Pr(\text{more than one arrival in } [t, t + \Delta t)) = o(\Delta t). \tag{8.30}$$

Surprisingly enough, these rather general properties are enough to exactly specify the distribution of the counting process as shown next.

Consider the PMF of the counting process at time $t + \Delta t$. In particular, consider finding the probability of the event $\{X(t + \Delta t) = 0\}$.

$$P_X(0; t + \Delta t) = \Pr(\text{no arrivals in } [0, t + \Delta t))$$

$$= \Pr(\text{no arrivals in } [0, t)) \Pr(\text{no arrivals in } [t, t + \Delta t))$$

$$= P_X(0; t)[1 - \lambda \Delta t + o(\Delta t)] \tag{8.31}$$

Subtracting $P_X(0; t)$ from both sides and dividing by Δt results in

$$\frac{P_X(0; t + \Delta t) - P_X(0; t)}{\Delta t} = -\lambda P_X(0; t) + \frac{o(\Delta t)}{\Delta t} P_X(0; t). \tag{8.32}$$

Passing to the limit as $\Delta t \to 0$ gives the first order differential equation

$$\frac{d}{dt} P_X(0; t) = -\lambda P_X(0; t). \tag{8.33}$$

The solution to this equation is of the general form

$$P_X(0; t) = c \exp(-\lambda t) u(t) \tag{8.34}$$

for some constant c. The constant c is found to be equal to unity by using the fact that at time zero, the number of arrivals must be zero; that is, $P_X(0; 0) = 1$. Hence,

$$P_X(0; t) = \exp(-\lambda t) u(t). \tag{8.35}$$

---

[4]The notation $o(x)$ refers to an arbitrary function of x that goes to zero as $x \to 0$ in a faster than linear fashion. That is, some function $g(x)$ is said to be $o(x)$ if $\lim_{x \to 0} \frac{g(x)}{x} = 0$.

The rest of the PMF for the random process $X(t)$ can be specified in a similar manner. We find the probability of the general event $\{X(t + \Delta t) = i\}$ for some integer $i > 0$.

$$P_X(i; t + \Delta t) = \Pr(i \text{ arrivals in } [0, t)) \Pr(\text{no arrivals in } [t, t + \Delta t))$$

$$+ \Pr(i - 1 \text{ arrivals in } [0, t)) \Pr(\text{one arrival in } [t, t + \Delta t))$$

$$+ \Pr(\text{less than } i - 1 \text{ arrivals in } [0, t))$$

$$\times \Pr(\text{more than one arrival in } [t, t + \Delta t))$$

$$= P_X(i; t)[1 - \lambda \Delta t + o(\Delta t)] + P_X(i - 1; t)[\lambda \Delta t + o(\Delta t)]$$

$$+ \sum_{j=0}^{i-2} P_X(i; t) o(\Delta t) \tag{8.36}$$

As before, subtracting $P_X(i; t)$ from both sides and dividing by Δt results in

$$\frac{P_X(i; t + \Delta t) - P_X(i; t)}{\Delta t} = -\lambda P_X(i; t) + \lambda P_X(i - 1; t) + \sum_{j=0}^{i} P_X(i; t) \frac{o(\Delta t)}{\Delta t}. \tag{8.37}$$

Passing to the limit as $\Delta \to 0$ gives another first order differential equation,

$$\frac{d}{dt} P_X(i; t) + \lambda P_X(i; t) = \lambda P_X(i - 1; t). \tag{8.38}$$

It is fairly straightforward to solve this set of differential equations. For example, for $i = 1$, we have

$$\frac{d}{dt} P_X(1; t) + \lambda P_X(1; t) = \lambda P_X(0; t) = \lambda e^{-\lambda t} u(t), \tag{8.39}$$

together with the initial condition that $P_X(1; 0) = 0$. The solution to this equation can be shown to be

$$P_X(1; t) = \lambda t e^{-\lambda t} u(t). \tag{8.40}$$

It is left as an exercise for the reader (see Exercise 8.21) to verify that the general solution to the family of differential equations specified in Equation 8.38 is

$$P_X(i; t) = \frac{(\lambda t)^i}{i!} e^{-\lambda t} u(t). \tag{8.41}$$

Starting with the three properties made about the nature of this counting process at the start of this section, we have demonstrated that $X(t)$ follows a Poisson distribution, hence this process is referred to as a *Poisson counting process*. Starting with the PMF for the Poisson counting process specified in Equation 8.41, one can easily find the mean and autocorrelation functions for this process. First, the mean function is given by

$$\mu_X(t) = E[X(t)] = \sum_{i=0}^{\infty} i \frac{(\lambda t)^i}{i!} e^{-\lambda t} u(t) = \lambda t u(t). \tag{8.42}$$

In other words, the average number of arrivals in the interval $[0, t)$, is λt. This gives the parameter λ the physical interpretation as the average rate of arrivals or, as it is more commonly referred to, the *arrival rate* of the Poisson process. Another observation we can make from the mean process is that the Poisson counting process is not stationary.

The autocorrelation function can be calculated as follows:

$$R_{XX}(t_1, t_2) = E[X(t_1)X(t_2)] = E[X(t_1)\{X(t_1) + (X(t_2) - X(t_1))\}]$$

$$= E[X^2(t_1)] + E[X(t_1)\{X(t_2) - X(t_1)\}]. \tag{8.43}$$

To simplify the second expression, we use the independent increments property of the Poisson counting process. Assuming that $t_1 < t_2$, then $X(t_1)$ represents the number of arrivals in the interval $[0, t_1)$, while $X(t_2) - X(t_1)$ is the number of arrivals in the interval $[t_1, t_2)$. Since these two intervals are nonoverlapping, the number of arrivals in the two intervals are independent. Therefore,

$$R_{XX}(t_1, t_2) = E[X^2(t_1)] + E[X(t_1)]E[X(t_2) - X(t_1)]$$

$$= \text{Var}(X(t_1)) + \mu_X(t_1)\mu_X(t_2) = \lambda t_1 + \lambda^2 t_1 t_2. \tag{8.44}$$

This can be written more concisely in terms of the autocovariance function,

$$C_{XX}(t_1, t_2) = \text{Var}(X(t_1)) = \lambda t_1. \tag{8.45}$$

If $t_2 < t_1$, then the roles of t_1 and t_2 need to be reversed. In general for the Poisson counting process, we have

$$C_{XX}(t_1, t_2) = \lambda \min(t_1, t_2). \tag{8.46}$$

Another feature that can be extracted from the PMF of the Poisson counting process is the distribution of the interarrival time. That is, let T be the time at which the first arrival occurs. We seek the distribution of the random variable T. The CDF of T can be found as

$$F_T(t) = \Pr(T \le t) = \Pr(\text{at least one arrival in } [0, t))$$

$$= 1 - \Pr(\text{no arrivals in } [0, t)) = [1 - e^{-\lambda t}]u(t). \tag{8.47}$$

Hence it follows that the arrival time is an exponential random variable with a mean value of $E[T] = 1/\lambda$. The PDF of T is

$$f_T(t) = \lambda e^{-\lambda t} u(t). \tag{8.48}$$

We could get the same result starting from any point in time. That is, we do not need to measure the time to the next arrival starting from time zero. Picking any arbitrary point in time t_o, we could define T to be the time until the first arrival after time t_o. Using the same reasoning as before, we would arrive at the same exponential distribution. If we pick t_o to be the time of a specific arrival and then define T to be the time to the next arrival, then T is interpreted as an inter-arrival time. Hence, we conclude that the time between successive arrivals in the Poisson counting process follows an exponential distribution with a mean of $1/\lambda$.

The Poisson counting process can be represented as a sum of randomly shifted unit step functions. That is, let S_i be the time of the ith arrival. Then

$$X(t) = \sum_{i=1}^{\infty} u(t - S_i). \tag{8.49}$$

The random variables, S_i, are sometimes referred to as *points* of the Poisson process. Many other related random processes can be constructed by replacing the unit step functions with alternative functions. For example, if the step function is replaced by a delta function, the *Poisson impulse process* results, which is expressed as

$$X(t) = \sum_{i=1}^{\infty} \delta(t - S_i). \tag{8.50}$$

A sample realization of this process is shown in Figure 8.9.

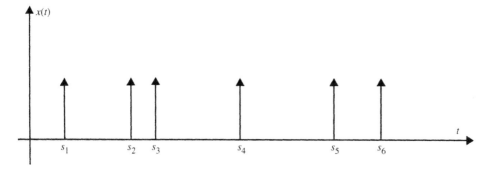

Figure 8.9 A sample realization of the Poisson impulse process.

8.7 Engineering Application: Shot Noise in a *p-n* Junction Diode

Both the Poisson counting process and the Poisson impulse process can be viewed as special cases of a general class of processes referred to as *shot noise processes*. Given an arbitrary waveform $h(t)$ and a set of Poisson points, S_i, the shot noise process is constructed as

$$X(t) = \sum_{i=1}^{\infty} h(t - S_i). \tag{8.51}$$

As an example of the physical origin of such a process, consider the operation of a *p-n* junction diode. When a forward bias voltage is applied to the junction, a current is generated. This current is not constant, but actually consists of discrete holes from the *p* region and electrons from the *n* region that have sufficient energy to overcome the potential barrier at the junction. Carriers do not cross the junction in a steady deterministic fashion; rather, each passage is a random event that might be modeled as a Poisson point process. The arrival rate of that Poisson process would be dependent on the bias voltage across the junction. As a carrier crosses the junction, it produces a current pulse, which we represent with some pulse shape $h(t)$, such that the total area under $h(t)$ is equal to the charge in an electron, q. Hence, the total current produced by the *p-n* junction diode can be modeled as a shot noise process.

   To start with, we compute the mean function of a shot noise process. However, upon examining Equation 8.51, it is not immediately obvious how to take the expected value for an arbitrary pulse shape $h(t)$. There are several ways to achieve this goal. One approach is to divide the time axis into infinitesimal intervals of length Δt. Then, define a sequence of Bernoulli random variables V_n such that $V_n = 1$ if a point occurred within the interval $[n\Delta t, (n + 1)\Delta t)$ and $V_n = 0$ if no points occurred in the same interval. Since the intervals are taken to be infinitesimal, the probability of having more than one point in a single interval is negligible. Furthermore, from the initial assumptions that led to the Poisson process, the distribution of the V_n is given by

$$\Pr(V_n = 1) = \lambda \Delta t \quad \text{and} \quad \Pr(V_n = 0) = 1 - \lambda \Delta t. \tag{8.52}$$

The shot noise process can be approximated by

$$X(t) \cong \sum_{n=0}^{\infty} V_n h(t - n\Delta t). \tag{8.53}$$

In the limit as $\Delta t \to 0$, the approximation becomes exact. Using this alternative representation of the shot noise process, calculation of the mean function is straightforward.

$$E[X(t)] \cong \sum_{n=0}^{\infty} E[V_n]h(t - n\Delta t) = \lambda \sum_{n=0}^{\infty} h(t - n\Delta t)\Delta t \qquad (8.54)$$

Note that in this calculation, the fact that $E[V_n] = \lambda \Delta t$ was used. Passing to the limit as $\Delta t \to 0$ results in

$$\mu_X(t) = \lambda \int_0^{\infty} h(t - u)\, du = \lambda \int_0^t h(v)\, dv. \qquad (8.55)$$

Strictly speaking, the mean function of the shot noise process is not a constant, and hence the process is not stationary. However, in practice, the current pulse will be time limited. Suppose the current pulse, $h(t)$, has a time duration of t_h. That is, for $t > t_h$, $h(t)$ is essentially equal to zero. For the example of the *p-n* junction diode, the time duration of the current pulse is the time it takes the carrier to pass through the depletion region. For most devices, this number may be a small fraction of a nanosecond. Then for any $t > t_h$,

$$\mu_X(t) = \lambda \int_0^{\infty} h(v)\, dv = \lambda q = \text{constant.} \qquad (8.56)$$

For example, using the fact that the charge on an electron is 1.6×10^{-19} C, if carriers made transitions at an average rate of 10^{15} per second (1 per femtosecond), then the average current produced in the diode would be 0.16 mA.

Next, we seek the autocorrelation (or autocovariance) function of the shot noise process. The same procedure used to calculate the mean function can also be used here.

$$R_{XX}(t, t + \tau) = E[X(t)X(t + \tau)] = \sum_{n=0}^{\infty} \sum_{m=0}^{\infty} E[V_n V_m]h(t - n\Delta t)h(t + \tau - m\Delta t)$$

$$= \sum_{n=0}^{\infty} E[V_n^2]h(t - n\Delta t)h(t + \tau - n\Delta t)]$$

$$+ \sum_{n=0}^{\infty} E[V_n]h(t - n\Delta t) \sum_{m \neq n}^{\infty} E[V_m]h(t + \tau - m\Delta t)$$

$$= \lambda \sum_{n=0}^{\infty} h(t - n\Delta t)h(t + \tau - n\Delta t)]\Delta t$$

$$+ \lambda^2 \sum_{n=0}^{\infty} h(t - n\Delta t)\Delta t \sum_{m \neq n}^{\infty} h(t + \tau - m\Delta t)\Delta t$$

$$= \sum_{n=0}^{\infty} h(t - n\Delta t)h(t + \tau - n\Delta t)[\lambda\Delta t - (\lambda\Delta t)^2]$$

$$+ \lambda^2 \sum_{n=0}^{\infty} h(t - n\Delta t)\Delta t \sum_{m=0}^{\infty} h(t + \tau - m\Delta t)\Delta t \qquad (8.57)$$

Passing to the limit as $\Delta t \to 0$, we note that the term involving $(\lambda\Delta t)^2$ is negligible compared to $\lambda\Delta t$. The resulting limit then takes the form

$$R_{XX}(t, t + \tau) = \lambda \int_0^{\infty} h(t - u)h(t + \tau - u)\, du + \lambda^2 \int_0^{\infty} h(t - u)\, du \int_0^{\infty} h(t + \tau - u)\, du$$

$$= \lambda \int_0^t h(v)h(v + \tau)\, dv + \lambda^2 \int_0^t h(v)\, dv \int_0^{t+\tau} h(v)\, dv. \qquad (8.58)$$

Note that the last term (involving the product of integrals) is just the product of the mean function evaluated at time t and the mean function evaluated at time $t + \tau$. Hence, we have

$$R_{XX}(t, t + \tau) = \lambda \int_0^t h(v)h(v + \tau)\, dv + \mu_X(t)\mu_X(t + \tau), \qquad (8.59)$$

or equivalently, in terms of the autocovariance function,

$$C_{XX}(t, t + \tau) = \lambda \int_0^t h(v)h(v + \tau)\, dv. \qquad (8.60)$$

As with the mean function, it is seen that the autocovariance function is a function of not only τ but also t. Again, for sufficiently large t, the upper limit in the preceding integral will be much longer than the time duration of the pulse, $h(t)$. Hence, for $t > t_h$,

$$C_{XX}(t, t + \tau) = C_{XX}(\tau) = \lambda \int_0^{\infty} h(v)h(v + \tau)\, dv, \qquad (8.61)$$

or

$$R_{XX}(t, t + \tau) = R_{XX}(\tau) = \lambda \int_0^{\infty} h(v)h(v + \tau)\, dv + \mu_X^2. \qquad (8.62)$$

We say that the shot noise process is asymptotically WSS. That is, after waiting a sufficiently long period of time, the mean and autocorrelation functions will be invariant to time shifts. In this case, the phrase "sufficiently long time" may mean a small fraction of a nanosecond! So for all practical purposes, the process is WSS. Also, it is noted that the width of the autocovariance function is t_h. That is, if $h(t)$ is time limited to a duration of t_h, then $C_{XX}(\tau)$ is zero for $|\tau| > t_h$.

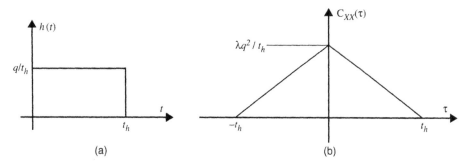

Figure 8.10 A square current pulse (a) and the corresponding autocovariance function (b).

This relationship is illustrated in Figure 8.10, assuming $h(t)$ is a square pulse, and implies that any samples of the shot noise process that are separated by more than t_h will be uncorrelated.

Finally, in order to characterize the PDF of the shot noise process, consider the approximation to the shot noise process given in Equation 8.53. At any fixed point in time, the process $X(t)$ can be viewed as the linear combination of a large number of independent Bernoulli random variables. By virtue of the central limit theorem, this sum can be very well approximated by a Gaussian random variable. Since the shot noise process is WSS (at least in the asymptotic sense) and is a Gaussian random process, then the process is also stationary in the strict sense. Also, samples spaced by more than t_h are independent.

EXAMPLE 8.24: Consider a shot noise process in a p-n junction diode where the pulse shape is square as illustrated in Figure 8.10. The mean current is $\mu_X = \lambda q$, which is presumably the desired signal we are trying to measure. The fluctuation of the shot noise process about the mean, we view as the unwanted disturbance, or noise. It would be interesting to measure the ratio of the power in the desired part of the signal to the power in the noise part of the signal. The desired part has a time-average power of $\mu_X^2 = (\lambda q)^2$, while the noise part has a power of $\sigma_X^2 = C_{XX}(0) = \lambda q^2/t_h$. The signal-to-noise ratio (SNR) is then

$$\mathrm{SNR} = \frac{\mu_X^2}{\sigma_X^2} = \frac{(\lambda q)^2}{\lambda q^2/t_h} = \lambda t_h.$$

We write this in a slightly different form,

$$\mathrm{SNR} = \lambda t_h = \lambda q \left(\frac{t_h}{q}\right) = \mu_X \left(\frac{t_h}{q}\right).$$

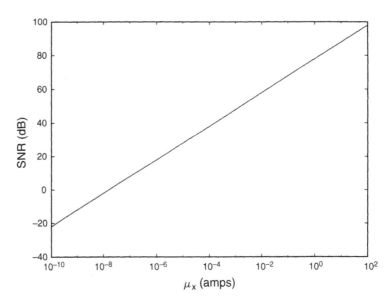

Figure 8.11   Signal-to-noise ratio in a shot noise process for an example *p-n* junction diode.

> For example, if the pulse duration were $t_h = 10$ picoseconds, the SNR
> as it depends on the strength of the desired part of the signal would be
> as illustrated in Figure 8.11. It is noted that the SNR is fairly strong until
> we try to measure signals that are below a microamp.

In Chapter 10, Power Spectral Density, we will view random processes in the
frequency domain. Using the frequency domain tools we will develop in that
chapter, it will become apparent that the noise power in the shot noise process is
distributed over a very wide bandwidth (about 100 GHz for the previous example).
Typically, our measuring equipment would not respond to that wide of a frequency
range, and so the amount of noise power we actually see would be much less than
that presented in Example 8.24 and would be limited by the bandwidth of our
equipment.

EXAMPLE 8.25: In this example, we provide some MATLAB code
to generate a sample realization of a shot noise process. We chose
to use a current pulse shape of the form $h(t) = t \exp(-t^2)$, but the
reader could easily modify this to use other pulse shapes as well.
A typical realization is shown in Figure 8.12. Note that after a short initial
transient period, the process settles into a steady state behavior.

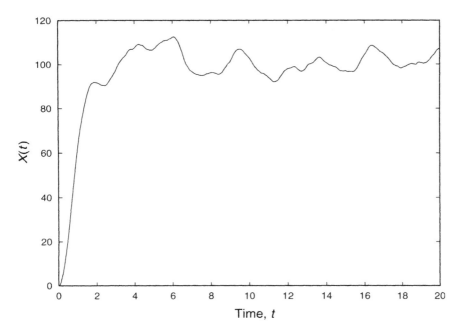

Figure 8.12 A typical realization of a shot noise process.

```
dt=0.001;                               % Time sample interval.
t=[0:dt:20];                            % Time axis.
v=rand(size(t))<0.2;                    % Impulse process.
h=t.*exp(-t.^2);                        % Pulse shape.
x=conv(v,h);                            % Shot noise process.
plot(t,x(1:length(t)))                  % Plot results.
xlabel('time, t'); ylabel('X(t)')
```

Exercises

8.1 A random process $X(t)$ consists of three-member functions: $x_1(t) = 1$, $x_2(t) = -3$, and $x_3(t) = \sin(2\pi t)$. Each member function occurs with equal probability.

   (a) Find the mean function, $\mu_X(t)$.
   (b) Find the autocorrelation function, $R_{X,X}(t_1, t_2)$.
   (c) Is the process WSS? Is it stationary in the strict sense?

8.2 A random process $X(t)$ has the following member functions: $x_1(t) = -2\cos(t)$, $x_2(t) = -2\sin(t)$, $x_3(t) = 2[\cos(t)+\sin(t)]$, $x_4(t) = [\cos(t)-\sin(t)]$, and $x_5(t) = [\sin(t) - \cos(t)]$.

(a) Find the mean function, $\mu_X(t)$.

(b) Find the autocorrelation function, $R_{X,X}(t_1, t_2)$.

(c) Is the process WSS? Is it stationary in the strict sense?

8.3 Let a discrete random process $X[n]$ be generated by repeated tosses of a fair die. Let the values of the random process be equal to the results of each toss.

(a) Find the mean function, $\mu_X[n]$.

(b) Find the autocorrelation function, $R_{X,X}(k_1, k_2)$.

(c) Is the process WSS? Is it stationary in the strict sense?

8.4 A discrete random process, $X[n]$, is generated by repeated tosses of a coin. Let the occurrence of a head be denoted by 1 and that of a tail by -1. A new discrete random process is generated by $Y[2n] = X[n]$ for $n = 0, \pm1, \pm2, \dots$ by and $Y[n] = X[n+1]$ for $\pm n$ for n odd. Find the autocorrelation function for $Y[n]$.

8.5 Let $X[n]$ be a wide sense stationary, discrete random process with autocorrelation function $R_{XX}[n]$, and let c be a constant.

(a) Find the autocorrelation function for the discrete random process $Y[n] = X[n] + c$.

(b) Are $X[n]$ and $Y[n]$ independent? Uncorrelated? Orthogonal?

8.6 A wide sense stationary, discrete random process $X[n]$ has an autocorrelation function of $R_{XX}[k]$. Find the expected value of $Y[n] = (X[n + m] - X[n - m])^2$, where m is an arbitrary integer.

8.7 A random process is given by $X(t) = A\cos(\omega t) + B\sin(\omega t)$, where A and B are independent zero mean random variables.

(a) Find the mean function, $\mu_X(t)$.

(b) Find the autocorrelation function, $R_{X,X}(t_1, t_2)$.

(c) Under what conditions (on the variances of A and B) is $X(t)$ WSS?

8.8 Show by example that the random process $Z(t) = X(t) + Y(t)$ may be a wide sense stationary process even though the random processes $X(t)$ and

$Y(t)$ are not. *Hint:* Let $A(t)$ and $B(t)$ be independent, wide sense stationary random processes with zero means and identical autocorrelation functions. Then let $X(t) = A(t)\sin(t)$ and $Y(t) = B(t)\cos(t)$. Show that $X(t)$ and $Y(t)$ are not wide sense stationary. Then show that $Z(t)$ is wide sense stationary.

8.9 Let $X(t) = A(t)\cos(\omega_0 t + \theta)$, where $A(t)$ is a wide sense stationary random process independent of θ, and let θ be a random variable distributed uniformly over $[0, 2\pi)$. Define a related process $Y(t) = A(t)\cos(\omega_0 + \omega_1)t + \theta)$. Show that $X(t)$ and $Y(t)$ are stationary in the wide sense but that the cross-correlation $R_{XY}(t, t + \tau)$, between $X(t)$ and $Y(t)$, is not a function of τ only and, hence, $Z(t) = X(t) + Y(t)$ is not stationary in the wide sense.

8.10 Let $X(t)$ be a modified version of the random telegraph process. The process switches between the two states $X(t) = 1$ and $X(t) = -1$, with the time between switches following exponential distributions $f_T(s) = \lambda\exp(-\lambda)u(s)$. Also, the starting state is determined by flipping a biased coin so that $\Pr(X(0) = 1) = p$ and $\Pr(X(0) = -1) = 1 - p$.

(a) Find $\Pr(X(t) = 1)$ and $\Pr(X(t) = -1)$.
(b) Find the mean function, $\mu_X(t)$.
(c) Find the autocorrelation function, $R_{X,X}(t_1, t_2)$.
(d) Is this process WSS?

8.11 Let $s(t)$ be a periodic square wave as illustrated in the accompanying figure. Suppose a random process is created according to $X(t) = s(t - T)$, where T is a random variable uniformly distributed over $(0, 1)$.

(a) Find the probability mass function of $X(t)$.
(b) Find the mean function, $\mu_X(t)$.
(c) Find the autocorrelation function, $R_{X,X}(t_1, t_2)$.
(d) Is this process WSS?

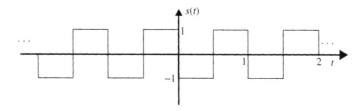

8.12 Let $s(t)$ be a periodic triangle wave as illustrated in the accompanying figure. Suppose a random process is created according to $X(t) = s(t - T)$, where T is a random variable uniformly distributed over $(0, 1)$.

(a) Find the probability density function of $X(t)$.

(b) Find the mean function, $\mu_X(t)$.

(c) Find the autocorrelation function, $R_{X,X}(t_1, t_2)$.

(d) Is this process WSS?

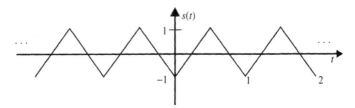

8.13 Let a random process consist of a sequence of pulses with the following properties: (i) the pulses are rectangular and of equal duration, Δ (with no "dead" space in between pulses), (ii) the pulse amplitudes are equally likely to be ± 1, (iii) all pulse amplitudes are statistically independent, and (iv) the various members of the ensemble are not synchronized.

(a) Find the mean function, $\mu_X(t)$.

(b) Find the autocorrelation function, $R_{X,X}(t_1, t_2)$.

(c) Is this process WSS?

8.14 A random process is defined by $X(t) = \exp(-At)u(t)$, where A is a random variable with PDF, $f_A(a)$.

(a) Find the PDF of $X(t)$ in terms of $f_A(a)$.

(b) If A is an exponential random variable, with $f_A(a) = e^{-a}u(a)$, find $\mu_X(t)$ and $R_{X,X}(t_1, t_2)$. Is the process WSS?

8.15 Let W_n be an IID sequence of zero-mean Gaussian random variables with variance σ_W^2. Define a discrete time random process $X[n] = pX[n-1] + W_n$, $n = 1, 2, 3, \ldots$, where $X[0] = W_0$ and p is a constant.

(a) Find the mean function, $\mu_X[n]$.

(b) Find the autocorrelation function, $R_{X,X}[n_1, n_2]$.

8.16 Let $X(t)$ and $Y(t)$ be two jointly wide sense stationary Gaussian random processes with zero means and with autocorrelation and crosscorrelation functions denoted as $R_{XX}(\tau)$, $R_{YY}(\tau)$, $R_{XY}(\tau)$. Determine the crosscorrelation function between $X^2(t)$ and $Y^2(t)$.

8.17 If $X(t)$ is a wide sense stationary Gaussian random process, find the cross-correlation between $X(t)$ and $X^3(t)$ in terms of the autocorrelation function $R_{XX}(\tau)$.

8.18 Two zero mean discrete random processes, $X[n]$ and $Y[n]$, are statistically independent. Let a new random process be $Z[n] = X[n] + Y[n]$. Let the autocorrelation functions for $X[n]$ and $Y[n]$ be

$$R_{XX}[k] = \left(\frac{1}{2}\right)^{|k|}, \quad R_{YY}[k] = \left(\frac{1}{3}\right)^{|k|}.$$

Find $R_{ZZ}[k]$. Plot all three autocorrelation functions (you may want to use MATLAB to help).

8.19 Consider a discrete time wide sense stationary random processes whose autocorrelation funcion is of the form

$$R_{XX}[k] = a^{|k|}, \quad where \quad |a| < 1.$$

Assume this process has zero-mean. Is the process ergodic in the mean?

8.20 Let $X(t)$ be a wide sense stationary random process that is ergodic in the mean and the autocorrelation. However, $X(t)$ is not zero-mean. Let $Y(t) = CX(t)$, where C is a random variable independent of $X(t)$ and C is not zero-mean. Show that $Y(t)$ is not ergodic in the mean or the autocorrelation.

8.21 Prove that the family of differential equations

$$\frac{d}{dt}P_X(0;t) + \lambda P_X(0;t) = 0,$$

$$\frac{d}{dt}P_X(i;t) + \lambda P_X(i;t) = \lambda P_X(i-1;t), \quad i = 1,2,3,\dots,$$

leads to the Poisson distribution

$$P_X(i;t) = \frac{(\lambda t)^i}{i!}e^{-\lambda t}.$$

8.22 Consider a Poisson counting process with arrival rate λ.

(a) Suppose it is observed that there is exactly one arrival in the time interval $[0, t_o)$. Find the PDF of that arrival time.

(b) Now suppose there were exactly two arrivals in the time interval $[0, t_o)$. Find the joint PDF of those two arrival times.

Can you extend these results to an arbitrary number, n, of arrivals?

8.23 Let $X(t)$ be a Poisson counting process with arrival rate λ. Find $\Pr(N(t) = k|N(t + \tau) = m)$, where $\tau > 0$ and $m \geq k$.

8.24 Let $X(t)$ be a WSS random process with mean μ_X and autocorrelation function $R_{XX}(\tau)$. Consider forming a new process according to

$$Y(t) = \frac{X(t + t_o) - X(t)}{t_o}.$$

(a) Find the mean function of $Y(t)$.

(b) Find the autocorrelation function of $Y(t)$. Is $Y(t)$ WSS?

8.25 Let $X_i(t)$, $i = 1, 2, \ldots, n$ be a sequence of independent Poisson counting processes with arrival rates λ_i. Show that the sum of all of these Poisson processes,

$$X(t) = \sum_{i=1}^{n} X_i(t),$$

is itself a Poisson process. What is the arrival rate of the sum process?

8.26 A workstation is used until it fails and is then sent out for repair. The time between failures, or the length of time the workstation functions until it needs repair, is a random variable T. Assume the times between failures, $T_1, T_2, \ldots T_n$, of the workstations available are independent random variables that are identically distributed. For $t > 0$, let the number of workstations that have failed be $N(t)$.

(a) If the time between failures of each workstation has an exponential PDF, then what type of process is $N(t)$?

(b) Assume that you have just purchased 10 new workstations and that each has a 90-day warranty. If the mean time between failures (MTBF) is 250 days, what is the probability that at least one workstation will fail before the end of the warranty period?

8.27 Suppose the arrival of calls at a switchboard is modeled as a Poisson process with the rate of calls per minute being $\lambda_a = 0.1$.

(a) What is the probability that the number of calls arriving in a 10-minute interval is less than 10?

(b) What is the probability that the number of calls arriving in a 10-minute interval is less than 10 if $\lambda_a = 10$?

(c) Assuming $\lambda_a = 0.1$, what is the probability that one call arrives during the first 10-minute interval and two calls arrive during the second 10-minute interval?

8.28 Model lightning strikes to a power line during a thunderstorm as a Poisson impulse process. Suppose the number of lightning strikes in time interval t has a mean rate of arrival given by s, which is one strike per 3 minutes.

(a) What is the expected number of lightning strikes in 1 minute? in 10 minutes?

(b) What is the average time between lightning strikes?

8.29 Suppose the power line in the previous problem has an impulse response that may be approximated by $h(t) = te^{-at}u(t)$, where $a = 10\sec^{-1}$.

(a) What does the shot noise on the power line look like? Sketch a possible member function of the shot noise process.

(b) Find the mean function of the shot noise process.

(c) Find the autocorrelation function of the shot noise process.

8.30 A shot noise process with random amplitudes is defined by

$$X(t) = \sum_{i=1}^{\infty} A_i h(t - S_i),$$

where the S_i are a sequence of points from a Poisson process and the A_i are IID random variables that are also independent of the Poisson points.

(a) Find the mean function of $X(t)$.

(b) Find the autocorrelation function of $X(t)$.

MATLAB Exercises

8.31 You are given a member function of a random process as $y(t) = 10\sin(2\pi t + \pi/2)$ where the amplitude is in volts. Quantize the amplitude of $y(t)$ into 21 levels with the intervals ranging from -10.5 to 10.5 in 1-volt steps. Consider 100 periods of $y(t)$ and let t take on discrete values given by nt_s where $t_s = 5$ msec. Construct a histogram of $y(t)$.

8.32 Write a MATLAB program to generate a Bernoulli process $X[n]$ for which each time instant of the process is a Bernoulli random variable with $\Pr(X[n] = 1) = 0.1$ and $\Pr(X[n] = 0) = 0.9$. Also, the process is IID (i.e., $X[n]$ is independent of $X[m]$ for all $m \neq n$). Once you have created the program to simulate $X[n]$, then create a counting process $Y[n]$ that counts the number of occurrences of $X[m] = 1$ in the interval $m \in [0, n]$. Plot member functions of each of the two processes.

8.33 Let W_n, $n = 0, 1, 2, 3, \ldots$ be a sequence of IID zero-mean Gaussian random variables with variance $\sigma_W^2 = 1$.

(a) Write a MATLAB program to generate the process

$$X[n] = \tfrac{1}{2}X[n-1] - \tfrac{1}{4}X[n-2] - \tfrac{1}{4}X[n-3] + W_n,$$

where $X[0] = W_0$ and $X[n] = 0$ for $n < 0$.

(b) Estimate the mean function of this process by generating a large number of realizations of the random process and computing the sample mean.

(c) Compute the time averaged mean of the process from a single realization. Does this seem to give the same result as the ensemble mean estimated in part (b)?

8.34 A certain random process is created as a sum of a large number, n, of sinusoids with random frequencies and random phases,

$$X(t) = \sum_{k=1}^{n} \cos(2\pi F_k t + \theta_k),$$

where the random phases θ_k are IID and uniformly distributed over $(0, 2\pi)$, and the random frequencies are given by $F_k = f_o + f_d \cos(\beta_k)$, where the β_k are IID and uniformly distributed over $(0, 1)$. (*Note:* These types of processes occur in the study of wireless communication systems.) For this exercise, we will take the constants f_o and f_d to be $f_o = 25$ Hz and $f_d = 10$ Hz, while we will let the number of terms in the sum be $n = 32$.

(a) Write a MATLAB program to generate realizations of this random process.

(b) Assuming the process is stationary and ergodic, use a single realization to estimate the first order PDF of the process, $f_X(x)$.

(c) Assuming the process is stationary and ergodic, use a single realization to estimate the mean of the process, μ_X.

(d) Assuming the process is stationary and ergodic, use a single realization to estimate the autocorrelation function of the process, $R_{XX}(\tau)$.

8.35 Write a MATLAB program to generate a shot noise process with

$$h(t) = b(at)\exp(-at)u(t),$$

where $a = 10^{12} \text{ sec}^{-1}$ and the constant b is chosen so that $q = \int h(t)\,dt$. For this program, assume that carriers cross the depletion region at a rate of 10^{13} per second. Plot a member function of this random process.

Markov Processes

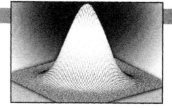

In this chapter we study a class of random processes that possess a certain characteristic that could crudely be described as memoryless. These processes appear in numerous applications, including queuing systems, computer communications networks, biological systems, and a wide array of other applications. As a result of their frequent occurrence, these processes have been studied extensively and a wealth of theory exists to solve problems related to these processes. We make no attempt to give an exhaustive treatment here, but rather present some of the fundamental concepts involving Markov processes.

9.1 Definition and Examples of Markov Processes

DEFINITION 9.1: A random process, $X(t)$, is said to be a Markov process if for any time instants, $t_1 < t_2 < \cdots < t_n < t_{n+1}$, the random process satisfies

$$F_X(X(t_{n+1}) \leq x_{n+1} \mid X(t_n) = x_n, X(t_{n-1}) = x_{n-1}, \ldots, X(t_1) = x_1)$$

$$= F_X(X(t_{n+1}) \leq x_{n+1} \mid X(t_n) = x_n). \qquad (9.1)$$

To understand this definition, we interpret t_n as the present time so that t_{n+1} represents some point in the future and $t_1, t_2, \ldots, t_{n-1}$ represent various points in the past. The Markovian property then states that given the present, the future is independent of the past. Or, in other words, the future of the random process depends only on where it is now and not on how it got there.

EXAMPLE 9.1: A classical example of a continuous time Markov process is the Poisson counting process studied in the previous chapter. Let $X(t)$ be a Poisson counting process with rate λ. Then its probability mass function satisfies

$$\Pr(X(t_{n+1}) = x_{n+1} \mid X(t_n) = x_n, X(t_{n-1}) = x_{n-1}, \ldots, X(t_1) = x_1)$$

$$= \begin{cases} 0 & x_{n+1} < x_n \\ \dfrac{(\lambda(t_{n+1} - t_n))^{x_{n+1}-x_n}}{(x_{n+1} - x_n)!} e^{-\lambda(t_{n+1}-t_n)} & x_{n+1} \geq x_n \end{cases}.$$

Clearly, this is independent of $\{X(t_{n-1}) = x_{n-1}, \ldots, X(t_1) = x_1\}$. In fact, the Markovian property must be satisfied because of the independent increments assumption of the Poisson process.

To start with, we will focus our attention on discrete-valued Markov processes in discrete time, better known as Markov chains. Let $X[k]$ be the value of the process at time instant k. Since the process is discrete-valued, $X[k] \in \{x_1, x_2, x_3, \ldots\}$ and we say that if $X[k] = x_n$, then the process is in state n at time k. A Markov chain is described statistically by its transition probabilities which are defined as follows.

DEFINITION 9.2: Let $X[k]$ be a Markov chain with states $\{x_1, x_2, x_3, \ldots\}$, then the probability of transitioning from state i to state j in one time instant is

$$p_{i,j} = \Pr(X[k+1] = j | X[k] = i). \tag{9.2}$$

If the Markov chain has a finite number of states, n, then it is convenient to define a *transition probability matrix*,

$$P = \begin{bmatrix} p_{1,1} & p_{1,2} & \cdots & p_{1,n} \\ p_{2,1} & p_{2,2} & \cdots & p_{2,n} \\ & & & \\ p_{n,1} & p_{n,2} & \cdots & p_{n,n} \end{bmatrix}. \tag{9.3}$$

One can encounter processes where the transition probabilities vary with time and hence need to be explicitly written as a function of k (e.g., $p_{i,j,k}$), but we do not consider such processes in this text and henceforth it is assumed that transition probabilities are independent of time.

EXAMPLE 9.2: Suppose every time a child buys a kid's meal at his favorite fast food restaurant, he receives one of four superhero action figures. Naturally, the child wants to collect all four action figures and so he regularly eats lunch at this restaurant in order to complete the collection. This process can be described by a Markov chain. In this case, let $X[k] \in \{0, 1, 2, 3, 4\}$ be the number of different action figures that the child has collected after purchasing k meals. Assuming each meal contains one of the four superheroes with equal probability and that the action figure in any meal is independent of what is contained in any previous or future meals, then the transition probability matrix easily works out to be

$$P = \begin{bmatrix} 0 & 1 & 0 & 0 & 0 \\ 0 & 1/4 & 3/4 & 0 & 0 \\ 0 & 0 & 1/2 & 1/2 & 0 \\ 0 & 0 & 0 & 3/4 & 1/4 \\ 0 & 0 & 0 & 0 & 1 \end{bmatrix}.$$

Initially (before any meals are bought), the process starts in state 0 (the child has no action figures). When the first meal is bought, the Markov chain must move to state 1 since no matter which action figure is contained in the meal, the child will now have one superhero. Hence, $p_{0,1} = 1$ and $p_{0,j} = 0$ for all $j \neq 1$. If the child has one distinct action figure, when he buys the next meal he has a 25 percent chance of receiving a duplicate and a 75 percent chance of getting a new action figure. Hence, $p_{1,1} = 1/4$, $p_{1,2} = 3/4$, and $p_{1,j} = 0$ for $j \neq 1, 2$. Similar logic is used to complete the rest of the matrix. The child might be interested in knowing the average number of lunches he needs to buy until his collection is completed. Or, maybe the child has saved up only enough money to buy 10 lunches and wants to know what his chances are of completing the set before running out of money. We will develop the theory needed to answer such questions.

The transition process of a Markov chain can also be illustrated graphically using a state diagram. Such a diagram is illustrated in Figure 9.1 for the Markov chain in Example 9.2. In the figure, each directed arrow represents a possible transition and the label on each arrow represents the probability of making that transition. Note that for this Markov chain, once we reach state 4, we remain there forever. This type of state is referred to as an *absorbing state*.

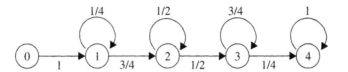

Figure 9.1 State diagram for the Markov chain of Example 9.2.

EXAMPLE 9.3: (The Gambler's Ruin Problem) Suppose a gambler plays a certain game of chance (e.g., blackjack) against the "house." Every time the gambler wins the game, he increases his fortune by one unit (say, a dollar) and every time he loses, his fortune decreases by one unit. Suppose the gambler wins each game with probability p and loses with probability $q = 1 - p$. Let X_n represent the amount of the gambler's fortune after playing the game n times. If the gambler ever reaches the state $X_n = 0$, the gambler is said to be "ruined" (he has lost all of his money). Assuming that the outcome of each game is independent of all others, the sequence x_n, $n = 0, 1, 2, \ldots$ forms a Markov chain. The state transition matrix is of the form

$$
P = \begin{bmatrix}
1 & 0 & 0 & 0 & 0 & \cdots \\
q & 0 & p & 0 & 0 & \cdots \\
0 & q & 0 & p & 0 & \cdots \\
0 & 0 & q & 0 & p & \cdots \\
\cdots & & & \cdots & & \cdots
\end{bmatrix}.
$$

The state transition diagram for the gambler's ruin problem is shown in Figure 9.2. One might then be interested in determining how long it

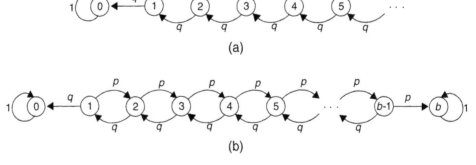

Figure 9.2 State transition diagram for Example 9.3 (The Gambler's Ruin Problem), with one absorbing state (a) and with two absorbing states (b).

might take before the gambler is ruined (enters the zero state). Is ruin inevitable for any p, or if the gambler is sufficiently proficient at the game, can he avoid ruin indefinitely? A more realistic alternative to this model is one where the house also has a finite amount of money. Suppose the gambler starts with d dollars and the house has $b-d$ dollars so that between the two competitors there is a total of b dollars in the game. Now if the gambler ever gets to the state 0 he is ruined, while if he gets to the state b he has "broken the bank" (i.e., the house is ruined). Now the Markov chain has two absorbing states as shown in part (b) of the figure. It would seem that sooner or later the gambler must have a run of bad luck sufficient to send him to the 0 state (i.e., ruin) or a run of good luck which will cause him to enter the state b (i.e., break the bank). It would be interesting to find the probabilities of each of these events.

The previous example is one of a class of Markov chains known as *random walks*. Random walks are often used to describe the motion of a particle. Of course, there are many applications that can be described by a random walk that do not involve the movement of a particle, but it is helpful to think of such a particle when describing such a Markov chain. In one dimension, a random walk is a Markov chain whose states are the integers and whose transition probabilities satisfy $p_{i,j} = 0$ for any $j \neq i-1, i, i+1$. In other words, at each time instant, the state of the Markov chain can either increase by one, stay the same, or decrease by one. If $p_{i,i+1} = p_{i,i-1}$, then the random walk is said to be symmetric, whereas if $p_{i,i+1} \neq p_{i,i-1}$ the random walk is said to have drift. Often the state space of the random walk will be a finite range of integers, $n, n+1, n+1, \ldots, m-1, m$ (for $m > n$), in which case the states n and m are said to be boundaries, or barriers. The gambler's ruin problem is an example of a random walk with absorbing boundaries, where $p_{n,n} = p_{m,m} = 1$. Once the particle reaches the boundary, it is absorbed and remains there forever. We could also construct a random walk with reflecting boundaries, in which case $p_{n,n+1} = p_{m,m-1} = 1$. That is, whenever the particle reaches the boundary, it is always reflected back to the adjacent state.

EXAMPLE 9.4: (A Queueing System) A common example of Markov chains (and Markov processes in general) is that of queueing systems. Consider, for example, a taxi stand at a busy airport. A line of taxis, which for all practical purposes can be taken to be infinitely long, is available to serve travelers. Customers wanting a taxi enter a queue and are given a taxi on a first come, first serve basis. Suppose it takes one unit of time (say, a minute) for the customer at the head of the queue to load himself and his luggage into a taxi. Hence, during each unit of time, one customer in the queue receives service and leaves the queue while some random number of new customers enter the end of the queue. Suppose at each time instant, the number of new customers arriving for service is described by a discrete distribution $(p_0, p_1, p_2, \ldots)$, where p_k is the probability of k new customers. For such a system, the transition probability matrix of the Markov chain would look like

$$
P = \begin{bmatrix}
p_0 & p_1 & p_2 & p_3 & \cdots \\
p_0 & p_1 & p_2 & p_3 & \cdots \\
0 & p_0 & p_1 & p_2 & \cdots \\
0 & 0 & p_0 & p_1 & \cdots \\
\cdots & & \cdots & & \cdots
\end{bmatrix}.
$$

The manager of the taxi stand might be interested in knowing the probability distribution of the queue length. If customers have to wait too long, they may get dissatisfied and seek other forms of transportation.

EXAMPLE 9.5: (A Branching Process) Branching processes are commonly used in biological studies. Suppose a certain species of organism has a fixed lifespan (one time unit). At the end of its lifespan, the nth organism in the species produces a number of offspring described by some random variable, Y_n, whose sample space is the set of nonnegative integers. Also, assume that the number of offspring produced by each organism is independent and identically distributed (IID). Then, X_k, the number of organisms in the species during the kth generation is a random variable that depends only on the number of organisms in the previous generation. In particular, if $X_k = i$, then $X_{k+1} = \sum_{n=1}^{i} Y_n$. The transition probability is then

$$
p_{i,j} = \Pr(X_{k+1} = j \mid X_k = i) = \Pr\left(\sum_{n=1}^{i} Y_n = j \right).
$$

Let $H_Y(z)$ be the probability generating function of the random variable Y_n (i.e., $H_Y(z) = \sum_{i=0}^{\infty} \Pr(Y_n = i) z^i$). Then, due to the IID nature of the Y_n, $\sum_{n=1}^{i} Y_n$ will have a probability generating function given by $[H_Y(z)]^i$. Hence, the transition probability, $p_{i,j}$, will be given by the coefficient of z^j in the power series expansion of $[H_Y(z)]^i$.

EXAMPLE 9.6: (A Genetic Model) Suppose a gene contains n units. Of these units, i are mutant and $n - i$ are normal. Every time a cell doubles, the n units double and each of the two cells receives a gene composed of n units. After doubling, there is a pool of $2n$ units of which $2i$ are mutant. These $2n$ units are grouped into two sets of n units randomly. As we trace a single line of descent, the number of mutant units in each gene forms a Markov chain. Define the kth gene to be in state i (where $X_k = i$) if it is composed of i mutant and $n - i$ normal units. It is not difficult to show that, given $X_k = i$,

$$p_{i,j} = \Pr(X_{k+1} = j \mid X_k = i) = \frac{\binom{2i}{j}\binom{2n - 2i}{n - j}}{\binom{2n}{n}}.$$

9.2 Calculating Transition and State Probabilities in Markov Chains

The state transition probability matrix of a Markov chain gives the probabilities of transitioning from one state to another in a single time unit. It will be useful to extend this concept to longer time intervals.

DEFINITION 9.3: The n-step transition probability for a Markov chain is

$$p_{i,j}^{(n)} = \Pr(X_{k+n} = j \mid X_k = i). \tag{9.4}$$

Also, define an n-step transition probability matrix $P^{(n)}$ whose elements are the n-step transition probabilities just described in Equation 9.4.

Given the one-step transition probabilities, it is straightforward to calculate higher order transition probabilities using the following result.

THEOREM 9.1: (Chapman-Kolmogorov Equation)

$$p_{i,j}^{(n)} = \sum_{k} p_{i,k}^{(m)} p_{k,j}^{(n-m)}, \quad \text{for any } m = 0, 1, 2, \dots, n \tag{9.5}$$

PROOF: First, condition on the event that in the process of transitioning from state i to state j, the Markov chain passes through state k at some intermediate point in time. Then, using the principle of total probability,

$$\Pr(X_{l+n}=j\,|\,X_l=i)=\sum_k \Pr(X_{l+n}=j\,|\,X_l=i,\ X_{l+m}=k)\Pr(X_{l+m}=k\,|\,X_k=i). \quad (9.6)$$

Using the Markov property, the expression reduces to the desired form:

$$\Pr(X_{l+n}=j\,|\,X_l=i)=\sum_k \Pr(X_{l+n}=j\,|\,X_{l+m}=k)\Pr(X_{l+m}=k\,|\,X_k=i). \quad (9.7)$$

$\blacksquare$

This result can be written in a more compact form using transition probability matrices. It is easily seen that the Chapman-Kolmogorov equations can be written in terms of the n-step transition probability matrices as

$$P^{(n)}=P^{(m)}P^{(n-m)}. \quad (9.8)$$

Then, starting with the fact that $P^{(1)}=P$, it follows that $P^{(2)}=P^{(1)}P^{(1)}=P^2$, and using induction, it is established that

$$P^{(n)}=P^n. \quad (9.9)$$

Hence, we can find the n-step transition probability matrix through matrix multiplication. If n is large, it may be more convenient to compute P^n via eigen-decomposition. The matrix P can be expanded as $P=U\Lambda U^{-1}$, where Λ is the diagonal matrix of eigenvalues and U is the matrix whose columns are the corresponding eigenvectors. Then

$$P^n=U\Lambda^n U^{-1}. \quad (9.10)$$

Another quantity of interest is the probability distribution of the Markov chain at some time instant k. If the initial probability distribution of the Markov chain is known, then the distribution at some later point in time can easily be found. Let $\pi_j(k)=\Pr(X_k=j)$ and $\pi(k)$ be the row vector whose jth element is $\pi_j(k)$. Then

$$\pi_j(k)=\Pr(X_k=j)=\sum_i \Pr(X_k=j\,|\,X_0=i)\Pr(X_0=i)=\sum_i p_{i,j}^{(k)}\pi_i(0), \quad (9.11)$$

or in vector form,

$$\pi(k)=\pi(0)P^k. \quad (9.12)$$

EXAMPLE 9.7: (Continuation of Example 9.2) Recall in Example 9.2 the child who purchased kid's meals at his favorite restaurant in order to collect a set of four superhero action figures. Initially, before any meals are purchased, the child has no action figures and so the initial probability distribution is $\pi(0) = 1,0,0,0,0$). Repeated application of Equation 9.12 with the probability transition matrix given in Example 9.2 results in

$$\pi(1) = (0,1,0,0,0),$$

$$\pi(2) = (0,1/4,3/4,0,0),$$

$$\pi(3) = (0,1/16,9/16,3/8,0),$$

$$\pi(4) = (0,1/64,21/64,9/16,3/32),$$

$$\pi(5) = (0,1/256,45/256,75/128,15/64),$$

$$\pi(6) = (0,1/1024,93/1024,135/256,195/512),$$

and so on. It is to be expected that if the child buys enough meals, he will eventually complete the collection (i.e., get to state 4) with probability approaching unity. This can be easily verified analytically by calculating the limiting form of P^k as $k \to \infty$. Recall that for this example, P is a triangular matrix and hence its eigenvalues are simply the diagonal entries. Thus, the diagonal matrix of eigenvalues is

$$\Lambda = \begin{bmatrix} 0 & 0 & 0 & 0 & 0 \\ 0 & 1/4 & 0 & 0 & 0 \\ 0 & 0 & 1/2 & 0 & 0 \\ 0 & 0 & 0 & 3/4 & 0 \\ 0 & 0 & 0 & 0 & 1 \end{bmatrix}.$$

It should be clear that $\lim_{k \to \infty} \Lambda^k$ is a matrix with all zero entries except the one in the lower-right corner, which is equal to one. Using MATLAB (or some other math package) to calculate the corresponding matrix of eigenvectors, it is found that

$$\lim_{k \to \infty} P^k = U \left(\lim_{k \to \infty} \Lambda^k \right) U^{-1} = \begin{bmatrix} 0 & 0 & 0 & 0 & 1 \\ 0 & 0 & 0 & 0 & 1 \\ 0 & 0 & 0 & 0 & 1 \\ 0 & 0 & 0 & 0 & 1 \\ 0 & 0 & 0 & 0 & 1 \end{bmatrix}.$$

Then, using the initial distribution of $\pi(0) = (1,0,0,0,0)$, the state distribution as $k \to \infty$ works out to be

$$\pi = \lim_{k \to \infty} \pi(k) = \lim_{k \to \infty} \pi(0) P^k = [0\,0\,0\,0\,1].$$

In Example 9.6, it was seen that as $k \to \infty$, the k-step transition probability matrix approached that of a matrix whose rows were all identical. In that case, the limiting product $\lim_{k \to \infty} \pi(0)P^k$ is the same regardless of the initial distribution $\pi(0)$. Such a Markov chain is said to have a unique steady state distribution, π. It should be emphasized that not all Markov chains have a steady state distribution. For example, the Poisson counting process of Example 9.1 clearly does not, since any counting process is a monotonic, nondecreasing function of time and, hence, it is expected that the distribution should skew towards larger values as time progresses.

This concept of a steady state distribution can be viewed from the perspective of stationarity. Suppose that at time k, the process has some distribution, $\pi(k)$. The distribution at the next time instant is then $\pi(k+1) = \pi(k)P$. If $\pi(k) = \pi(k+1)$, then the process has reached a point where the distribution is stationary (independent of time). This stationary distribution, π, must satisfy the relationship

$$\pi = \pi P. \tag{9.13}$$

In other words, π (if it exists) is the left eigenvector of the transition probability matrix P, that corresponds to the eigenvalue $\lambda = 1$. The next example shows that this eigenvector is not always unique.

EXAMPLE 9.8: (The Gambler's Ruin Revisited) Suppose a certain gambler has \$5 and plays against another player (the house). The gambler decides that he will play until he either doubles his money or loses it all. Suppose the house has designed this game of chance so that the gambler will win with probability $p = 0.45$ and the house will win with probability $q = 0.55$. Let X_k be the amount of money the gambler has after playing the game k times. The transition probability matrix for this Markov chain is

$$P = \begin{bmatrix}
1 & 0 & 0 & 0 & 0 & 0 & 0 & 0 & 0 & 0 & 0 \\
0.55 & 0 & 0.45 & 0 & 0 & 0 & 0 & 0 & 0 & 0 & 0 \\
0 & 0.55 & 0 & 0.45 & 0 & 0 & 0 & 0 & 0 & 0 & 0 \\
0 & 0 & 0.55 & 0 & 0.45 & 0 & 0 & 0 & 0 & 0 & 0 \\
0 & 0 & 0 & 0.55 & 0 & 0.45 & 0 & 0 & 0 & 0 & 0 \\
0 & 0 & 0 & 0 & 0.55 & 0 & 0.45 & 0 & 0 & 0 & 0 \\
0 & 0 & 0 & 0 & 0 & 0.55 & 0 & 0.45 & 0 & 0 & 0 \\
0 & 0 & 0 & 0 & 0 & 0 & 0.55 & 0 & 0.45 & 0 & 0 \\
0 & 0 & 0 & 0 & 0 & 0 & 0 & 0.55 & 0 & 0.45 & 0 \\
0 & 0 & 0 & 0 & 0 & 0 & 0 & 0 & 0.55 & 0 & 0.45 \\
0 & 0 & 0 & 0 & 0 & 0 & 0 & 0 & 0 & 0 & 1
\end{bmatrix}.$$

This matrix has (two) repeated eigenvalues of $\lambda = 1$, and the corresponding eigenvectors are $[1\,0\,0\,0\,0\,0\,0\,0\,0\,0\,0]$ and $[0\,0\,0\,0\,0\,0\,0\,0\,0\,0\,1]$. Note that any linear combination of these will also be an eigenvector. Hence, any vector of the form $[p\,0\,0\,0\,0\,0\,0\,0\,0\,1 - p]$ is a left eigenvector of P and therefore there is no unique stationary distribution for this Markov chain. For this example, the limiting form of the state distribution of the Markov chain depends on the initial distribution. The limiting form of P^k can be easily found to be

$$\lim_{k \to \infty} P^k = \begin{bmatrix} 1 & 0\,0\,0\,0\,0\,0\,0\,0\,0 & 0 \\ 0.9655 & 0\,0\,0\,0\,0\,0\,0\,0\,0 & 0.0345 \\ 0.9233 & 0\,0\,0\,0\,0\,0\,0\,0\,0 & 0.0767 \\ 0.8717 & 0\,0\,0\,0\,0\,0\,0\,0\,0 & 0.1283 \\ 0.8087 & 0\,0\,0\,0\,0\,0\,0\,0\,0 & 0.1913 \\ 0.7317 & 0\,0\,0\,0\,0\,0\,0\,0\,0 & 0.2683 \\ 0.6376 & 0\,0\,0\,0\,0\,0\,0\,0\,0 & 0.3624 \\ 0.5225 & 0\,0\,0\,0\,0\,0\,0\,0\,0 & 0.4775 \\ 0.3819 & 0\,0\,0\,0\,0\,0\,0\,0\,0 & 0.6181 \\ 0.2101 & 0\,0\,0\,0\,0\,0\,0\,0\,0 & 0.7899 \\ 0 & 0\,0\,0\,0\,0\,0\,0\,0\,0 & 1 \end{bmatrix}.$$

Using the initial distribution $\pi(0) = [0\,0\,0\,0\,0\,1\,0\,0\,0\,0\,0]$ (that is, the gambler starts off in state 5), then it is seen that the steady state distribution is $\lim_{k \to \infty} \pi(k) = [0.7317\,0\,0\,0\,0\,0\,0\,0\,0\,0\,0.2683]$. So, when the gambler starts with \$5, he has about a 73 percent chance of losing all of his money and about a 27 percent chance of doubling his money.

As seen in Example 9.7, with some Markov chains, the limiting form of P^k (as $k \to \infty$) does not necessarily converge to a matrix whose rows are all identical. In that case, the limiting form of the state distribution will depend on the starting distribution. In the case of the gambler's ruin problem, we probably could have guessed this behavior. If the gambler had started with very little money, we would expect him to end up in the state of ruin with very high probability; whereas, if the gambler was very wealthy and the house had very little money, we would expect a much greater chance of the gambler eventually breaking the house. Accordingly, our intuition tells us that the probability distribution of the gambler's ultimate state should depend on the starting state.

In general, there are several different manners in which a Markov chain's state distribution can behave as $k \to \infty$. In some cases, $\lim_{k \to \infty} \pi(k)$ does not exist. Such would be the case when the process tends to oscillate between two or more states. A second possibility, as in Example 9.7, is that $\lim_{k \to \infty} \pi(k)$ does in fact converge to a fixed distribution, but the form of this limiting distribution depends on the starting distribution. The last case is when $\lim_{k \to \infty} \pi(k) = \pi$. That is, the state distribution converges to some fixed distribution, π, and the form of π is independent of the starting distribution. Here, the transition probability matrix, P, will have a single (not repeated) eigenvalue at $\lambda = 1$, and the corresponding eigenvector (properly normalized) will be the steady state distribution, π. Furthermore, the limiting form of P^k will be one whose rows are all identical and equal to the steady state distribution, π. In the next section, we look at some conditions that must be satisfied for the Markov chain to achieve a unique steady state distribution.

EXAMPLE 9.9: In this example, we provide the MATLAB code to simulate the distribution of the queue length of the taxi stand described in Example 9.4. For this example, we take the number of arrivals per time unit, X, to be a Poisson random variable whose PMF is

$$P_X(k) = \frac{\lambda^k e^{-\lambda}}{k!}.$$

Recall that in the taxi stand, one customer is served per time unit (assuming there is at least one customer in the queue waiting to be served). The following code can be used to estimate and plot the PMF of the queue length. The average queue length was also calculated to be 3.36 customers for an arrival rate of $\lambda = 0.85$ customers per time unit. Figure 9.3 shows a histogram of the PMF of the queue length for the same arrival rate.

```
N=10000;                                    % Length of simulation.
a=0.85;                                     % Arrival rate.
k=[0:10];
Poisson=zeros(size(k));                     % Calculate Poisson PMF.
for m=k
  Poisson(m+1)=a.^m*exp(-a)./factorial(m);
end
queue(1)=0;                                 % Initial queue size.
for n=1:N
  x=rand(1);
  arrivals=sum(x>cumsum(Poisson));          % Poisson RV.
  departures=queue(n)>0;
  queue(n+1)=queue(n)+arrivals-departures;  % Current queue length.
end
mean_queue_length=sum(queue)/length(queue)  % Compute average queue
                                            length.
bins=[0:25]
```

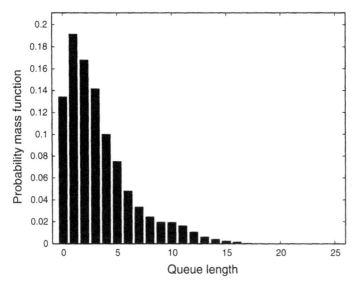

Figure 9.3   Histogram of the queue length for the Taxi stand of Example 9.4 assuming a Poisson arrival process with an average arrival rate of 0.85 arrivals per time unit.

```
y=hist(queue,bins);
PMF=y/N;                                        % Estimate PMF.
bar(bins,PMF)                                   % Plot results.
axis([min(bins)-1 max(bins)+1 0 1.1*max(PMF)])
```

9.3 Characterization of Markov Chains

Using the methods presented in the previous sections, calculating the steady state distribution of a Markov chain requires performing an eigendecomposition of the transition probability matrix, P. If the number of states in the Markov chain is large (or infinite), performing the required linear algebra may be difficult (or impossible). Hence, it would be useful to seek alternative methods to determine if a steady state distribution exists, and if so, to calculate it. To develop the necessary theory, we must first proceed through a sequence of definitions and classifications of the states of a Markov chain.

DEFINITION 9.4: State j is *accessible* from state i if for some finite n, $p_{i,j}^{(n)} > 0$. This simply means that if the process is in state i, it is possible for the process to get to state j in a finite amount of time. Furthermore, if state j is accessible from state i and state i is accessible from state j, then the states i and j are said to *communicate*. It is common to use the shorthand notation $i \leftrightarrow j$ to represent the relationship "state i communicates with state j."

The states of any Markov chain can be divided into sets or classes of states where all the states in a given class communicate with each other. It is possible for the process to move from one communicating class of states to another, but once that transition is made, the process can never return to the original class. If it did, the two classes would communicate with each other and hence would be a part of a single class.

DEFINITION 9.5: A Markov chain for which all of the states are part of a single communicating class is called an *irreducible* Markov chain. Also, the corresponding transition probability matrix is called an irreducible matrix.

The examples in Section 9.1 can be used to help illustrate these concepts. For both processes in Examples 9.1 and 9.2, none of the states communicate with any other states. This is a result of the fact that both processes are counting processes and it is therefore impossible to go backward in the chain of states. Hence, if $j > i$, state j is accessible from state i but state i is not accessible from state j. As a result, for any counting process, all states form a communication class of their own. That is, the number of classes is identical to the number of states. In the gambler's ruin problem of Example 9.3, the two absorbing states do not communicate with any other state since it is impossible ever to leave an absorbing state, while all the states in between communicate with each other. Hence, this Markov chain has three communicating classes. The two absorbing states each form a class to themselves, while the third class consists of all the states in between.

The queueing system (i.e., taxi stand) of Example 9.4 represents a Markov chain where all states communicate with each other, and hence that Markov chain is irreducible. For the branching process of Example 9.5, all states communicate with each other except the state 0, which represents the extinction of the species, which presumably is an absorbing state. The genetic model of Example 9.6 is similar to the gambler's ruin problem in that the Markov chain has two absorbing states at the endpoints, while everything in between forms a single communicating class.

DEFINITION 9.6: The *period* of state i, $d(i)$, is the greatest common divisor of all integers $n \geq 1$ such that $p_{i,i}^{(n)} > 0$. Stated another way, $d(i)$ is the period of state i if any transition from state i to itself must occur in a number of steps that is a multiple of $d(i)$. Furthermore, a Markov chain for which all states have a period of $d(i) = 1$ is called an *aperiodic* Markov chain.

Most Markov chains are aperiodic. The class of random walks is an exception. Suppose a Markov chain defined on the set of integers has transition probabilities that satisfy

$$p_{i,j} = \begin{cases} p & j = i+1 \\ 1-p & j = i-1 \\ 0 & \text{otherwise} \end{cases} . \tag{9.14}$$

Then each state will have a period of 2. If we add absorbing boundaries to the random walk, then the absorbing states are not periodic because for the absorbing states, $p_{i,i}^{(n)} = 1$ for all n and hence the period of an absorbing state is 1. It is left as an exercise for the reader (see Exercise 9.18) to establish the fact that the period is a property of a class of communicating states. That is, if $i \leftrightarrow j$ then $d(i) = d(j)$ and hence all states in the same class must have the same period.

DEFINITION 9.7: Let $f_{i,i}^{(n)}$ be the probability that given a process is in state i, the first return to state i will occur in exactly n steps. Mathematically,

$$f_{i,i}^{(n)} = \Pr(X_{k+n} = i, \ X_{k+m} \neq i \quad \text{for} \quad m = 1, 2, \dots, n-1 \,|\, X_k = i). \tag{9.15}$$

Also, define $f_{i,i}$ to be the probability that the process will eventually return to state i. The probability of eventual return is related to the first return probabilities by

$$f_{i,i} = \sum_{n=1}^{\infty} f_{i,i}^{(n)}. \tag{9.16}$$

It should be noted that the first return probability $f_{i,i}^{(n)}$ is not the same thing as the n-step transition probability, but the two quantities are related. To develop this relationship, it is observed that

$$p_{i,i}^{(n)} = \sum_{m=0}^{n} \Pr(\{X_{k+n} = i\}, \{\text{first return to state } i \text{ occurs in } m \text{ steps}\} \,|\, X_k = i) \tag{9.17}$$

$$= \sum_{m=0}^{n} \Pr(X_{k+n} = i \,|\, X_{k+m} = i) \Pr(\text{first return to state } i \text{ occurs in } m \text{ steps} \,|\, X_k = i) \tag{9.18}$$

$$= \sum_{m=0}^{n} p_{i,i}^{(n-m)} f_{i,i}^{(m)}, \quad n = 1, 2, 3, \dots . \tag{9.19}$$

In the Equations 9.17–9.19, $p_{i,i}^{(0)}$ is taken to be equal to 1 and $f_{i,i}^{(0)}$ is taken to be equal to 0. Given the n-step transition probabilities, $p_{i,i}^{(n)}$, one could solve the preceding

system of equations for the first return probabilities. However, since the previous
system is a convolution, this set of equations may be easier to solve using frequency
domain techniques. Define the generating functions[1]

$$P_{i,i}(z) = \sum_{n=0}^{\infty} p_{i,i}^{(n)} z^n, \tag{9.20}$$

$$F_{i,i}(z) = \sum_{n=0}^{\infty} f_{i,i}^{(n)} z^n. \tag{9.21}$$

It is left as an exercise for the reader (see Exercise 9.19) to demonstrate that these
two generating functions are related by

$$P_{i,i}(z) - 1 = P_{i,i}(z) F_{i,i}(z). \tag{9.22}$$

This relationship provides an easy way to compute the first return probabilities
from the n-step transition probabilities. Note that if the transition probability matrix
is known, then calculating the generating function, $P_{i,i}(z)$, is straightforward. Recall
that $P^n = U \Lambda^n U^{-1}$, and hence if $[U]_{i,j}$ is the element in the ith row and jth column
of U, then

$$p_{i,i}^{(n)} = \sum_{j} [U]_{i,j} \Lambda^n [U^{-1}]_{j,i}. \tag{9.23}$$

Forming the generating function from this equation results in

$$P_{i,i}(z) = \sum_{n=0}^{\infty} p_{i,i}^{(n)} z^n = \sum_{n=0}^{\infty} \sum_{j} [U]_{i,j} (\Lambda z)^n [U^{-1}]_{j,i}$$

$$= \sum_{j} [U]_{i,j} \left(\sum_{n=0}^{\infty} (\Lambda z)^n \right) [U^{-1}]_{j,i}$$

$$= \sum_{j} [U]_{i,j} \left(\sum_{n=0}^{\infty} (\Lambda z)^n \right) [U^{-1}]_{j,i}$$

$$= \sum_{j} [U]_{i,j} (I - \Lambda z)^{-1} [U^{-1}]_{j,i}. \tag{9.24}$$

In other words, $P_{i,i}(z)$ is the element in the ith row and ith column of the matrix
$U(I - \Lambda z)^{-1} U^{-1}$.

---

[1] Note that these generating functions are not necessarily probability generating functions
because the sequences involved are not necessarily probability distributions.

DEFINITION 9.8: The ith state of a Markov chain is *transient* if $f_{i,i} < 1$ and *recurrent* if $f_{i,i} = 1$. Since $f_{i,i}$ represents the probability of the process eventually returning to state i given that it is in state i, the state is transient if there is some nonzero probability that it will never return, and the state is recurrent if the process must eventually return with probability 1.

THEOREM 9.2: State i of a Markov chain is recurrent if and only if

$$\sum_{n=1}^{\infty} p_{i,i}^{(n)} = \infty. \tag{9.25}$$

Since this sum represents the expected number of returns to state i, it follows that a state is recurrent if and only if the expected number of returns is infinite.

PROOF: First, note that $f_{i,i} = \sum_{n=1}^{\infty} f_{i,i}^{(n)} = \lim_{z \to 1} F_{i,i}(z)$. From Equation 9.22, this would imply that

$$\lim_{z \to 1} P_{i,i}(z) - 1 = \lim_{z \to 1} P_{i,i}(z) f_{i,i}. \tag{9.26}$$

As a result,

$$\lim_{z \to 1} P_{i,i}(z) = \sum_{n=1}^{\infty} p_{i,i}^{(n)} = \frac{1}{1 - f_{i,i}}. \tag{9.27}$$

If state i is transient, then $f_{i,i} < 1$ and hence,

$$\sum_{n=1}^{\infty} p_{i,i}^{(n)} < \infty,$$

whereas if state i is recurrent,

$$f_{i,i} = 1 \quad \text{and} \quad \sum_{n=1}^{\infty} p_{i,i}^{(n)} = \infty.$$

■

We leave it to the reader (see Exercise 9.21) to verify that recurrence is a class property. That is, if one state in a communicating class is recurrent, then all are recurrent, and if one is transient, then all are transient.

EXAMPLE 9.10: Consider a random walk on the integers (both positive and negative) that initially starts at the origin ($X_0 = 0$). At each time instant, the process either increases by 1 with probability p or decreases by 1 with probability $1 - p$:

$$p_{i,j} = \begin{cases} p & \text{if } j = i + 1 \\ 1 - p & \text{if } j = i - 1 \\ 0 & \text{otherwise} \end{cases}.$$

First note that this Markov chain is periodic with period 2 and hence $p_{i,i}^{(n)} = 0$ for any odd n. For even n, given the process is in state i, the process will be in state i again n time instants later if during those time instants the process increases $n/2$ times and decreases $n/2$ times. This probability follows a binomial distribution so that

$$p_{i,i}^{(n)} = \binom{n}{n/2} p^{n/2}(1 - p)^{n/2}, \quad \text{for even } n.$$

To determine if the states of this random walk are recurrent or transient, we must determine whether or not the series

$$\sum_{n=1}^{\infty} p_{i,i}^{(2n)} = \sum_{n=1}^{\infty} \binom{2n}{n} (p(1 - p))^n$$

converges. To help make this determination, the identity

$$\sum_{n=1}^{\infty} \binom{2n}{n} x^n = \frac{1}{\sqrt{1 - 4x}} - 1, \quad |x| < 1/4$$

is used. This identity can easily be verified by expanding the binomial on the right-hand side in powers of x and confirming (after a little algebra) that the coefficients of the power series expansion do take on the desired form. Applying this identity results in

$$\sum_{n=1}^{\infty} p_{i,i}^{(2n)} = \frac{1}{\sqrt{1 - 4p(1 - p)}} - 1.$$

Note that for a probability p, $4p(1 - p) \leq 1$ with equality if and only if $p = 1/2$. Hence, the series converges and all states are transient if $p \neq 1/2$, while if $p = 1/2$ the series diverges and all states are recurrent.

DEFINITION 9.9: The *mean time to first return* for a recurrent state i of a Markov chain is

$$\mu_i = \sum_{n=1}^{\infty} n f_{i,i}^{(n)}. \tag{9.28}$$

If the state is transient, then the mean time to first return must be infinite, since with some nonzero probability, the process will never return.

DEFINITION 9.10: A recurrent state is referred to as *null recurrent* if $\mu_i = \infty$, while the state is *positive recurrent* if $\mu_i < \infty$.

The mean time to first return of a recurrent state is related to the steady state probability of the process being in that state. To see this, define a sequence of random variables $T_1, T_2, T_3, \ldots$, where T_m represents the time between the $(m-1)$th and mth returns to the state i. That is, suppose that $X_k = i$ for some time instant k which is sufficiently large so that the process has pretty much reached steady state. The process then returns to state i at time instants $k+T_1, k+T_1+T_2, k+T_1+T_2+T_3$, and so on. Over some period of time where the process visits state i exactly n times, the fraction of time the process spends in state i can be written as

$$\text{fraction of time process is in state } i = \frac{n}{\displaystyle\sum_{j=1}^{n} T_j} = \frac{1}{\dfrac{1}{n}\displaystyle\sum_{j=1}^{n} T_j}. \tag{9.29}$$

As $n \to \infty$ (assuming the process is ergodic), the left-hand side of the previous equation becomes the steady state probability that the process is in state i, π_i. Furthermore, due to the law of large numbers, the denominator of the right-hand side converges to μ_i. This proves the following key result.

THEOREM 9.3: For an irreducible, aperiodic, recurrent Markov Chain, the steady state distribution is unique and is given by

$$\pi_i = \frac{1}{\mu_i}. \tag{9.30}$$

Note that if a state is positive recurrent then $\pi_i > 0$, while if a state is null recurrent then $\pi_i = 0$. Note that for any transient state, $\mu_i = \infty$ and as a result, $\pi_i = 0$.

EXAMPLE 9.11: Continuing with the random walk from the previous example, the generating function for the n-step transition probabilities is found to be

$$P_{i,i}(z) = \sum_{n=0}^{\infty} p_{i,i}^{(n)} z^n = \sum_{n=0}^{\infty} \binom{2n}{n} (p(1-p))^n z^{2n} = \frac{1}{\sqrt{1 - 4p(1-p)z^2}}.$$

Using the relationship $P_{i,i}(z) - 1 = P_{i,i}(z)F_{i,i}(z)$, the generating function for the first return probabilities is

$$F_{i,i}(z) = 1 - \sqrt{1 - 4p(1-p)z^2}.$$

Since the random walk is recurrent only for $p = 1/2$, we consider only that case so that $F_{i,i}(z) = 1 - \sqrt{1 - z^2}$. The mean time to first return can be found directly from the generating function.

$$\mu_i = \sum_{n=1}^{\infty} n f_{i,i}^{(n)} = \lim_{z \to 1} \frac{d}{dz} F_{i,i}(z) = \lim_{z \to 1} \frac{z}{\sqrt{1 - z^2}} = \infty$$

Thus, when the transition probabilities of the random walk are balanced so that all states are recurrent, then the mean time to first return is infinite and, in fact, all states are null recurrent.

9.4 Continuous Time Markov Processes

In this section, we investigate Markov processes where the time variable is continuous. In particular, most of our attention will be devoted to the so-called birth-death processes which are a generalization of the Poisson counting process studied in the previous chapter. To start with, consider a random process $X(t)$ whose state space is either finite or countable infinite so that we can represent the states of the process by the set of integers, $X(t) \in \{\ldots, -3, -2, -1, 0, 1, 2, 3, \ldots\}$. Any process of this sort that is a Markov process has the interesting property that the time between any change of states is an exponential random variable. To see this, define T_i to be the time between the ith and the $(i+1)$th change of state and let $h_i(t)$ be the complement to its CDF, $h_i(t) = \Pr(T_i > t)$. Then, for $t > 0, s > 0$,

$$h_i(t+s) = \Pr(T_i > t+s) = \Pr(T_i > t+s, T_i > s) = \Pr(T_i > t+s \,|\, T_i > s) \Pr(T_i > s).$$
$$(9.31)$$

Due to the Markovian nature of the process, $\Pr(T_i > t+s \,|\, T_i > s) = \Pr(T_i > t)$, and hence the previous equation simplifies to

$$h_i(t+s) = h_i(t)h_i(s). \qquad (9.32)$$

The only function which satisfies this type of relationship for arbitrary t and s is an exponential function of the form $h_i(t) = e^{-\rho_i t}$ for some constant ρ_i.

Furthermore, for this function to be a valid probability, the constant ρ_i must not be negative. From this, the PDF of the time between change of states is easily found to be $f_{T_i}(t) = \rho_i e^{-\rho_i t} u(t)$.

As with discrete time Markov chains, the continuous time Markov process can be described by its transition probabilities.

DEFINITION 9.11: Define $p_{i,j}(t) = \Pr(X(t_o + t) = j | X(t_o) = i)$ to be the transition probability for a continuous time Markov process. If this probability does not depend on t_o, then the process is said to be a *homogeneous* Markov process.

Unless otherwise stated, we assume for the rest of this chapter that all continuous time Markov processes are homogeneous. The transition probabilities, $p_{i,j}(t)$, are somewhat analogous to the n-step transition probabilities used in the study of discrete time processes and as a result, these probabilities satisfy a continuous time version of the Chapman-Kolmogorov equations:

$$p_{i,j}(t + s) = \sum_k p_{i,k}(t) p_{k,j}(s), \quad \text{for} \quad t, s > 0. \tag{9.33}$$

One of the most commonly studied class of continuous time Markov processes is the birth-death process. These processes get their name from applications in the study of biological systems, but they are also commonly used in the study of queueing theory, and many other applications. The birth-death process is similar to the discrete time random walk studied in the previous section in that when the process changes states, it either increases by 1 or decreases by 1. As with the Poisson counting process, the general class of birth-death processes can be described by the transition probabilities over an infinitesimal period of time, Δt. For a birth-death process,

$$p_{i,j}(\Delta t) = \begin{cases} \lambda_i \Delta t + o(\Delta t) & \text{if } j = i + 1 \\ \mu_i \Delta t + o(\Delta t) & \text{if } j = i - 1 \\ 1 - (\lambda_i + \mu_i)\Delta t + o(\Delta t) & \text{if } j = i \\ o(\Delta t) & \text{if } j \neq i - 1, i, i + 1 \end{cases} \tag{9.34}$$

The parameter λ_i is called the birth rate, while μ_i is the death rate when the process is in state i. In the context of queueing theory, λ_i and μ_i are referred to as the arrival and departure rates, respectively.

Similar to what was done with the Poisson counting process, by letting $s = \Delta t$ in Equation 9.33 and then applying the infinitesimal transition probabilities, a set of differential equations can be developed that will allow us to solve for the general transition probabilities. From Equation 9.33,

$$p_{i,j}(t + \Delta t) = \sum_k p_{i,k}(t)p_{k,j}(\Delta t)$$

$$= (\lambda_{j-1}\Delta t)p_{i,j-1}(t) + (1 - (\lambda_j + \mu_j)\Delta t)p_{i,j}(t) + (\mu_{j+1}\Delta t)p_{i,j+1}(t) + o(\Delta t). \quad (9.35)$$

Rearranging terms and dividing by Δt produces

$$\frac{p_{i,j}(t + \Delta t) - p_{i,j}(t)}{\Delta t} = \lambda_{j-1}p_{i,j-1}(t) - (\lambda_j + \mu_j)p_{i,j}(t) + \mu_{j+1}p_{i,j+1}(t) + \frac{o(\Delta t)}{\Delta t}. \tag{9.36}$$

Finally, passing to the limit as $\Delta t \to 0$ results in

$$\frac{d}{dt}p_{i,j}(t) = \lambda_{j-1}p_{i,j-1}(t) - (\lambda_j + \mu_j)p_{i,j}(t) + \mu_{j+1}p_{i,j+1}(t). \tag{9.37}$$

This set of equations is referred to as the *forward Kolmogorov equations*. One can follow a similar procedure (see Exercise 9.24) to develop a slightly different set of equations known as the *backward Kolmogorov equations*:

$$\frac{d}{dt}p_{i,j}(t) = \lambda_i p_{i+1,j}(t) - (\lambda_i + \mu_i)p_{i,j}(t) + \mu_i p_{i-1,j}(t). \tag{9.38}$$

For all but the simplest examples, it is very difficult to find a closed form solution for this system of equations. However, the Kolmogorov equations can lend some insight into the behavior of the system. For example, consider the steady state distribution of the Markov process. If a steady state exists, we would expect that as $t \to \infty$, $p_{i,j}(t) \to \pi_j$ independent of i and also that $dp_{i,j}(t)/(dt) \to 0$. Plugging these simplifications into the forward Kolmogorov equations leads to

$$\lambda_{j-1}\pi_{j-1} - (\lambda_j + \mu_j)\pi_j + \mu_{j+1}\pi_{j+1} = 0. \tag{9.39}$$

These equations are known as the *global balance equations*. From them, the steady state distribution can be found (if it exists). The solution to the balance equations is surprisingly easy to obtain. First, we rewrite the difference equation in the more symmetric form

$$\lambda_j\pi_j - \mu_{j+1}\pi_{j+1} = \lambda_{j-1}\pi_{j-1} - \mu_j\pi_j. \tag{9.40}$$

Next, assume that the Markov process is defined on the states $j = 0, 1, 2, \ldots$. Then the previous equation must be adjusted for the end point $j = 0$ (assuming $\mu_0 = 0$ which merely states that there can be no deaths when the population's size is zero) according to

$$\lambda_0\pi_0 - \mu_1\pi_1 = 0. \tag{9.41}$$

Combining Equations 9.40 and 9.41 results in

$$\lambda_j \pi_j - \mu_{j+1}\pi_{j+1} = 0, \quad j = 0, 1, 2, \ldots, \tag{9.42}$$

which leads to the simple recursion

$$\pi_{j+1} = \frac{\lambda_j}{\mu_{j+1}}\pi_j, \quad j = 0, 1, 2, \ldots, \tag{9.43}$$

whose solution is given by

$$\pi_j = \pi_0 \prod_{i=1}^{j} \frac{\lambda_{i-1}}{\mu_i}, \quad j = 1, 2, 3, \ldots. \tag{9.44}$$

This gives the π_j in terms of π_0. In order to determine π_0, the constraint that the π_j must form a distribution is imposed.

$$\sum_{j=0}^{\infty} \pi_j = 1 \Rightarrow \pi_0 = \frac{1}{1 + \sum_{j=1}^{\infty}\prod_{i=1}^{j}\frac{\lambda_{i-1}}{\mu_i}}. \tag{9.45}$$

This completes the proof of the following theorem.

THEOREM 9.4: For a Markov birth-death process with birth rate λ_n, $n = 0, 1, 2, \ldots$, and death rate μ_n, $n = 1, 2, 3, \ldots$, the steady state distribution is given by

$$\pi_k = \lim_{t \to \infty} p_{i,k}(t) = \frac{\prod_{i=1}^{k}\frac{\lambda_{i-1}}{\mu_i}}{1 + \sum_{j=1}^{\infty}\prod_{i=1}^{j}\frac{\lambda_{i-1}}{\mu_i}}. \tag{9.46}$$

If the series in the denominator diverges, then $\pi_k = 0$ for any finite k. This indicates that a steady state distribution does not exist. Likewise, if the series converges, the π_k will be nonzero, resulting in a well-behaved steady state distribution.

EXAMPLE 9.12: (The M/M/1 Queue) In this example, we consider the birth-death process with constant birth rate and constant death rate. In particular, we take

$$\lambda_n = \lambda, \quad n = 0, 1, 2, \ldots \quad \text{and} \quad \mu_0 = 0, \ \mu_n = \mu, \quad n = 1, 2, 3, \ldots.$$

This model is commonly used in the study of queueing systems and, in that context, is referred to as the M/M/1 queue. In this nomenclature, the first "M" refers to the arrival process as being Markovian, the second "M" refers to the departure process as being Markovian, and the "1" is the number of servers. So this is a single server queue, where the interarrival time of new customers is an exponential random variable with mean $1/\lambda$ and the service time for each customer is exponential with mean $1/\mu$. For the M/M/1 queueing system, $\lambda_{i-1}/\mu_i = \lambda/\mu$ for all i so that

$$1 + \sum_{j=1}^{\infty} \prod_{i=1}^{j} \frac{\lambda_{i-1}}{\mu_i} = \sum_{j=0}^{\infty} \left(\frac{\lambda}{\mu} \right)^j = \frac{1}{1 - \lambda/\mu} \quad \text{for } \lambda < \mu.$$

The resulting steady state distribution of the queue size is then

$$\pi_k = \frac{(\lambda/\mu)^k}{\dfrac{1}{1 - \lambda/\mu}} = (1 - \lambda/\mu)(\lambda/\mu)^k, \quad k = 0, 1, 2, \ldots, \quad \text{for } \lambda < \mu.$$

Hence, if the arrival rate is less than the departure rate, the queue size will have a steady state. It makes sense that if the arrival rate is greater than the departure rate, then the queue size will tend to grow without bound.

EXAMPLE 9.13: (The M/M/∞ Queue) Next suppose the last example is modified so that there are an infinite number of servers available to simultaneously provide service to all customers in the system. In that case, there are no customers ever waiting in line, and the process $X(t)$ now counts the number of customers in the system (receiving service) at time t. As before, we take the arrival rate to be constant $\lambda_n = \lambda$, but now the departure rate needs to be proportional to the number of customers in service, $\mu_n = n\mu$. In this case, $\lambda_{i-1}/\mu_i = \lambda/(i\mu)$ and

$$1 + \sum_{j=1}^{\infty} \prod_{i=1}^{j} \frac{\lambda_{i-1}}{\mu_i} = 1 + \sum_{j=1}^{\infty} \prod_{i=1}^{j} \frac{\lambda}{i\mu} = 1 + \sum_{j=1}^{\infty} \frac{(\lambda/\mu)^j}{j!} = e^{\lambda/\mu}.$$

Note that the series converges for any λ and μ, and hence the M/M/∞ queue will always have a steady state distribution given by

$$\pi_k = \frac{(\lambda/\mu)^k}{k!} e^{-\lambda/\mu}.$$

EXAMPLE 9.14: This example demonstrates one way to simulate the M/M/1 queueing system of Example 9.12. One realization of this process as produced by the code that follows is illustrated in Figure 9.4. In generating the figure, we use an average arrival rate of $\lambda = 20$ customers per hour and an average service time of $1/\mu = 2$ minutes. This leads to the condition $\lambda < \mu$ and the M/M/1 queue exhibits stable behavior. The reader is encouraged to run the program for the case when $\lambda > \mu$ to observe the unstable behavior (the queue size will tend to grow continuously over time).

```
a=20;                    % Arrival rate (customers/hour).
b=30;                    % Departure rate (1/b=avg service
                         time).
N=25;                    % Number of arrivals in simulation.
X=-log(rand(1,N))/a;     % Random interarrival times.
X=cumsum(X);             % Random arrival times.
Y=-log(rand(1,N))/b;     % Service times for each customer.
serv_start=X(1);         % First customer starts service
                         % immediately upon arrival.
```

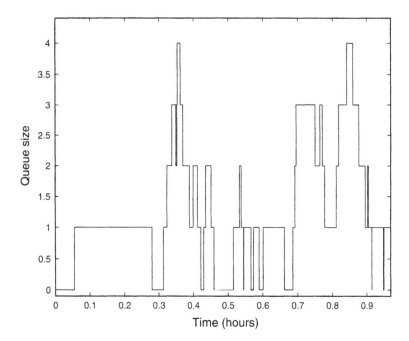

Figure 9.4 Simulated realization of the birth-death process for M/M/1 queueing system of Example 9.12.

```
Z(1)=serv_start+Y(1);              % Departure time of first
                                   customer.
for k=2:N                          % kth customer.
   serv_start=max([Z(k-1), X(k)]); % Beginning of service
                                   time.
   Z(k)=serv_start+Y(k);           % End of service time.
end

% Construct data to plot graph of queue size vs. time.
xaxis=[0, X(1)];                   % Vector of points for
                                   the M/M/1.
                                   % Birth-death process
yaxis=[0, 0];                      % Vector of queue sizes at
                                   points in preceding vector.
qs=1;                              % Current queue size.
X=X(2:length(X));
while length(X)>0
   if X(1)<Z(1)                    % Next point is arrival.
      qs=qs+1;                     % Increase queue size.
      xaxis=[xaxis xaxis(length(xaxis)) X(1)];
      yaxis=[yaxis qs qs];
      X=X(2:length(X));
   else                            % Next point is departure.
      qs=qs-1; % decrease queue size
      xaxis=[xaxis xaxis(length(xaxis)) Z(1)];
      yaxis=[yaxis qs qs];
      Z=Z(2:length(Z));
   end
end
plot(xaxis,yaxis)                  % Plot realization of
                                   % birth-death process.

xlabel('time (hours)');
ylabel('queue size')
```

If the birth-death process is truly modeling the size of a population of some organism, then it would be reasonable to consider the case when $\lambda_0 = 0$. That is, when the population size reaches zero, no further births can occur. In that case, the species is extinct and the state $X(t) = 0$ is an absorbing state. A fundamental question would then be, Is extinction a certain event, and if not what is the probability of the process being absorbed into the state of extinction? Naturally the answer to this question would depend on the starting population size. Let q_i be the probability that the process eventually enters the absorbing state, given that it is initially in state i. Note that if the process is currently in state i, after the next transition, the birth-death process must be either in state $i - 1$ or state $i + 1$. The time to the next birth, B_i, is a random variable with an exponential distribution with a mean of $1/\lambda_i$, while the time to the next death is an exponential random variable, D_i, with a mean of $1/\mu_i$. Hence, the process will transition to state $i + 1$ if $B_i < D_i$, otherwise it will transition to state $i - 1$. The reader can easily verify that $\Pr(B_i < D_i) = \lambda_i/(\lambda_i + \mu_i)$. The absorption probability can then be written as

$$
\begin{aligned}
q_i &= \Pr\,(\text{absorption} \mid \text{in state } i) \\
&= \Pr\,(\text{absorption, next state is } i + 1 | \text{in state } i) \\
&\quad + \Pr\,(\text{absorption, next state is } i - 1 | \text{ in state } i) \\
&= \Pr\,(\text{absorption} \mid \text{in state } i + 1)\,\Pr\,(\text{next state is } i + 1 | \text{ in state } i) \\
&\quad + \Pr\,(\text{absorption} \mid \text{in state } i - 1)\,\Pr\,(\text{next state is } i - 1 | \text{ in state } i) \\
&= q_{i+1}\frac{\lambda_i}{\lambda_i + \mu_i} + q_{i-1}\frac{\mu_i}{\lambda_i + \mu_i}, \quad i = 1, 2, 3, \ldots.
\end{aligned}
\tag{9.47}
$$

This provides a recursive set of equations that can be solved to find the absorption probabilities. To solve this set of equations, we rewrite them as

$$
q_{i+1} - q_i = \frac{\mu_i}{\lambda_i}(q_i - q_{i-1}), \quad i = 1, 2, 3, \ldots.
\tag{9.48}
$$

After applying this recursion repeatedly and using the fact that $q_0 = 1$,

$$
q_{i+1} - q_i = (q_1 - 1)\prod_{j=1}^{i}\frac{\mu_j}{\lambda_j}.
\tag{9.49}
$$

Summing this equation from $i = 1, 2, \ldots, n$ results in

$$
q_{n+1} - q_1 = (q_1 - 1)\sum_{i=1}^{n}\prod_{j=1}^{i}\frac{\mu_j}{\lambda_j}.
\tag{9.50}
$$

Next, suppose that the series on the right-hand side of the previous equation diverges as $n \to \infty$. Since the q_i are probabilities, the left-hand side of the equation must be bounded, which implies that $q_1 = 1$. Then from Equation 9.49, it is determined that q_n must be equal to one for all n. That is, if

$$
\sum_{i=1}^{\infty}\prod_{j=1}^{i}\frac{\mu_j}{\lambda_j} = \infty,
\tag{9.51}
$$

then absorption will eventually occur with probability 1 regardless of the starting state. If $q_1 < 1$ (absorption is not certain), then the preceding series must converge to a finite number. It is expected in that case that as $n \to \infty$, $q_n \to 0$. Passing to the limit as $n \to \infty$ in Equation 9.50 then allows a solution for q_1 of the form

$$q_1 = \frac{\sum_{i=1}^{\infty} \prod_{j=1}^{i} \frac{\mu_j}{\lambda_j}}{1 + \sum_{i=1}^{\infty} \prod_{j=1}^{i} \frac{\mu_j}{\lambda_j}}. \tag{9.52}$$

Furthermore, the general solution for the absorption probability is

$$q_n = \frac{\sum_{i=n}^{\infty} \prod_{j=1}^{i} \frac{\mu_j}{\lambda_j}}{1 + \sum_{i=1}^{\infty} \prod_{j=1}^{i} \frac{\mu_j}{\lambda_j}}. \tag{9.53}$$

EXAMPLE 9.15: Consider a population model where both the birth and death rates are proportional to the population, $\lambda_n = n\lambda$, $\mu_n = n\mu$. For this model,

$$\sum_{i=1}^{\infty} \prod_{j=1}^{i} \frac{\mu_j}{\lambda_j} = \sum_{i=1}^{\infty} \prod_{j=1}^{i} \frac{\mu}{\lambda} = \sum_{i=1}^{\infty} \left(\frac{\mu}{\lambda}\right)^i = \frac{\mu/\lambda}{1 - \mu/\lambda} = \frac{\mu}{\lambda - \mu} \quad \text{for} \quad \lambda > \mu.$$

Hence, if $\lambda < \mu$, the series diverges and the species will eventually reach extinction with probability 1. If $\lambda > \mu$,

$$\sum_{i=n}^{\infty} \prod_{j=1}^{i} \frac{\mu_j}{\lambda_j} = \sum_{i=n}^{\infty} \left(\frac{\mu}{\lambda}\right)^i = \frac{(\mu/\lambda)^n}{1 - \mu/\lambda},$$

and the absorption (extinction) probabilities are

$$q_n = \left(\frac{\mu}{\lambda}\right)^n, \quad n = 1, 2, 3, \ldots.$$

Continuous time Markov processes do not necessarily need to have a discrete amplitude as in the previous examples. In the following, we discuss a class of continuous time, continuous amplitude Markov processes. To start with, it is noted that for any time instants $t_0 < t_1 < t_2$, the conditional PDF of a Markov process must satisfy the Chapman-Kolmogorov equation

$$f(x_2, t_2 | x_0, t_0) = \int_{-\infty}^{\infty} f(x_2, t_2 | x_1, t_1) f(x_1, t_1 | x_0, t_0) \, dx_1. \tag{9.54}$$

This is just the continuous amplitude version of Equation 9.33. Here we use the notation $f(x_2, t_2 | x_1, t_1)$ to represent the conditional probability density of the process $X(t_2)$ at the point x_2 conditioned on $X(t_1) = x_1$. Next, suppose we interpret these time instants as $t_0 = 0$, $t_1 = t$, and $t_2 = t + \Delta t$. In this case, we interpret $x_2 - x_1 = \Delta x$ as the the infinitesimal change in the process that occurs during the infinitesimal time instant Δt and $f(x_2, t_2 | x_1, t_1)$ is the PDF of that increment.

Define $\Phi_{\Delta x}(\omega)$ to be the characteristic function of $\Delta x = x_2 - x_1$:

$$\Phi_{\Delta x}(\omega) = E[e^{j\omega \Delta x}] = \int_{-\infty}^{\infty} e^{j\omega(x_2 - x_1)} f(x_2, t + \Delta t | x_1, t) \, dx_2. \tag{9.55}$$

We note that the characteristic function can be expressed in a Taylor series as

$$\Phi_{\Delta x}(\omega) = \sum_{k=0}^{\infty} \frac{M_k(x_1, t)}{k!} (j\omega)^k, \tag{9.56}$$

where $M_k(x_1, t) = E[(x_2 - x_1)^k | (x_1, t)]$ is the kth moment of the increment Δx. Taking inverse transforms of this expression, the conditional PDF can be expressed as

$$f(x_2, t + \Delta t | x_1, t) = \sum_{k=0}^{\infty} \frac{M_k(x_1, t)}{k!} (-1)^k \frac{\partial^k}{\partial x_2^k} (\delta(x_2 - x_1)). \tag{9.57}$$

Inserting this result into the Chapman-Kolmogorov equation, Equation 9.54, results in

$$f(x_2, t + \Delta t | x_0, t_0) = \sum_{k=0}^{\infty} \frac{(-1)^k}{k!} \int_{-\infty}^{\infty} M_k(x_1, t) \frac{\partial^k}{\partial x_2^k} \delta(x_2 - x_1) f(x_1, t | x_0, t_0) \, dx_1$$

$$= \sum_{k=0}^{\infty} \frac{(-1)^k}{k!} \frac{\partial^k}{\partial x_2^k} [M_k(x_2, t) f(x_2, t | x_0, t_0)]$$

$$= f(x_2, t | x_0, t_0) + \sum_{k=1}^{\infty} \frac{(-1)^k}{k!} \frac{\partial^k}{\partial x_2^k} [M_k(x_2, t) f(x_2, t | x_0, t_0)]. \tag{9.58}$$

Subtracting $f(x_2, t|x_0, t_0)$ from both sides of this equation and dividing by Δt results in

$$\frac{f(x_2, t + \Delta t|x_0, t_0) - f(x_2, t|x_0, t_0)}{\Delta t} = \sum_{k=1}^{\infty} \frac{(-1)^k}{k!} \frac{\partial^k}{\partial x_2^k} \left[\frac{M_k(x_2, t)}{\Delta t} f(x_2, t|x_0, t_0)\right].$$

$$(9.59)$$

Finally, passing to the limit as $\Delta t \to 0$ results in the partial differential equation

$$\frac{\partial}{\partial t} f(x, t|x_0, t_0) = \sum_{k=1}^{\infty} \frac{(-1)^k}{k!} \frac{\partial^k}{\partial x^k} [K_k(x, t)f(x, t|x_0, t_0)], \qquad (9.60)$$

where the function $K_k(x, t)$ is defined as

$$K_k(x, t) = \lim_{\Delta t \to 0} \frac{E[(X(t + \Delta t) - X(t))^k | X(t)]}{\Delta t}. \qquad (9.61)$$

For many processes of interest, the PDF of an infinitesimal increment can be accurately approximated from its first few moments, and hence we take $K_k(x, t) = 0$ for $k > 2$. For such processes, the PDF must satisfy

$$\frac{\partial}{\partial t} f(x, t|x_0, t_0) = -\frac{\partial}{\partial x} (K_1(x, t)f(x, t|x_0, t_0)) + \frac{1}{2} \frac{\partial^2}{\partial x^2} (K_2(x, t)f(x, t|x_0, t_0)). \quad (9.62)$$

This is known as the (one-dimensional) *Fokker-Planck equation* and is used extensively in diffusion theory to model the dispersion of fumes, smoke, and similar phenomena.

In general, the Fokker-Planck equation is notoriously difficult to solve and doing so is well beyond the scope of this text. Instead, we consider a simple special case where the functions $K_1(x, t)$ and $K_2(x, t)$ are constants, in which case the Fokker Planck equation reduces to

$$\frac{\partial}{\partial t} f(x, t|x_0, t_0) = -2c \frac{\partial}{\partial x} (f(x, t|x_0, t_0)) + D \frac{\partial^2}{\partial x^2} (f(x, t|x_0, t_0)), \qquad (9.63)$$

where in diffusion theory, D is known as the coefficient of diffusion and c is the drift. This equation is used in models that involve the diffusion of smoke or other pollutants in the atmosphere, the diffusion of electrons in a conductive medium, the diffusion of liquid pollutants in water and soil, and the diffusion of plasmas. This equation can be solved in several ways. Perhaps one of the easiest methods is to use Fourier transforms. This is explored further in the exercises where the reader is asked to show that (taking $x_0 = 0$ and $t_0 = 0$) the solution to this diffusion equation is

$$f(x, t|x_0 = 0, t_0 = 0) = \frac{1}{\sqrt{4\pi Dt}} \exp\left(-\frac{(x - 2ct)^2}{4Dt}\right). \qquad (9.64)$$

That is, the PDF is Gaussian with a mean and variance that changes with time. The behavior of this process is explored in the next example.

EXAMPLE 9.16: In this example, we model the diffusion of smoke from a forest fire that starts in a National Park at time $t = 0$ and location $x = 0$. The smoke from the fire drifts in the positive x direction due to wind blowing at 10 miles per hour, and the diffusion coefficient is 1 square mile per hour. The probability density function is given in Equation 9.64. We provide a three-dimensional rendition of this function in Figure 9.5 using the following MATLAB program.

```
c=10;                           % Drift.
D=1;                            % Diffusion coefficient.
tpoints=[0.25, 0.5, 1, 1.5, 2]; % Time samples.
x=[0:0.1:50];                   % x-axis.
```

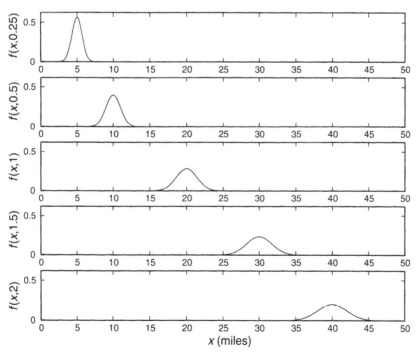

Figure 9.5 Observations of the PDF at different time instants showing the drift and dispersion of smoke for Example 9.16.

```
for k=1:length(tpoints)
    t=tpoints(k);                           % Set t.
    pdf(k,:)=exp(-(x-2*c*t).^2/(4*D*t))/sqrt(4*pi*D*t);
                                            % f(x,t).
    subplot(5,1,k)
    plot(x,pdf(k,:))                        % Plot PDF.
    axis([0 max(x) 0 1.1*max(max(pdf))])
    s=num2str(t);
    leftstr='f(x,';
    rightstr=')';
    txt=[leftstr s rightstr];
    ylabel(txt)
end
xlabel('x (miles)')
```

9.5 Engineering Application: A Computer Communication Network

Consider a local area computer network where a cluster of nodes is connected by a common communications line. Suppose for simplicity that these nodes occasionally need to transmit a message of some fixed length (or, number of packets). Also, assume that the nodes are synchronized so that time is divided into slots, each of which is sufficiently long to support one packet. In this example, we consider a random access protocol known as *slotted Aloha*. Messages (packets) are assumed to arrive at each node according to a Poisson process. Assuming there are a total of n nodes, the packet arrival rate at each node is assumed to be λ/n so that the total arrival rate of packets is fixed at λ packets/slot. In slotted Aloha, every time a new packet arrives at a node, that node attempts to transmit that packet during the next slot. During each slot, one of three events can occur: (1) no node attempts to transmit a packet, in which case the slot is said to be idle; (2) exactly one node attempts to transmit a packet, in which case the transmission is successful; or (3) more than one node attempts to transmit a packet, in which case a collision is said to have occurred.

All nodes involved in a collision will need to retransmit their packets, but if they all retransmit during the next slot, then they will continue to collide and the packets will never be successfully transmitted. All nodes involved in a collision are said to be backlogged until their packet is successfully transmitted. In the slotted Aloha protocol, each backlogged node chooses to transmit during the next slot with probability p (and hence chooses not to transmit during the next slot with probability $1 - p$). Viewed in an alternative manner, every time a collision occurs, each node involved waits a random amount of time until they attempt retransmission, where that random time follows a geometric distribution.

This computer network can be described by a Markov chain, X_k = number of backlogged nodes at the end of the kth slot. To start with, we evaluate the transition probabilities of the Markov chain, $p_{i,j}$. Assuming that there are an infinite number of nodes (or a finite number of nodes, each of which could store an arbitrary number of backlogged packets in a buffer), we note that

$$\Pr(m \text{ backlogged nodes attempt to transmit}|X_k = n) = \binom{n}{m} p^m (1 - p)^{n-m}, \quad (9.65)$$

$$\Pr(m \text{ new arrivals}|X_k = n) = \frac{\lambda^m}{m!} e^{-\lambda}. \quad (9.66)$$

Using these equations, it is straightforward to determine that the transition probabilities are given by

$$p_{i,j} = \begin{cases} 0 & \text{for } j < i - 1 \\ ip(1 - p)^{i-1} e^{-\lambda} & \text{for } j = i - 1 \\ (1 + \lambda(1 - p)^i - ip(1 - p)^{i-1}) e^{-\lambda} & \text{for } j = i \\ (1 - (1 - p)^i) \lambda e^{-\lambda} & \text{for } j = i + 1 \\ \dfrac{\lambda^{j-i}}{(j - i)!} e^{-\lambda} & \text{for } j > i + 1 \end{cases} \quad (9.67)$$

In order to get a feeling for the steady state behavior of this Markov chain, we define the drift of the chain in state i as

$$d_i = E[X_{k+1}|X_k = i] - i. \quad (9.68)$$

Given that the chain is in state i, if the drift is positive, then the number of backlogged nodes will tend to increase; whereas, if the drift is negative, the number of backlogged nodes will tend to decrease. Crudely speaking, a drift of zero represents some sort of equilibrium for the Markov chain. Given the preceding transition probabilities a, the drift works out to be

$$d_i = \lambda - (1 - p)^{i-1} e^{-\lambda} [ip + \lambda(1 - p)]. \quad (9.69)$$

Assuming that $p \ll 1$, then we can use the approximations $(1 - p) \approx 1$ and $(1 - p)^i \approx e^{-ip}$ to simplify the expression for the drift:

$$d_i \approx \lambda - g(i) e^{-g(i)}, \quad \text{where } g(i) = \lambda + ip. \quad (9.70)$$

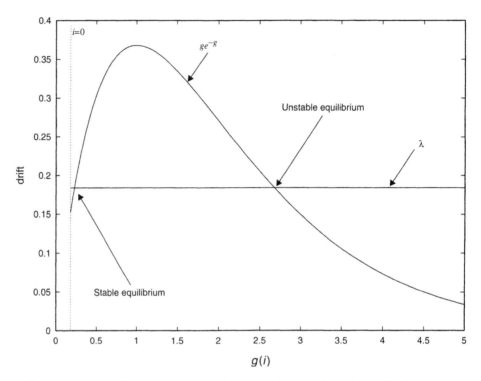

Figure 9.6 Arrival rate and successful transmission rate for a slotted Aloha system.

The parameter $g(i)$ has the physical interpretation of the average number of transmissions per slot given that there are i backlogged states. To understand the significance of this result, the two terms in the expression for the drift are plotted in Figure 9.6. The first term, λ, has the interpretation of the average number of new arrivals per slot, while the second term, $g\exp(-g)$, is the average number of successful transmissions per slot or the average departure rate. For a very small number of backlogged states, the arrival rate is greater than the departure rate and the number of backlogged states tends to increase. For moderate values of i, the departure rate is greater than the arrival rate and the number of backlogged states tends to decrease. Hence, the drift of the Markov chain is such that the system tends to stabilize around the point marked stable equilibrium in Figure 9.6. This is the first point where the two curves cross. Note, however, that for very large i, the drift becomes positive again. If the number of backlogged states ever becomes large enough to push the system to the right of the point marked unstable equilibrium in the figure, then the number of backlogged nodes will tend to grow without bound and the system will become unstable.

Note that the value of λ represents the throughput of the system. If we try to use a value of λ that is greater than the peak value of $g\exp(-g)$, then the drift will always be positive and the system will be unstable from the beginning. This maximum throughput occurs when $g(i) = 1$ and has a value of $\lambda_{max} = 1/e$. By choosing an arrival rate less than λ_{max}, we can get the system to operate near the stable equilibrium, but sooner or later, we will get a string of bad luck and the system will drift into the unstable region. The lower the arrival rate, the longer it will take (on average) for the system to become unstable, but at any arrival rate, the system will eventually reach the unstable region. Hence, slotted Aloha is inherently an unstable protocol. As a result, various modifications have been proposed that exhibit stable behavior.

9.6 Engineering Application: A Telephone Exchange

Consider a base station in a cellular phone system. Suppose calls arrive at the base station according to a Poisson process with some arrival rate λ. These calls are initiated by a mobile unit within the cell served by that base station. Furthermore, suppose each call has a duration that is an exponential random variable with some mean, $1/\mu$. The base station has some fixed number of channels, m, that can be used to service the demands of the mobiles in its cell. If all m channels are being used, any new call that is initiated cannot be served and the call is said to be blocked. We are interested in calculating the probability that when a mobile initiates a call, the customer is blocked.

Since the arrival process is memoryless and the departure process is memoryless, the number of calls being serviced by the base station at time t, $X(t)$, is a birth-death Markov process. Here, the arrival rate and departure rates (given there are n channels currently being used) are given by

$$\lambda_n = \begin{cases} \lambda & 0 \leq n < m \\ 0 & n = m \end{cases}, \quad \mu_n = n\mu, \ 0 \leq n \leq m. \tag{9.71}$$

The steady state distribution of this Markov process is given by Equation 9.46. For this example, the distribution is found to be

$$\pi_n = \frac{\displaystyle\prod_{i=1}^{n} \frac{\lambda_{i-1}}{\mu_i}}{1 + \displaystyle\sum_{j=1}^{m} \prod_{i=1}^{j} \frac{\lambda_{i-1}}{\mu_i}} = \frac{\dfrac{1}{n!}\left(\dfrac{\lambda}{\mu}\right)^n}{\displaystyle\sum_{j=0}^{m} \frac{1}{j!}\left(\frac{\lambda}{\mu}\right)^j}. \tag{9.72}$$

The blocking probability is just the probability that when a call is initiated, it finds the system in state m. In steady state, this is given by π_m and the resulting blocking probability is the so-called Erlang-B formula,

$$\Pr(\text{blocked call}) = \frac{\dfrac{1}{m!}\left(\dfrac{\lambda}{\mu}\right)^m}{\displaystyle\sum_{j=0}^{m}\dfrac{1}{j!}\left(\dfrac{\lambda}{\mu}\right)^j}. \tag{9.73}$$

This equation is plotted in Figure 9.7 for several values of m. The horizontal axis is the ratio of λ/μ which is referred to in the telephony literature as the traffic intensity. As an example of the use of this equation, suppose a certain base station had 60 channels available to service incoming calls. Furthermore, suppose each user initiated calls at a rate of 1 call per 3 hours and calls had an average duration

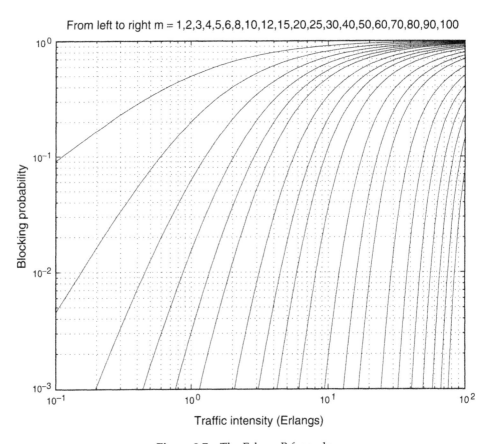

Figure 9.7 The Erlang-B formula.

of 3 minutes (0.05 hours). If a 2 percent probability of blocking is desired, then from Figure 9.7 we determine that the system can handle a traffic intensity of approximately 50 Erlangs. Note that each user generates an intensity of

$$\frac{\lambda}{\mu} = \frac{1/(3 \text{ hour})}{1/(0.05 \text{ hour})} = \frac{1}{60} \text{ Erlangs.} \tag{9.74}$$

Hence, a total of $50 * 60 = 3,000$ mobile users per cell could be supported while still maintaining a 2 percent blocking probability.

Concluding Remarks

This chapter has covered a class of random process that can be described as memoryless, with emphasis on Markov Processes. While the material is difficult, it is nonetheless important in that it is applicable to numerous applications such as queuing systems and computer communications networks, as illustrated by the last two engineering applications examples. The reader is encouraged to master this material by completing as many of the exercises that follow as possible.

Exercises

9.1 Consider a two-state Markov chain with a general transition probability matrix

$$P = \begin{bmatrix} 1-p & p \\ q & 1-p \end{bmatrix},$$

where $0 < p, q < 1$. Find an expression for the n-step transition probability matrix, P^n.

9.2 For the general two-state Markov chain of Exercise 9.1, suppose the states are called 0 and 1. Furthermore, suppose $\Pr(X_0 = 0) = s$ and $\Pr(X_0 = 1) = 1 - s$.

(a) Find $\Pr(X_1 = 0, X_2 = 1)$.
(b) Find $\Pr(X_1 = 1 | X_0 = 0, X_2 = 0)$.
(c) Find $\Pr(X_2 = X_1)$. Is it the same as $\Pr(X_1 = X_0)$?

9.3 For a Markov chain, prove or disprove the following statement:

$$\Pr(X_k = i_k | X_{k+1} = i_{k+1}, \ X_{k+2} = i_{k+2}, \ldots, X_{k+m} = i_{k+m})$$
$$= \Pr(X_k = i_k | X_{k+1} = i_{k+1}).$$

9.4 A square matrix P is called a stochastic matrix if all of its elements satisfy $0 \leq p_{i,j} \leq 1$ and, furthermore,

$$\sum_j p_{i,j} = 1 \quad \text{for all } i.$$

Every stochastic matrix is the transition probability matrix for some Markov chain; however, not every stochastic matrix is a valid two-step transition probability matrix. Prove that a 2×2 stochastic matrix is a valid two-step transition probability matrix for a two-state Markov chain if and only if the sum of the diagonal elements is greater than or equal to 1.

9.5 A random waveform is generated as follows. The waveform starts at 0 voltage. Every t_s seconds, the waveform switches to a new voltage level. If the waveform is at a voltage level of 0 volts, it may move to +1 volt with probability p, or it may move to -1 volt with probability $q = 1-p$. Once the waveform is at +1 (or -1), the waveform will return (with probability 1) to 0 volts at the next switching instant.

(a) Model this process as a Markov Chain. Describe the states of the system and give the transition probability matrix.

(b) Determine whether each state is periodic or aperiodic. If periodic, determine the period of each state.

(c) For each instant of time, determine the PDF for the value of the waveform.

9.6 Model the diffusion of electrons and holes across a potential barrier in an electronic device as follows. We have n black balls (electrons) in urn A and n whiteballs (holes) in urn B. An experimental outcome selects randomly one ball from each urn. The ball from urn A is placed in urn B and that from urn B is placed in A. Let the state of the process be the number of black balls in urn A. (By knowing the number of black balls in urn A, we know the composition of both urns.) Let k denote the state of the process. Find the transition probabilities, $p_{i,j}$.

9.7 A PCM waveform has the two states +1 and 0. Suppose the transition matrix is

$$P = \begin{bmatrix} 0.5 & 0.5 \\ 0.25 & 0.75 \end{bmatrix}.$$

The initial value of the waveform is determined by the flip of a coin, with the outcome of a head corresponding to +1 and a tail to 0.

(a) What is the probability that the waveform will be at $+1$ after one step if the coin is a fair coin?

(b) Find the same probability if the coin is biased such that a head occurs with probability $1/3$.

(c) Repeat the problem for two steps.

9.8 A student takes this course at period 1 on Monday, Wednesday, and Friday. Period 1 starts at 7:25 A.M. Consequently, the student sometimes misses class. The student's attendance behavior is such that she attends class depending only on whether or not she went to the last class. If she attended class on one day, then she will go to class the next time it meets with probability $1/2$. If she did not go to one class, then she will go to the next class with probability $3/4$.

(a) Find the transition matrix P.

(b) Find the probability that if she went to class on Wednesday, she will attend class on Friday.

(c) Find the probability that if she went to class on Monday, she will attend class on Friday.

(d) Does the Markov chain described by this transition matrix have a steady state distribution? If so, find that distribution.

9.9 A three-state Markov chain has the following transition matrix:

$$P = \begin{bmatrix} 0.25 & 0.5 & 0.25 \\ 0.4 & 0.6 & 0 \\ 1 & 0 & 0 \end{bmatrix}.$$

(a) Does this Markov chain have a unique steady state probability vector? If so, find it.

(b) What is the approximate value of $p_{1,3}^{(100)}$? What interpretation do you give to this result?

(c) What is the probability that after the third step you are in state 3 if the initial state probability vector is $(1/3 \; 1/3 \; 1/3)$?

9.10 The three letters C, A, and T represent the states of a word-generating system. Let the initial state probability vector be $(1/3 \; 1/3 \; 1/3)$ for the three letters, respectively. The transition matrix is given as

$$\begin{array}{c} \\ \\ P = \begin{array}{c} C \\ A \\ T \end{array} \end{array} \begin{array}{ccc} C & A & T \\ \begin{bmatrix} 0.1 & 0.7 & 0.2 \\ 0.6 & 0.1 & 0.3 \\ 0.1 & 0.8 & 0.1 \end{bmatrix} \end{array}.$$

What is the probability of generating a proper three-letter English word after two transitions from the initial state?

9.11 Two students play the following game. Two dice are tossed. If the sum of the numbers showing is less than 7, student A collects a dollar from student B. If the total is greater than 7, then student B collects a dollar from student A. If a 7 appears, then the student with the fewest dollars collects a dollar from the other. If the students have the same amount, then no dollars are exchanged. The game continues until one student runs out of dollars. Let student A's number of dollars represent the states. Let each student start with 3 dollars.

(a) What is the transition matrix, P?
(b) If student A reaches state 0 or 6, then he stays there with probability 1. What is the probability that student B loses in 3 tosses of the dice?
(c) What is the probability that student A loses in 5 or fewer tosses?

9.12 A biologist would like to estimate the size of a certain population of fish. A sequential approach is proposed whereby a member of the population is sampled at random, tagged, and then returned. This process is repeated until a member is drawn that has been previously tagged. If desired, we could then begin tagging again with a new kind of tag. Let M be the trial at which the first previously tagged fish is sampled and N be the total population size. This process can be described in terms of a Markov chain where X_k is the number of successive untagged members observed. That is, $X_k = k$ for $k = 1, 2, \ldots, M - 1$ and $X_M = 0$.

(a) For a fixed $N = n$, find the form of the transition probability matrix.
(b) Find $\Pr(M = m | X_0 = 0)$ for $m = 2, 3, 4, \ldots, n$.

9.13 A person with a contagious disease enters the population. Every day he either infects a new person (which occurs with probability p) or his symptoms appear and he is discovered by health officials (which occurs with probability $1 - p$). Assuming all infected persons behave in the same manner, compute the probability distribution of the number of infected but undiscovered people in the population at the time of the first discovery of the disease.

9.14 Let X_n be the sum of n independent rolls of a fair (cubicle) die.

(a) Find $\lim_{n \to \infty} \Pr(X_n$ is a multiple of 3).
(b) Find $\lim_{n \to \infty} \Pr(X_n$ is a multiple of 5).

9.15 A certain three-state Markov chain has a transition probability matrix given by

$$P = \begin{bmatrix} 0.4 & 0.5 & 0.1 \\ 0.05 & 0.7 & 0.25 \\ 0.05 & 0.5 & 0.45 \end{bmatrix}.$$

Determine if the Markov chain has a unique steady state distribution or not. If it does, find that distribution.

9.16 Suppose a process can be considered to be in one of two states (let's call them state A and state B), but the next state of the process depends not only on the current state but also on the previous state. We can still describe this process using a Markov chain, but we will now need four states. The chain will be in state (X, Y), $X, Y \in \{A, B\}$ if the process is currently in state X and was previously in state Y.

(a) Show that the transition probability matrix of such a four-state Markov chain must have zeros in at least half of its entries.

(b) Suppose that the transition probability matrix is given by

$$(A, A) \quad (A, B) \quad (B, A) \quad (B, B)$$

$$P = \begin{matrix} (A, A) \\ (A, B) \\ (B, A) \\ (B, B) \end{matrix} \begin{bmatrix} 0.8 & 0.2 & 0 & 0 \\ 0 & 0 & 0.4 & 0.6 \\ 0.6 & 0.4 & 0 & 0 \\ 0 & 0 & 0.1 & 0.9 \end{bmatrix}.$$

Find the steady state distribution of the Markov chain.

(c) What is the steady state probability that the underlying process is in state A?

9.17 For a Markov chain with each of the transition probability matrices shown in (a)–(c), find the communicating classes and the periodicity of the various states.

(a) $\begin{bmatrix} 0 & 0 & 1 & 0 \\ 1 & 0 & 0 & 0 \\ \frac{1}{2} & \frac{1}{2} & 0 & 0 \\ \frac{1}{3} & \frac{1}{3} & \frac{1}{3} & 0 \end{bmatrix}$,
(b) $\begin{bmatrix} 0 & 1 & 0 & 0 \\ 0 & 0 & 0 & 1 \\ 0 & 1 & 0 & 0 \\ \frac{1}{3} & 0 & \frac{2}{3} & 0 \end{bmatrix}$,
(c) $\begin{bmatrix} 0 & 1 & 0 & 0 \\ \frac{1}{2} & 0 & 0 & \frac{1}{2} \\ 0 & 0 & 0 & 1 \\ 0 & \frac{1}{2} & \frac{1}{2} & 0 \end{bmatrix}.$

9.18 Prove that if $i \leftrightarrow j$, then $d(i) = d(j)$, and hence all states in the same class must have the same period.

9.19 Demonstrate that the two generating functions defined in Equations 9.20 and 9.21 are related by

$$P_{i,i}(z) - 1 = P_{i,i}(z)F_{i,i}(z).$$

9.20 Define the generating functions

$$P_{i,j}(z) = \sum_{n=0}^{\infty} p_{i,j}^{(n)} z^n \quad \text{and} \quad F_{i,j}(z) = \sum_{n=0}^{\infty} f_{i,j}^{(n)} z^n.$$

(a) Show that $P_{i,j}(z) = F_{i,j}(z)P_{j,j}(z)$.
(b) Prove that if state j is a transient state, then for all i,

$$\sum_{n=1}^{\infty} p_{i,j}^{(n)} < \infty.$$

9.21 Verify that recurrence is a class property. That is, if one state in a communicating class is recurrent then all are recurrent, and if one is transient then all are transient.

9.22 Suppose a Bernoulli trial results in a success with probability p and a failure with probability $1 - p$. Suppose the Bernoulli trial is repeated indefinitely with each repitition independent of all others. Let X_n be a "success runs" Markov chain where X_n represents the number of most recent consecutive successes that have been observed at the nth trial. That is, $X_n = m$ if trial numbers $n, n - 1, n - 2, \ldots, n - m + 1$ were all successes but trial number $n - m$ was a failure. Note that $X_n = 0$ if the nth trial was a failure.

(a) Find an expression for the one-step transition probabilities, $p_{i,j}$.
(b) Find an expression for the n-step first return probabilities for state $0, f_{0,0}^{(n)}$.
(c) Prove that state 0 is recurrent for any $0 < p < 1$. Note that since all states communicate with one another, this result together with the result of the previous exercise is sufficient to show that all states are recurrent.

9.23 Find the steady state distribution of the success runs Markov chain described in Exercise 9.22.

9.24 Derive the *backward Kolmogorov equations*,

$$\frac{d}{dt} p_{i,j}(t) = \lambda_i p_{i+1,j}(t) - (\lambda_i + \mu_i) p_{i,j}(t) + \mu_i p_{i-1,j}(t).$$

9.25 In this problem, you will demonstrate that the Gaussian PDF in Equation 9.64 is in fact the solution to the diffusion Equation 9.63. To do this, we will use frequency domain methods. Define, $\Phi(\omega, t) = E[e^{j\omega X(t)}] = \int_{-\infty}^{\infty} f(x, t|x_0 = 0, t_0 = 0)e^{j\omega x} \, dx$ to be the time varying characteristic function of the random process $X(t)$.

(a) Starting from the diffusion Equation 9.63, show that the characteristic function must satisfy

$$\frac{\partial}{\partial t} \Phi(\omega, t) = (2cj\omega - D\omega^2)\Phi(\omega, t).$$

Also, determine the appropriate initial condition for this differential equation. That is, find $\Phi(\omega, 0)$.

(b) Solve the first order differential equation in part (a) and show that the characteristic function is of the form

$$\Phi(\omega, t) = \exp(-D\omega^2 + 2cj\omega).$$

(c) From the characteristic function, find the resulting PDF given by Equation 9.64.

9.26 A communication system sends data in the form of packets of fixed length. Noise in the communication channel may cause a packet to be received incorrectly. If this happens, then the packet is retransmitted. Let the probability that a packet is received incorrectly be q.

(a) Determine the average number of transmissions that are necessary before a packet is received correctly. Draw a state diagram for this problem.

(b) Let the the transmission time be T_t seconds for a packet. If the packet is received incorrectly, then a message is sent back to the transmitter stating that the message was received incorrectly. Let the time for sending such a message be T_a. Assume that if the packet is received correctly that we do not send an acknowledgment. What is the average time for a successful transmission? Draw a state diagram for this problem.

(c) Now suppose there are three nodes. The packet is to be sent from node 1 to node 2 to node 3 without an error. The probability of the packets being received incorrectly at each node is the same and is q. The transmission time is T_t and the time to acknowledge that a packet is received incorrectly is T_a. Draw a state diagram for this problem. Determine the average time for the packet to reach node 3 correctly.

MATLAB Exercises

9.27 On the first day of the new year it is cloudy. What is the probability that it is sunny on 4 July if the following transition matrix applies?

$$
\begin{array}{cccc}
 & \text{sunny} & \text{cloudy} & \text{rainy} \\
\begin{array}{c} \text{sunny} \\ \text{cloudy} \\ \text{rainy} \end{array} &
\left[\begin{array}{ccc}
0.7 & 0.2 & 0.1 \\
0.3 & 0.2 & 0.5 \\
0.3 & 0.3 & 0.4
\end{array}\right]
\end{array}
$$

How much does your answer change if it is a leap year?

9.28 Determine which of the transition matrices (a)–(g) represents a regular Markov chain. Find the steady state distribution for the regular matrices. Note a Markov chain is regular if some power of the transition matrix has only positive (nonzero) entries. This implies that a regular chain has no periodic states.

(a) $\begin{bmatrix} 1/3 & 2/3 \\ 5/6 & 1/6 \end{bmatrix}$, (b) $\begin{bmatrix} 0 & 1 \\ 1/4 & 3/4 \end{bmatrix}$, (c) $\begin{bmatrix} 0 & 1 \\ 1 & 0 \end{bmatrix}$,

(d) $\begin{bmatrix} 1/5 & 4/5 \\ 1 & 0 \end{bmatrix}$, (e) $\begin{bmatrix} 1/2 & 1/2 & 0 \\ 0 & 1/2 & 1/2 \\ 1/3 & 1/3 & 1/3 \end{bmatrix}$, (f) $\begin{bmatrix} 1/3 & 0 & 2/3 \\ 0 & 1 & 0 \\ 0 & 1/5 & 4/5 \end{bmatrix}$,

(g) $\begin{bmatrix} 1/2 & 1/4 & 1/4 \\ 1/3 & 2/3 & 0 \\ 0 & 1/4 & 3/4 \end{bmatrix}$.

9.29 Write a MATLAB program to simulate a three-state Markov chain with the transition probability matrix

$$
P = \begin{bmatrix} 1/2 & 1/4 & 1/4 \\ 1/2 & 0 & 1/2 \\ 1/4 & 1/4 & 1/2 \end{bmatrix}.
$$

Assuming that the process starts in the third state, generate a sequence of 500 states. Estimate the steady state probability distribution, π, using the sequence you generated. Does it agree with the theoretical answer? Does the steady state distribution depend on the starting state of the process?

9.30 Write a MATLAB program to simulate the M/M/1 queueing system. If you like, you may use the program provided in Example 9.14. Use your program to estimate the average amount of time a customer spends waiting in line for service. Assume an arrival rate of $\lambda = 15$ customers/hour and an average service time of $1/\mu = 3$ minutes. Note, that if a customer arrives to find no others in the system, the waiting time is zero.

9.31 Modify the program of Example 9.14 to simulate the M/M/∞ queue of Example 9.13. Based on your simulation results, estimate the PMF of the number of customers in the system. Compare your results with the analytical results found in Example 9.13.

Power Spectral Density

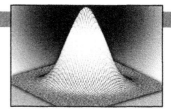

In the study of deterministic signals and systems, frequency domain techniques (e.g., Fourier transforms) provide a valuable tool that allows the engineer to gain significant insights into a variety of problems. In this chapter, we develop frequency domain tools for studying random processes. This will prepare us for the study of random processes in linear systems in the next chapter.

For a deterministic continuous signal, $x(t)$, the Fourier transform is used to describe its spectral content. In this text, we write the Fourier transform as[1]

$$X(f) = F[x(t)] = \int_{-\infty}^{\infty} x(t)e^{-j2\pi ft}\, dt, \tag{10.1}$$

and the corresponding inverse transform is

$$x(t) = F^{-1}[X(f)] = \int_{-\infty}^{\infty} X(f)e^{j2\pi ft}\, df. \tag{10.2}$$

For discrete time signals, we could use a discrete Fourier transform (DFT) or a z-transform. The Fourier transform, $X(f)$, is referred to as the spectrum of $x(t)$ since it describes the spectral contents of the signal. In general, $X(f)$ is a complex function of frequency, and hence we also speak of an amplitude (magnitude) spectrum, $|X(f)|$, and a phase spectrum, $\angle X(f)$. In order to study random processes in

---

[1]Even though we use an upper case letter to represent a Fourier transform, it is not necessarily random. Clearly, the Fourier transform of a nonrandom signal is also not random. While this is inconsistent with our previous notation of using upper case letters to represent random quantities, this notation of using upper case letters to represent Fourier transforms is so common in the literature, we felt it necessary to retain this convention. The context should make it clear whether a function of frequency is random or not.

the frequency domain, we seek a similar quantity that will describe the spectral characteristics of a random process.

The most obvious thing to do would be to try to define the Fourier transform of a random process as perhaps

$$X(f) = \int_{-\infty}^{\infty} X(t)e^{-j2\pi ft}\, dt = F[X(t)]; \tag{10.3}$$

however, this leads to several problems. First of all, there are problems with existence. Since $X(t)$ is a random process, there is not necessarily any guarantee that the integral exists for every possible realization, $x(t)$. That is, not every realization of the random process may have a Fourier transform. Even for processes that are well-behaved in the sense that every realization has a well-defined Fourier transform, we are still left with the problem that $X(f)$ is itself a random process. In Chapter 8, we described the temporal characteristics of random processes in terms of deterministic functions such as the mean function and the autocorrelation function. In a similar way, we seek a deterministic description of the spectral characteristics of a random process. The power spectral density (PSD) function, which is defined in the next section, will play that role.

Note that (10.3) uses an uppercase X to denote both the Fourier transform and the random variable. As noted on the previous page, we have elected to do this only because it is so common in the literature to use an uppercase letter to denote the Fourier transform.

10.1 Definition of Power Spectral Density

To start with, for a random process $X(t)$, define a truncated version of the random process as

$$X_{t_o}(t) = \begin{cases} X(t) & |t| \leq t_o \\ 0 & |t| > t_o \end{cases}. \tag{10.4}$$

The energy of this random process is

$$E_{X_{t_o}} = \int_{-t_o}^{t_o} X^2(t)\, dt = \int_{-\infty}^{\infty} X_{t_o}^2(t)\, dt, \tag{10.5}$$

and hence the time averaged power is

$$P_{X_{t_o}} = \frac{1}{2t_o} \int_{-\infty}^{\infty} X_{t_o}^2(t)\, dt = \frac{1}{2t_o} \int_{-\infty}^{\infty} |X_{t_o}(f)|^2\, df. \tag{10.6}$$

The last equality is obtained using Parseval's theorem. The quantity $X_{t_o}(f)$ is the Fourier transform of $X_{t_o}(t)$. Since the random process has been truncated to a finite time interval, there will generally not be any problem with the existence of the Fourier transform. Note that $P_{X_{t_o}}$ is a random variable and so to get the ensemble averaged power, we must take an expectation,

$$\overline{P_{X_{t_o}}} = E\left[P_{X_{t_o}}\right] = \frac{1}{2t_o} \int_{-\infty}^{\infty} E\left[|X_{t_o}(f)|^2\right] df. \tag{10.7}$$

The power in the (untruncated) random process $X(t)$ is then found by passing to the limit as $t_o \to \infty$,

$$\overline{P_X} = \lim_{t_o \to \infty} \frac{1}{2t_o} \int_{-\infty}^{\infty} E\left[|X_{t_o}(f)|^2\right] df = \int_{-\infty}^{\infty} \lim_{t_o \to \infty} \frac{E\left[|X_{t_o}(f)|^2\right]}{2t_o} df. \tag{10.8}$$

Define $S_{XX}(f)$ to be the integrand in Equation 10.8. That is, let

$$S_{XX}(f) = \lim_{t_o \to \infty} \frac{E\left[|X_{t_o}(f)|^2\right]}{2t_o}. \tag{10.9}$$

Then, the average power in the process can be expressed as

$$\overline{P_X} = \int_{-\infty}^{\infty} S_{XX}(f)\, df. \tag{10.10}$$

Hence, this function of frequency which we have referred to simply as $S_{XX}(f)$ has the property that when integrated over all frequency, the total power in the process is obtained. In other words, $S_{XX}(f)$ has the units of power per unit frequency and so it is the power density function of the random process in the frequency domain. Thus, the quantity $S_{XX}(f)$ is given the name power spectral density (PSD). In summary, we have the following definition of PSD.

DEFINITION 10.1: For a random process $X(t)$, the *power spectral density* (PSD) is defined as

$$S_{XX}(f) = \lim_{t_o \to \infty} \frac{E\left[|X_{t_o}(f)|^2\right]}{2t_o}, \tag{10.11}$$

where $X_{t_o}(f)$ is the Fourier transform of the truncated version of the process as described in Equation 10.4.

Several properties of the PSD function should be evident from Definition 10.1 and from the development that lead to that definition:

(1) $S_{XX}(f)$ is a real function. $\hspace{5cm}$ (10.12a)

(2) $S_{XX}(f)$ is a nonnegative function. $\hspace{4cm}$ (10.12b)

(3) $S_{XX}(f)$ is an even function.   (10.12c)

(4) The average power in a random process is given by $\overline{P_X} = \int_{-\infty}^{\infty} S_{XX}(f)\,df$.

(10.12d)

EXAMPLE 10.1: As a simple example, consider a sinusoidal process $X(t) = A\,\sin(\omega_0 t + \Theta)$ with random amplitude and phase. Assume the phase is uniform over $[0, 2\pi)$ and independent of the amplitude that we take to have an arbitrary distribution. Since each realization of this process is a sinusoid at frequency f_0, we would expect that all of the power in this process should be located at $f = f_0$ (and $f = -f_0$). Mathematically, we have

$$X_{t_0}(t) = A\,\sin(\omega_0 t + \Theta)\,\text{rect}\left(\frac{t}{2t_0}\right),$$

where rect(t) is a square pulse of unit height and unit width and centered at $t = 0$. The Fourier transform of this truncated sinusoid works out to be

$$X_{t_0}(f) = -jt_0 A e^{j\Theta}\text{sinc}(2(f - f_0)t_0) + jt_0 A e^{-j\Theta}\text{sinc}(2(f + f_0)t_0),$$

where the "sinc" function is $\text{sinc}(x) = \sin(\pi x)/(\pi x)$. We next calculate the expected value of the magnitude squared of this function.

$$E\left[|X_{t_0}(f)|^2\right] = E[A^2]t_0^2\{\text{sinc}^2(2(f - f_0)t_0) + \text{sinc}^2(2(f + f_0)t_0)\}$$

The PSD function for this random process is then

$$S_{XX}(f) = \lim_{t_0 \to \infty} \frac{E\left[|X_{t_0}(f)|^2\right]}{2t_0}$$

$$= \lim_{t_0 \to \infty} \frac{E[A^2]t_0}{2}\{\text{sinc}^2(2(f - f_0)t_0) + \text{sinc}^2(2(f + f_0)t_0)\}.$$

To calculate this limit, we observe that as t_0 gets large, the function $g(f) = t_0\text{sinc}^2(2ft_0)$ becomes increasingly narrower and taller. Hence, we could view the limit as an infinitely tall, infinitely narrow pulse. This is one way to define a delta function. One property of a delta function that is not necessarily shared by the function under consideration is that $\int \delta(f)\,df = 1$. Hence, the limiting form of $g(t)$ will have to be a scaled (in amplitude) delta function. To figure out what the scale factor needs to be, the integral of $g(f)$ is calculated:

$$\int_{-\infty}^{\infty} g(f)\,df = \int_{-\infty}^{\infty} t_0\text{sinc}^2(2ft_0)\,df = \frac{1}{2}\int_{-\infty}^{\infty} \text{sinc}^2(u)\,du = \frac{1}{2}.$$

Hence,

$$\lim_{t_o \to \infty} t_o \text{sinc}^2(2ft_o) \frac{1}{2} \delta(f).$$

The resulting PSD is then simplified to

$$S_{XX}(f) = \frac{E[A^2]}{4} \{\delta(f - f_o) + \delta(f + f_o)\}.$$

This is consistent with our intuition. The power in a sinusoid with amplitude A is $A^2/2$. Thus, the average power in the sinusoidal process is $E[A^2]/2$. This power is evenly split between the two points $f = f_o$ and $f = -f_o$.

One important lesson to learn from the previous example is that even for very simplistic random processes, it can be quite complicated to evaluate the PSD using the definition given in Equation 10.11. The next section presents a very important result that allows us to greatly simplify the process of finding the PSD of many random processes.

10.2 The Wiener-Khintchine-Einstein Theorem

THEOREM 10.1 (Wiener-Khintchine-Einstein): For a wide sense stationary (WSS) random process $X(t)$ whose autocorrelation function is given by $R_{XX}(\tau)$, the PSD of the process is

$$S_{XX}(f) = F(R_{XX}(\tau)) = \int_{-\infty}^{\infty} R_{XX}(\tau)e^{-j2\pi f\tau} d\tau. \qquad (10.13)$$

In other words, the autocorrelation function and PSD form a Fourier transform pair.

PROOF: Starting from the definition of PSD,

$$E\left[|X_{t_o}(f)|^2\right] = E\left[\int_{-t_o}^{t_o}\int_{-t_o}^{t_o} X(t)X(s)e^{-j2\pi f(t-s)} dt\, ds\right]$$

$$= \int_{-t_o}^{t_o}\int_{-t_o}^{t_o} E[X(t)X(s)]e^{-j2\pi f(t-s)} dt\, ds = \int_{-t_o}^{t_o}\int_{-t_o}^{t_o} R_{XX}(t,s)e^{-j2\pi f(t-s)} dt\, ds.$$

$$(10.14)$$

Using the assumption that the process is WSS, the autocorrelation function is only a function of a single time variable, $t - s$. Hence, this expression is rewritten as

$$E\left[|X_{t_o}(f)|^2\right] = \int_{-t_o}^{t_o} \int_{-t_o}^{t_o} R_{XX}(t - s)e^{-j2\pi f(t-s)}\, dt\, ds. \tag{10.15}$$

It is noted that the preceding integrand is a function of only a single variable; therefore, with the appropriate change of variables, the double integral can be reduced to a single integral. The details are given in the following.

The region of integration is a square in the s-t plane of width $2t_o$ centered at the origin. Consider an infinitesimal strip bounded by the lines $t - s = \tau$ and $t - s = \tau + d\tau$.

This strip is illustrated in Figure 10.1. Let $a(\tau)$ be the area of that strip that falls within the square region of integration. A little elementary geometry reveals that

$$a(\tau) = \begin{cases} 2t_o\left(1 - \dfrac{|\tau|}{2t_o}\right)d\tau & \text{for } |\tau| < 2t_o \\[2mm] 0 & \text{for } |\tau| > 2t_o \end{cases}. \tag{10.16}$$

To obtain the preceding result, one must neglect edge terms that contribute expressions that are quadratic in the infinitesimal $d\tau$. Since the integrand in Equation 10.15 is a function of only $t - s$, it is constant (and equal to $R_{XX}(\tau)e^{-j2\pi\tau}$) over the entire strip. The double integral over the strip can therefore be written as the value of the integrand multiplied by the area of the strip. The double integral over the entire

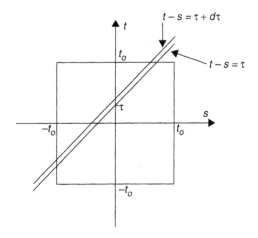

Figure 10.1   Illustration of the change of variables for the double integral in Equation 10.15.

square can be written as a sum of the integrals over all the strips that intersect the square:

$$\int_{-t_o}^{t_o}\int_{-t_o}^{t_o} R_{XX}(t-s)e^{-j2\pi f(t-s)}\,dt\,ds = \sum_{\text{strips}} R_{XX}(\tau)e^{-j2\pi\tau}a(\tau). \tag{10.17}$$

Passing to the limit as $d\tau \to 0$, the sum becomes an integral resulting in

$$E\left[|X_{t_o}(f)|^2\right] = 2t_o \int_{-2t_o}^{2t_o} R_{XX}(\tau)e^{-j2\pi\tau}\left(1-\frac{|\tau|}{2t_o}\right)\,d\tau. \tag{10.18}$$

The PSD function for the random process $X(t)$ is then

$$S_{XX}(f) = \lim_{t_o\to\infty}\frac{E\left[|X_{t_o}(f)|^2\right]}{2t_o} = \lim_{t_o\to\infty}\int_{-2t_o}^{2t_o}\left(1-\frac{|\tau|}{2t_o}\right)R_{XX}(\tau)e^{-j2\pi\tau}\,d\tau. \tag{10.19}$$

Passing to the limit as $t_o \to \infty$ then gives the desired result in Equation 10.13. ■

While most of the random processes we deal with are WSS, for those that are not, Theorem 10.1 needs to be adjusted since the autocorrelation function for a nonstationary process would be a function of two time variables. For nonstationary processes the Wiener-Khintchine-Einstein theorem is written as

$$S_{XX}(f) = \int_{-\infty}^{\infty}\langle R_{XX}(t,t+\tau)\rangle e^{-j2\pi f\tau}\,d\tau, \tag{10.20}$$

where in this case $\langle\rangle$ represents a time average with respect to the time variable t. We leave it as an exercise for the reader to prove this more general version of the theorem.

EXAMPLE 10.2: Let us revisit the random sinusoidal process, $X(t) = A\sin(\omega_o t + \Theta)$, of Example 10.1. This time the PSD function will be calculated by first finding the autocorrelation function.

$$R_{XX}(t,t+\tau) = E[X(t)X(t+\tau)] = E[A^2\sin(\omega_o t + \Theta)\sin(\omega_o(t+\tau)+\Theta)]$$

$$= \frac{1}{2}E[A^2]E[\cos(\omega_o\tau) - \cos(\omega_o(2t+\tau)+2\Theta)] = \frac{1}{2}E[A^2]\cos(\omega_o\tau)$$

The autocorrelation function is only a function of τ and thus the PSD is simply the Fourier transform of the autocorrelation function,

$$S_{XX}(f) = \int_{-\infty}^{\infty} R_{XX}(\tau)e^{-j2\pi f\tau}\,d\tau = \frac{1}{2}E[A^2]F[\cos(\omega_0\tau)]$$

$$= \frac{1}{4}E[A^2]\{\delta(f - f_0) + \delta(f + f_0)\}.$$

This is exactly the same result that was obtained in Example 10.1 using the definition of PSD, but in this case the result was obtained with much less work.

EXAMPLE 10.3: Now suppose we have a sinusoid with a random amplitude but a fixed phase, $X(t) = A\sin(\omega_0 t + \theta)$. Here the autocorrelation function is

$$R_{XX}(t, t + \tau) = E[X(t)X(t + \tau)] = E[A^2\sin(\omega_0 t + \theta)\sin(\omega_0(t + \tau) + \theta)]$$

$$= \frac{1}{2}E[A^2]E[\cos(\omega_0\tau) - \cos(\omega_0(2t + \tau) + 2\theta)]$$

$$= \frac{1}{2}E[A^2]\cos(\omega_0\tau) + \frac{1}{2}E[A^2]\cos(\omega_0(2t + \tau) + 2\theta).$$

In this case, the process is not WSS and so we must take a time average of the autocorrelation before we take the Fourier transform.

$$\langle R_{XX}(t, t + \tau)\rangle = \left\langle \frac{1}{2}E[A^2]\cos(\omega_0\tau) + \frac{1}{2}E[A^2]\cos(\omega_0(2t + \tau) + 2\theta)\right\rangle$$

$$= \frac{1}{2}E[A^2]\cos(\omega_0\tau) + \frac{1}{2}E[A^2]\langle\cos(\omega_0(2t + \tau) + 2\theta)\rangle = \frac{1}{2}E[A^2]\cos(\omega_0\tau)$$

The time-averaged autocorrelation is exactly the same as the autocorrelation in the previous example, and hence the PSD of the sinusoid with random amplitude and fixed phase is exactly the same as the PSD of the sinusoid with random amplitude and random phase.

EXAMPLE 10.4: Next, consider a modified version of the random telegraph signal of Example 8.4. In this case, the process starts at $X(0) = 1$ and switches back and forth between $X(t) = 1$ and $X(t) = -1$, with the switching times being dictated by a Poisson point process with rate λ.

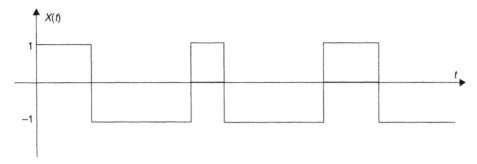

Figure 10.2 A sample realization for the random telegraph signal of Example 10.4.

A sample realization is shown in Figure 10.2. To find the PSD we first find the autocorrelation function.

$$R_{XX}(t, t + \tau) = E[X(t)X(t + \tau)]$$

$$= (1)\Pr(\text{even number of switches in } [t, t + \tau))$$

$$+ (-1)\Pr(\text{odd number of switches in } [t, t + \tau)).$$

The number of switches in a contiguous interval follows a Poisson distribution, and hence

$$R_{XX}(t, t + \tau) = \sum_{m \text{ even}} \frac{(\lambda|\tau|)^m}{m!} e^{-\lambda|\tau|} - \sum_{m \text{ odd}} \frac{(\lambda|\tau|)^m}{m!} e^{-\lambda|\tau|}$$

$$= \sum_{m=0}^{\infty} \frac{(-\lambda|\tau|)^m}{m!} e^{-\lambda|\tau|} = e^{-2\lambda|\tau|}.$$

Since this is a function of only τ, we directly take the Fourier transform to find the PSD:

$$S_{XX}(f) = F[e^{-2\lambda|\tau|}] = \frac{1/\lambda}{1 + (\pi f / \lambda)^2} = \frac{\lambda}{\lambda^2 + (\pi f)^2}.$$

The autocorrelation function and PSD for this random telegraph signal are illustrated in Figure 10.3.

EXAMPLE 10.5: To illustrate how some minor changes in a process can affect its autocorrelation and PSD, let us return to the random telegraph process as it was originally described in Example 8.4. As in the previous

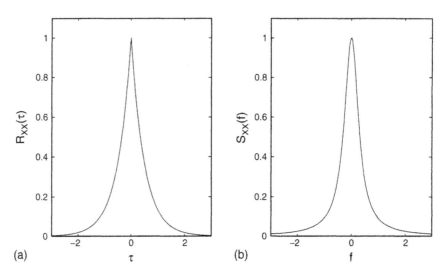

Figure 10.3   Autocorrelation function (a) and PSD (b) for the random telegraph signal of Example 10.4.

example, the process switches back and forth between two values as dictated by an underlying Poisson point process; however, now the two values of the process are $X(t) \in \{0,1\}$ instead of $X(t) \in \{+1, -1\}$. Also, the process starts at $X(0) = 0$ instead of $X(0) = 1$. Noting that the product $X(t)X(t + \tau)$ is equal to zero unless both $\{X(t) = 1\}$ and $\{X(t + \tau) = 1\}$ are true, the autocorrelation function is calculated as

$$R_{XX}(t, t+\tau) = E[X(t)X(t+\tau)] = \Pr(\{X(t)=1\} \cap \{X(t+\tau)=1\})$$

$$= \Pr(\text{odd number of switches in } [0, t))$$

$$\times \Pr(\text{even number of switches in } [t, t+\tau))$$

$$= \left(\sum_{m \text{ odd}} \frac{(\lambda t)^m}{m!} e^{-\lambda t} \right) \left(\sum_{m \text{ even}} \frac{(\lambda \tau)^m}{m!} e^{-\lambda t} \right) = \left(\frac{1}{2} - \frac{1}{2}e^{-2\lambda t} \right) \left(\frac{1}{2} + \frac{1}{2}e^{-2\lambda \tau} \right).$$

The last step was accomplished using some of the results obtained in Example 8.6 and assumes that τ is positive. If, on the other hand, τ is negative, then it turns out that

$$R_{XX}(t, t + \tau) = \left(\frac{1}{2} - \frac{1}{2}e^{-2\lambda(t+\tau)} \right) \left(\frac{1}{2} + \frac{1}{2}e^{2\lambda \tau} \right).$$

Clearly, this process is not stationary since the autocorrelation function is a function of both t and τ. Hence, before the Fourier transform is taken, the time average of the autocorrelation function must be computed.

$$\langle R_{XX}(t,t+\tau)\rangle = \left(\frac{1}{2} - \frac{1}{2}e^{-2\lambda t}\right)\left(\frac{1}{2} + \frac{1}{2}e^{-2\lambda\tau}\right) = \frac{1}{4} + \frac{1}{4}e^{-2\lambda\tau}, \quad \text{for } \tau > 0,$$

$$\langle R_{XX}(t,t+\tau)\rangle = \left(\frac{1}{2} - \frac{1}{2}e^{-2\lambda(t+\tau)}\right)\left(\frac{1}{2} + \frac{1}{2}e^{2\lambda\tau}\right) = \frac{1}{4} + \frac{1}{4}e^{2\lambda\tau}, \quad \text{for } \tau < 0$$

In summary, the autocorrelation function can be concisely expressed as

$$\langle R_{XX}(t,t+\tau)\rangle = \frac{1}{4} + \frac{1}{4}e^{-2\lambda|\tau|}.$$

The PSD function is then found to be

$$S_{XX}(f) = F\left[\frac{1}{4} + \frac{1}{4}e^{-2\lambda|\tau|}\right] = \frac{1}{4}\delta(f) + \frac{1}{4}\frac{\lambda}{\lambda^2 + (\pi f)^2}.$$

There are two differences between this result and that of the Example 10.4. First the total power (integral of PSD) in this process is $1/2$ the total power in the process of the previous example. This is easy to see since when $X(t) \in \{0,1\}$, $E[X^2(t)] = 1/2$, while when $X(t) \in \{+1,-1\}$, $E[X^2(t)] = 1$. Second, in this example, there is a delta function in the PSD which was not present in the previous example. This is due to the fact that the mean of the process in this example was (asymptotically) equal to $1/2$, whereas in the previous example it was zero. It is left as an exercise for the reader to determine if the initial conditions of the random process would have any effect on the PSD. That is, if the process started at $X(0) = 1$ and everything else remained the same, would the PSD change?

DEFINITION 10.2: The *cross spectral density* between two random processes, $X(t)$ and $Y(t)$, is the Fourier transform of the cross correlation function:

$$S_{XY}(f) = F[R_{XY}(\tau)] = \int_{-\infty}^{\infty} R_{XY}(\tau)e^{-j2\pi f\tau}d\tau. \tag{10.21}$$

The cross spectral density does not have a physical interpretation nor does it share the same properties as the PSD function. For example, $S_{XY}(f)$ is not necessarily real since $R_{XY}(\tau)$ is not necessarily even. The cross spectral density function does possess a form of symmetry known as *Hermitian symmetry*[2],

$$S_{XY}(f) = S_{YX}(-f) = S_{XY}^*(-f). \tag{10.22}$$

---

[2]Here and throughout the text, the superscript * refers to the complex conjugate.

This property follows from the fact that $R_{XY}(\tau) = R_{YX}(-\tau)$. The proof of this property is left to the reader.

10.3 Bandwidth of a Random Process

Now that we have an analytical function that describes the spectral content of a signal, it is appropriate to talk about the bandwidth of a random process. As with deterministic signals, there are many definitions of bandwidth. Which definition is used depends on the application and sometimes on personal preference. Several definitions of bandwidth are given next. To understand these definitions, it is helpful to remember that when measuring the bandwidth of a signal (whether random or deterministic), only positive frequencies are measured. Also, we tend to classify signals according to where their spectral contents lies. Those signals for which most of the power is at or near direct current (d.c.) are referred to as *lowpass* signals, while those signals whose PSD is centered around some nonzero frequency, $f = f_o$ are referred to as *bandpass* processes.

DEFINITION 10.3: For a lowpass process, the absolute bandwidth, B_{abs}, is the largest frequency for which the PSD is nonzero. That is, B_{abs} is the smallest value of B such that $S_{XX}(f) = 0$ for all $f > B$. For a bandpass process, let B_L be the largest value of B such that $S_{XX}(f) = 0$ for all $0 < f < B$, and similarly let B_R be the smallest value of B such that $S_{XX}(f) = 0$ for all $B < f$. Then $B_{\text{abs}} = B_R - B_L$. In summary, the absolute bandwidth of a random process is the width of the band that contains all frequency components. The concept of absolute bandwidth is illustrated in Figure 10.4.

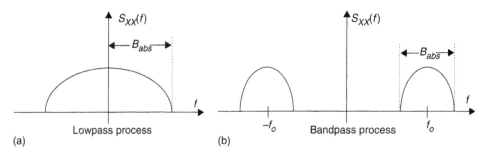

Figure 10.4 Measuring the absolute bandwidth of a lowpass (a) and bandpass (b) process.

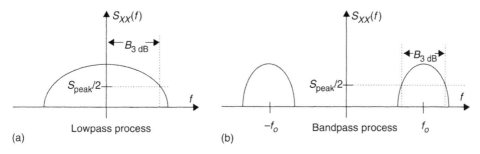

Figure 10.5 Measuring the 3-dB bandwidth of a lowpass (a) and bandpass (b) process.

DEFINITION 10.4: The *3-dB bandwidth* (or *half power* bandwidth), $B_{3\,dB}$, is the width of the frequency band where the PSD is within 3 dB of its peak value everywhere within the band. Let S_{peak} be the maximum value of the PSD. Then for a lowpass signal, $B_{3\,dB}$ is the largest value of B for which $S_{XX}(f) > S_{peak}/2$ for all frequencies such that $0 < f < B$. For a bandpass process, $B_{3\,dB} = B_R - B_L$, where $S_{XX}(f) > S_{peak}/2$ for all frequencies such that $B_L < f < B_R$ and it is assumed that the peak value occurs within the band. The concept of 3-dB bandwidth is illustrated in Figure 10.5.

DEFINITION 10.5: The *root-mean-square* (RMS) *bandwidth*, B_{rms}, of a lowpass random process is given by

$$B_{rms}^2 = \frac{\int_0^\infty f^2 S_{XX}(f)\,df}{\int_0^\infty S_{XX}(f)\,df}. \tag{10.23}$$

This measure of bandwidth is analogous to using standard deviation as a measure of the width of a PDF. For bandpass processes, this definition is modified according to

$$B_{rms}^2 = \frac{4\int_0^\infty (f - f_o)^2 S_{XX}(f)\,df}{\int_0^\infty S_{XX}(f)\,df}, \tag{10.24}$$

where

$$f_o = \frac{\int_0^\infty f S_{XX}(f)\,df}{\int_0^\infty S_{XX}(f)\,df}. \tag{10.25}$$

It is left as an exercise for the reader to figure out why the factor of 4 appears in the preceding definition.

EXAMPLE 10.6: Consider the random telegraph process of Example 10.4 where the PSD was found to be

$$S_{XX}(f) = \frac{\lambda}{\lambda^2 + (\pi f)^2}.$$

The absolute bandwidth of this process is $B_{abs} = \infty$. This can be seen from the picture of the PSD in Figure 10.2. To find the 3-dB bandwidth, it is noted that the peak of the PSD occurs at $f = 0$ and has a value of $S_{peak} = \lambda^{-1}$. The 3-dB bandwidth is then the value of f for which $S_{XX}(f) = 1/(2\lambda)$. This is easily found to be $B_{3\,dB} = \lambda/\pi$. Finally, the RMS bandwidth of this process is infinite since

$$\int_0^\infty \frac{\lambda f^2}{\lambda^2 + (\pi f)^2}\, df = \infty.$$

10.4 Spectral Estimation

The problem of estimating the PSD of a random process has been the topic of extensive research over the past several decades. Many books are dedicated to this topic alone and hence we cannot hope to give a complete treatment of the subject here; however, some fundamental concepts are introduced in this section that will provide the reader with a basic understanding of the problem and some rudimentary solutions. Spectral estimators are generally grouped into two classes, *parametric* and *nonparametric*. A parametric estimator assumes a certain model for the random process with several unknown parameters and then attempts to estimate the parameters. Given the model parameters, the PSD is then computed analytically from the model. On the other hand, a nonparametric estimator makes no assumptions about the nature of the random process and estimates the PSD directly. Since parametric estimators take advantage of some prior knowledge of the nature of the process, it would be expected that these estimators are more accurate. However, in some cases, prior knowledge may not be available, in which case a nonparametric estimator may be more appropriate. We start with a description of some basic techniques for nonparametric spectral estimation.

10.4.1 Nonparametric Spectral Estimation

Suppose we observe a random process, $X(t)$, over some time interval $(-t_o, t_o)$ (or a discrete time process $X[n]$ over some time interval $[0, n_o - 1]$) and we wish to estimate its PSD function. Two approaches immediately come to mind. The first method we will refer to as the direct method, or the *periodogram*. It is based on the definition of PSD in Equation 10.11. The second method we will refer to as the indirect method, or the *correlation method*. The basic idea here is to estimate the autocorrelation function and then take the Fourier transform of the estimated autocorrelation to form an estimate of the PSD. We first describe the correlation method. In all of the discussion on spectral estimation to follow, it is assumed that the random processes are WSS.

An estimate of the autocorrelation function of a continuous time random process can be formed by taking a time average of the particular realization observed:

$$\widehat{R}_{XX}(\tau) = \left\langle X\left(t - \frac{\tau}{2}\right) X\left(t + \frac{\tau}{2}\right)\right\rangle = \frac{1}{2t_o - |\tau|} \int_{t_o + \frac{|\tau|}{2}}^{t_o - \frac{|\tau|}{2}} X\left(t - \frac{\tau}{2}\right) X\left(t + \frac{\tau}{2}\right) dt.$$

$$(10.26)$$

It is not difficult to show that this estimator is unbiased (i.e., $E[\widehat{R}_{XX}(\tau)] = R_{XX}(\tau)$), but at times it is not a particularly good estimator, especially for large values of τ. The next example illustrates this fact.

EXAMPLE 10.7: Consider the random telegraph process of Example 10.4. A sample realization of this process is shown in Figure 10.6, along with the estimate of the autocorrelation function. For convenience, the true autocorrelation is shown as well. Note that the estimate matches quite well for small values of τ, but as $|\tau| \to t_o$, the estimate becomes very bad.

In order to improve the quality of the autocorrelation estimate, it is common to introduce a windowing function to suppress the erratic behavior of the estimate at large values of τ. This is particularly important when estimating the PSD because the wild behavior at large values of $|\tau|$ will distort the estimate of the

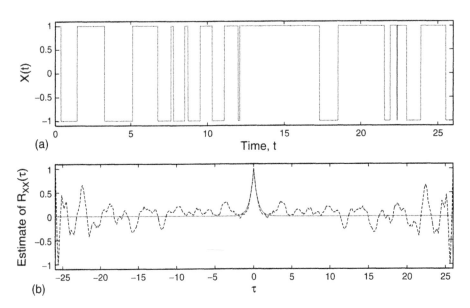

Figure 10.6 A sample realization of the random telegraph signal (a) and the estimate of the autocorrelation function based on that realization (b). The dotted line is the true autocorrelation function.

PSD at all frequencies once the Fourier transform of the autocorrelation estimate is taken.

DEFINITION 10.6: For a WSS random process $X(t)$, the windowed estimate of the autocorrelation function using a windowing function $w(t)$ is given by

$$\widehat{R}_{XX}^{(w)}(\tau) = \frac{w(\tau)}{2t_o - |\tau|} \int_{t_o + \frac{|\tau|}{2}}^{t_o - \frac{|\tau|}{2}} X\left(t - \frac{\tau}{2}\right) X\left(t + \frac{\tau}{2}\right) dt. \tag{10.27}$$

There are many possible windowing functions that can be used. The previous autocorrelation estimate (without the windowing function) can be viewed as a windowed estimate with a rectangular window,

$$w(t) = \text{rect}\left(\frac{t}{4t_o}\right). \tag{10.28}$$

Another option would be to use a triangular window,

$$w(t) = \text{tri}\left(\frac{t}{2t_o}\right) = \begin{cases} 1 - \dfrac{|t|}{2t_o} & |t| < t_o \\ 0 & |t| > t_o \end{cases}. \tag{10.29}$$

This would lead to the autocorrelation estimate,

$$\widehat{R}_{XX}^{(\text{tri})}(\tau) = \frac{1}{2t_o} \int_{t_o+\frac{|\tau|}{2}}^{t_o-\frac{|\tau|}{2}} X\left(t - \frac{\tau}{2}\right) X\left(t + \frac{\tau}{2}\right) dt. \tag{10.30}$$

While this estimator is biased, the mean-squared error in the estimate will generally be smaller than when the rectangular window is used. Much of the classical spectral estimation theory focuses on how to choose an appropriate window function to satisfy various criteria.

EXAMPLE 10.8: The autocorrelation function of the random telegraph signal is once again estimated, this time with the windowed autocorrelation estimator using the triangular window. The sample realization, as well as the autocorrelation estimate, are shown in Figure 10.7. Note this time that the behavior of the estimate for large values of τ is more controlled.

Once an estimate of the autocorrelation can be found, the estimate of the PSD is obtained through Fourier transformation.

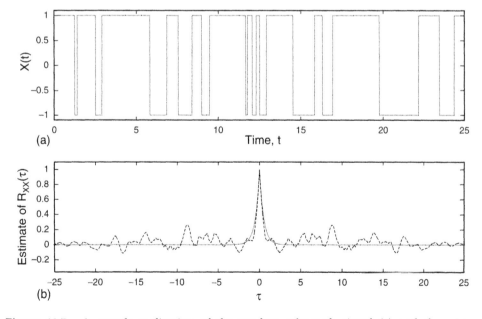

Figure 10.7 A sample realization of the random telegraph signal (a) and the windowed estimate of the autocorrelation function (using a triangular window) based on that realization (b).

DEFINITION 10.7: For a WSS random process $X(t)$, the correlation-based estimate (with windowing function $w(t)$) of the PSD is given by

$$\widehat{S}_{XX}^{(w)}(f) = F\left[\widehat{R}_{XX}^{(w)}(\tau)\right] = \int_{-\infty}^{\infty} \widehat{R}_{XX}^{(w)}(\tau) e^{-j2\pi f \tau} d\tau = \int_{-2t_o}^{2t_o} \widehat{R}_{XX}^{(w)}(\tau) e^{-j2\pi f \tau} d\tau. \quad (10.31)$$

EXAMPLE 10.9: The PSD estimates corresponding to the autocorrelation estimates of the previous example are illustrated in Figure 10.8. There the correlation based PSD estimates are plotted and compared with the true PSD. Note that when no windowing is used, the PSD estimate tends to overestimate the true PSD. Another observation is that it appears from these results that the PSD estimates could be improved by smoothing. We will elaborate on that shortly.

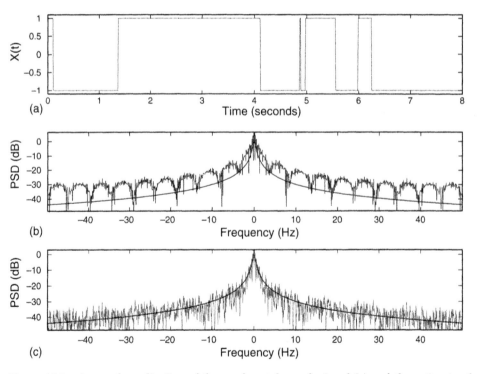

Figure 10.8 A sample realization of the random telegraph signal (a) and the estimate of the PSD function (b, c). Plot (b) is for the unwindowed estimator, while plot (c) is for the triangular windowed estimator. For both PSD plots, the smooth thick line is the true PSD.

The next approach we consider for estimating the PSD of a random process is to directly use the definition of PSD in Equation 10.11. This approach is referred to as the periodogram estimate.

DEFINITION 10.8: Given an observation of the process $X(t)$ over an interval $(-t_0, t_0)$, $X_{t_0}(t)$, the *periodogram* estimate of the PSD is

$$\widehat{S}_{XX}^{(p)}(f) = \frac{1}{2t_0} \left| X_{t_0}(f) \right|^2. \tag{10.32}$$

THEOREM 10.2: The periodogram estimate of the PSD is equivalent to the autocorrelation-based estimate with a triangular window. That is,

$$\widehat{S}_{XX}^{(p)}(f) = \widehat{S}_{XX}^{(tri)}(f). \tag{10.33}$$

PROOF: The proof of this theorem is a fairly elementary exercise in manipulating the properties of Fourier transforms. Recall that for any two signals, $x(t)$ and $y(t)$, the product of their spectra forms a transform pair with the convolution of the two signals. That is, $F[x(t) * y(t)] = X(f)Y(f)$. Applying this to Equation 10.32 results in

$$\left| X_{t_0}(f) \right|^2 = F[X_{t_0}(\tau) * X_{t_0}(-\tau)] = F\left[\int_{-\infty}^{\infty} X_{t_0}(u) X_{t_0}(u - \tau)\, du \right]$$

$$= F\left[\int_{-t_0}^{t_0} X_{t_0}(t) X_{t_0}(t + \tau)\, dt \right],$$

$$\frac{1}{2t_0} \left| X_{t_0}(f) \right|^2 = F\left[\frac{1}{2t_0} \int_{-t_0}^{t_0} X_{t_0}(t) X_{t_0}(t + \tau)\, dt \right] = F\left[\widehat{R}_{XX}^{(tri)}(\tau) \right] = \widehat{S}_{XX}^{(tri)}(f). \tag{10.34}$$

■

An example of the periodogram was given in plot (c) of Figure 10.8. At the time, it was referred to as the correlation-based estimate with a triangular windowing function. Now it is clear that the two are the same. It was mentioned in Example 10.9 that the quality of the periodogram might be improved by smoothing the PSD estimate. This can be accomplished by convolving $\widehat{S}_{XX}^{(p)}(f)$ with some smoothing function, $\widetilde{w}(f)$.

DEFINITION 10.9: The smoothed periodogram with smoothing function $\widetilde{w}(f)$ is given by

$$\widehat{S}_{XX}^{(wp)}(f) = \widetilde{w}(f) * \widehat{S}_{XX}^{(p)}(f). \tag{10.35}$$

The smoothed periodogram can be viewed in terms of the correlation-based estimate as well. Note that if $w(\tau) = F^{-1}[\widetilde{w}(f)]$, then

$$\widehat{S}_{XX}^{(wp)}(f) = F[w(\tau)] * F[\widehat{R}_{XX}^{(tri)}(\tau)] = F[w(\tau)\widehat{R}_{XX}^{(tri)}(\tau)]. \qquad (10.36)$$

Hence, the smoothed periodogram is nothing more than the windowed correlation-based estimate with a window that is the product of $w(t)$ and the triangular window. This seems to indicate that there would be some potential benefit to using windowing functions other than what has been presented here. The reader is referred to the many books on spectral estimation in the literature for discussions of other possibilities.

In all of the spectral estimators presented thus far, an ensemble average was estimated using a single realization. A better estimate could be obtained if several independent realizations of the random process were observed and a sample average were used to replace the ensemble average. Even though we may be able to observe only a single realization, it may still be possible to achieve the same effect. This is done by breaking the observed time waveform into segments and treating each segment as an independent realization of the random process. The periodogram is computed on each segment and then the resulting estimates are averaged.

EXAMPLE 10.10: Figure 10.9 compares the periodogram estimate of the PSD of the random telegraph signal with and without segmentation. In plot (b) no segmentation is used, while in plot (c) the data is segmented into $M = 8$ frames. A periodogram is computed for each frame and the results are then averaged. Note the improvement in the PSD estimate when the segmentation is used. Also note that there is a slight bias appearing in the segmented estimate. This is most noticeable at the higher frequencies. This bias will get worse as more segments are used. There is a trade-off in wanting to use a large value of M to reduce the "jitter" in the estimate and wanting to use a small value of M to keep the bias to a minimum. The following MATLAB functions were used to implement the periodogram estimates with and without seqmentation. These same functions can be used to estimate the PSD of any input signal.

```
function [Shat, f]=Periodogram(x,dx)
% This function computes the periodogram estimate of the PSD
% of the input signal. The vector x contains the samples of the
% input while dx indicates the time interval between samples.
Nx=length(x);
```

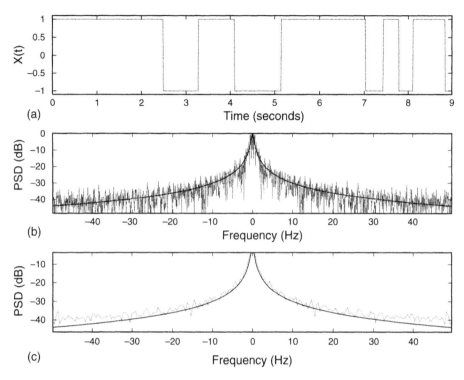

Figure 10.9 A sample realization of the random telegraph signal (a) and the periodogram estimate of the PSD function (b, c). Plot (b) is for the unsegmented data, while plot (c) is for when the data is segmented into $M = 8$ frames. For both PSD plots, the smooth thick line is the true PSD.

```
Rhat=conv(x,fliplr(x))/Nx;
Nr=length(Rhat);
Shat=fft(Rhat);
Shat=fftshift(dx*abs(Shat));
Nf=(Nr-1)/2; df=1/(dx*Nr);
f=[-Nf:Nf]*df;
function [S,f]=EPrdgm(x,dx,M)
% This function computes the periodogram estimate of the PSD
% of the input signal by breaking the signal into M frames and
% and performing a periodogram estimate on each frame and then
% averaging the results. The vector x contains the samples of
% the signal, while dx is the sampling interval. S is the
% estimated PSD and f is a vector of frequency samples that
% gives the frequency scale to be used when plotting S.
```

```
Nx=length(x);
Nframe=floor(Nx/M);                          % frame length
S=zeros(1,2*Nframe-1);
for m=1:M
  xm=x((((m-1)*Nframe+1):(m*Nframe)));
  [Stemp,f]=Periodogram(xm,dx);
  S=S+Stemp;
end
S=S/M;
```

10.4.2 Parametric Spectral Estimation

In parametric spectral estimation, a general model of the data is assumed, which usually contains one or more unknown parameters. Given the general model, the PSD can be calculated analytically. The problem then becomes one of estimating the unknown parameters and plugging the result into the analytic form of the PSD. To provide an example of how this general approach works, we present a specific class of random process models.

DEFINITION 10.10: Given a process $X[n]$ with known statistics, a new process, $Y[n]$, is formed according to the difference equation:

$$a_0 Y[n] = \sum_{i=1}^{p} a_i Y[n-i] + \sum_{i=0}^{q} b_i X[n-i]. \tag{10.37}$$

This process is referred to as an *autoregressive moving average process* (ARMA). As special cases, if all of the a_i are equal to zero (except a_0, which is usually set equal to unity), then Equation 10.37 simplifies to

$$Y[n] = \sum_{i=0}^{q} b_i X[n-i], \tag{10.38}$$

and the process is referred to as a *moving average* (MA) process. The notation MA(q) is used to refer to a qth order moving average process. If all of the b_i are equal to zero except for b_0, then the difference equation becomes

$$a_0 Y[n] = \sum_{i=0}^{p} a_i Y[n-i] + b_0 X[n], \tag{10.39}$$

and the process is referred to as an *autoregressive* (AR) process. The notation $AR(p)$ is used to refer to a pth order autoregressive process. For the general case, the notation $ARMA(p, q)$ is used.

To demonstrate the basic principles of parametric estimation, suppose it was determined that a certain random process, $Y[n]$, is well modeled by an $AR(1)$ model,

$$Y[n] = a_1 Y[n-1] + X[n], \tag{10.40}$$

where $X[n]$ is a IID random process with zero-mean and a variance of σ_X^2. It is noted that

$$Y[n+1] = a_1 Y[n] + X[n+1], \tag{10.41}$$

$$Y[n+2] = a_1^2 Y[n] + a_1 X[n+1] + X[n+2], \tag{10.42}$$

$$Y[n+3] = a_1^3 Y[n] + a_1^2 X[n] + a_1 X[n+1] + X[n+2], \tag{10.43}$$

and in general,

$$Y[n+k] = a_1^k Y[n] + \sum_{i=0}^{k-1} a_1^{k-1-i} X[n+i]. \tag{10.44}$$

Using this expression, the autocorrelation function of the $AR(1)$ process can be computed.

$$
\begin{aligned}
R_{YY}[n, n+k] &= E[Y[n]Y[n+k]] \\
&= a_1^k E[Y^2[n]] + \sum_{i=0}^{k-1} a_1^{k-1-i} E[Y[n]X[n+i]] = a_1^k R_{YY}[n, n] \tag{10.45}
\end{aligned}
$$

The last step is accomplished using the fact that $Y[n]$ is independent of $X[n+i]$ for $i > 0$. The expression $R_{YY}[n, n]$ is calculated according to

$$R_{YY}[n, n] = E[(a_1 Y[n-1] + X[n])^2] = a_1^2 R_{YY}[n-1, n-1] + \sigma_X^2. \tag{10.46}$$

Assuming that the process $Y[n]$ is WSS[3], this recursion becomes

$$R_{YY}[0] = a_1^2 R_{YY}[0] + \sigma_X^2 \Rightarrow R_{YY}[0] = \frac{\sigma_X^2}{1 - a_1^2}. \tag{10.47}$$

---

[3]It will be shown in the next chapter that this is the case provided that $X[n]$ is WSS.

Hence, the autocorrelation function of the AR(1) process is

$$R_{YY}[k] = \frac{\sigma_X^2}{1 - a_1^2} a_1^{|k|}. \tag{10.48}$$

Assuming that the samples of this discrete time process are taken at a sampling interval of Δt, the PSD of this process works out to be

$$S_{YY}(f) = \frac{\Delta t \sigma_X^2}{\left|1 - a_1 e^{-j2\pi f \Delta t}\right|^2}. \tag{10.49}$$

For this simple AR(1) model, the PSD can be expressed as a function of two unknown parameters, a_1 and σ_X^2. The problem of estimating the PSD then becomes one of estimating the two parameters and then plugging the result into the general expression for the PSD. In many cases, the total power in the process may be known, which eliminates the need to estimate σ_X^2. Even if that is not the case, the value of σ_X^2 is just a multiplicative factor in the expression for PSD and does not change the shape of the curve. Hence, in the following, we focus attention on estimating the parameter, a_1.

Since we know the AR(1) model satisfies the recursion of Equation 10.40, the next value of the process can be predicted from the current value according to

$$\widehat{Y}[n+1] = \hat{a}_1 Y[n]. \tag{10.50}$$

This is known as *linear prediction* since the predictor of the next value is a linear function of the current value. The error in this estimate is

$$E = Y[n+1] - \widehat{Y}[n+1] - Y[n+1] - \hat{a}_1 Y[n]. \tag{10.51}$$

Typically, we choose as an estimate of a_1 the value of $\hat{a}_1$ that makes the linear predictor as good as possible. Usually, "good" is interpreted as minimizing the mean-square error, which is given by

$$E[E^2] = E[(Y[n+1] - \hat{a}_1 Y[n])^2] = R_{YY}(0)(1 + \hat{a}_1^2) - 2\hat{a}_1 R_{YY}(1). \tag{10.52}$$

Differentiating the mean-square error with respect to $\hat{a}_1$ and setting equal to zero results in

$$2\hat{a}_1 R_{YY}(0) - 2R_{YY}(1) = 0 \Rightarrow \hat{a}_1 = \frac{R_{YY}(1)}{R_{YY}(0)}. \tag{10.53}$$

Of course, we don't know what the autocorrelation function is. If we did, we would not need to estimate the PSD. So, the preceding ensemble averages must be replaced with sample averages, and the minimum mean-square error (MMSE) linear prediction coefficient is given by

$$\hat{a}_1 = \frac{\widehat{R}_{YY}(1)}{\widehat{R}_{YY}(0)} = \frac{\sum_{n=-n_0}^{n_0-1} y[n]y[n+1]}{\sum_{n=-n_0}^{n_0} y^2[n]}. \tag{10.54}$$

Note that we have used a lower case $y[n]$ in Equation 10.54 since we are dealing with a single realization rather than the ensemble $Y[n]$.

EXAMPLE 10.11: In this example, we use the AR(1) model to estimate the PSD of the random telegraph process. Clearly, the AR(1) model does not describe the random telegraph process; however, the autocorrelation function of the random telegraph signal is a two-sided exponential, as is the autocorrelation function of the AR(1) process. As a consequence, we expect this model to give good results. The results are shown in Figure 10.10. Notice how nicely the estimated PSD matches the actual PSD.

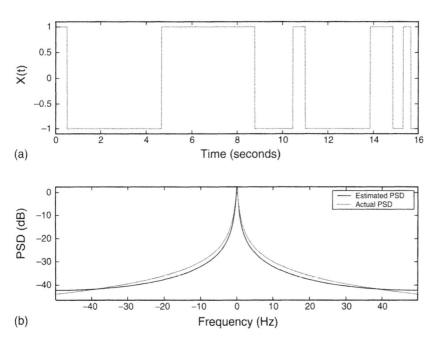

(a)

(b)

Figure 10.10 A sample realization of the random telegraph signal (a) and the parametric estimate of the PSD function based on the AR(1) model (b).

In the previous example, the results were quite good because the functional form of the PSD of the AR(1) process nicely matched the functional form of the true PSD. If the fit had not been so good, it might have been necessary to move to a higher order AR(p) model. In the exercises at the end of the chapter (see Exercises 10.19 and 10.20), the reader is led through the problem of finding the MMSE linear prediction coefficients for a general AR(p) model. The problem of analytically finding the PSD of the AR(p) process is dealt with in the next chapter (also see Exercise 10.18).

10.5 Thermal Noise

The most commonly encountered source of noise in electronic systems is that caused by thermal agitation of electrons in any conductive material, which is commonly referred to as *thermal noise*. Unlike shot noise, thermal noise does not require the presence of a direct current and thus is always present. We will not delve into the underlying thermodynamics to derive a model for this type of noise, but rather will just summarize some of the important results. Nyquist's theorem states that for a resistive element with an impedance of r ohms (Ω), at a temperature of t_k (measured in degrees Kelvin), the mean-square voltage of the thermal noise measured in an incremental frequency band of width Δf centered at frequency f is found to be

$$E[V^2(t)] = v_{\text{rms}}^2 = 4kt_k r \Delta f \left[\frac{h|f|/kt_k}{\exp(h|f|/kt_k) - 1} \right] \text{volts}^2, \qquad (10.55)$$

where

h = Planck's constant = 6.2×10^{-34} J-sec;
k = Boltzman's constant = 1.38×10^{-23} J/°K;
t_k = absolute temperature = $273 + °C$.

Typically, a practical resistor is modeled as a Thevenin equivalent circuit, as illustrated in Figure 10.11, consisting of a noiseless resistor in series with a noise source with a mean-square value as specified in the previous equation. If this noisy resistor were connected to a resistive load of impedance r_L, the average power delivered to the load would be

$$\overline{P_L} = \frac{v_{\text{rms}}^2 r_L}{(r + r_L)^2}. \qquad (10.56)$$

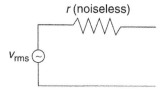

Figure 10.11 A Thevenin equivalent circuit for a noisy resistor.

The power delivered to the load is maximized when the source and the load impedance are matched (i.e., $r_L = r$). It is common to refer to the maximum power that can be delivered to a load as the *available power*. For a noisy resistor, the available power (in a bandwidth of Δf) is

$$\overline{P} = \frac{v_{\text{rms}}^2}{4r} = kt_k \Delta f \left[\frac{h|f|/kt_k}{\exp(h|f|/kt_k) - 1} \right] \text{watts.} \qquad (10.57)$$

The power spectral density of the thermal noise in the resistor is then

$$S_{NN}(f) = \frac{1}{2} kt_k \left[\frac{h|f|/kt_k}{\exp(h|f|/kt_k) - 1} \right]. \qquad (10.58)$$

The extra factor of $1/2$ is due to the fact that our PSD function is a two-sided function of frequency, and so the actual power in a given frequency band is evenly split between the positive and negative frequencies. Note that the power available to a load and the resulting PSD are independent of the impedance of the resistor, r.

This PSD function is plotted in Figure 10.12 for several different temperatures. Note that for frequencies that are of interest in most applications (except optical, infrared, etc.), the PSD function is essentially constant. It is straightforward (and left as an exercise to the reader) to show that this constant is given by

$$S_{NN}(f) = \frac{1}{2} kt_k = \frac{N_o}{2}, \qquad (10.59)$$

where we have defined the constant $N_o = kt_f$. At $t_k = 298°\text{K}^4$, the parameter N_o takes on a value of $N_o = 4.11 \times 10^{-21} \text{ W/Hz} = -173.86 \text{ dBm/Hz}$.

---

[4]Most texts use $t_k = 290°\text{K}$ as "room temperature"; however, this corresponds to a fairly chilly room ($17°\text{C} \approx 63°\text{F}$). On the other hand $t_k = 298°\text{K}$ is a more balmy environment ($25°\text{C} \approx 77°\text{F}$). These differences would change the value of N_o by only a small fraction of a dB.

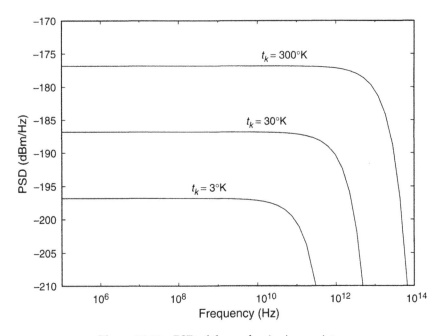

Figure 10.12   PSD of thermal noise in a resistor.

It is common to use this simpler function as a model for the PSD of thermal noise. Because the PSD contains equal power at all frequencies, this noise model is referred to as *white noise* (analogous to white light, which contains all frequencies). The corresponding autocorrelation function is

$$R_{NN}(\tau) = \frac{N_o}{2}\delta(\tau). \tag{10.60}$$

It should be pointed out that the noise model of Equation 10.59 is a mathematical approximation to the actual PSD. There is no such thing as truly white noise, since such a process (if it existed) would have infinite power and would destroy any device we tried to measure it with. However, this mathematical model is simple, easy to work with, and serves as a good approximation to thermal noise for most applications.

In addition to modeling thermal noise as having a flat PSD, it can also be shown that the first order characteristics of thermal noise can be well approximated with a zero-mean Gaussian process. We say that thermal noise is zero-mean white Gaussian noise (WGN) with a (two-sided) PSD of $N_o/2$. While thermal noise is the most common source of noise in electronic devices, there are other sources as well. Shot noise was discussed at the end of Chapter 8. In addition, one may encounter flicker noise, which occurs primarily in active devices; burst or popcorn noise, which is found in integrated circuits and some discrete transistors; avalanche noise, which is produced by avalanche breakdown in a p-n junction; as well as several other types of noise. For the purposes of this text, we will stick with the white Gaussian model of thermal noise.

10.6 Engineering Application: PSDs of Digital Modulation Formats

In this section, we evaluate the PSD of a class of signals that might be used in a digital communications system. Suppose we have a sequence of data symbols $\{B_k\}$ that we wish to convey across some communication medium. We can use the data symbols to determine the amplitude of a sequence of pulses that we would then transmit across the medium (e.g., a twisted copper pair, or an optical fiber). This is known as pulse amplitude modulation (PAM). If the pulse amplitudes are represented by the sequence of random variables $\{\ldots, A_{-2}, A_{-1}, A_0, A_1, A_2, \ldots\}$ and the basic pulse shape is given by the waveform $p(t)$, then the transmitted signal might be of the form

$$S(t) = \sum_{k=-\infty}^{\infty} A_k p(t - kt_s - \Theta), \tag{10.61}$$

where t_s is the symbol interval (that is, one pulse is launched every t_s seconds) and Θ is a random delay, which we take to be uniformly distributed over $[0, t_s)$ and independent of the pulse amplitudes. The addition of the random delay in the model makes the process $S(t)$ WSS. This is not necessary and the result we will obtain would not change if we did not add this delay, but it does slightly simplify the derivation.

If the data symbols are drawn from an alphabet of size 2^n symbols, then each symbol can be represented by an n-bit word, and hence the data rate of the digital communication system is $r = n/t_s$ bits/second. The random process $S(t)$ used to represent this data has a certain spectral content, and thus requires a communications channel with a bandwidth adequate to carry that spectral content. It would be interesting to see how the required bandwidth relates to the data rate. Toward that end, we seek to determine the PSD of the PAM signal $S(t)$. We will find the PSD by first computing the autocorrelation function of $S(t)$ and then converting this to PSD via the Wiener-Khintchine-Einstein theorem.

Using the definition of autocorrelation, the autocorrelation function of the PAM signal is given by

$$R_{SS}(t, t+\tau) = E[S(t)S(t+\tau)] = E\left[\sum_{k=-\infty}^{\infty} \sum_{m=-\infty}^{\infty} A_k A_m p(t - kt_s - \Theta)(p(t + \tau - mt_s - \Theta))\right]$$

$$= \sum_{k=-\infty}^{\infty} \sum_{m=-\infty}^{\infty} E[A_k A_m] E[p(t - kt_s - \Theta)p(t + \tau - mt_s - \Theta)]$$

$$= \frac{1}{t_s} \sum_{k=-\infty}^{\infty} \sum_{m=-\infty}^{\infty} E[A_k A_m] \int_0^{t_s} p(t - kt_s - \theta)p(t + \tau - mt_s - \theta)]d\theta. \qquad (10.62)$$

To simplify notation, we define $R_{AA}[n]$ to be the autocorrelation function of the sequence of pulse amplitudes. Note that we are assuming the sequence is stationary (at least in the wide sense). Going through a simple change of variables ($v = t - kt_s - \theta$) then results in

$$R_{SS}(t, t+\tau) = \frac{1}{t_s} \sum_{k=-\infty}^{\infty} \sum_{m=-\infty}^{\infty} R_{AA}[m - k] \int_{t-(k+1)t_s}^{t-k_s} p(v)p(v + \tau - (m - k)t_s)\, dv. \qquad (10.63)$$

Finally, we go through one last change of variables ($n = m - k$) to produce

$$R_{SS}(t, t+\tau) = \frac{1}{t_s} \sum_{k=-\infty}^{\infty} \sum_{n=-\infty}^{\infty} R_{AA}[n] \int_{t-(k+1)t_s}^{t-k_s} p(v)p(v + \tau - nt_s)\, dv$$

$$= \frac{1}{t_s} \sum_{n=-\infty}^{\infty} R_{AA}[n] \sum_{k=-\infty}^{\infty} \int_{t-(k+1)t_s}^{t-kt_s} p(v)p(v + \tau - nt_s)\, dv$$

$$= \frac{1}{t_s} \sum_{n=-\infty}^{\infty} R_{AA}[n] \int_{-\infty}^{\infty} p(v)p(v + \tau - nt_s)\, dv. \qquad (10.64)$$

To aid in taking the Fourier transform of this expression, we note that the integral in this equation can be written as a convolution of $p(t)$ with $p(-t)$:

$$\int_{-\infty}^{\infty} p(v)p(v + \tau - nt_s)\, dv = p(t)^* p(-t)|_{t=\tau-nt_s}. \qquad (10.65)$$

Using the fact that convolution in the time domain becomes multiplication in the frequency domain along with the time reversal and time shifting properties of Fourier transforms (see Appendix C, Review of Linear Time Invariant Systems), the transform of this convolution works out to be

$$F\left[\int_{-\infty}^{\infty} p(v)p(v+\tau - nt_s)\,dv\right] = |P(f)|^2 e^{-j2\pi nft_s},$$ (10.66)

where $P(f) = F[p(t)]$ is the Fourier transform of the pulse shape used. With this result, the PSD of the PAM signal is found by taking the transform of Equation 10.64, resulting in

$$S_{SS}(f) = \frac{|P(f)|^2}{t_s} \sum_{n=-\infty}^{\infty} R_{AA}[n] e^{-j2\pi nft_s}.$$ (10.67)

It is seen from the previous equation that the PSD of a PAM signal is the product of two terms, the first of which is the magnitude squared of the pulse shapes spectrum, while the second term is essentially the PSD of the discrete sequence of amplitudes. As a result, we can control the spectral content of our PAM signal by carefully designing a pulse shape with a compact spectrum and also by introducing memory into the sequence of pulse amplitudes.

EXAMPLE 10.12: To start with, suppose the pulse amplitudes are an IID sequence of random variables that are equally likely to be $+1$ or -1. In that case, $R_{AA}[n] = \delta[n]$ and the PSD of the sequence of amplitudes is

$$\sum_{n=-\infty}^{\infty} R_{AA}[n] e^{-j2\pi nft_s} = 1.$$

In this case, $S_{SS}(f) = |P(f)|^2/t_s$ and the spectral shape of the PAM signal is completely determined by the spectral content of the pulse shape. Suppose we use as a pulse shape a square pulse of height a and width t_s,

$$p(t) = a\mathrm{rect}(t/t_s) \longleftrightarrow P(f) = at_s\mathrm{sinc}(ft_s).$$

The PSD of the resulting PAM signal is then $S_{SS}(f) = a^2 t_s \mathrm{sinc}^2 f(t_s)$. Note that the factor $a^2 t_s$ is the energy in each pulse sent, E_p. A sample realization of this PAM process along with a plot of the PSD is given in Figure 10.13. Most of the power in the process is contained in the main lobe, which has a bandwidth of $1/t_s$ (equal to the data rate), but there is also a nontrivial amount of power in the sidelobes, which die off very slowly. The high-frequency content can be attributed to the instantaneous jumps in the process. These frequency sidelobes can

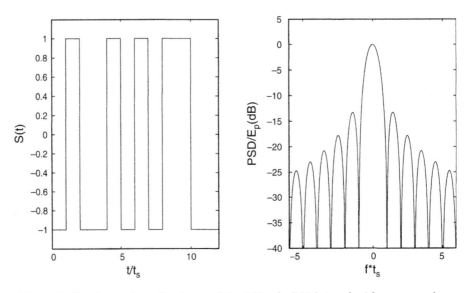

Figure 10.13   A sample realization and the PSD of a PAM signal with square pulses.

be suppressed by using a pulse with a smoother shape. Suppose, for example, we used a pulse that was a half cycle of a sinusoid of height a,

$$p(t) = a \cos\left(\frac{\pi t}{t_s}\right) \text{rect}\left(\frac{t}{t_s}\right) \longleftrightarrow P(f)$$

$$= \frac{at_s}{2}\left[\text{sinc}\left(ft_s - \frac{1}{2}\right) + \text{sinc}\left(ft_s + \frac{1}{2}\right)\right] = \frac{at_s}{2\pi}\frac{\cos(\pi ft_s)}{\frac{1}{4} - (ft_s)^2}.$$

The resulting PSD of the PAM signal with half-sinusoidal pulse shapes is then

$$S_{SS}(f) = \frac{a^2 t_s}{4\pi^2}\frac{\cos^2(\pi ft_s)}{\left[\frac{1}{4} - (ft_s)^2\right]^2}.$$

In this case, the energy in each pulse is $E_p = a^2 t_s/2$. As shown in Figure 10.14, the main lobe is now 50 percent wider than it was with square pulses, but the sidelobes decay much more rapidly.

EXAMPLE 10.13: In this example, we show how the spectrum of the PAM signal can also be manipulated by adding memory to the sequence of pulse amplitudes. Suppose the data to be transmitted

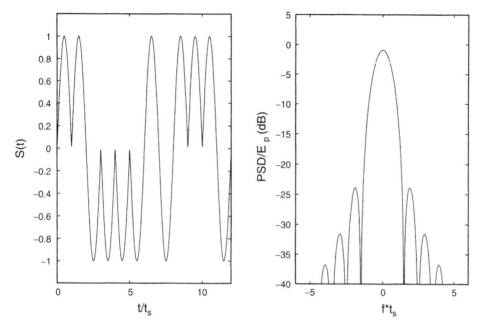

Figure 10.14 A sample realization and the PSD of a PAM signal with half-sinusoidal pulses.

$\{\ldots, B_{-2}, B_{-1}, B_0, B_1, B_2, \ldots\}$ is an IID sequence of Bernoulli random variables, $B_k \in \{+1, -1\}$. In the previous example, we formed the pulse amplitudes according to $A_k = B_k$. Suppose instead that we formed these amplitudes according to $A_k = B_k + B_{k-1}$. Now the pulse amplitudes can take on three values (even though each pulse still carries only 1 bit of information. This is known as *duobinary* precoding. The resulting autocorrelation function for the sequence of pulse amplitudes is

$$R_{AA}[n] = E[A_k A_{k+n}] = E[(B_k + B_{k-1})(B_{k+n} + B_{k+n-1}]) = \begin{cases} 2 & n = 0 \\ 1 & n = \pm 1 \\ 0 & \text{otherwise} \end{cases} .$$

The PSD of this sequence of pulse amplitudes is then

$$\sum_{n=-\infty}^{\infty} R_{AA}[n] e^{-j2\pi f t_s} = 2 + e^{j2\pi f t_s} + e^{-j2\pi n f t_s}$$

$$= 2 + 2\cos(2\pi f t_s) = 4\cos^2(\pi f t_s).$$

This expression then multiplies whatever spectral shape results from the pulse shape chosen. The PSD of duobinary PAM with square pulses

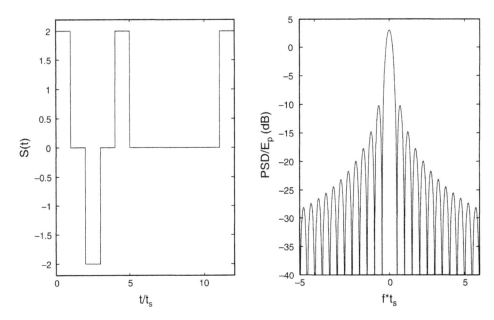

Figure 10.15 A sample realization and the PSD of a PAM signal with duobinary precoding and square pulses.

is illustrated in Figure 10.15. In this case, the duobinary precoding has the benefit of suppressing the frequency sidelobes without broadening the main lobe.

EXAMPLE 10.14: The following MATLAB code creates a realization of a binary PAM signal where the pulse amplitudes are either $+1$ or -1. In this example, a half-sinusoidal pulse shape is used, but the code is written so that it is easy to change the pulse shape (just change the sixth line where the pulse shape is assigned to the variable p). After a realization of the PAM signal is created, the PSD of the resulting signal is estimated using the segmented periodogram technique given in Example 10.10. The resulting PSD estimate is shown in Figure 10.16. Note the agreement between the estimate and the actual PSD shown in Figure 10.14. The reader is encouraged to try running this program with different pulse shapes to see how the pulse shape changes the spectrum of the PAM signal.

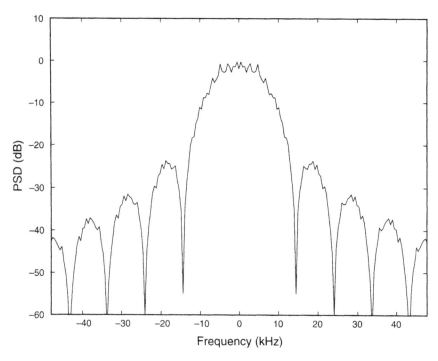

Figure 10.16 An estimate of the PSD of a PAM signal with half-sinusoidal pulse shapes.

```
N=100;                                % Number of bit intervals in
                                      realization.
Ns=19;                                % Number of time samples per bit.
Rb=9600;                              % Bit rate (bits/sec).
dt=1/(Rb*Ns);                         % Time between samples.
t=([1:Ns]-0.5)/Ns;                    % Time axis for pulse shape.
p=sin(pi*t);                          % Pulse shape.
Ep=p*p'*dt;                           % Energy per pulse.
X(1:Ns:(Ns*(N-1)+1))                  % Random data bits.
  =sign(rand(1,N)-0.5);
X=conv(X,p);                          % PAM signal with pulse shape
                                      added.
M=10;                                 % Number of segments.
[S,f]=EPrdgm(X,dt,M);                 % (Normalized) PSD estimate.
plot(f/1000,10*log10(abs(S/Ep)))      % Plot results.
axis([-5*Rb/1000 5*Rb/1000 -60 10])
xlabel('frequency (kHz)')
ylabel('PSD (dB)')
```

Exercises

10.1 Consider a random process of the form

$$X(t) = b\cos(2\pi\Psi t + \Theta),$$

where b is a constant, Θ is a uniform random variable over $[0, 2\pi)$, and Ψ is a random variable that is independent of Θ and has a PDF, $f_\Psi(\psi)$. Find the PSD, $S_{XX}(f)$ in terms of $f_\Psi(\psi)$. In so doing, prove that for any $S(f)$ that is a valid PSD function, we can always construct a random process with PSD equal to $S(f)$.

10.2 Let $X(t) = A\cos(\omega t) + B\sin(\omega t)$ where A and B are independent, zero-mean, identically distributed non-Gaussian random variables (IID). Show that $X(t)$ is wide sense stationary (WSS), but not strict sense stationary. *Hint:* For the latter case, consider $E[X^3(t)]$. Recall the discussion prior to Example 8.23. If A and B *are* Gaussian, with zero mean and (IID), then $X(t)$ *is* strict sense stationary.

10.3 Let $X(t) = \sum_{n=1}^{N} a_n \cos(\omega_n t + \theta_n)$, where all of the ω_n are nonzero constants, the a_n are constants, and the θ_n are IID random variables, each uniformly distributed over $[0, 2\pi)$.

(a) Determine the autocorrelation function of $X(t)$.

(b) Determine the power spectral density of $X(t)$.

10.4 Let $X(t) = \sum_{n=1}^{\infty} [A_n \cos(n\omega t) + B_n \sin(n\omega t)]$ be a random process, where A_n and B_n are random variables such that $E[A_n] = E[B_n] = 0$, $E[A_n B_m] = 0$, $E[A_n A_m] = \delta_{n,m} E[A_n^2]$, and $E[B_n B_m] = \delta_{n,m} E[B_n^2]$ for all m and n, where $\delta_{n,m}$ is the Kronecker delta function. This process is sometimes used as a model for random noise.

(a) Find the time-varying autocorrelation function $R_{XX}(t, t + \tau)$.

(b) If $E[B_n^2] = E[A_n^2]$, is this process WSS?

10.5 Find the power spectral density for a process for which $R_{XX}(\tau) = 1$ for all τ.

10.6 Suppose $X(t)$ is a stationary, zero-mean Gaussian random process with PSD, $S_{XX}(f)$.

(a) Find the PSD of $Y(t) = X^2(t)$ in terms of $S_{XX}(f)$.

(b) Sketch the resulting PSD if $S_{XX}(f) = \text{rect}\left(\dfrac{f}{2B}\right)$.

(c) Is $Y(t)$ WSS?

10.7 Consider a random sinusoidal process of the form $X(t) = b\cos(2\pi ft + \Theta)$, where Θ has an arbitrary PDF, $f_\Theta(\theta)$. Determine analytically how the PSD of $X(t)$ depends on $f_\Theta(\theta)$. Give an intuitive explanation for your result.

10.8 Let $s(t)$ be a deterministic periodic waveform with period t_o. A random process is constructed according to $X(t) = s(t - T)$, where T is a random variable uniformly distributed over $[0, t_o)$. Show that the random process $X(t)$ has a line spectrum and write the PSD of $X(t)$ in terms of the Fourier Series coefficients of the periodic signal $s(t)$.

10.9 A sinusoidal signal of the form $X(t) = b\cos(2\pi f_o t + \Theta)$ is transmitted from a fixed platform. The signal is received by an antenna on a mobile platform that is in motion relative to the transmitter, with a velocity of V relative to the direction of signal propagation between the transmitter and receiver. Hence, the received signal experiences a Doppler shift and (ignoring noise in the receiver) is of the form.

$$Y(t) = b\cos\left(2\pi f_o\left(1 + \frac{V}{c}\right)t + \Theta\right),$$

where c is the speed of light. Find the PSD of the received signal if V is uniformly distributed over $(-v_o, v_o)$. Qualitatively, what does the Doppler effect do to the PSD of the sinusoidal signal?

10.10 Two zero-mean, discrete random processes, $X[n]$ and $Y[n]$, are statistically independent and have autocorrelation functions given by $R_{XX}[k] = (1/2)^k$ and $R_{YY}[k] = (1/3)^k$. Let a new random process be $Z[n] = X[n] + Y[n]$.

(a) Find $R_{ZZ}[k]$. Plot all three autocorrelation functions.
(b) Determine all three power spectral density functions analytically and plot the power spectral densities.

10.11 Let $S_{XX}(f)$ be the PSD function of a WSS discrete time process $X[n]$. Recall that one way to obtain this PSD function is to compute $R_{XX}[n] = E[X[k]X[k+n]]$ and then take the DFT of the resulting autocorrelation function. Determine how to find the average power in a discrete time random process directly from the PSD function, $S_{XX}(f)$.

10.12 A binary phase shift keying signal is defined according to

$$X(t) = \cos\left(2\pi f_c t + B[n]\frac{\pi}{2}\right) \quad \text{for} \quad nT \leq t < (n+1)T,$$

for all n, and $B[n]$ is a discrete time, Bernoulli random process that has values of $+1$ or -1.

(a) Determine the autocorrelation function for the random process $X(t)$. Is the process WSS?

(b) Determine the power spectral density of $X(t)$.

10.13 Develop a formula to compute the RMS bandwidth of a random process, $X(t)$, directly from its autocorrelation function, $R_{XX}(\tau)$.

10.14 A random process has a PSD function given by

$$S(f) = \frac{1}{\left(1 + \left(\frac{f}{B}\right)^2\right)^3}.$$

(a) Find the absolute bandwidth.

(b) Find the 3-dB bandwidth.

(c) Find the RMS bandwidth.

Can you generalize your result to a spectrum of the form

$$S(f) = \frac{1}{\left(1 + \left(\frac{f}{B}\right)^2\right)^N},$$

where N is an integer greater than 1?

10.15 A random process has a PSD function given by

$$S(f) = \frac{f^2}{\left(1 + \left(\frac{f}{B}\right)^2\right)^3}.$$

(a) Find the absolute bandwidth.

(b) Find the 3-dB bandwidth.

(c) Find the RMS bandwidth.

Can you generalize your result to a spectrum of the form

$$S(f) = \frac{f^2}{\left(1 + \left(\frac{f}{B}\right)^2\right)^N},$$

where N is an integer greater than 2?

10.16 Let $X(t)$ be a random process whose PSD is shown in the accompanying figure. A new process is formed by multiplying $X(t)$ by a carrier to produce

$$Y(t) = X(t)\cos(\omega_o t + \Theta),$$

where Θ is uniform over $[0, 2\pi)$ and independent of $X(t)$. Find and sketch the PSD of the process $Y(t)$.

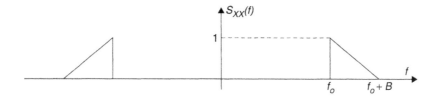

10.17 Consider the linear prediction random process $X[n] = (1/2)X[n-1]+E[n]$, $n = 1, 2, 3, \ldots$, where $X[0] = 0$ and $E[n]$ is a zero-mean, IID random process. Find the mean and autocorrelation functions for $X[n]$. Is $X[n]$ WSS?

10.18 Consider an AR(2) process described by the recursion

$$Y[n] = a_1 Y[n-1] + a_2 Y[n-2] + X[n],$$

where $X[n]$ is an IID random process with zero-mean and variance σ_X^2.

(a) Show that the autocorrelation function of the AR(2) process satisfies the difference equation,

$$R_{YY}[k] = a_1 R_{YY}[k-1] + a_2 R_{YY}[k-2], \quad k = 2, 3, 4, \ldots.$$

(b) Show that the first two terms in the autocorrelation function satisfy

$$(1 - a_1^2 - a_2^2)R_{YY}[0] - 2a_1 a_2 R_{YY}[1] = \sigma_X^2,$$

and

$$(1 - a_2)R_{YY}[1] = a_1 R_{YY}[0].$$

From these two equations, solve for $R_{YY}[0]$ and $R_{YY}[1]$ in terms of a_1, a_2, and σ_X^2.

(c) Using the difference equation in part (a) together with the initial conditions in part (b), find a general expression for the autocorrelation function of an AR(2) process.

(d) Use your result in part (c) to find the PSD of an AR(2) process.

10.19 Suppose we use an AR(2) model to predict the next value of a random process based on observations of the two most recent samples. That is, we form

$$\widehat{Y}[n+1] = a_1 Y[n] + a_2 Y[n-1].$$

(a) Derive an expression for the mean-square estimation error,

$$E[\varepsilon^2] = E[(Y[n+1] - \widehat{Y}[n+1])^2].$$

(b) Find the values of the prediction coefficients, a_1 and a_2, that minimize the mean-square error.

10.20 Extend the results of Exercise 10.19 to a general AR(p) model. That is, suppose we wish to predict the next value of a random process by forming a linear combination of the p most recent samples:

$$\widehat{Y}[n+1] = \sum_{k=1}^{p} a_k Y[n-k+1].$$

Find an expression for the values of the prediction coefficients that minimize the mean-square prediction error.

10.21 A random process $Y[n]$ is found to obey the AR(2) model. That is, $Y[n] = a_1 Y[n-1] + a_2 Y[n-2] + X[n]$. Find expressions for the autocorrelation function and PSD of $Y[n]$. Can you generalize your results to an arbitrary AR(p) model?

10.22

(a) Prove that the expression for the PSD of thermal noise in a resistor converges to the constant $N_o/2 = kt_k/2$ as $f \to 0$.

(b) Assuming a temperature of 298°K, find the range of frequencies over which thermal noise has a PSD that is within 99 percent of its value at $f = 0$.

(c) Suppose we had a very sensitive piece of equipment that was able to accurately measure the thermal noise across a resistive element. Furthermore, suppose our equipment could respond to a range of frequencies that spanned 50 MHz. Find the power (in watts) and the root mean square (RMS) voltage (in volts) that we would measure across a 75 Ω resistor. Assume the equipment had a load impedance matched to the resistor.

10.23 Suppose two resistors of impedance r_1 and r_2 are placed in series and held at different physical temperatures, t_1 and t_2. We would like to model this series combination of noisy resistors as a single noiseless resistor, with an impedance of $r = r_1 + r_2$, together with a noise source with an effective temperature of t_e. In short, we want the two models shown in the accompanying figure to be equivalent. Assuming the noise produced by the two resistors is independent, what should t_e, the effective noise temperature, of the series combination of resistors be? If the two resistors are held at the same physical temperature, is the effective temperature equal to the true common temperature of the resistors?

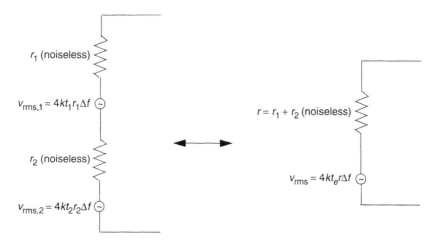

10.24 Repeat Exercise 10.26 for a parallel combination of resistors.

MATLAB Exercises

10.25

(a) Create a random process $X[n]$ where each sample of the random process is an IID, Bernoulli random variable equally likely to be ± 1. Form a new process according to the MA(2) model $Y[n] = X[n] - \frac{1}{2}X[n-1] + \frac{1}{4}X[n-2]$. Assume $X[n] = 0$ for $n < 0$.

(b) Compute the time average autocorrelation function $\langle Y[n]Y[n+k]\rangle$ from a single realization of this process.

(c) Compute the ensemble average autocorrelation function $E[Y[n]Y \times [n + k]]$ from several realizations of this process. Does the process appear to be ergodic in the autocorrelation?

(d) Estimate the PSD of this process using the periodogram method.

10.26

(a) Create a random process $X[n]$ where each sample of the random process is an IID, Bernoulli random variable equally likely to be ± 1. Form a new process according to the AR(2) model $Y[n] = \frac{1}{2}Y[n - 1] - \frac{1}{4}Y \times [n - 2] + X[n]$. Assume $Y[n] = 0$ for $n < 0$.

(b) Compute the time average autocorrelation function $\langle Y[n]Y[n+k]\rangle$ from a single realization of this process.

(c) Compute the ensemble average autocorrelation function $E[Y[n]Y \times [n+k]]$ from several realizations of this process. Does the process appear to be ergodic in the autocorrelation?

(d) Estimate the PSD of this process using the periodogram method.

10.27

(a) For the process in Exercise 10.29, find a parametric estimate of the PSD by using an AR(1) model. Compare the resulting PSD estimate with the nonparametric estimate found in Exercise 10.29(d). Explain any differences you see.

(b) Again referring to the process in Exercise 10.29, find a parametric estimate of the PSD this time by using an AR(2) model. Compare the resulting PSD estimate with the nonparametric estimate found in Exercise 10.29(d). Explain any differences you see.

10.28

(a) Write a MATLAB program to create a realization of a binary PAM signal with square pulses. You can accomplish this with a simple modification to the program as given in Example 10.14. Call this signal $x(t)$.

(b) We can create a frequency shift keying (FSK) signal according to

$$y(t) \cos\left(2\pi f_c t + \frac{\pi}{2t_s}\int_0^t x(u)\, du\right), \qquad (10.68)$$

where t_s is the duration of the square pulses in $x(t)$ and f_c is the carrier frequency. Write a MATLAB program to create a 10-msec realization of this FSK signal assuming $t_s = 100\ \mu$ sec and $f_c = 20$ kHz.

(c) Using the segmented periodogram, estimate the PSD of the FSK signal you created in part (b).

(d) Estimate the 30-dB bandwidth of the FSK signal. That is, find the bandwidth where the PSD is down 30 dB from its peak value.

10.29 Construct a signal-plus-noise random sequence using 10 samples of

$$X[n] = \cos(2\pi nf_0 t_s) + N[n],$$

where $N[n]$ is a sequence of zero-mean, unit variance, IID Gaussian random variables, $f_0 = 0.1/t_s = 100$ kHz, and $t_s = 1$ μsec is the time between samples of the process.

(a) Calculate the periodogram estimate of the PSD, $S_{XX}(f)$.

(b) Calculate a parametric estimate of the PSD using AR models with $p = 1, 2, 3$, and 5. Compare the parametric estimates with the periodogram. In your opinion, which order AR model is the best fit?

(c) Repeat parts (a) and (b) using 100 samples instead of 10.

10.30 Construct a signal-plus-noise random sequence using 10 samples of

$$X[n] = \cos(2\pi nf_1 t_s) + \cos(2\pi nf_2 t_s) + N[n],$$

where $N[n]$ is a sequence of zero-mean, unit variance, IID Gaussian random variables, $f_1 = 0.1/t_s = 100$ kHz, $f_1 = 0.4/t_s = 400$ kHz, and $t_s = 1$ μsec is the time between samples of the process.

(a) Calculate the periodogram estimate of the PSD, $S_{XX}(f)$.

(b) Calculate a parametric estimate of the PSD using AR models with $p = 3, 4, 5, 6$, and 7. Compare the parametric estimates with the periodogram. In your opinion, which order AR model is the best fit?

(c) Repeat parts (a) and (b) using 100 samples instead of 10.

Random Processes in Linear Systems

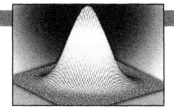

11

In this chapter, we consider the response of both continuous and discrete linear systems to random processes, such as a signal plus noise. We develop statistical descriptions of the output of linear systems with random inputs by viewing the systems in both the time domain and the frequency domain. Two engineering application sections at the end of this chapter demonstrate how filters can be optimized for the purpose of enhancing signal-to-noise ratio and also for the purpose of prediction or smoothing of a signal in noise. It is assumed that the reader is familiar with the study of linear time invariant (LTI) systems. A brief overview is provided in Appendix C, Review of Linear Time–Invariant Systems, for those needing a refresher.

11.1 Continuous Time Linear Systems

Consider an LTI system described by an impulse response $h(t)$ or a transfer function $H(f)$. If a random process, $X(t)$, is input to this system, the output will also be random and is given by the convolution integral

$$Y(t) = \int_{-\infty}^{\infty} X(u)h(t-u)\,du. \tag{11.1}$$

We would like to statistically describe the output of the system. Ultimately, the joint PDF of any number of samples of the output would be nice. In general, this is a very difficult problem and hence we have to be satisfied with a simpler description. However, if the input process is Gaussian, then the output process will also be

Gaussian since any linear processing of Gaussian random variables (processes) produces new Gaussian random variables (processes). In that case, to completely describe the output of the system, we need merely to compute the mean and the autocovariance (or autocorrelation) function of the output. Even if the processes involved are not Gaussian, the mean and autocorrelation functions will serve as a good start toward describing the process. Hence, our first goal will be to specify the mean and autocorrelation functions of the output of an LTI system with a random input.

To start with, consider the mean function of the output:

$$\mu_Y(t) = E[Y(t)] = E\left[\int_{-\infty}^{\infty} X(u)h(t-u)\,du\right]$$

$$= \int_{-\infty}^{\infty} E[X(u)]h(t-u)\,du = \int_{-\infty}^{\infty} \mu_X(u)h(t-u)\,du. \tag{11.2}$$

Hence the output mean is the convolution of the input mean process with the impulse response of the system. For the special case when the input is wide sense stationary (WSS) and the input mean function is therefore constant, the output mean function becomes

$$\mu_Y(t) = \mu_X \int_{-\infty}^{\infty} h(t-u)\,du = \mu_X \int_{-\infty}^{\infty} h(s)\,ds = \mu_X H(0). \tag{11.3}$$

Note that the mean function of the output is also constant provided the input mean is constant.

The autocorrelation function of the output is calculated in a similar manner.

$$R_{YY}(t_1, t_2) = E[Y(t_1)Y(t_2)] = E\left[\left(\int_{-\infty}^{\infty} X(u)h(t_1-u)\,du\right)\left(\int_{-\infty}^{\infty} X(v)h(t_2-v)\,dv\right)\right]$$

$$= \int_{-\infty}^{\infty}\int_{-\infty}^{\infty} E[X(u)X(v)]h(t_1-u)h(t_2-v)\,du\,dv$$

$$= \int_{-\infty}^{\infty}\int_{-\infty}^{\infty} R_{XX}(u,v)h(t_1-u)h(t_2-v)\,du\,dv. \tag{11.4}$$

For WSS inputs, this expression can be simplified a little by using the fact that $R_{XX}(u,v) = R_{XX}(v-u)$. The output autocorrelation function is then

$$R_{YY}(t_1, t_2) = \int_{-\infty}^{\infty}\int_{-\infty}^{\infty} R_{XX}(v-u)h(t_1-u)h(t_2-v)\,du\,dv. \tag{11.5}$$

Although it may not appear like it from this expression, here the output autocorrelation function is also a function of time difference only. To see this, perform the change of variables $s = t_1 - u$ and $w = t_2 - v$. Then Equation 11.5 becomes

$$R_{YY}(t_1, t_2) = \int_{-\infty}^{\infty} \int_{-\infty}^{\infty} R_{XX}(t_2 - t_1 + s - w)h(s)h(w) \, dw \, ds. \qquad (11.6)$$

Now it is clear that $R_{YY}(t_1, t_2) = R_{YY}(t_2 - t_1)$. To write this result in a more compact form, note that the inner integral in Equation 11.6 can be expressed as

$$\int_{-\infty}^{\infty} R_{XX}(t_2 - t_1 + s - w)h(w) \, dw = R_{XX}(t) * h(t)\big|_{t=t_2-t_1+s}, \qquad (11.7)$$

where $*$ denotes convolution. Let $g(t) = R_{XX}(t) * h(t)$, then the output autocorrelation can be expressed as

$$R_{YY}(t_2 - t_1) = \int_{-\infty}^{\infty} g(t_2 - t_1 + s)h(s) \, ds = g(t) * h(-t)\big|_{t=t_2-t_1}. \qquad (11.8)$$

Putting all these results together, we get

$$R_{YY}(\tau) = R_{XX}(\tau) * h(\tau) * h(-\tau). \qquad (11.9)$$

Hence, the output autocorrelation function is found by a double convolution. The presence of the double convolution in Equation 11.9 begs for an equivalent frequency domain representation. Taking Fourier transforms of both sides gives an expression for the power spectral density (PSD) of the output of the filter in terms of the input PSD:

$$S_{YY}(f) = S_{XX}(f)H(f)H*(f) = S_{XX}(f)|H(f)|^2. \qquad (11.10)$$

The term $|H(f)|^2$ is sometimes referred to as the *power transfer function* because it describes how the power is transferred from the input to the output of the system. In summary, we have shown the following results.

THEOREM 11.1: Given an LTI system with impulse response $h(t)$ or transfer function $H(f)$ and a random input process $X(t)$, the mean and autocorrelation functions of the output process, $Y(t)$, can be described by

$$\mu_Y(t) = \mu_X(t) * h(t), \qquad (11.11a)$$

$$R_{YY}(t_1, t_2) = \int_{-\infty}^{\infty} \int_{-\infty}^{\infty} R_{XX}(u, v)h(t_1 - u)h(t_2 - v) \, du \, dv. \qquad (11.11b)$$

Furthermore, if $X(t)$ is WSS, then $Y(t)$ is also WSS with

$$\mu_Y = \mu_X H(0), \tag{11.12a}$$

$$R_{YY}(\tau) = R_{XX}(\tau) * h(\tau) * h(-\tau), \tag{11.12b}$$

$$S_{YY}(f) = S_{XX}(f)|H(f)|^2. \tag{11.12c}$$

At times it is desirable to specify the relationship between the input and output of a filter. Toward that end, we can calculate the cross correlation function between the input and output.

$$R_{XY}(t_1, t_2) = E[X(t_1)Y(t_2)] = E\left[X(t_1) \int_{-\infty}^{\infty} X(u)h(t_2 - u)\, du\right]$$

$$= \int_{-\infty}^{\infty} E[X(t_1)X(u)]h(t_2 - u)\, du = \int_{-\infty}^{\infty} R_{XX}(t_1, u)h(t_2 - u)\, du$$

$$= \int_{-\infty}^{\infty} R_{XX}(t_1, t_2 - v)h(v)\, dv \tag{11.13}$$

If $X(t)$ is WSS, then this simplifies to

$$R_{XY}(\tau) = \int_{-\infty}^{\infty} R_{XX}(\tau - v)h(v)\, dv = R_{XX}(\tau) * h(\tau). \tag{11.14}$$

In a similar manner, it can be shown that

$$R_{YX}(\tau) = R_{XX}(\tau) * h(-\tau). \tag{11.15}$$

In terms of cross spectral densities, these equations can be written as

$$S_{XY}(f) = S_{XX}(f)H(f) \quad \text{and} \quad S_{YX}(f) = S_{XX}(f)H * (f). \tag{11.16}$$

EXAMPLE 11.1: White Gaussian noise, $N(t)$ with a PSD of $S_{NN}(f) = N_0/2$ is input to an resistor, capacitor (RC) lowpass filter. Such a filter will have a transfer function and impulse response given by

$$H(f) = \frac{1}{1 + j2\pi fRC} \quad \text{and} \quad h(t) = \frac{1}{RC} \exp\left(-\frac{t}{RC}\right) u(t),$$

respectively. If the input noise is zero-mean, $\mu_N = 0$, then the output process will also be zero-mean, $\mu_Y = 0$. Also

$$S_{YY}(f) = S_{NN}(f)|H(f)|^2 = \frac{N_0/2}{1 + (2\pi fRC)^2}.$$

Using inverse Fourier transforms, the output autocorrelation is found to be

$$R_{YY}(\tau) = \frac{N_o}{4RC} \exp\left(-\frac{|\tau|}{RC}\right).$$

EXAMPLE 11.2: Suppose we wish to convert a white noise process from continuous time to discrete time using a sampler. Since white noise has infinite power, it cannot be sampled directly and must be filtered first. Suppose for simplicity we use an ideal lowpass filter of bandwidth B to perform the sampling so that the system is as illustrated in Figure 11.1. Let $N_f(t)$ be the random process at the output of the lowpass filter. This process has a PSD of

$$S_{N_f N_f}(f) = \frac{N_0}{2}\mathrm{rect}\left(\frac{f}{2B}\right) = \begin{cases} N_0/2 & |f| < B \\ 0 & \text{otherwise} \end{cases}.$$

The corresponding autocorrelation function is

$$R_{N_f N_f}(\tau) = N_0 B \,\mathrm{sinc}(2B\tau).$$

If the output of the filter is sampled every t_o seconds, the discrete time noise process will have an autocorrelation of $R_{NN}[k] = N_0 B \,\mathrm{sinc}(2kBt_o)$. If the discrete time output process $N[n]$ is to be white, then we want all samples to be uncorrelated. That is, we want $R_{NN}[k] = 0$ for all $k \neq 0$. Recall that the sinc function has nulls whenever its argument is an integer. Thus, the discrete time process will be white if (and only if) $2Bt_o$ is an integer. In terms of the sampling rate, $f_o = 1/t_o$, for the discrete time process to be white, the sampling rate must be $f_o = 2B/m$ for some integer m.

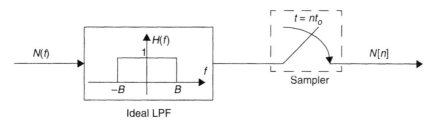

Figure 11.1 Block diagram of a sampling system to convert white noise from continuous time to discrete time.

11.2 Discrete Time Linear Systems

The response of a discrete time linear system to a (discrete time) random process is found using virtually identical techniques to those used with continuous time systems. As such, we do not repeat the derivations here, but rather summarize the relevant results. We start with a linear system described by the difference equation

$$\sum_{i=0}^{p} a_i Y[n-i] = \sum_{k=0}^{q} b_k X[n-k],$$ (11.17)

where $X[n]$ is the input to the system and $Y[n]$ is the output. The reader might recognize this system as producing an autoregressive moving average process (ARMA) process as described in Section 10.4. This system can be described by a transfer function expressed using either z-transforms or discrete Fourier transforms (DFT) as

$$H(z) = \frac{\sum_{k=0}^{q} b_k z^{-k}}{\sum_{i=0}^{p} a_i z^{-i}} \quad \text{or} \quad H(f) = \frac{\sum_{k=0}^{q} b_k e^{-j2\pi kf}}{\sum_{i=0}^{p} a_i e^{-j2\pi if}}.$$ (11.18)

If the DFT is used, it is understood that the frequency variable f is actually a normalized frequency (normalized by the sampling rate). The system can also be described in terms of a discrete time impulse response, $h[n]$, which can be found through either an inverse z-transform or an inverse DFT. The following results apply to any discrete time system described by an impulse response, $h[n]$, and transfer function, $H(f)$.

THEOREM 11.2: Given a discrete time LTI system with impulse response $h[n]$, transfer function $H(f)$, and a random input process $X[n]$, the mean and autocorrelation functions of the output process, $Y[n]$, can be described by

$$\mu_Y[n] = \mu_X[n] * h[n],$$ (11.19a)

$$R_{YY}[n_1, n_2] = \sum_{k_1=-\infty}^{\infty} \sum_{k_2=-\infty}^{\infty} R_{XX}[k_1, k_2] h[n_1 - k_1] h[n_2 - k_2].$$ (11.19b)

Furthermore, if $X[n]$ is WSS, then $Y[n]$ is also WSS with

$$\mu_Y = \mu_X H(0),$$ (11.20a)

$$R_{YY}[n] = R_{XX}[n] * h[n] * h[-n],$$ (11.20b)

$$S_{YY}(f) = S_{XX}(f)|H(f)|^2.$$ (11.20c)

Again, it is emphasized, that the frequency variable in the PSD of a discrete time process is to be interpreted as frequency normalized by the sampling rate.

EXAMPLE 11.3: A discrete time Gaussian white noise process has zero-mean and an autocorrelation function of $R_{XX}[n] = \sigma^2 \delta[n]$. This process is input to a system described by the difference equation

$$Y[n] = aY[n-1] + bX[n].$$

Note that this produces an AR(1) process as the output. The transfer function and impulse response of this system are

$$H(f) = \frac{b}{1 - ae^{-j2\pi f}} \quad \text{and} \quad h[n] = ba^n u[n],$$

respectively, assuming that $|a| < 1$. The autocorrelation and PSD functions of the output process are

$$R_{YY}[n] = \frac{b^2 a^{|n|}}{1 - a^2} \quad \text{and} \quad S_{YY}(f) = \frac{\sigma^2 b^2}{|1 - ae^{-j2\pi f}|^2} = \frac{\sigma^2 b^2}{1 + a^2 - 2a\cos(2\pi f)},$$

respectively.

11.3 Noise Equivalent Bandwidth

Consider an ideal lowpass filter with a bandwidth B whose transfer function is shown in Figure 11.2. Suppose white Gaussian noise with PSD $N_0/2$ is passed through this filter. The total output power would be $P_o = N_0 B$. For an arbitrary lowpass filter, the output noise power would be

$$P_o = \frac{N_0}{2} \int_{-\infty}^{\infty} |H(f)|^2 \, df. \tag{11.21}$$

One way to define the bandwidth of an arbitrary filter is to construct an ideal lowpass filter that produces the same output power as the actual filter. This results in the following definition of bandwidth known as *noise equivalent bandwidth*.

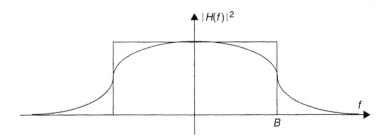

Figure 11.2 Power transfer function of an arbitrary and ideal lowpass filter. $B = B_{\text{neq}}$ if areas under two curves are equal.

DEFINITION 11.1: The *noise equivalent bandwidth* of a lowpass filter with transfer function $H(f)$ is

$$B_{\text{neq}} = \frac{1}{2|H(0)|^2} \int_{-\infty}^{\infty} |H(f)|^2 \, df = \frac{1}{|H(0)|^2} \int_{0}^{\infty} |H(f)|^2 \, df. \tag{11.22}$$

This definition needs to be slightly adjusted for bandpass filters. If the center of the passband is taken to be at some frequency, f_o, then the noise equivalent bandwidth is

$$B_{\text{neq}} = \frac{1}{2|H(f_o)|^2} \int_{-\infty}^{\infty} |H(f)|^2 \, df = \frac{1}{|H(f_o)|^2} \int_{0}^{\infty} |H(f)|^2 \, df. \tag{11.23}$$

EXAMPLE 11.4: Consider the RC lowpass filter whose transfer function is

$$H(f) = \frac{1}{1 + j2\pi f RC}.$$

The noise equivalent bandwidth of this filter is

$$B_{\text{neq}} = \int_{0}^{\infty} \frac{1}{1 + (2\pi f RC)^2} \, df = \frac{1}{2\pi RC} \tan^{-1}(u) \Big|_{0}^{\infty} = \frac{1}{4RC}.$$

In addition to using the noise equivalent bandwidth, the definitions in Section 10.3 presented for calculating the bandwidth of a random process can also be applied to find the bandwidth of a filter. For example, the absolute bandwidth and the RMS bandwidth of this filter are both infinite, thus the 3-dB (half power) bandwidth of this filter is

$$B_{3\text{dB}} = \frac{1}{2\pi RC},$$

which for this example is slightly smaller than the noise equivalent bandwidth.

11.4 Signal-to-Noise Ratios

Often the input to a linear system will consist of signal plus noise, namely,

$$X(t) = S(t) + N(t), \tag{11.24}$$

where the signal part can be deterministic or a random process. We can invoke linearity to show that the mean process of the output can be viewed as a sum of the mean due to the signal input alone plus the mean due to the noise input alone. That is,

$$\mu_Y(t) = \mu_S(t) * h(t) + \mu_N(t) * h(t). \tag{11.25}$$

In most cases, the noise is taken to be zero-mean, in which case the mean at the output is due to the signal part alone.

When calculating the autocorrelation function of the output, we cannot invoke superposition since autocorrelation is not a linear operation. First, we calculate the autocorrelation function of the signal plus noise input.

$$R_{XX}(t_1, t_2) = E[(S(t_1) + N(t_1))(S(t_2) + N(t_2))]. \tag{11.26}$$

If the signal and noise part are independent, which is generally a reasonable assumption, and the noise is zero-mean, then this autocorrelation becomes

$$R_{XX}(t_1, t_2) = R_{SS}(t_1, t_2) + R_{NN}(t_1, t_2), \tag{11.27}$$

or, assuming all processes involved are WSS,

$$R_{XX}(\tau) = R_{SS}(\tau) + R_{NN}(\tau). \tag{11.28}$$

As a result, the PSD of the output can be written as

$$S_{YY}(f) = S_{XX}(f)|H(f)|^2 = S_{SS}(f)|H(f)|^2 + S_{NN}(f)|H(f)|^2, \tag{11.29}$$

which is composed of two terms, namely that due to the signal and that due to the noise. We can then calculate the output power due to the signal part and the output power due to the noise part.

DEFINITION 11.2: The *signal-to-noise ratio* (SNR) for a signal comprised of the sum of a desired (signal) part and a noise part is defined as the ratio of the power of the signal part to the power (variance) of the noise part. That is, for $X(t) = S(t) + N(t)$,

$$\text{SNR} = \frac{E[S^2(t)]}{E[N^2(t)]} = \frac{R_{SS}(0)}{R_{NN}(0)} = \frac{\int_{-\infty}^{\infty} S_{SS}(f)\,df}{\int_{-\infty}^{\infty} S_{NN}(f)\,df}. \tag{11.30}$$

EXAMPLE 11.5: Suppose the input to the RC lowpass filter of the previous example consists of a sinusoidal signal plus white noise. That is, let the input be $X(t) = S(t) + N(t)$, where $N(t)$ is white Gaussian noise as in the previous example and $S(t) = a\cos(\omega_0 t + \Theta)$, where Θ is a uniform random variable over $[0, 2\pi)$ that is independent of the noise. The output can be written as $Y(t) = S_o(t) + N_o(t)$, where $S_o(t)$ is the output due to the sinusoidal signal input and $N_o(t)$ is the output due to the noise. The signal output can be expressed as

$$S_o(t) = a|H(f_0)|\cos(\omega_0 t + \angle H(f_0) + \Theta),$$

and the power in this sinusoidal signal is

$$R_{S_o S_o}(0) = \frac{a^2|H(f_0)|^2}{2}.$$

From the results of Example 11.1, the noise power at the output is

$$R_{N_o N_o}(0) = \frac{N_o}{4RC}.$$

Hence, the SNR of the output of the RC lowpass filter is

$$\text{SNR} = \frac{2a^2 RC|H(f_0)|^2}{N_o} = \frac{2a^2 RC}{N_o(1 + (2\pi f_0 RC)^2)}.$$

Suppose we desire to adjust the RC time constant (or, equivalently, adjust the bandwidth) of the filter so that the output SNR is optimized. Differentiating with respect to the quantity RC, setting equal to zero and solving the resulting equation produces the optimum time constant

$$RC_{\text{opt}} = \frac{1}{2\pi f_0}.$$

Stated another way, the 3-dB frequency of the RC filter is set equal to the frequency of the input sinusoid in order to optimize output SNR. The resulting optimum SNR is

$$\text{SNR}_{\text{opt}} = \frac{a^2}{2\pi N_o f_o}.$$

11.5 The Matched Filter

Suppose we are given an input process consisting of a (known, deterministic) signal plus an independent white noise process (with a PSD of $N_o/2$). It is desired to filter out as much of the noise as possible while retaining the desired signal. The general system is shown in Figure 11.3. The input process $X(t) = s(t) + N(t)$ is to be passed through a filter with impulse response $h(t)$ that produces an output process $Y(t)$. The goal here is to design the filter to maximize the SNR at the filter output. Due to the fact that the input process is not necessarily stationary, the output process may not be stationary and hence the output SNR may be time varying. We must therefore specify at what point in time we want the SNR to be maximized. Picking an arbitrary sampling time, t_o, for the output process, we desire to design the filter such that the SNR is maximized at time $t = t_o$.

Let Y_o be the value of the output of the filter at time t_o. This random variable can be expressed as

$$Y_o = Y(t_o) = s(t) * h(t)\big|_{t=t_o} + N_Y(t_o), \tag{11.31}$$

where $N_Y(t)$ is the noise process out of the filter. The power in the signal and noise parts, respectively, is given by

$$\text{signal power} = \left[s(t) * h(t)\big|_{t=t_o} \right]^2 = \left[\int_0^\infty h(u)s(t_o - u)\, du \right]^2, \tag{11.32}$$

$$\text{noise power} = \frac{N_o}{2} \int_{-\infty}^\infty |H(f)|^2\, df = \frac{N_o}{2} \int_{-\infty}^\infty h^2(t)\, dt. \tag{11.33}$$

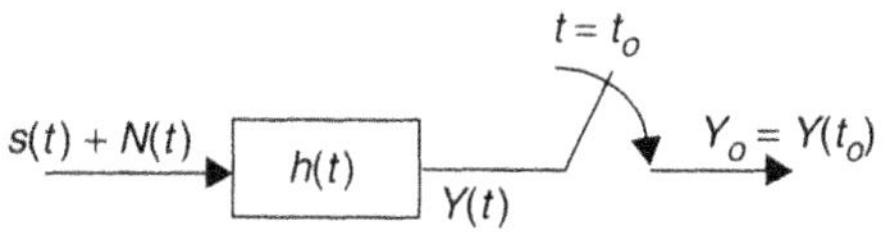

Figure 11.3 Linear system for filtering noise from a desired signal.

The SNR is then expressed as the ratio of these two quantities,

$$\text{SNR} = \frac{2}{N_0} \frac{\left[\int_0^\infty h(u)s(t_o - u)\, du \right]^2}{\int_{-\infty}^\infty h^2(t)\, dt}. \tag{11.34}$$

We seek the impulse response (or, equivalently, the transfer function) of the filter that maximizes the SNR as given in Equation 11.34. To simplify this optimization problem, we use Schwarz's inequality, which states that

$$\left| \int_{-\infty}^\infty x(t)y(t)\, dt \right|^2 \leq \int_{-\infty}^\infty |x(t)|^2\, dt \int_{-\infty}^\infty |y(t)|^2\, dt, \tag{11.35}$$

where equality holds if and only if $x(t) \propto y(t)$[1]. Applying this result to the expression for SNR produces an upper bound on the SNR:

$$\text{SNR} \leq \frac{2}{N_0} \frac{\int_{-\infty}^\infty |h(t)|^2\, dt \int_{-\infty}^\infty |s(t_o - t)|^2\, dt}{\int_{-\infty}^\infty h^2(t)\, dt} = \frac{2}{N_0} \int_{-\infty}^\infty |s(t_o - t)|^2\, dt$$

$$= \frac{2}{N_0} \int_{-\infty}^\infty |s(t)|^2\, dt = \frac{2E_s}{N_o}, \tag{11.36}$$

where E_s is the energy in the signal $s(t)$. Furthermore, this maximum SNR is achieved when $h(t) \propto s(t_o - t)$. In terms of the transfer function, this relationship is expressed as $H(f) \propto S^*(f)e^{-j2\pi f t_o}$. The filter that maximizes the SNR is referred to as a *matched filter* since the impulse response is matched to that of the desired signal. These results are summarized in the following theorem.

THEOREM 11.3: If an input to an LTI system characterized by an impulse response, $h(t)$, is given by $X(t) = s(t) + N(t)$ where $N(t)$ is a white noise process, then a matched filter will maximize the output SNR at time t_o. The impulse response and transfer function of the matched filter are given by

$$h(t) = s(t_o - t) \quad \text{and} \quad H(f) = S^*(f)e^{-j2\pi f t_o}. \tag{11.37}$$

Furthermore, if the white noise at the input has a PSD of $S_{NN}(f) = N_0/2$ then the optimum SNR produced by the matched filter is

$$\text{SNR}_{\text{max}} = \frac{2E_s}{N_o}, \tag{11.38}$$

where E_s is the energy in the signal $s(t)$.

---

[1] The notation $x(t) \propto y(t)$ means that $x(t)$ is proportional to $y(t)$.

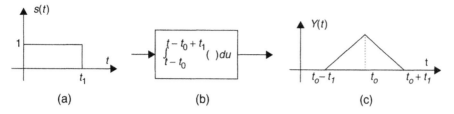

(a)   (b)   (c)

Figure 11.4 A square pulse (a), the corresponding matched filter (b), and the response of the matched filter to the square pulse (c).

EXAMPLE 11.6: A certain communication system transmits a square pulse given by

$$s(t) = \begin{cases} 1 & 0 \le t < t_1 \\ 0 & \text{otherwise} \end{cases}.$$

This signal is received in the presence of white noise at a receiver producing the received process $R(t) = s(t) + N(t)$. The matched filter that produces the optimum SNR at time t_o for this signal has an impulse response of the form

$$h(t) = s(t_o - t) = \begin{cases} 1 & t_o - t_1 < t \le t_o \\ 0 & \text{otherwise} \end{cases}.$$

The output of the matched filter is then given by

$$Y(t) = h(t) * R(t) = \int_{-\infty}^{\infty} h(t - u)R(u)\, du$$

$$= \int_{-\infty}^{\infty} s(u + t_o - t)R(u)\, du = \int_{t-t_o}^{t-t_o+t_1} R(u)\, du.$$

Hence, the matched filter for a square pulse is just a finite time integrator. The matched filter simply integrates the received signal for a period of time equal to the width of the pulse. When sampled at the correct point in time, the output of this integrator will produce the maximum SNR. The operation of this filter is illustrated in Figure 11.4.

EXAMPLE 11.7: In this example, we expand on the results of the previous example and consider a sequence of square pulses with random (binary) amplitudes as might be transmitted in a typical communication system. Suppose this signal is corrupted by white

Gaussian noise and we must detect the transmitted bits. That is, we must determine whether each pulse sent has a positive or negative amplitude. Plot (a) in Figure 11.5 shows both the square pulse train and the same signal corrupted by noise. Note that by visually observing the signals, it is very difficult to make out the original signal from the noisy version. We attempt to clean up this signal by passing it through the matched filter from Example 11.6. In the absence of noise, we would expect to see a sequence of overlapping triangular pulses. The matched filter output both with and without noise is illustrated in plot (b) of Figure 11.5. Notice that a great deal of noise has been eliminated by the matched filter. To detect the data bits, we would sample the matched filter output at the end of each bit interval (shown by circles in the plot) and use the sign of the sample to be the estimate of the transmitted data bit. In this example, all of our decisions would be correct. The MATLAB

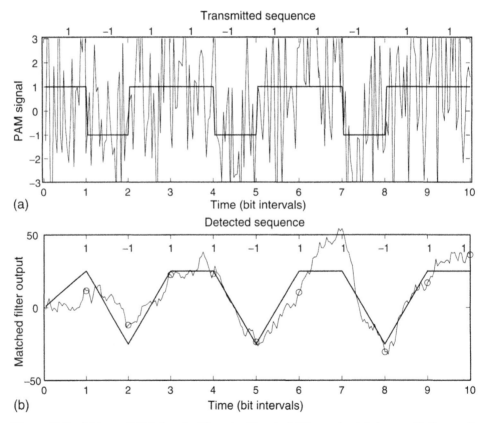

Figure 11.5　A binary PAM signal with and without additive noise (a) along with the result of passing both signals through a matched filter (b).

code used to generate these signals follows. This is just a modified version of the code used to generate the PAM signal in Example 10.14.

```
N=10;                           % Number of bit intervals.
Ns=25;                          % Number of time samples per bit.
t=([1:Ns]-0.5)/Ns;              % Time axis for pulse shape.
p=ones(size(t));                % Square pulse shape.
d=sign(rand(1,N)-0.5);          % Random data bits.
X(1:Ns:(Ns*(N-1)+1))=d;
X=conv(X,p);                    % PAM signal with pulse shape.
sigma=sqrt(Ns/10);              % Noise strength.
noise=sigma*randn(size(X));     % Gaussian noise.
R=X+noise;
subplot(2,1,1)                  % Plot clean and noisy signals.
x_axis=[1:length(R)]/Ns;
plot(x_axis,R, x_axis,X)
axis([0 N -3 3])
h=fliplr(p);                    % Matched filter impulse response.
z=conv(R,h);                    % Noisy output of matched filter.
z2=conv(X,h);                   % Noise-free MF output.
zs=z(Ns*[1:N]);                 % Sample matched filter outputs.
subplot(2,1,2)                  % Plot matched filter output.
x_axis=[1:length(z)]/Ns;
plot(x_axis,z,'-',x_axis,z2,'-',[1:N],zs,'o')
```

11.6 The Wiener Filter

In this section, we consider another filter design problem that involves removing the noise from a sum of a desired signal plus noise. In this case, the desired signal is also a random process (rather than a known, deterministic signal as in the last section) and the goal here is to estimate the desired part of the signal plus noise. In its most general form, the problem is stated as follows. Given a random process $X(t)$, we want to form an estimate $Y(t)$ of some other zero-mean process $Z(t)$ based on observation of some portion of $X(t)$. We require the estimator to be linear. That is, we will obtain $Y(t)$ by filtering $X(t)$. Hence,

$$Y(t) = \int_{-\infty}^{\infty} h(t-u)X(u)\,du. \tag{11.39}$$

We want to design the filter to minimize the mean square error

$$E[\varepsilon^2(t)] = E[(Z(t) - Y(t))^2]. \tag{11.40}$$

In this section, we will consider the special case where the observation consists of the process we are trying to estimate plus independent noise. That is $X(t) = Z(t) + N(t)$. We observe $X(t)$ for some time interval $t \in (t_1, t_2)$ and based on that observation, we will form an estimate of $Z(t)$. Consider a few special cases:

- Case I. If $(t_1, t_2) = (-\infty, t)$, then we have a filtering problem in which we must estimate the present based on the entire past. We may also have $(t_1, t_2) = (t - t_o, t)$, in which case we have a filtering problem where we must estimate the present based on the most recent past.
- Case II. If $(t_1, t_2) = (-\infty, \infty)$, then we have a smoothing problem where we must estimate the present based on a noisy version of the past, present, and future.
- Case III. If $(t_1, t_2) = (-\infty, t - t_o)$, then we have a prediction problem where we must estimate the future based on the past and present.

All of these cases can be cast in the same general framework and a single result will describe the optimal filter for all cases. In order to derive the optimal filter, it is easier to view the problem in discrete time and then ultimately pass to the limit of continuous time. Hence, we reformulate the problem in discrete time. Given an observation of the discrete time process $X[n] = Z[n] + N[n]$ over some time interval $n \in [n_1, n_2]$, we wish to design a filter $h[n]$ such that the linear estimate

$$Y[n] = \sum_{k=n_1}^{n_2} h[n-k]X[k] \tag{11.41}$$

minimizes the mean square error $E[\varepsilon^2[n]] = E[(Z[n] - Y[n])^2]$.

The filter $h[n]$ can be viewed as a sequence of variables. We seek to jointly optimize with respect to each variable in that sequence. This can be done by differentiating with respect to each variable and setting the resulting equations equal to zero:

$$\frac{d}{dh[m]} E[\varepsilon^2[n]] = 2E\left[\varepsilon[n]\frac{d}{dh[m]}\varepsilon[n]\right] = 0, \quad m \in [n - n_2, n - n_1]. \tag{11.42}$$

Noting that

$$\frac{d}{dh[m]}\varepsilon[n] = \frac{d}{dh[m]}(Z[n] - Y[n]) = -\frac{d}{dh[m]}Y[n] = -X[n - m], \tag{11.43}$$

the system of equations to solve becomes

$$E[\varepsilon[n]X[n - m]] = 0, \quad \text{for} \quad m \in [n - n_1, n - n_2]. \tag{11.44}$$

Equivalently, this can be rewritten as

$$E[\varepsilon[n]X[m]] = 0, \quad \text{for} \quad m \in [n_1, n_2].$$ (11.45)

In summary, the filter that minimizes the mean square error will cause the observed data to be orthogonal to the error. This is known as the *orthogonality principle*. Applying the orthogonality principle, we have

$$E[\varepsilon[n]X[m]] = E\left[\left(Z[n] - \sum_{k=n_1}^{n_2} h[n-k]X[k]\right)X[m]\right] = 0.$$ (11.46)

Assuming all the processes involved are jointly WSS, these expectations can be written in terms of autocorrelation and cross correlation functions as

$$\sum_{k=n_1}^{n_2} h[n-k]R_{XX}[k-m] = R_{ZX}[n-m], \quad m \in [n_1, n_2],$$ (11.47)

or equivalently,

$$\sum_{i=n-n_2}^{n-n_1} h[i]R_{XX}[k-i] = R_{ZX}[k], \quad k \in [n-n_2, n-n_1].$$ (11.48)

These equations are known as the *Wiener-Hopf equations*, the *normal equations*, or the *Yule-Walker* equations. The resulting filter found by solving this system of equations is known as the *Wiener* filter.

A similar result can be found for continuous time systems by applying the orthogonality principle in continuous time. Given an observation of $X(t)$ over the time interval (t_1, t_2), the orthogonality principle states that the filter that minimizes the mean square prediction error will satisfy

$$E[\varepsilon(t)X(s)] = 0, \quad \text{for} \quad s \in (t_1, t_2).$$ (11.49)

This produces the continuous-time version of the Wiener-Hopf equation,

$$\int_{t-t_2}^{t-t_1} h(v)R_{XX}(\tau - v)\,dv = R_{ZX}(\tau), \quad \tau \in (t - t_2, t - t_1).$$ (11.50)

The techniques used to solve the Wiener-Hopf equation depend on the nature of the observation interval. For example, consider the smoothing problem where the observation interval is $(t_1, t_2) = (-\infty, \infty)$. In that case, the Wiener-Hopf equation becomes

$$\int_{-\infty}^{\infty} h(v)R_{XX}(\tau - v)\,dv = R_{ZX}(\tau).$$ (11.51)

The left-hand side of the equation is a convolution and the integral equation can easily be solved using Fourier transforms. Taking a Fourier transform of both sides of the equation results in

$$H(f)S_{XX}(f) = S_{ZX}(f) \Rightarrow H(f) = \frac{S_{ZX}(f)}{S_{XX}(f)}. \tag{11.52}$$

Note also that if the noise is zero-mean and independent of $Z(t)$, then $R_{ZX}(\tau) = R_{ZZ}(\tau)$ and $R_{XX}(\tau) = R_{ZZ}(\tau) + R_{NN}(\tau)$. The transfer function of the Wiener filter for the smoothing problem then becomes

$$H(f) = \frac{S_{ZZ}(f)}{S_{ZZ}(f) + S_{NN}(f)}. \tag{11.53}$$

EXAMPLE 11.8: Suppose the desired signal $Z(t)$ has a spectral density of

$$S_{ZZ}(f) = \frac{1}{1+f^2}$$

and the noise is white with a PSD of $S_{NN}(f) = 1$. Then the Wiener filter for the smoothing problem has the form

$$H(f) = \frac{\dfrac{1}{1+f^2}}{\dfrac{1}{1+f^2} + 1} = \frac{1}{2+f^2}.$$

The corresponding impulse response is

$$h(t) = \frac{\pi}{\sqrt{2}} \exp(-\sqrt{8}\pi |t|).$$

Note that this filter is not causal. This is due to the nature of the smoothing problem, whereby we estimate the present based on past, present, and future.

Next, consider the filtering problem where the observation interval is $(t_1, t_2) = (-\infty, t)$. In this case, the Wiener-Hopf equation becomes

$$\int_0^\infty h(v)R_{XX}(\tau - v)\, dv = R_{ZX}(\tau), \quad \tau \in (0, \infty) \tag{11.54}$$

It is emphasized now that the left-hand side of the equation is not a convolution since the lower limit of the integral is not $-\infty$. The resulting integral equation is much trickier to solve than in the case of the smoothing problem. In order to develop a procedure for solving this general equation, consider the special case when $R_{XX}(\tau) = \delta(\tau)$. In that case, the preceding integral equation becomes

$$h(\tau) = R_{ZX}(\tau) \quad \text{for} \quad \tau > 0. \tag{11.55}$$

Because we are estimating the present based on observing the past, the filter must be causal and thus its impulse response must be zero for negative time. Thus, for the special case when $R_{XX}(\tau) = \delta(\tau)$, the Wiener filter is $h(\tau) = R_{ZX}(\tau)u(\tau)$.

This example in itself is not very interesting since we would not expect $X(t)$ to be white, but it does help to find the general solution to the Wiener-Hopf equation. First, before estimating $Z(t)$, suppose we pass the input $X(t)$ through a causal filter with a transfer function $1/G(f)$. Call the output $\widetilde{X}(t)$. If $G(f)$ is chosen such that $|G(f)|^2 = S_{XX}(f)$, then the process $\widetilde{X}(t)$ will be a white process and the filter is called a *whitening filter*. We can then use the result of the previous special case to estimate $Z(t)$ based on the white process $\widetilde{X}(t)$. Hence, we are designing the Wiener filter in two stages as illustrated in Figure 11.6.

To find the impulse response of the second filter, we start with the result that $h^2(\tau) = R_{Z\widetilde{X}}(\tau)U(\tau)$. Also, since $\widetilde{X}(t)$ can be written as $\widetilde{X}(t) = X(t) * h_1(t)$, then

$$S_{Z\widetilde{X}}(f) = S_{ZX}(f)H_1^*(f). \tag{11.56}$$

The resulting quantities needed to form the second filter are then

$$S_{Z\widetilde{X}}(f) = \frac{S_{ZX}(f)}{G^*(f)} \longleftrightarrow R_{Z\widetilde{X}}(\tau) = F^{-1}\left\{\frac{S_{ZX}(f)}{G^*(f)}\right\}. \tag{11.57}$$

To construct the whitening filter, we need to find a $G(f)$ such that (1) $H_1(f) = 1/G(f)$ is causal, and (2) $|G(f)|^2 = S_{XX}(f)$. The procedure for doing this is known as *spectral factorization*. Since $S_{XX}(f)$ is a PSD and thus is an even function of f, it will factor in the form

$$S_{XX}(f) = G(f)G * (f), \tag{11.58}$$

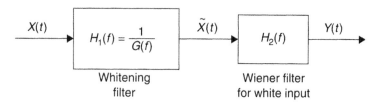

Figure 11.6 Constructing the Wiener filter for the filtering problem as a cascade of two filters.

where half of the poles and zeros are assigned to $G(f)$ and the other half are assigned to $G^*(f)$. In order to be sure that $H_1(f)$ is causal, we assign to $G(f)$ those zeros in the upper half-plane. As will be shown in the next example, it is also important to assign poles from the upper half-plane to $G(f)$ as well.

EXAMPLE 11.9: For this example, let $X(t) = Z(t) + N(t)$, where $N(t)$ is white noise with a spectral density of $S_{NN}(f) = 1$ and independent of $Z(t)$, which has a spectral density of

$$S_{ZZ}(f) = \frac{3}{1 + (2\pi f)^2}.$$

Note also that

$$S_{ZX}(f) = S_{ZZ}(f) \quad \text{and} \quad S_{XX}(f) = S_{ZZ}(f) + S_{NN}(f) = \frac{3}{1 + (2\pi f)^2} + 1$$

$$= \frac{4 + (2\pi f)^2}{1 + (2\pi f)^2} = \frac{(2 + j2\pi f)(2 - j2\pi f)}{(1 + j2\pi f)(1 - j2\pi f)}.$$

A pole-zero plot for the PSD function $S_{XX}(f)$ is shown in Figure 11.7. We assign the poles and zeros in the upper half-plane to the function $G(f)$ (and hence the poles and zeros in the lower half-plane go to $G^*(f)$). This results in

$$G(f) = \frac{2 + j2\pi f}{1 + j2\pi f} \implies H_1(f) = \frac{1 + j2\pi f}{2 + j2\pi f} = 1 - \frac{1}{2 + j2\pi f}.$$

The corresponding impulse response is

$$h_1(t) = \delta(t) - e^{-2t} u(t).$$

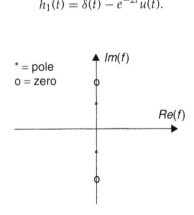

Figure 11.7 Pole-zero plot for the PSD function of Example 11.9.

As desired, the whitening filter is causal. To find the form of the second filter, we first calculate the cross spectral density between $\tilde{X}(t)$ and $Z(t)$:

$$S_{Z\tilde{X}}(f) = S_{ZX}(f)H_1^*(f) = \frac{3}{1 + (2\pi f)^2} \frac{1 - j2\pi f}{2 - j2\pi f}$$

$$= \frac{3}{(1 + j2\pi f)(2 - j2\pi f)} = \frac{1}{1 + j2\pi f} + \frac{1}{2 - j2\pi f}.$$

Taking an inverse Fourier transform, the cross correlation function is

$$R_{Z\tilde{X}}(\tau) = e^{-\tau}u(\tau) + e^{2\tau}u(-\tau).$$

The impulse response of the second filter is then given by

$$h_2(\tau) = R_{Z\tilde{X}}(\tau)u(\tau) = e^{-\tau}u(\tau).$$

When the actual Wiener filter is implemented, there is no reason the filter has to be implemented as a cascade of two filters. We did this for ease of determining the filter. Combining these two filters into one produces

$$H(f) = H_1(f)H_2(f) = \left(\frac{1 + j2\pi f}{2 + j2\pi f}\right)\left(\frac{1}{1 + j2\pi f}\right)$$

$$= \frac{1}{2 + j2\pi f} \longleftrightarrow h(t) = e^{-2t}u(t).$$

It can be easily verified that this filter does indeed satisfy the Wiener-Hopf equation.

Note, we also could have chosen $G(f) = (2 + j2\pi f)/(1 - j2\pi f)$, and $H_1(f) = 1/G(f)$ would still have been causal:

$$H_1(f) = \frac{1 - j2\pi f}{2 + j2\pi f} = -1 + \frac{3}{2 + j2\pi f} \longleftrightarrow h_1(t) = -\delta(t) + 3e^{-2t}u(t).$$

In this case,

$$S_{Z\tilde{X}}(f) = S_{ZX}(f)H_1^*(f) = \frac{3}{1 + (2\pi f)^2} \frac{1 + j2\pi f}{2 - j2\pi f}$$

$$= \frac{3}{(1 - j2\pi f)(2 - j2\pi f)} = \frac{3}{1 - j2\pi f} + \frac{-3}{2 - j2\pi f}.$$

$$\longleftrightarrow R_{Z\tilde{X}}(\tau) = 3e^{\tau}u(-\tau) - 3e^{2\tau}u(-\tau)$$

This leads to $h_2(t) = 0$!! Thus, we see it is important to assign both poles and zeros from the upper half-plane to $G(f)$.

Finally, we consider the prediction problem where we wish to estimate the value of $Z(t)$ based on observing $X(t)$ over the time interval $(-\infty, t - t_0)$. Applying the orthogonality principle, the appropriate form of the Wiener-Hopf equation for the prediction problem becomes

$$\int_{t_0}^{\infty} h(v) R_{XX}(\tau - v)\, dv = R_{ZX}(\tau), \quad \tau \in (0, \infty) \tag{11.59}$$

This equation is solved using the same technique as with the filtering problem. First, the input is passed through a whitening filter and then Equation 11.59 is solved for the case when the input process is white. The procedure for finding the whitening filter is exactly the same as before. The solution to Equation 11.59 when $R_{XX}(\tau) = \delta(\tau)$ is

$$h(\tau) = R_{ZX}(\tau) u(\tau - t_0). \tag{11.60}$$

In summary, the solution to the prediction problem is found by following these steps:

- Step 1. Factor the input PSD according to $S_{XX}(f) = 1/(G(f) G^*(f))$, where $G(f)$ contains all poles and zeros of $S_{XX}(f)$ that are in the upper half-plane. The whitening filter is then specified by $H_1(f) = 1/G(f)$. Call $\tilde{X}(t)$ the output of the whitening filter when $X(t)$ is input.
- Step 2. Calculate the cross correlation function, $R_{Z\tilde{X}}(\tau)$. The second stage of the Wiener filter is then specified by $h_2(\tau) = R_{Z\tilde{X}}(\tau) u(\tau - t_0)$.
- Step 3. The overall Wiener filter is found by combining these two filters, $H(f) = H_1(f) H_2(f)$.

It should be noted that the filtering problem can be viewed as a special case of the prediction problem when $t_0 = 0$, thus this summary applies to the prediction problem as well.

EXAMPLE 11.10: In this example, we repeat the filter design of Example 11.9 for the case of the prediction problem. As before, we pick the whitening filter to be of the form

$$H_1(f) = \frac{1 + j2\pi f}{2 + j2\pi f} \longleftrightarrow h_1(t) = \delta(t) - e^{-2t} u(t).$$

As before, the resulting cross correlation function is then

$$R_{Z\tilde{X}}(\tau) = e^{-\tau} u(\tau) + e^{2\tau} u(-\tau).$$

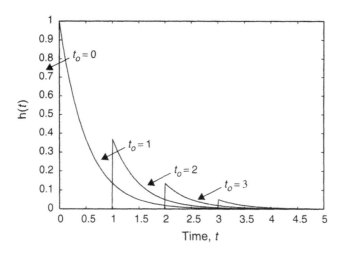

Figure 11.8 Impulse response of the Wiener prediction filter for Example 11.10.

Assuming that $t_o > 0$ so that the problem is indeed one of prediction, the impulse response of the second filter is

$$h_2(t) = R_{Z\widetilde{X}}(t)u(t - t_o) = e^{-t}u(t - t_o).$$

The impulse response of the Wiener prediction filter is then

$$h(t) = h_1(t) * h_2(t) = [\delta(t) - e^{-2t}u(t)] * [e^{-t}u(t - t_o)] = e^{-2t+t_o}u(t - t_o).$$

This agrees with the result of the previous example when $t_o = 0$. Figure 11.8 illustrates the impulse response of the Wiener prediction filter for several values of t_o.

11.7 Bandlimited and Narrowband Random Processes

A random processes is said to be *bandlimited* if all of its frequency components are limited to some bandwidth, B. Specifically, if a random process $X(t)$ has a PSD function with an absolute bandwidth of B, then the process is said to be bandlimited to B Hz. For many bandlimited random processes, the frequency components are clustered at or near direct current (d.c.). Such a process is referred to as a *lowpass*

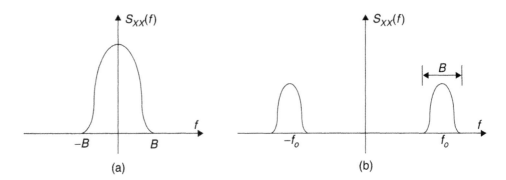

Figure 11.9 The PSD functions of a lowpass (a), and a bandpass (b) random process.

random process. If, on the other hand, the frequency components of a random process are removed from d.c. and reside in some nonzero frequency band, the process is called a *bandpass* process. These distinctions are illustrated in Figure 11.9. For a bandpass process, in addition to the bandwidth, the location of the frequency band where the PSD is nonzero must also be specified. In the figure, the parameter f_o describes that location. While often f_o is taken to be the center of the band as depicted in the figure, this does not have to be the case. The parameter f_o can be chosen to be any convenient frequency within the band. In any event, f_o is referred to as the *center frequency* of the band (even though it may not really be in the center). Finally, if a bandpass random process has a center frequency that is large compared to its bandwidth, $f_o \gg B$, then the process is said to be *narrowband*.

Narrowband random processes frequently are found in the study of communications systems. For example, a commercial FM radio broadcast system uses channels with bandwidths of 200 kHz, which are located near 100 MHz. Hence the center frequency of an FM radio signal is about 500 times greater than its bandwidth. In the U.S. digital cellular system, 30-kHz channels are used at frequencies near 900 MHz. In that case, the center frequencies are on the order of 30,000 times the bandwidth.

From studies of deterministic signals, the reader is probably aware that working with bandpass signals can be rather cumbersome. Trigonometric functions pop up everywhere and lead to seemingly endless usage of various identities. On the other hand, working with lowpass signals is often much simpler. To ease the complexity of working with bandpass signals, various representations have been formulated that allow bandpass signals to be decomposed into combinations of related lowpass signals. In this section, we will focus on the most common of those decompositions, which is valid for narrowband signals. Generalizations of the following results are available for signals that are bandpass but not necessarily narrowband but will not be covered here.

Suppose a random process $Z(t)$ is narrowband. Then $Z(t)$ can be expressed in terms of two lowpass processes $X(t)$ and $Y(t)$ according to

$$Z(t) = X(t)\cos(\omega_o t) - Y(t)\sin(\omega_o t). \tag{11.61}$$

The two processes $X(t)$ and $Y(t)$ are referred to as the *inphase* (I) and *quadrature* (Q) components of $Z(t)$. Although it is not proven here, the equality in the previous equation is in the mean-square sense. That is,

$$E[\{Z(t) - (X(t)\cos(\omega_o t) - Y(t)\sin(\omega_o t))\}^2] = 0. \tag{11.62}$$

This Cartesian representation of the narrowband random process can also be replaced by a polar representation of the form

$$Z(t) = R(t)\cos(\omega_o t + \Theta(t)), \tag{11.63}$$

where $R(t)$ is called the *real envelope* of $Z(t)$ and $\Theta(t)$ is the *excess phase*. We next describe the relationship between the statistics of the I and Q components and the statistics of the original random process.

The I and Q components of a signal can be extracted using the system shown in Figure 11.10. The passbands of the lowpass filters need to be large enough to pass the desired components (i.e., $> B/2$ Hz) but not so large as to pass the double frequency components produced by the mixers (at and around $2f_o$ Hz). For narrowband signals where $f_o \gg B$, the filters can be very loosely designed, and hence we do not need to worry too much about the particular forms of these filters. To see how this network functions, consider the output of the upper mixer:

$$2Z(t)\cos(\omega_o t) = 2\{X(t)\cos(\omega_o t) - Y(t)\sin(\omega_o t)\}\cos(\omega_o t)$$

$$= X(t)\{1 + \cos(2\omega_o t)\} - Y(t)\sin(2\omega_o t). \tag{11.64}$$

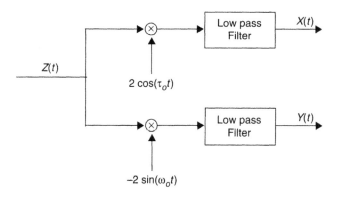

Figure 11.10 Network for decomposing a narrowband process into its I and Q components.

After passing this through the lowpass filter, the terms involving the double frequencies will be attenuated and the output of the upper LPF is indeed $X(t)$. Similar calculations reveal that $Y(t)$ is indeed the output of the lower branch.

Next we calculate PSDs involving the I and Q components. Consider first, multiplication of the process $Z(t)$ by a sinusoid. Let $A(t) = 2Z(t)\cos(\omega_0 t)$. The autocorrelation function of $A(t)$ is easily calculated to be

$$R_{AA}(t, t + \tau) = 2R_{ZZ}(\tau)\{\cos(\omega_0\tau) + \cos(2\omega_0 t + \omega_0\tau)\}. \tag{11.65}$$

Note that the process $A(t)$ is not WSS. In order to compute the PSD of a process that is not stationary, we must first take the time average of the autocorrelation function (with respect to t) so that the result will be a function of τ only. This time-averaged autocorrelation function works out to be

$$\langle R_{AA}(t, t + \tau)\rangle = 2R_{ZZ}(\tau)\{\cos(\omega_0\tau) + \langle\cos(2\omega_0 t + \omega_0\tau)\rangle\} = 2R_{ZZ}(\tau)\cos(\omega_0\tau). \tag{11.66}$$

At this point, the PSD of $A(t)$ can then be found through Fourier transformation to be

$$S_{AA}(f) = S_{ZZ}(f - f_0) + S_{ZZ}(f + f_0). \tag{11.67}$$

Recall that the process $Z(t)$ was assumed to be narrowband. That is, its PSD has components near f_0 and $-f_0$. After shifting by f_0, the term $S_{ZZ}(f - f_0)$ has components near d.c. and also near $2f_0$. The components near d.c. will pass through the filter while those at and around $2f_0$ will be attenuated. Similarly, $S_{ZZ}(f + f_0)$ will have terms near d.c. that will pass through the filter and terms near $-2f_0$ that will not. This is illustrated in Figure 11.11. For notational convenience, let LP{} denote the

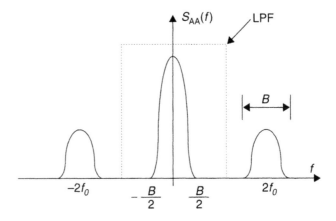

Figure 11.11   PSD of the input to the LPF in the I branch.

lowpass part of a quantity. Then the PSD of the inphase component of a narrowband process can be written in terms of the PSD of the original process as

$$S_{XX}(f) = \text{L. P.}\{S_{ZZ}(f - f_o) + S_{ZZ}(f + f_o)\}. \tag{11.68}$$

Following a similar set of steps, the PSD of the Q component is found to be identical to the I component. That is,

$$S_{YY}(f) = S_{XX}(f) = \text{L. P.}\{S_{ZZ}(f - f_o) + S_{ZZ}(f + f_o)\}. \tag{11.69}$$

The cross spectral density can also be calculated in a manner similar to the PSDs. The result is

$$S_{XY}(f) = j\,\text{L. P.}\{S_{ZZ}(f - f_o) - S_{ZZ}(f + f_o)\}, \tag{11.70}$$

It is noted that if the PSD function $S_{ZZ}(f)$ is symmetric about $f = f_o$, then the cross spectral density works out to be zero. In that case, the I and Q components are orthogonal (since $R_{XY}(\tau) = 0$). Furthermore, if the process $Z(t)$ is zero-mean, then the I and Q components will also be zero-mean. In this case, the I and Q components are then uncorrelated. Finally, if in addition $Z(t)$ is a Gaussian random process, then the I and Q components are also statistically independent. In summary, we have proven the results of the following theorem.

THEOREM 11.4: For a narrowband process $Z(t)$, the PSDs involving the I and Q components $X(t)$ and $Y(t)$ are given by

$$S_{YY}(f) = S_{XX}(f) = \text{L. P.}\{S_{ZZ}(f - f_o) + S_{ZZ}(f + f_o)\}, \tag{11.71a}$$

$$S_{XY}(f) = j\,\text{L. P.}\{S_{ZZ}(f - f_o) - S_{ZZ}(f + f_o)\}. \tag{11.71b}$$

If $Z(t)$ is a zero-mean Gaussian random process and its PSD is symmetric about $f = f_o$, then the I and Q components are statistically independent.

EXAMPLE 11.11: Suppose zero-mean white Gaussian noise with a PSD of $N_o/2$ is passed through an ideal BPF with a bandwidth of B Hz to produce the narrowband noise process $Z(t)$ as shown in Figure 11.12. The I and Q components will then have a PSD that is given by

$$S_{YY}(f) = S_{XX}(f) = N_o\,\text{rect}(f/B).$$

The corresponding autocorrelation functions are

$$R_{XX}(\tau) = R_{YY}(\tau) = N_o B\,\text{sinc}(B\tau).$$

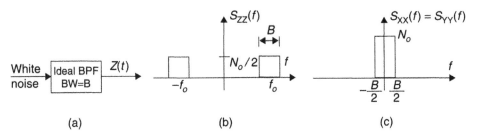

Figure 11.12   Generation of a narrowband noise process (a), its PSD function (b), and the PSD function of the I and Q components (c).

> Since the PSD of $Z(t)$ is symmetric about $f - f_o$, the cross PSD is zero and hence the I and Q components are independent.

11.8 Complex Envelopes

When working with narrowband random processes, it is convenient to combine the I and Q components into a single lowpass random process whose real part is the I component and whose imaginary part is the Q component. The resulting random process is a complex lowpass random process.

DEFINITION 11.3: For a narrowband process, $Z(t)$, with I and Q components, $X(t)$ and $Y(t)$, respectively, the *complex envelope*, $G_z(t)$, is defined as

$$G_z(t) = X(t) + jY(t). \tag{11.72}$$

With this definition, the random process is written in terms of its complex envelope according to

$$Z(t) = Re[G_z(t)e^{j\omega_o t}]. \tag{11.73}$$

The properties developed in the previous section for the I and Q components of a narrowband random process can be used to determine equivalent properties for the complex envelope. To be consistent with the definitions for complex random variables given in Chapter 5, we define the autocorrelation function of a complex random process as follows.

DEFINITION 11.4: For any complex random process $G(t)$, the autocorrelation function is defined as[2]

$$R_{GG}(t, t + \tau) = \frac{1}{2}E[G(t)G^*(t + \tau)]. \tag{11.74}$$

If $G(t)$ represents the complex envelope of a narrowband random process and the I and Q components are jointly WSS, then this autocorrelation function will be a function of only τ. Also, the corresponding PSD can be found through Fourier transformation:

$$S_{GG}(f) = F[R_{GG}(\tau)]. \tag{11.75}$$

Using this definition together with the properties developed in the previous section, the autocorrelation function for a complex envelope is found to be

$$R_{GG}(\tau) = \frac{1}{2}E[(X(t) + jY(t))(X(t + \tau) - jY(t + \tau))]$$

$$= \frac{1}{2}R_{XX}(\tau) + \frac{1}{2}R_{YY}(\tau) + \frac{j}{2}R_{YX}(\tau) - \frac{j}{2}R_{XY}(\tau). \tag{11.76}$$

For the case where the I and Q components are orthogonal, this reduces to

$$R_{GG}(\tau) = \frac{1}{2}R_{XX}(\tau) + \frac{1}{2}R_{YY}(\tau) = R_{XX}(\tau) = R_{YY}(\tau). \tag{11.77}$$

The corresponding PSD is then

$$S_{GG}(f) = S_{XX}(f) = S_{YY}(f) = \text{L. P.}\{S_{ZZ}(f - f_o) + S_{ZZ}(f + f_o)\}. \tag{11.78}$$

Hence, the complex envelope has the same PSD and autocorrelation function as the I and Q components. It is left as an exercise for the reader to show that the autocorrelation and PSD functions of the original narrowband process can be found from those of the complex envelope according to

$$S_{ZZ}(f) = \frac{1}{2}S_{GG}(f - f_o) + \frac{1}{2}S_{GG}(-f - f_o), \tag{11.79}$$

$$R_{ZZ}(\tau) = Re[R_{GG}(\tau)e^{j\omega_o\tau}]. \tag{11.80}$$

---

[2]The same disclaimer must be made here as in Definition 5.13. Many texts do not include the factor of 1/2 in the definition of the autocorrelation function for complex random processes, and hence the reader should be aware that there are two different definitions prevalent in the literature.

11.9 Engineering Application: An Analog Communication System

A block diagram of a simple amplitude modulation (AM) analog communication system is shown in Figure 11.13. A message (usually voice or music) is represented as a random process $X(t)$ with some bandwidth B. This message is modulated onto a carrier using amplitude modulation. The resulting AM signal is of the form

$$S_{AM}(t) = [A_o + X(t)]\cos(\omega_c t + \Theta). \tag{11.81}$$

In AM, a bias, A_o, is added to the message and the result forms the time-varying amplitude of a radio frequency (RF) carrier. In order to allow for envelope detection at the receiver, the bias must satisfy $A_o > \max|X(t)|$. Some example waveforms are shown in Figure 11.14. Note that due to the process of modulation, the AM signal now occupies a bandwidth of $2B$. The modulated RF signal is transmitted using an antenna and propagates through the environment to the receiver where it is picked up with a receiving antenna.

To study the effects of noise on the AM system, we ignore other sources of corruption (e.g., interference, distortion) and model the received signal as simply a noisy version of the transmitted signal,

$$R(t) = S_{AM}(t) + N_w(t), \tag{11.82}$$

where $N_w(t)$ is a Gaussian white noise process with a PSD of $N_0/2$. To limit the effects of noise, the received signal is passed through a BPF whose passband is chosen to allow the AM signal to pass while removing as much of the noise as possible. This receiver filter is taken to be an ideal bandpass filter whose bandwidth is $2B$. The output of the filter is then modeled as an AM signal plus narrowband noise:

$$Z(t) = S_{AM}(t) + N_x(t)\cos(\omega_c t + \Theta) - N_y(t)\sin(\omega_c t + \Theta)$$

$$= [A_o + X(t) + N_x(t)]\cos(\omega_c t + \Theta) - N_y(t)\sin(\omega_c t + \Theta). \tag{11.83}$$

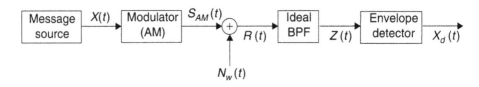

Figure 11.13 A block diagram of an amplitude modulation (AM) communications system.

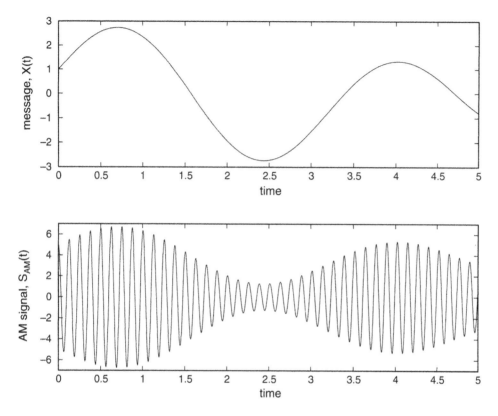

Figure 11.14 A sample message and the corresponding AM signal.

The envelope detector is a device that outputs the real envelope of the input. Hence, for the preceding input, the demodulated output will be

$$X_d(t) = \sqrt{(A_o + X(t) + N_x(t))^2 + (N_y(t))^2}. \qquad (11.84)$$

In its normal operating mode, the desired portion of the filter output will generally be much stronger than the noise. In this case, we observe that most of the time $A_o + X(t) + N_x(t) \gg N_y(t)$ so that the demodulated output can be well approximated by

$$X_d(t) \approx A_o + X(t) + N_x(t). \qquad (11.85)$$

The DC component can easily be removed and hence $X(t)$ represents the desired component of the demodulated output and $N_x(t)$ is the undesired (noise) component. The power in the desired component depends on the message and we simply write it as $E[X^2(t)]$. The I-component of the narrowband noise has a spectral density equal to N_0 for $|f| < B$ (and zero otherwise). Hence, the noise power at the demodulator output is $2N_0 B$. The resulting signal-to-noise ratio at the output of the AM systems is

$$\text{SNR} = \frac{E[X^2(t)]}{2N_0 B}. \tag{11.86}$$

It is common to express this SNR in terms of the transmitted power required to support the AM modulator. For an AM signal of the form in Equation 11.81, the transmitted power is

$$P_T = E[(A_o + X(t))^2 \cos^2(\omega_c t + \Theta)] = E[(A_o + X(t))^2] E[\cos^2(\omega_c t + \Theta)], \tag{11.87}$$

assuming the carrier phase is independent of the message. If the message is a zero-mean random process (which is usually the case), this expression simplifies to

$$P_T = \frac{A_0^2 + E[X^2(t)]}{2}. \tag{11.88}$$

Using this relationship, the SNR of the AM system can then be expressed as

$$\text{SNR} = \frac{E[X^2(t)]}{A_0^2 + E[X^2(t)]} \frac{P_T}{N_0 B}. \tag{11.89}$$

The factor,

$$\eta_{AM} = \frac{E[X^2(t)]}{A_0^2 + E[X^2(t)]}, \tag{11.90}$$

is known as the *modulation efficiency* of AM and is usually expressed as a percentage. Note that due to the requirement that $A_o > \max |X(t)|$, the modulation efficiency of AM must be less than 50 percent (which would occur for square wave messages). For a sinusoidal message, the modulation efficiency would be no more than 33 percent while for a typical voice or music signal, the modulation efficiency might be much smaller.

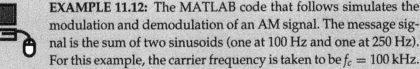

EXAMPLE 11.12: The MATLAB code that follows simulates the modulation and demodulation of an AM signal. The message signal is the sum of two sinusoids (one at 100 Hz and one at 250 Hz). For this example, the carrier frequency is taken to be $f_c = 100$ kHz. We've added noise to the AM signal to produce a typical received signal as shown in plot (a) of Figure 11.15. To demodulate the signal, we decompose the received signal into its I and Q components using the technique illustrated in Figure 11.9. The lowpass filter we used was a second-order Butterworth filter with a cutoff frequency of 400 Hz, but the particular form of the filter

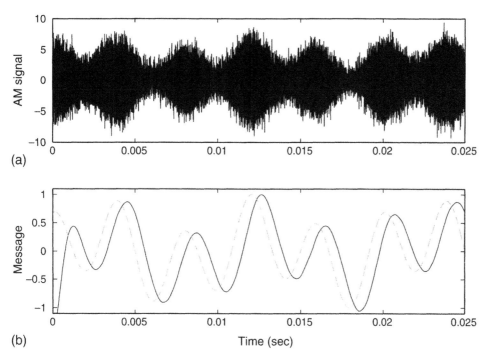

Figure 11.15 An AM signal corrupted by noise (a) along with a comparison of the original message and the demodulated message (b).

is not crucial. Once these components are found, the real envelope is computed according to $R(t) = \sqrt{X^2(t) + Y^2(t)}$. Our estimate of the message is then a scaled version of this envelope with the DC component removed. Plot (b) in Figure 11.15 shows the original message (dashed) and the recovered message (solid).

```
fc=100000;                          % Carrier frequency.
dt=1/(4*fc);                        % Sampling interval.
t=[0:dt:0.025];
f1=100; f2=250;
m=sin(2*pi*f1*t)+2*cos(2*pi*f2*t);  % Message signal (two
                                      tones).
s=(m+1+max(abs(m))).*cos(2*pi*      % AM signal.
  (fc*t+rand(1)));
s=s+randn(1,length(s));             % Noisy AM signal.
subplot(2,1,1)
plot(t,s); ylabel('AM signal');
```

```
f0=400;
a=2*pi*f0/sqrt(2);
h=exp(-a*t).*sin(a*t);        % Second-order lowpass filter
                             % with cut-off frequency at f0.

temp=s.*cos(2*pi*fc*t);
x=conv(temp,h)*dt;            % I component.
temp=s.*sin(2*pi*fc*t);
y=conv(temp,h)*dt;            % Q component.
r=sqrt(x.^2+y.^2);            % Real envelope.
r=r(1:length(t));
mhat=r-sum(r)/length(r);
subplot(2,1,2)
plot(t,mhat/max(mhat),'b',t,m/max(m),'g--')
axis([0 max(t) -1.1 1.1])
ylabel('Message'); xlabel('time (sec)')
```

Exercises

11.1 A white noise process, $X(t)$, with a power spectral density (PSD) of $S_{XX}(f) = N_o/2$ is passed through a finite time integrator whose output is given by

$$Y(t) = \int_{t-t_o}^{t} X(u)\, du.$$

Find the following:

(a) the PSD of the output process,

(b) the total power in the output process,

(c) the noise equivalent bandwidth of the integrator (filter).

11.2 A certain linear time-invariant system has an input/output relationship given by

$$Y(t) = \frac{X(t) - X(t - t_o)}{t_o}.$$

(a) Find the output autocorrelation, $R_{YY}(\tau)$, in terms of the input autocorrelation, $R_{XX}(\tau)$.

(b) Find the output PSD, $S_{YY}(f)$, in terms of the input PSD, $S_{XX}(f)$.

(c) Does your answer to part (b) make sense in the limit as $t_o \to 0$?

11.3 The output $Y(t)$ of a linear filter is c times the input $X(t)$. Show that $R_{YY}(\tau) = c^2 R_{XX}(\tau)$.

11.4 The output $Y(t)$ of a filter is given in terms of its input $X(t)$ by $Y(t) = X(t) + X(t - t_o) + X(t - 2t_o)$.

(a) Determine $R_{YY}(\tau)$ as a function of $R_{XX}(\tau)$.

(b) Find $E[Y^2(t)]$.

11.5 Is the following function a valid discrete-time autocorrelation function? Justify your answer.

$$R_{XX}[k] = \begin{cases} 1 & k = 1 \\ \dfrac{3}{4} & k = \pm 3 \\ 0 & \text{otherwise} \end{cases}.$$

11.6 A discrete random sequence $X[n]$ is the input to a discrete linear filter $h[n]$. The output is $Y[n]$. Let $Z[n] = X[n + i] - Y[n]$. Find $E[Z^2[n]]$ in terms of the autocorrelation functions for $X[n]$ and $Y[n]$ and the cross correlation function between $X[n]$ and $Y[n]$.

11.7 The unit impulse response of a discrete linear filter is $h[n] = a^n u[n]$, where $|a| < 1$. The autocorrelation function for the input random sequence is

$$R_{XX}[k] = \begin{cases} 1 & k = 0 \\ 0 & \text{otherwise} \end{cases}.$$

Determine the cross correlation function between the input and output random sequences.

11.8 Find the PSD of a discrete random sequence with the following autocorrelation function: $R_{XX}[k] = a(b^{|k|})$, where $|b| < 1$.

11.9 A discrete time linear filter has a unit pulse response $h[n]$. The input to this filter is a random sequence with uncorrelated samples. Show that the output power spectral density is real and nonnegative.

11.10 Consider a nonlinear device such that the output is $Y(t) = aX^2(t)$, where the input $X(t)$ consists of a signal plus a noise component, $X(t) = S(t) + N(t)$.

Determine the mean and autocorrelation function for $Y(t)$ when the signal $S(t)$ and the noise $N(t)$ are both Gaussian random processes and wide sense stationary (WSS) with zero mean, and $S(t)$ is independent of $N(t)$.

11.11 Calculate the spectrum for $Y(t)$ in Exercise 11.10 if

$$S_{SS}(f) = \frac{A^2}{4}[\delta(f + f_c) + \delta(f - f_c)],$$

and

$$S_{NN}(f) = \begin{cases} \dfrac{N_o}{2} & f_c - \dfrac{B}{2} < |f| < f_c + \dfrac{B}{2} \\ 0 & \text{otherwise} \end{cases}.$$

11.12 If the input to a linear filter is a random telegraph process with c zero crossings per second and an amplitude A, determine the output power spectral density. The filter impulse response is $h(t) = b\exp(-at)\,u(t)$.

11.13 The input to a linear filter is a random process with the following autocorrelation function:

$$R_{XX}(\tau) = \frac{A\omega_o}{\pi} \frac{\sin(\omega_o\tau)}{\omega_o\tau}.$$

The impulse response of the filter is of the same form and is

$$h(t) = \frac{\omega_1}{\pi} \frac{\sin(\omega_1 t)}{\omega_1 t}.$$

Determine the autocorrelation function of the filter output for $\omega_o \geq \omega_1$ and for $\omega_o < \omega_1$.

11.14 The power spectrum at the input to an ideal bandpass filter is

$$S_{XX}(f) = \frac{a}{1 + (f/f_o)^2}.$$

Let the transfer function for the ideal bandpass filter be

$$H(f) = \begin{cases} b & \text{for } f_1 < |f| < f_2 \\ 0 & \text{otherwise} \end{cases}.$$

Determine the autocorrelation function of the output. You may have to make a reasonable approximation to obtain a simplified form. Assume that $f_1 - f_2 \gg f_o$ and $(f_2 - f_1)/f_2 \ll 1$.

11.15 A random process with a PSD of $S_{XX}(f) = 1/(1 + f^2)$ is input to a filter. The filter is to be designed so that the output process is white (constant PSD). This filter is called a *whitening filter*.

(a) Find the transfer function of the whitening filter for this input process. Be sure that the filter is causal.

(b) Sketch a circuit that will realize this transfer function.

11.16 White Gaussian noise is input to an RC lowpass filter.

(a) At what sampling instants is the output independent of the input at time $t = t_1$?

(b) At what sampling instants is the output independent of the output at time $t = t_1$?

(c) Repeat parts (a) and (b) if the filter is replaced by the finite time integrator of Exercise 11.1.

11.17

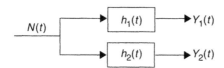

A white Gaussian noise process, $N(t)$, is input to two filters with impulse responses, $h_1(t)$ and $h_2(t)$, as shown in the accompanying figure. The corresponding outputs are $Y_1(t)$ and $Y_2(t)$, respectively.

(a) Derive an expression for the cross correlation function of the two outputs, $R_{Y_1 Y_2}(\tau)$.

(b) Derive an expression for the cross spectral density of the two outputs, $S_{Y_1 Y_2}(\tau)$.

(c) Under what conditions (on the filters) are the two outputs independent when sampled at the same instants in time? That is, when are $Y_1(t_o)$ and $Y_2(t_o)$ independent? Express your constraints in terms of the impulse responses of the filters and also in terms of their transfer functions.

(d) Under what conditions (on the filters) are the two outputs independent when sampled at different instants in time? That is, when are $Y_1(t_1)$ and $Y_2(t_2)$ independent for arbitrary t_1 and t_2? Express your constraints in terms of the impulse responses of the filters and also in terms of their transfer functions.

11.18 If the inputs to two linear filters $h_1(t)$ and $h_2(t)$ are $X_1(t)$ and $X_2(t)$, respectively, show that the cross correlation between the outputs $Y_1(t)$ and $Y_2(t)$

of the two filters is

$$R_{Y_1 Y_2}(\tau) = \int_{-\infty}^{\infty} \int_{-\infty}^{\infty} h_1(\alpha) h_2(\beta) R_{X_1 X_2}(\tau + \alpha - \beta)(d\alpha) \, d\beta.$$

11.19 Determine the noise equivalent bandwidth for a filter with impulse response $h(t) = b \exp(-at) u(t)$.

11.20 A filter has the following transfer function:

$$H(f) = \frac{4}{10 + j2\pi f}$$

Determine the ratio of the noise equivalent bandwidth for this filter to its 3-dB bandwidth.

11.21 For the highpass RC network in the accompanying figure, let $X(t) = A \sin(\omega_c t + \Theta) + N(t)$, where $N(t)$ is white, WSS, Gaussian noise and Θ is a random variable uniformly distributed over $[0, 2\pi)$. Assuming zero initial conditions:

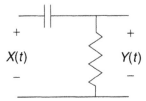

(a) Find the output mean and variance.
(b) Find and plot the autocorrelation function of the output.
(c) Find and plot the output PSD.
(d) Find the output signal-to-noise ratio.

11.22 A parallel resistor, inductor, capacitor (RLC) network is driven by an input current source of $X(t) = A \sin(\omega_c t + \Theta) + N(t)$, where $N(t)$ is white, WSS noise with zero mean. The output is the voltage across the network. The phase Θ is a random variable uniformly distributed over $[0, 2\pi)$.

(a) Find the output power spectrum by first computing the output autocorrelation function and then transforming.
(b) Check the result of part (a) by using Equation 11.12c.
(c) Determine the output signal-to-noise ratio and optimize the bandwidth to maximize the signal-to-noise ratio. Assume ω_c differs from the center frequency of the RLC filter.

Hints: You may have to calculate the autocorrelation function as a function of t and τ and then let t go to infinity to find the steady state output. There are several conditions you may want to consider; for example, the filter may be overdamped, critically damped, or underdamped. It may also have an initial voltage on the capacitor and a current through the inductor. State your assumption about these conditions.

11.23 Suppose you want to learn the characteristics of a certain filter. A white noise source with an amplitude of 15 watts/Hz is connected to the input of the filter. The power spectrum of the filter output is measured and found to be

$$S_{YY}(f) = \frac{30}{(2\pi f)^2 + 10^2}.$$

(a) What is the bandwidth (3-dB) of the filter?

(b) What is the attenuation (or gain) at zero frequency?

(c) Show one possible (i.e., real, causal) filter that could have produced this output PSD.

11.24 A one-sided exponential pulse, $s(t) = \exp(-t)u(t)$, plus white noise is input to the finite time integrator of Exercise 11.1. Adjust the width of the integrator, t_o, so that the output SNR is maximized at $t = t_o$.

11.25

(a) Determine the impulse response of the filter matched to the pulse shape in the accompanying figure. Assume that the filter is designed to maximize the output SNR at time $t = 3t_o$.

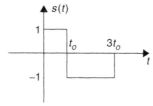

(b) Sketch the output of the matched filter designed in part (a) when the signal $s(t)$ is at the input.

11.26 Find the impulse response and transfer function of a filter matched to a triangular waveform as shown in the accompanying figure when the noise is stationary and white with a power spectrum of $N_o/2$.

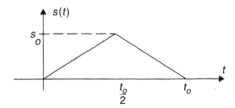

11.27 A known deterministic signal, $s(t)$, plus colored (not white) noise, $N(t)$, with a PSD of $S_{NN}(f)$ is input to a filter. Derive the form of the filter that maximizes the SNR at the output of the filter at time $t = t_0$. To make this problem simpler, you do not need to insist that the filter is causal.

11.28* Suppose we observe a random process $Z(t)$ (without any noise) over a time interval $(-\infty, t)$. Based on this observation, we wish to predict the value of the same random process at time $t + t_0$. That is, we desire to design a filter with impulse response, $h(t)$, whose output will be an estimate of $Z(t + t_0)$:

$$Y(t) = \widehat{Z}(t + t_0) = \int_0^\infty h(u)Z(t - u)\, du.$$

(a) Find the Wiener-Hopf equation for the optimum (in the minimum mean square error (MMSE) sense) filter.

(b) Find the form of the Wiener filter if $R_{ZZ}(\tau) = \exp(-|\tau|)$.

(c) Find an expression for the mean square error $E[\varepsilon^2(t)] = E[(Z(t + t_0) - Y(t))^2]$.

11.29 (Adapted form Leon-Garcia, 7.49) Suppose we are allowed to observe a random process $Z(t)$ at two points in time, t_0 and t_2. Based on these observations we would like to estimate $Z(t)$ at time $t = t_1$ where $t_0 < t_1 < t_2$. We can view this as an interpolation problem. Let our estimator be a linear combination of the two observations,

$$Y(t_1) = \widehat{Z}(t_1) = aZ(t_0) + bZ(t_2).$$

(a) Use the orthogonality principle to find the MMSE estimator.

(b) Find an expression for the mean square error of the MMSE estimator.

11.30 (Roughly adapted from Papoulis 14.1 and 14.5) Suppose we are allowed to observe a random process $Z(t)$ at two points in time, t_0 and t_1. Based on

---

*Adapted from Leon-Garcia, 7.59

those observations, we would like to estimate $Z(t)$ at time $t = t_2$ where $t_0 < t_1 < t_2$. We can view this as a prediction problem. Let our estimator be a linear combination of the two observations,

$$Y(t_2) = \widehat{Z}(t_2) = aZ(t_0) + bZ(t_1).$$

(a) Use the orthogonality principle to find the MMSE estimator.

(b) Suppose that $R_{ZZ}(\tau) = c\exp(-b|\tau|)$ for positive constants b and c. Show that in this case, the sample at time $t = t_0$ is not useful for predicting the value of the process at time $t = t_2$ (given we have observed the process at time $t = t_1 > t_0$). In other words, show that $a = 0$.

11.31 The sum of two independent random processes with power spectral densities

$$S_{SS}(f) = \frac{2}{50 + (2\pi f)^2} \quad \text{and} \quad S_{NN}(f) = 40$$

are input to an linear time invariant (LTI) filter.

(a) Determine the Wiener smoothing filter. That is, find the impulse response of the filter that produces an output that is an MMSE estimate of $S(t)$.

(b) Find the PSD of the filtered signal plus noise.

11.32 The sum of two independent random sequences with autocorrelation functions

$$R_{SS}[k] = \frac{10}{1 - \left(\dfrac{1}{10}\right)^2}\left(\frac{1}{10}\right)^{|k|} \quad \text{and} \quad R_{NN}[k] = \frac{100}{1 - \left(\dfrac{1}{4}\right)^2}\left(\frac{1}{4}\right)^{|k|}$$

are input to an LTI filter.

(a) Determine the Wiener smoothing filter. That is, find the impulse response of the filter that produces an output that is an MMSE estimate of $S[n]$.

(b) Find the PSD of the filtered signal plus noise.

(c) Find the input SNR and an estimate of the output SNR. Discuss whether or not the Wiener filter improves the SNR.

Hint: Compare the spectra of the Wiener filter, the signal, and the noise by plotting each on the same graph.

MATLAB Exercises

11.33 (a) Construct a signal plus noise random sequence using 10 samples of

$$X[n] = \cos(2\pi f_o n t_s) + N[n],$$

where $N[n]$ is generated using randn in MATLAB and $f_o = 0.1/t_s$. Design a discrete time matched filter for the cosine signal. Filter the signal plus noise sequence with the matched filter. At what sample value does the filter output peak?

(b) Construct a signal plus noise random sequence using 10 samples of

$$X[n] = \cos(2\pi f_1 n t_s) + 10 \cos(2\pi f_2 n t_s) + N[n],$$

where $N[n]$ is generated using randn in MATLAB, $f_1 = 0.1/t_s$, and $f_2 = 0.4/t_s$. Design a discrete time matched filter for the f_2 cosine signal. Filter the signal plus noise sequence with the matched filter. At what sample value does the filter output peak?

11.34 If a random sequence has an autocorrelation function $R_{XX}[k] = 10\left(0.8^{|k|}\right)$, find the discrete time, pure prediction Wiener filter. That is, find the filter $h[n]$ such that when $X[n]$ is input, the output, $Y[n]$, will be an MMSE estimate of $X[n+1]$. Determine and plot the PSD for the random sequence and the power transfer function, $|H(f)|^2$, of the filter.

11.35 You have a random process with the correlation matrix

$$\begin{bmatrix} 1.0 & 0.3 & 0.09 & 0.027 \\ 0.3 & 1.0 & 0.3 & 0.09 \\ 0.09 & 0.3 & 1.0 & 0.3 \\ 0.027 & 0.09 & 0.3 & 1.0 \end{bmatrix}$$

Determine the pure prediction Wiener filter. That is, find the coefficients of the impulse response, $h = [h_0, h_1, h_2, 0, 0, 0, \ldots]$. Then determine the power spectrum of the Wiener filter. Are the results what you expected?

11.36 Generate a 100-point random sequence using `randn(1,100)` in MATLAB. Use a first-order autoregressive (AR) filter to filter this random process. That is, the filter is

$$A(z) = \frac{1}{1 + a_1 z^{-1}}.$$

Let $a_1 = -0.1$. Use the filtered data to obtain an estimate for the first-order prediction Wiener filter. Compare the estimated filter coefficient with the true value.

Simulation Techniques

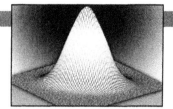

With the increasing computational power of very inexpensive computers, simulation of various systems is becoming very common. Even when a problem is analytically tractable, sometimes it is easier to write a quick program to simulate the desired results. However, there is a certain art to building good simulations, and many times avoidable mistakes have led to incorrect simulation results. This chapter aims at helping the reader to build a basic knowledge of some common techniques used for simulation purposes. Most of the results presented in this chapter are just applications of material covered in previous chapters, so there is nothing fundamentally new here. Nevertheless, armed with some of the basic simulation principles presented in this chapter, the reader should be able to develop simulation tools with confidence.

12.1 Computer Generation of Random Variables

In this section, we study techniques used to generate random numbers. However, we must start with a disclaimer. Most of the techniques used in practice to generate so-called random numbers will actually generate a completely deterministic sequence of numbers. So, what is actually random about these random number generators? Strictly speaking, nothing! Rather, when we speak of *computer-generated* random numbers, we are usually creating a sequence of numbers that have certain statistical properties that make them behave like random numbers, but in fact they are not really random at all. Such sequences of numbers are more appropriately referred to as *pseudorandom* numbers.

12.1.1 Binary Pseudorandom Number Generators

To start with, suppose we would like to simulate a sequence of independent and identically distributed (IID) Bernoulli random variables, $x_1, x_2, x_3, \ldots$. One way to do this would be to grab a coin and start flipping it and observe the sequence of heads and tails, which could then be mapped to a sequence of 1s and 0s. One drawback to this approach is that it is very time consuming. If our application demanded a sequence of length 1 million, not many of us would have the patience to flip the coin that many times. Naturally, we seek to assign this task to a computer. So, to simulate an IID sequence of random variables, essentially we would like to create a computer program that will output a binary sequence with the desired statistical properties. For example, in addition to having the various bits in the sequence be statistically independent, we might also want 0s and 1s to be equally likely.

Consider the binary sequence generated by the linear feedback shift register (LFSR) structure illustrated in Figure 12.1. In that figure, the square boxes represent binary storage elements (i.e., flip-flops) while the adder is modulo-2 (i.e., an exclusive OR gate). Suppose the shift register is initially loaded with the sequence 1, 1, 1, 1. It is not difficult to show that the shift register will output the sequence

$$1, 1, 1, 1, 0, 0, 0, 1, 0, 0, 1, 1, 0, 1, 0, \ldots. \tag{12.1}$$

If the shift register were clocked longer, it would become apparent that the output sequence would be periodic with a period of 15 bits. While periodicity is not a desirable property for a pseudorandom number generator, if we are interested in generating short sequences (of length less than 15 bits), then the periodicity of this sequence generator would not come into play. If we are interested in generating longer sequences, we could construct a shift register with more stages so that the period of the resulting sequence would be sufficiently long that its periodicity would not be a concern.

The sequence generated by our LFSR does possess several desirable properties. First, the number of ones and zeros is almost equal (8 ones and 7 zeros is as close as we can get to "equally likely" with a sequence of length 15). Second, the autocorrelation function of this sequence is nearly equal to that of a truly random binary IID sequence (again, it is as close as we can possibly get with a sequence of period 15; see Exercise 12.1). Practically speaking, the sequence output by this completely deterministic device does a pretty good job of mimicking the behavior of an IID binary sequence. It also has the advantage of being repeatable. That is, if we load the shift register with the same initial contents, we can always reproduce the exact same sequence.

It should be noted that not all LFSRs will serve as good pseudorandom number generators. Consider for example the four-stage LFSR in Figure 12.2. This shift

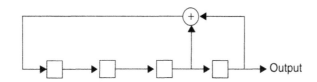

Figure 12.1 A four-stage, binary linear feedback shift register.

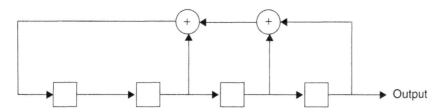

Figure 12.2 Another four-stage, binary linear feedback shift register.

register is only a slightly modified version of the one in Figure 12.1, yet when loaded with the sequence 1, 1, 1, 1, this shift register outputs a repeating sequence of all 1s (i.e., the period of the output sequence is 1). The only difference between the two shift registers is in the placement of the feedback tap connections. Clearly, the placement of these tap connections is crucial in creating a good pseudorandom sequence generator.

A general N-stage LFSR is shown in Figure 12.3. The feedback tap gains, g_i, $i = 0, 1, 2, \ldots, N$, are each either 0 or 1. A 1 represents the presence of a tap connection, while a 0 represents the absence of a tap connection. It is also understood that $g_0 = g_N = 1$. That is, there must always be connections at the first and last position. A specific LFSR can then be described by the sequence of tap connections. For example, the four stage LFSR in Figure 12.1 can be described by the sequence of tap connections $[g_0, g_1, g_2, g_3, g_4] = [1, 0, 0, 1, 1]$. It is also common to further simplify this shorthand description of the LFSR by converting the binary sequence of tap connections to an octal number. For example, the sequence $[1, 0, 0, 1, 1]$ becomes the octal number 23. Likewise, the LFSR in Figure 12.2 is described by $[g_0, g_1, g_2, g_3, g_4] = [1, 0, 1, 1, 1] \rightarrow 27$.

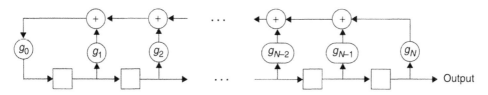

Figure 12.3 A general N-stage, binary linear feedback shift register.

An N-stage LFSR must necessarily generate a periodic sequence of numbers. At any point in time, the N-bit contents of the shift register can take on only one of 2^N possibilities. Given any starting state, the LFSR could at best cycle through all possible states, but sooner or later must return to its initial state (or some other state it has already been in). At that point, the output will then begin to repeat itself. Also, note that if the LFSR ever gets into the all zero state, it will remain in that state and output a sequence of all zeros from that point on. Hence, to get the maximum period out of a LFSR, we would like the shift register to cycle through all possible nonzero states exactly once before returning to its initial state. This will produce a periodic output sequence with a period of $2^N - 1$. Such a sequence is referred to as a *maximal length linear feedback shift register* (MLLFSR) sequence, or an m-sequence, for short.

To study the question of how to connect the feedback taps of an LFSR in order to produce an m-sequence requires a background in abstract algebra beyond the scope of this book. Instead, we include a short list in Table 12.1 describing a few feedback connections, in the octal format described, that will produce m-sequences. This list is not exhaustive in that it does not list all possible feedback connections for a given shift register length that lead to m-sequences.

As mentioned, m-sequences have several desirable properties in that they mimic those properties exhibited by truly random IID binary sequences. Some of the properties of m-sequences generated by an N-stage LFSR are summarized as follows:

- m-sequences are periodic with a period of $2^N - 1$.
- In one period, an m-sequence contains $2^{N/2}$ ones and $2^{N/2} - 1$ zeros. Hence, zeros and ones are almost equally likely.
- The autocorrelation function of m-sequences is almost identical to that of an IID sequence.
- Define a run of length n to be a sequence of either n consecutive ones or n consecutive zeros. An m-sequence will have one run of length N, one run of

Table 12.1   LFSR Feedback Connections for m-sequences

SR length, N	Feedback connections (in octal format)
2	7
3	13
4	23
5	45, 75, 67
6	103, 147, 155
7	211, 217, 235, 367, 277, 325, 203, 313, 345
8	435, 551, 747, 453, 545, 537, 703, 543

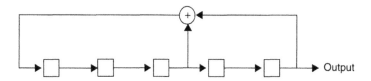

Figure 12.4 A five-stage LFSR with 45 (octal) feedback tap connections.

length $N-1$, two runs of length $N-2$, four runs of length $N-3$, eight runs of length $N-4,\ldots$, and 2^{N-2} runs of length 1.

M-sequences possess many other interesting properties that are not as relevant to their use as random number generators.

EXAMPLE 12.1: Suppose we wish to construct an m-sequence of length 31. Since the period of an m-sequence is $2^N - 1$, we will need an $N = 5$ stage shift register to do the job. From Table 12.1, there are three different feedback connections listed, any of which will work. Using the first entry in the table, the octal number 45 translates to the feedback tap connections (1,0,0,1,0,1). This describes the LFSR shown in Figure 12.4. Assuming this LFSR is initially loaded with the sequence (1,0,0,0,0), the resulting m-sequence will be

$$(0\,0\,0\,0\,1\,0\,0\,1\,0\,1\,1\,0\,0\,1\,1\,1\,1\,1\,0\,0\,0\,1\,1\,0\,1\,1\,1\,0\,1\,0\,1).$$

12.1.2 Nonbinary Pseudorandom Number Generators

Next suppose that it is desired to generate a sequence of pseudorandom numbers drawn from a nonbinary alphabet. One simple way to do this is to modify the output of the binary LFSRs discussed previously. For example, suppose we want a pseudorandom number generator that produces octal numbers (i.e, from an alphabet of size 8) that simulates a sequence of IID octal random variables where the eight possible outcomes are equally probable. One possible approach would be to take the 5-stage LFSR of Example 12.1, group the output bits in triplets (i.e., three at a time), and then convert each triplet to an octal number. Doing so (assuming the LFSR is loaded initially as in Example 12.1), one period of the resulting octal sequence is

$$(0\,2\,2\,6\,3\,7\,0\,6\,7\,2\,4\,1\,1\,3\,1\,7\,4\,3\,3\,5\,2\,0\,4\,5\,4\,7\,6\,1\,5\,6\,5). \qquad (12.2)$$

Note that each of the octal numbers $0, 1, 2, \ldots, 7$ occurs exactly four times in this sequence except the number 0, which occurs three times. This is as close as we can get to equally likely octal numbers with a sequence of length 31. The number of runs of various lengths in this sequence also matches what we might expect from a truly random IID sequence of equally likely octal numbers. For example, the probability of a run of length 2 occurring in a random sequence of octal numbers is $1/8$. Given a random sequence of length 31, the expected number of runs of length 2 is $31/8 = 3.875$. This pseudorandom sequence has three runs of length 2. The expected number of runs of length 3 is $31/64 = 0.4844$. The pseudorandom sequence has no runs of length 3.

It should be noted that since the octal sequence in Equation 12.2 is generated by the underlying feedback structure of Example 12.1, there should be a recursive formula that can be used to generate the sequence. In this case, the recursion is fairly complicated and not very instructive, but this leads us to the idea of generating pseudorandom sequences of nonbinary numbers using some recursive formula. One commonly used technique is the *power residue method* whereby a pseudorandom sequence is generated through a recursion of the form

$$x_k = ax_{k-1} \bmod q, \quad k = 1, 2, 3, \ldots, \tag{12.3}$$

where a and q are suitably chosen constants. Due to the modulo-q operation, the elements of the sequence are from the set $\{0, 1, 2, \ldots, q-1\}$. The first element of the sequence x_0 is called the seed and given the set $\{a, q, x_0\}$, the resulting sequence is completely deterministic. Note also that the resulting sequence must be periodic since once any element appears a second time in the sequence, the sequence will then repeat itself. Furthermore, since the elements of the sequence are from a finite alphabet with q symbols, the maximum period of the sequence is $q - 1$ (the period is not q because the element 0 must never appear or the sequence will produce all 0s after that and hence be of period 1). The next example shows that the desirable statistical properties of the sequence are dependent upon a careful selection of the numbers a and q.

EXAMPLE 12.2: First suppose that $\{a, q\} = \{4, 7\}$. Then the sequence produced has a period of 3 and assuming that the seed is $x_0 = 1$, the sequence is given by

$$(x_0, x_1, x_2, \ldots) = (1, 4, 2, 1, 4, 2, 1, 4, 2, \ldots).$$

This is not a particularly good result since with the selection of $q = 7$, we would hope for a period of $q - 1 = 6$. However, if we make a slight change so that $\{a, q\} = \{3, 7\}$, with the seed of $x_0 = 1$, the sequence becomes

$$(x_0, x_1, x_2, \ldots) = (1, 3, 2, 6, 4, 5, 1, 3, 2, 6, 4, 5, \ldots).$$

Now, as desired, the sequence has the maximal period of 6 and cycles through each of the integers from 1 through 6 exactly once each. As another example of a choice of $\{a, q\}$ that leads to a bad sequence, suppose we selected $\{a, q\} = \{4, 8\}$. Then the sequence produced (assuming $x_0 = 1$) would be

$$(x_0, x_1, x_2, \ldots) = (1, 4, 0, 0, 0, \ldots).$$

Clearly, we can get pseudorandom sequences with very different statistical properties depending on how we choose $\{a, q\}$. By choosing the number q to be very large, the resulting period of the pseudorandom sequence will also be very large, and hence the periodicity of the sequence will not become an issue. Most math packages and high-level programming languages have built in random number generators that use this method. Commonly, the parameters $\{a, q\} = \{7^5, 2^{31} - 1\}$ are used. This produces a sequence of length $2^{31} - 2$, which is over 2 billion. Furthermore, by normalizing the elements of the sequence by q, the resulting pseudorandom sequence has elements from the set $\{1/q, 2/q, \ldots, (q-1)/q\}$. With a very large choice for the value of q, for almost all practical purposes, the elements will appear to be drawn from the continuous interval $(0, 1)$. Hence, we have constructed a simple method to simulate random variables drawn from a uniform distribution.

12.1.3 Generation of Random Numbers from a Specified Distribution

Quite often we are interested in generating random variables that obey some distribution other than a uniform distribution. In this case, it is generally a fairly simple task to transform a uniform random number generator into one that follows some other distribution. Consider forming a monotonic increasing transformation $g()$ on a random variable X to form a new random variable Y. From the results of Chapter 4, the PDFs of the random variables involved are related by

$$f_Y(y) = \frac{f_X(x)}{dg(x)/dx}. \tag{12.4}$$

Given an arbitrary PDF, $f_X(x)$, the transformation $Y = g(X)$ will produce a uniform random variable Y if $dg/dx = f_X(x)$, or equivalently $g(x) = F_X(x)$. Viewing this result in reverse, if X is uniformly distributed over $(0, 1)$ and we want to create a new random variable, Y, with a specified distribution, $F_Y(y)$, the transformation $Y = F_Y^{-1}(X)$ will do the job.

EXAMPLE 12.3: Suppose we want to transform a uniform random variable into an exponential random variable with a PDF of the form

$$f_Y(y) = a \exp(-ay)u(y).$$

The corresponding CDF is

$$F_Y(y) = [1 - \exp(-ay)]u(y).$$

Hence, to transform a uniform random variable into an exponential random variable, we can use the transformation

$$Y = F_Y^{-1} = -\frac{\ln(1 - X)}{a}.$$

Note that if X is uniformly distributed over $(0, 1)$, then $1 - X$ will be uniformly distributed as well so that the slightly simpler transformation

$$Y = -\frac{\ln(X)}{a}$$

will also work.

This approach for generation of random variables works well provided that the CDF of the desired distribution is invertible. One notable exception where this approach will be difficult is the Gaussian random variable. Suppose, for example, we wanted to transform a uniform random variable, X, into a standard normal random variable, Y. The CDF in this case is the complement of a Q-function, $F_Y(y) = 1 - Q(y)$. The inverse of this function would then provide the appropriate transformation, $y = Q^{-1}(1 - x)$, or as with the previous example, we could simplify this to $y = Q^{-1}(x)$. The problem here lies with the inverse Q-function, which cannot be expressed in a closed form. One could devise efficient numerical routines to compute the inverse Q-function, but fortunately there is an easier approach.

An efficient method to generate Gaussian random variables from uniform random variables is based on the following 2×2 transformation. Let X_1 and X_2 be two independent uniform random variables (over the interval $(0,1)$). Then if two new random variables, Y_1 and Y_2, are created according to

$$Y_1 = \sqrt{-2\ln(X_1)}\cos(2\pi X_2), \tag{12.5a}$$

$$Y_2 = \sqrt{-2\ln(X_1)}\sin(2\pi X_2), \tag{12.5b}$$

then Y_1 and Y_2 will be independent standard normal random variables (see Example 5.24). This famous result is known as the *Box-Muller transformation* and is commonly used to generate Gaussian random variables. If a pair of Gaussian random variables is not needed, one of the two can be discarded. This method is particularly convenient for generating complex Gaussian random variables since it naturally generates pairs of independent Gaussian random variables. Note that if Gaussian random variables are needed with different means or variances, this can easily be accomplished through an appropriate linear transformation. That is, if $Y \sim N(0,1)$, then $Z = \sigma Y + \mu$ will produce $Z \sim N(\mu, \sigma^2)$.

12.1.4 Generation of Correlated Random Variables

Quite often it is desirable to create a sequence of random variables that are not independent, but rather have some specified correlation. Suppose we have a Gaussian random number generator that generates a sequence of IID standard normal random variables, $X = (X_1, X_2, \ldots, X_N)$ and it is desired to transform this set of random variables into a set of Gaussian random variables, $Y = (Y_1, Y_2, \ldots, Y_N)$, with some specified covariance matrix, C_Y. By using a linear transformation of the form $Y = AX$, the joint Gaussian distribution will be preserved. The problem of how to choose the transformation to produce the desired covariance matrix was covered in Chapter 6, Section 6.4A. Recall that to specify this transformation, an eigen decomposition of the covariance matrix is performed to produce $C_Y = Q\Lambda Q^T$, where Λ is the diagonal matrix of eigenvalues of C_Y, and Q is the corresponding matrix of eigenvectors. Then the matrix $A = Q\sqrt{\Lambda}$ will produce the vector Y with the correct covariance matrix C_Y. Once again, if a random vector with a nonzero mean vector is desired, this approach can be augmented as $Y = AX + B$, where B is the vector of appropriate means.

12.2 Generation of Random Processes

Next consider the problem of simulating a random process, $X(t)$, with a desired PSD, $S_{XX}(f)$. It is not feasible to create a continuous time random process with a computer. Fortunately, we can invoke the sampling theorem to represent the continuous time random process by its samples. Let $X_k = X(kT_s)$ be the kth sample of the random process taken at a sampling rate of $R_s = 1/T_s$. Then, provided the sampling rate is chosen to be at least twice the absolute bandwidth of $X(t)$ (i.e., twice the largest nonzero frequency component of $S_{XX}(f)$), the random process can be reproduced from its samples. Thus, the problem of generating a random process can be translated into one of creating a sequence of random variables. The question is how should the random variables be correlated in order to produce the correct PSD? Of course, the autocorrelation function, $R_{XX}(\tau)$, provides the answer to this question. If the random process is sampled at a rate of $R_s = 1/T_s$, then the kth and mth sample will have a correlation specified by $E[X_k X_m] = R_{XX}((k-m)T_s)$. Hence, if $\mathbf{X} = (X_1, X_2, \ldots, X_N)$ is a sequence of samples of the random process $X(t)$, the correlation matrix of these samples will have a Toeplitz structure (assuming $X(t)$ is stationary) of the form

$$\mathbf{R_{XX}} = \begin{bmatrix} R_{XX}(0) & R_{XX}(T_s) & R_{XX}(2T_s) & \ldots & R_{XX}(NT_s) \\ R_{XX}(-T_s) & R_{XX}(0) & R_{XX}(T_s) & \ldots & R_{XX}((N-1)T_s) \\ R_{XX}(-2T_s) & R_{XX}(-T_s) & R_{XX}(0) & \ldots & R_{XX}((N-2)T_s) \\ \ldots & \ldots & \ldots & & \ldots \\ R_{XX}(-NT_s) & R_{XX}(-(N-1)T_s) & R_{XX}(-(N-2)T_s) & \ldots & R_{XX}(0) \end{bmatrix}.$$

(12.6)

Once the appropriate correlation matrix is specified, the procedure developed in the last section can be used to generate the samples with the appropriate correlation matrix.

This approach will work fine provided that the number of samples desired is not too large. However, in many cases, we need to simulate a random process for a large time duration. In that case, the number of samples, N, needed becomes large and hence the matrix $\mathbf{R_{XX}}$ is also large. Performing the necessary eigendecomposition on this matrix then becomes a computationally intensive problem. The following sections look at some alternative approaches to this general problem that offer some computational advantages.

12.2.1 Frequency Domain Approach

If the random process to be simulated is a Gaussian random process, we can approach the problem by creating samples of the random process in the frequency domain. Suppose we wish to create a realization of the random process, $X(t)$, of time duration T_d, say over the interval $(0, T_d)$, and that we don't much care what happens to the process outside this interval. To start with, we produce a periodic

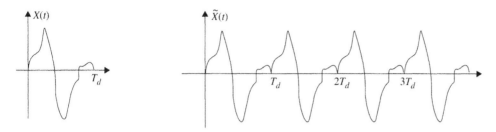

Figure 12.5 A realization of the random process $X(t)$ along with its periodic extension $\widetilde{X}(t)$.

signal $\widetilde{X}(t)$ by repeating $X(t)$ every T_d seconds as illustrated in Figure 12.5. Since $\widetilde{X}(t)$ is periodic, it has a Fourier series representation

$$\widetilde{X}(t) = \sum_k X_k e^{j2\pi kf_o t}, \quad f_o = \frac{1}{T_d}. \tag{12.7}$$

Note that due to the linearity of the Fourier series construction, if the X_k are zero-mean Gaussian random variables, then the resulting process $\widetilde{X}(t)$ will be a zero-mean Gaussian random process. Furthermore, the periodic random process $\widetilde{X}(t)$ has a line spectrum given by

$$S_{\widetilde{X},\widetilde{X}}(f) = \sum_k s_k^2 \delta(f - kf_o), \quad s_k^2 = E\left[|X_k|^2\right]. \tag{12.8}$$

The s_k can be chosen to shape the spectrum to any desired form. If the desired PSD of the random process is $S_{XX}(f)$, then we could pick $s_k^2 \propto S_{xx}(kf_o)$. The constant of proportionality can be chosen so that the total power in the process $\widetilde{X}(t)$ matches that of $X(t)$. In particular, suppose that $X(t)$ is bandlimited so that $S_{XX}(f) = 0$ for $|f| > W$. Then the number of terms in the Fourier series in Equation 12.7 is finite. Let

$$M = \left\lfloor \frac{W}{f_o} \right\rfloor = \lfloor WT_d \rfloor. \tag{12.9}$$

Then s_k will be nonzero only for $|k| \leq M$. Hence, we need to generate a total of $2M + 1$ random variables, $X_{-M}, X_{-M+1}, \ldots, X_{-1}, X_0, X_1, \ldots, X_M$. The variances of these random variables are chosen such that

$$s_k^2 = E\left[|X_k|^2\right] = \beta S_{XX}(kf_o), \quad \beta = \frac{\int_{-W}^{W} S_{XX}(f)\,df}{\sum_{k=-M}^{M} S_{XX}(kf_o)}. \tag{12.10}$$

In summary, the random process can be simulated by first generating a sequence of $2M + 1$ zero-mean complex Gaussian random variables. Each random variable should be scaled so that the variances are as specified in Equation 12.10. Samples of the random process in the time domain can be constructed for any desired time resolution, Δt, according to

$$X[i] = X(i\Delta t) = \sum_{k=-M}^{M} X_k (e^{j2\pi f_0 \Delta t})^{ik}. \tag{12.11}$$

If the random process is real, it is sufficient to generate the $M + 1$ random variables $X_0, X_1, \ldots, X_M$ independently and then form the remaining random variables $X_{-1}, X_{-2}, \ldots, X_{-M}$ using the conjugate symmetry relationship $X_{-k} = X_k^*$. In this case, X_0 must also be real so that a total $2M + 1$ real Gaussian random variables are needed (one for X_0 and two each for X_k, $k = 1, 2, \ldots, M$) to construct the random process, $X(t)$.

EXAMPLE 12.4: Suppose we wish to generate a 5-msec segment of a real zero-mean Gaussian random process with a PSD given by

$$S_{XX}(f) = \frac{1}{1 + (f/f_3)^4},$$

where $f_3 = 1$ kHz is the 3-dB frequency of the PSD. Strictly speaking, this process has an infinite absolute bandwidth. However, for sufficiently high frequencies there is minimal power present. For the purposes of this example, we (somewhat arbitrarily) take the bandwidth to be $W = 6f_3$ so that approximately 99.9 percent of the total power in the process is contained within $|f| < W$. Since we want to simulate a time duration of $T_d = 5$ msec, the number of Fourier series coefficients needed is given by $M = \lfloor WT_d \rfloor = 30$. Figure 12.6 shows a comparison of the actual PSD, $S_{XX}(f)$, with the discrete line spectrum approximation. Also, one realization of the random process generated by this method is shown in Figure 12.7. The MATLAB code used to create these plots follows.

```
% Set parameters.
I=sqrt(-1);
Td=5e-3; fo=1/Td; f3=1e3; dt=1e-5;
M=floor(6*f3*Td); m=[-M:M];

% Construct discrete samples of PSD.
x=[0:0.01:10]; psd=1./(1+x.^4);
```

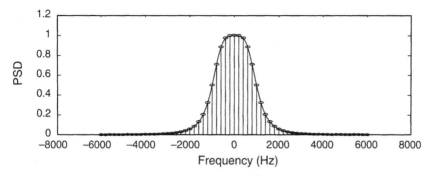

Figure 12.6 A comparison of the exact PSD along with the discrete approximation for Example 12.4.

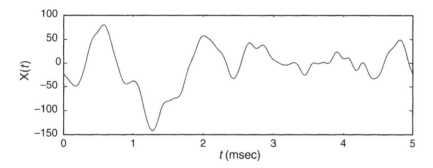

Figure 12.7 A realization of the random process of Example 12.4.

```
power=2*f3*sum(psd)*0.01;
s=1./(1+((m*fo)/f3).^4);
beta=power/sum(s);
s=beta*s;
% Construct "continuous" PSD.
f=[-8:0.01:8]*f3;
psd=1./(1+(f/f3).^4);

% Plot results.
subplot(2,1,1)
stem(m*fo,s/fo); hold on
plot(f,psd,'g'); hold off
axis([-8*f3 8*f3 0 1.2])
xlabel('frequency (Hz)'); ylabel('PSD');
```

```
% Generate frequency domain samples.
z0=randn(1); z0=z0*sqrt(s(M+1));
zplus=sqrt(s(M+2:2*M+1)/2).*(randn(1,M)+I*randn(1,M));
zminus=conj(fliplr(zplus));
z=[zminus z0 zplus];

% Create time domain process.
t=[0:dt:Td];
rp=zeros(1,length(t));
for m=-M:M
  rp=rp+z(m+M+1)*exp(I*2*pi*m*fo*t);
end;

% Plot results.
subplot(2,1,2)
plot(t*1000,real(rp))
xlabel('t (msec)'); ylabel('X(t)')
```

12.2.2 Time Domain Approach

A simple alternative to the previous frequency domain approach is to perform time domain filtering on a white Gaussian noise process as illustrated in Figure 12.8. Suppose white Gaussian noise with PSD, $S_{XX}(f) = 1$ is input to an LTI filter that can be described by the transfer function, $H(f)$. Then, from the results in Chapter 11, it is known that the PSD of the output process is $S_{YY}(f) = |H(f)|^2$. Hence, in order to create a Gaussian random process, $Y(t)$, with a prescribed PSD, $S_{YY}(f)$, we can construct a white process and pass this process through an appropriately designed filter. The filter should have a magnitude response that satisfies

$$|H(f)| = \sqrt{S_{YY}(f)}. \tag{12.12}$$

The phase response of the filter is irrelevant, and hence any convenient phase response can be given to the filter.

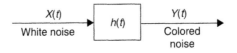

Figure 12.8   Time domain filtering to create a colored Gaussian random process.

This technique is particularly convenient when the prescribed PSD, $S_{YY}(f)$ can be written as the ratio of two polynomials in f. Then the appropriate transfer function can be found through spectral factorization techniques. If the desired PSD is a more complicated function of f, then designing and/or implementing a filter to produce that PSD may be difficult. In that case, it may be necessary to use an approximate filter design.

EXAMPLE 12.5: In this example, we design the filter needed to generate the random process specified in Example 12.4 using the time domain method. The PSD is factored as follows:

$$S(f) = \frac{1}{1+(f/f_3)^4} = \frac{f_3^4}{\left(f - f_3 e^{j\pi/4}\right)\left(f - f_3 e^{j3\pi/4}\right)\left(f - f_3 e^{j5\pi/4}\right)\left(f - f_3 e^{j7\pi/4}\right)}.$$

If the first two poles are associated with $H(f)$ (and the last two with $H^*(f)$), then the filter has a transfer function of

$$H(f) = \frac{f_3^2}{\left(f - f_3 e^{j\pi/4}\right)\left(f - f_3 e^{j3\pi/4}\right)} = \frac{f_3^2}{f^2 - j\sqrt{2}f_3 f - f_3^2},$$

which can be represented in the time domain as

$$h(t) = -2\omega_o e^{-\omega_o t}\sin(\omega_o t)u(t),$$

where $\omega_o = \sqrt{2}\pi f_3$. For the purposes of creating a random process with the desired spectrum, the negative sign in front of this impulse response is irrelevant and can be ignored. Hence, to produce the desired random process, we start by generating a white Gaussian random process and then convolve the input with the impulse response specified.

Once an appropriate analog filter has been designed, the filter must be converted to a discrete time form. If the sampling rate is taken to be sufficiently high, then the impulse response of the discrete time filter can be found by simply sampling the impulse response of the continuous time filter. This is the so-called impulse invariance transformation. However, because of aliasing that occurs in the process of sampling the impulse response, the frequency response of the digital filter will not be identical to that of the original analog filter unless the analog filter is absolutely bandlimited. Of course, if the analog filter is approximately bandlimited and if the sampling rate is sufficiently large, this aliasing can be kept to a minimum and the resulting digital filter will produce a very good approximation to the desired frequency response.

An alternative popular approach for producing a digital filter from an analog design is to use a bilinear transformation. That is, suppose we have an analog filter with transfer function $H_a(s)$, a digital approximation to this filter, $H_d(z)$, can be obtained (assuming a sampling frequency of f_s) according to

$$H_d(z) = H_a(s)\big|_{s=2f_s\left(\frac{1-z^{-1}}{1+z^{-1}}\right)}. \tag{12.13}$$

One advantage of the bilinear transformation is that if the analog filter is stable, then the digital filter will be stable as well. Note also that if the analog filter is an nth-order filter, then the order of the digital filter will be no more than n as well.

 EXAMPLE 12.6: In this example, we find the digital approximation to the analog filter designed in Example 12.5 using the bilinear approximation. From the results of that example, the analog filter was a second-order Butterworth filter whose transfer function (in terms of s) was given by

$$H_a(s) = \frac{\omega_3^2}{s^2 + \sqrt{2}\omega_3 s + \omega_3^2},$$

where $\omega_3 = 2\pi f_3$ is the 3-dB frequency of the filter in radians per second. After a little bit of algebraic manipulation, application of the bilinear transformation in Equation 12.13 results in

$$H_d(z) = \frac{b_0 + b_1 z^{-1} + b_2 z^{-2}}{a_0 + a_1 z^{-1} + a_2 z^{-2}},$$

where $b_0 = 1$, $b_1 = 2$, $b_2 = 1$, $a_0 = 1 + \sqrt{2}\gamma + \gamma^2$, $a_1 = 2 - 2\gamma^2$, $a_2 = 1 - \sqrt{2}\gamma + \gamma^2$, and $\gamma = f_s/(\pi f_3)$. Figure 12.9 shows a plot of the impulse response of this filter as well as one realization of the random process created by passing white Gaussian noise through this filter. Note that for this example, the impulse response of the filter lasts for about 1 msec (this makes sense since the bandwidth was 1 kHz). Therefore, when creating the filtered process, at least the first millisecond of output data should be discarded since the contents of the digital filter have not reached a statistical steady state until that point. The relevant MATLAB code follows.

```
I=sqrt(-1);
imp_len=150;     % Length of impulse response in samples.
sim_len=150;     % Length of simulation in samples.
```

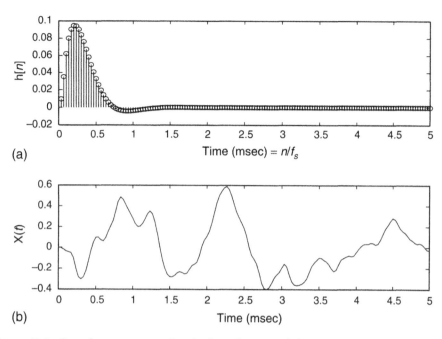

Figure 12.9 Impulse response and a single realization of the output of the filter designed in Example 12.6.

```
f3=1000;                        % 3-dB frequency of desired PSD.
fs=30000;                       % Sampling rate.
g=fs/(pi*f3);                   % Compute filter coefficients.
b0=1; b1=2; b2=1; b=[b0 b1 b2];
a0=1+sqrt(2)*g+g^2; a1=2-2*g^2;
a2=1-sqrt(2)*g+g^2; a=[a0 a1 a2];
x=zeros(1,imp_len); x(1)=1;     % Impulse.
y=filter(b,a,x);                % Impulse response of filter.
time_scale=[1:length(y)]/fs;
subplot(2,1,1)                  % Plot impulse response.
stem(time_scale*1000,y,'o')
xlabel('time(msec)=n/f_s')
ylabel('h[n]')
x=randn(1,sim_len);             % White Gaussian random process.
y=filter(b,a,x);                % Filtered process.
time_scale=[1:length(y)]/fs;
subplot(2,1,2)                  % Plot realization.
plot(time_scale*1000,y);
xlabel('time (msec)'); ylabel('X(t)');
```

One advantage of the time domain approach is that it is convenient for generating very long realizations of the random process. Once the filter is designed, the output process is created by performing a convolution of the impulse response of the filter with a white input sequence. The complexity of this operation is linear in the length of the sequence. However, the process of creating a long sequence can be broken into several smaller sequences. The smaller sequences can then be concatenated together to create a long sequence. There will be no discontinuities at the points of concatenation if the contents of the filter are stored after each realization and used as the starting contents for the next realization.

Some care must be taken when using the time domain method if the desired sampling rate of the process is much larger than the bandwidth of the filter. In this case, the poles of the digital filter will be close to the unit circle and the filter might exhibit stability problems. This is illustrated in Figure 12.10, where the magnitude of the poles of the digital filter from Example 12.6 are plotted as a function of the sampling rate. Note that when the rate at which the process is sampled becomes a few hundred times the bandwidth of the filter, the poles of the digital filter become very close to the unit circle. This problem can be easily avoided by creating a digital filter to create samples of the process at a lower rate (perhaps at several times the

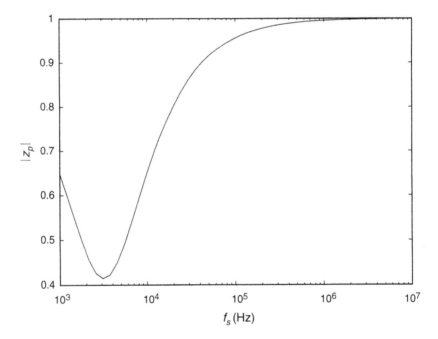

Figure 12.10 Magnitude of the filter poles as a function of sampling frequency for the filter designed in Example 12.6.

bandwidth of the filter so as to avoid aliasing) and then upsampling (through interpolation) the resulting process to any desired rate.

12.2.3 Generation of Gaussian White Noise

Generation of a white noise process is exceedingly common and also very simple. However, it is often the source of frequent mistakes among engineers, so we felt it worth making a few comments about computer generation of white Gaussian noise processes. The source of confusion in the generation of white noise is that one cannot represent white noise from its time domain samples. White noise has infinite power, therefore samples of a white noise process would require infinite variance. Alternatively, white noise has infinite bandwidth, so the Nyquist rate for recovering white noise from its samples would be infinite. In order to represent a "white" process in discrete time, we must invoke some form of prefiltering before sampling. This will limit the bandwidth such that a finite sampling rate can be used, and at the same time it will limit the power of the process such that the resulting samples of the filtered white process will have a finite variance.

Strictly speaking, once we filter the white noise it is no longer white, but this should not be of too much concern. In practice, there is no such thing as truly white noise. Recall that the white noise model was an approximation to a noise process that had a constant PSD over a very large (but not infinite) bandwidth. Any equipment we use to measure or receive the noise will automatically filter the process. With this in mind, we imagine a prefilter that has a bandwidth much larger than any bandwidth we are concerned with in the specific system we are simulating. The noise we create, although not truly white, will behave as though it were white for all practical purposes.

In order to simplify the process of creating the samples of the prefiltered white noise, it is common to employ an ideal lowpass prefilter with bandwidth W as illustrated in Figure 12.11. Now that the process is bandlimited, it can be represented by samples at any rate that satisfies $f_s \geq 2W$. Since the prefilter is ideal, the autocorrelation function of the filter output is easily calculated to be

$$R(\tau) = N_0 W \operatorname{sinc}(2W\tau). \tag{12.14}$$

Figure 12.11 A/D conversion process for white noise.

Note that since this sinc function has nulls at multiples of $1/2W$, the samples of the filtered process will be uncorrelated provided that the samples are spaced by any integer multiple of $1/2W$. In other words, if the sampling rate satisfies $f_s = 2W/n$ for any integer n, then the samples of the prefiltered white noise will be uncorrelated. By choosing $n = 1$ so that the sampling rate is exactly the Nyquist rate, $f_s = 2W$, the process can be recovered from the discrete samples and the samples are uncorrelated. For Gaussian noise, this implies that the filtered white noise can be represented by a sequence of independent, zero-mean, Gaussian random variables with variance of $\sigma^2 = N_o W$. Note that the variance of the samples and the rate at which they are taken are related by $\sigma^2 = N_o f_s/2$.

The lesson to be learned here is that if we wish to represent Gaussian white noise as a sequence of *independent* Gaussian random variables, then there is an implicit assumption about the nature of the prefiltering. Furthermore, to be consistent with this assumption, the variance of the samples must be adjusted when the sampling rate is changed. The variance and sampling rate cannot be selected independently.

12.3 Simulation of Rare Events

Quite often, we are interested in estimating the probability of some event, A. If analytically calculating this probability is too cumbersome, we can design a computer program to simulate the system in question and then observe whether or not the event occurs. By repeating this procedure many times, we can observe how often the event A occurs, and hence get an estimate of its probability through a relative frequency approach. The event A could be a bit error in a digital communications system—in which case we are interested in calculating bit error probability—or it could be a buffer overflow in a computer network or even something as extravagant as breaking the bank at the blackjack table.

12.3.1 Monte Carlo Simulations

In general, suppose we have the ability to recreate (simulate) the experiment an arbitrary number of times and define a sequence of Bernoulli random variables, X_i, that are defined according to

$$X_i = \begin{cases} 1 & \text{if } A \text{ occurs during the } i\text{th experiment} \\ 0 & \text{otherwise} \end{cases}. \tag{12.15}$$

Hence, X_i is simply an indicator function for the event A. If the experiments are independent, then the probability of the event A, p_A, can be estimated according to

$$\hat{p}_A = \frac{1}{n} \sum_{i=1}^{n} X_i. \tag{12.16}$$

This is nothing more than estimating the mean of an IID sequence of random variables. From the development of Chapter 7, we know that this estimator is unbiased and that as $n \to \infty$ the estimate converges (almost everywhere via the strong law of large numbers) to the true probability.

In practice, we do not have the patience to run our simulation an infinite number of times nor do we need to. At some point the accuracy of our estimate should be "good enough," but how many trials is enough? Some very concrete answers to this question can be obtained using the theory developed in Chapter 7. If the event A is fairly probable, then it will not take too many trials to get a good estimate of the probability, in which case runtime of the simulation is not really too much of an issue. However, if the event A is rare, then we will need to run many trials to get a good estimate of p_A. In the case where n gets large, we want to be sure not to make it any larger than necessary so that our simulation runtimes do not get excessive. Thus, the question of how many trials to run becomes important when simulating rare events.

Assuming n is large, the random variable $\hat{p}_A$ can be approximated as a Gaussian random variable via the central limit theorem. The mean and variance are $E[\hat{p}_A] = p_A$ and $\sigma_{\hat{p}_A}^2 = n^{-1} p_A (1 - p_A)$, respectively. One can then set up a confidence interval based on the desired accuracy. For example, suppose we wish to obtain an estimate that is within 1 percent of the true value with 90 percent probability. That is, we want to run enough trials to insure that

$$\Pr\left(|\hat{p}_A - p_A| < 0.01 p_A\right) = 0.9 = 1 - \alpha \tag{12.17}$$

From the results of Chapter 7, Section 7.5, we get

$$\varepsilon_{0.1} = 0.01 p_A = \frac{\sigma_X}{\sqrt{n}} c_{0.1} = \sqrt{\frac{p_A(1 - p_A)}{n}} c_{0.1}, \tag{12.18}$$

where the value of $c_{0.1}$ is taken from Table 7.1 as $c_{0.1} = 1.64$. Solving for n gives us an answer for how long to run the simulation:

$$n = \frac{(100 c_{0.1})^2 (1 - p_A)}{p_A} = \frac{(164)^2 (1 - p_A)}{p_A} \approx \frac{(164)^2}{p_A}. \tag{12.19}$$

Or in general, if we want the estimate, $\hat{p}_A$, to be within β percent of the true value (i.e., $|\hat{p}_A - p_A| < \beta p_A / 100$) with probability $1 - \alpha$, then the number of trials in the simulation should be chosen according to

$$n = \frac{\left(\frac{100c_\alpha}{\beta}\right)^2 (1 - p_A)}{p_A} \approx \frac{\left(\frac{100c_\alpha}{\beta}\right)^2}{p_A}. \tag{12.20}$$

This result is somewhat unfortunate because in order to know how long to run the simulation, we have to know the value of the probability we're trying to estimate in the first place. In practice, we may have a crude idea of what to expect for p_A, which we could then use to guide us in selecting the number of trials in the simulation. However, we can use this result to give us very specific guidance in how to choose the number of trials to run, even when p_A is completely unknown to us. Define the random variable N_A to be the number of occurrences of the event A in n trials; that is,

$$N_A = \sum_{i=1}^{n} X_i. \tag{12.21}$$

Note that $E[N_A] = np_A$. That is, the quantity np_A can be interpreted as the expected number of occurrences of the event A in n trials. Multiplying both sides of Equation 12.20 by p_A then produces

$$E[N_A] = \left(\frac{100c_\alpha}{\beta}\right)^2 (1 - p_A) \approx \left(\frac{100c_\alpha}{\beta}\right)^2. \tag{12.22}$$

Hence, one possible procedure to determine how many trials to run is to repeat the experiment for a random number of trials until the event A occurs some fixed number of times as specified by Equation 12.22. Let M_k be the random variable that represents the trial number of the kth occurrence of the event A. Then, one could form an estimate of p_A according to

$$\hat{p}_A = \frac{k}{M_k}. \tag{12.23}$$

It turns out that this produces a biased estimate; however a slightly modified form,

$$\hat{p}_A = \frac{k - 1}{M_k - 1}, \tag{12.24}$$

produces an unbiased estimate (see Exercise 7.12).

EXAMPLE 12.7: Suppose we wish to estimate the probability of an event that we expect to be roughly on the order of $p \sim 10^{-4}$.

Assuming we want 1 percent accuracy with a 90 percent confidence level, the number of trials needed will be

$$n = \frac{1}{p}\left(\frac{100c_\alpha}{\beta}\right)^2 = 10^4 \left(\frac{100 * 1.64}{1}\right)^2 = 268,960,000.$$

Alternatively, we need to repeat the simulation experiment until we observe the event

$$N_A = \left(\frac{100c_\alpha}{\beta}\right)^2 = 26,896 \text{ times.}$$

Assuming we do not have enough time available to repeat our simulation over 1/4 of a billion times, we would have to accept less accuracy. Suppose that due to time limitations we decide that we can only repeat our experiment 1 million times, then we can be sure that with 90 percent confidence, the estimate will be within the interval $(p - \varepsilon, p + \varepsilon)$, if ε is chosen according to

$$\varepsilon = \frac{\sqrt{p(1-p)}}{\sqrt{n}}c_\alpha \approx \frac{\sqrt{p}}{\sqrt{n}}c_\alpha = 1.64\frac{10^{-2}}{10^3} = 1.64 \times 10^{-5} = 0.164p.$$

With one million trials, we can only be 90 percent sure that the estimate is within 16.4 percent of the true value.

The preceding example demonstrates that using the Monte Carlo approach to simulating rare events can be very time consuming in that we may need to repeat our simulation experiments many times to get a high degree of accuracy. If our simulations are complicated, this may put a serious strain on our computational resources. The next section presents a novel technique, which when applied intelligently can substantially reduce the number of trials we may need to run.

12.3.2 Importance Sampling

The general idea behind importance sampling is fairly simple. In the Monte Carlo approach, we spend a large amount of time with many repetitions of an experiment while we are waiting for an event to occur that may happen only very rarely. In importance sampling, we skew the distribution of the underlying randomness in our experiment so that the "important" events happen more frequently. We can then use analytical tools to convert our distorted simulation results into an unbiased estimate of the probability of the event in which we are interested. To help present this technique, we first generalize the problem treated in Section 12.3.1. Suppose the simulation experiment consisted of creating a sequence of random variables, $X = (X_1, X_2, \ldots, X_m)$ according to some density, $f_X(x)$, and then observing whether or not some event A occurred which could be defined in terms of the X_i. For example, suppose X_i represents the number of messages that arrive at a certain node in a computer communications network at time instant i. Furthermore, suppose that it is known that the node requires a fixed amount of time to forward each message along to its intended destination and that the node has some finite buffer capacity for storing messages. The event A might represent the event that the node's buffer overflows, and hence a message is lost during the time interval $i = 1, 2, \ldots, m$. Ideally, the communications network has been designed so that this event is fairly uncommon, but it is desirable to quantify how often this overflow will occur. While it may be fairly straightforward to determine whether or not a buffer overflow has occurred given a specific sequence of arrivals, $X = x$, determining analytically the probability of buffer overflow may be difficult, so we decide to approach this via simulation. Let $\eta_A(X)$ be an indicator function for the event A. That is, let $\eta_A(x) = 1$ if x is such that the event A occurs and $\eta_A(x) = 0$ otherwise. Also, let $x^{(i)}$ be the realization of the random vector X that occurs on the ith trial of the simulation experiment. Then the Monte Carlo approach to estimating p_A is

$$\hat{p}_{A,MC} = \frac{1}{n} \sum_{i=1}^{n} \eta_A(x^{(i)}). \tag{12.25}$$

Now suppose instead that we generate a sequence of random variables, $Y = (Y_1, Y_2, \ldots, Y_m)$ according to a different distribution $f_Y(y)$, and form the estimate

$$\hat{p}_{A,IS} = \frac{1}{n} \sum_{i=1}^{n} \frac{f_X(y^{(i)})}{f_Y(y^{(i)})} \eta_A(y^{(i)}). \tag{12.26}$$

It is pretty straightforward (see Exercise 12.11) to establish that this estimator is also unbiased. By carefully choosing the density function, $f_Y(y)$, we may be able to drastically speed up the convergence of the series in Equation 12.26 relative to that in Equation 12.25.

The important step here is to decide how to choose the distribution of Y. In general, the idea is to choose a distribution of Y so that the $\{\eta_A(Y) = 1\}$ occurs more frequently than $\{\eta_A(X) = 1\}$. In other words, we want to choose a distribution so that the "important" event is sampled more often. It is common to employ the so-called twisted distribution, which calls on concepts taken from large deviation theory. But using these techniques are beyond the scope of this book. Instead, we take an admittedly ad hoc approach here and on a case-by-case basis we try to find a good (but not necessarily optimal) distribution. An example of using importance sampling is provided in the following application section.

12.4 Engineering Application: Simulation of a Coded Digital Communication System

In this section, we demonstrate use of the importance sampling technique outlined in the previous section in the simulation of a digital communications system with convolutional coding. A basic block diagram of the system is illustrated in Figure 12.12. A source outputs binary data, X_i, which is input to an encoder that adds redundancy to the data stream for the purposes of error protection. For this particular example, an encoder with a code rate of $1/2$ is used. Simply put, this means that for every one bit input, the convolutional encoder outputs two coded bits, $\{Y_{2i-1}, Y_{2i}\}$. To keep this example simple, the channel is modeled as one which randomly inverts bits with some probability p in a memoryless fashion. That is, what happens to one bit on the channel is independent of any of the other bits. Given a vector of bits input to the channel, $\mathbf{Y} = (Y_1, Y_2, \ldots, Y_n)$, the probability of observing a certain output of the channel $\mathbf{R} = (R_1, R_2, \ldots, R_n)$ is given by

$$\Pr(\mathbf{R}|\mathbf{Y}) = \prod_{i=1}^{n} \Pr\left(R_i|Y_i\right), \tag{12.27}$$

where

$$\Pr\left(R_i|Y_i\right) = \begin{cases} 1-p & \text{if } R_i = Y_i \\ p & \text{if } R_i = \overline{Y}_i \end{cases} . \tag{12.28}$$

The decoder then takes the received sequence output from the channel and determines what was the most likely data sequence.

For this example, it is not necessary to understand the workings of the encoder and decoder. We will just assume the existence of some computer subroutines that simulate their functions. Each trial of our simulation experiment will consist of randomly generating an IID sequence of equally likely data bits, passing them through the encoder, randomly corrupting some of the encoded bits according to the channel model we've developed, and then decoding the received sequence.

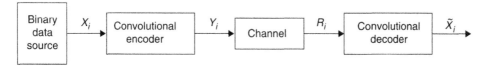

Figure 12.12 Block diagram of a digital communications system.

The decoded bits are then compared with the original bits to measure the decoded bit error probability. For the purposes of this example, it is assumed that data is transmitted in blocks of 50 information bits (which are encoded into blocks of 100 coded bits). We refer to each of these blocks as a frame. The channel is simulated by creating an error pattern $E = (E_1, E_2, \ldots, E_{100})$, where the E_i are a sequence of IID random variables with $\Pr(E_i = 1) = p$ and $\Pr(E_i = 0) = 1 - p$. Then

$$R = Y \oplus E. \tag{12.29}$$

The event $\{E_i = 1\}$ implies that the ith bit is inverted in the process of going through the channel, while $\{E_i = 0\}$ means the ith bit is received correctly.

Using the standard Monte Carlo approach, the decoded bit error rate is estimated according to

$$\hat{P}_{e,MC} = \frac{1}{mn} \sum_{j=1}^{m} \eta(x^{(j)}, \hat{x}^{(j)}, e), \tag{12.30}$$

where m is the number of packets transmitted in the simulation; n is the number of bits per packet; and the function $\eta(x^{(j)}, \hat{x}^{(j)}, e)$ counts the number of bit errors that occurred in the jth packet. If the channel error rate, p, is fairly high (e.g., a noisy channel), then the Monte Carlo approach will work quite nicely. However, if $p \ll 1$, then channel errors will be infrequent and the decoder will usually correct them. Hence, the decoded error probability will be very small and the Monte Carlo approach will require us to simulate an astronomical number of packets.

Alternatively, consider a simple importance sampling approach. To avoid simulating endless packets that ultimately end up error-free, we now create IID error patterns with $\Pr(E_i = 1) = q$, where q is some suitably chosen value larger than p. Note that any pattern $e = (e_1, e_2, \ldots, e_{100})$ that contains exactly w ones and $100 - w$ zeros will be generated with probability $\Pr(e) = q^w(1 - q)^{100-w}$. Let $w(e)$ be the number of ones in a particular error pattern, e. Then, our importance sampling estimate of the decoded error probability will be

$$\hat{P}_{e,IS} = \frac{1}{mn} \sum_{j=1}^{m} \frac{p^{w(e)}(1 - p)^{100-w(e)}}{q^{w(e)}(1 - q)^{100-w(e)}} \eta(x^{(j)}, \hat{x}^{(j)}, e). \tag{12.31}$$

Simulation results are shown in Figure 12.13, where the channel bit error probability is $p = 0.01$. Note that there are theoretical bounds that tell us that for this example, the actual probability of decoded bit error should be bounded by $1.97 \times 10^{-5} \le P_e \le 6.84 \times 10^{-5}$. To get fairly accurate results via the Monte Carlo approach, we would expect to have to simulate on the order of several hundred thousand packets. It is seen in Figure 12.13 that, indeed, even after simulating 10,000 packets, the estimated error probability has still not converged well. For the

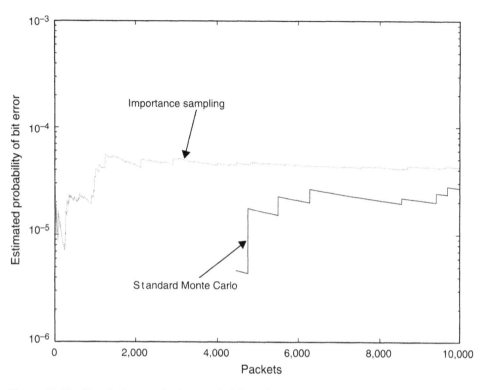

Figure 12.13 Simulation results for a coded digital communications system using standard Monte Carlo and importance sampling techniques.

importance sampling results, we used $q = 0.05$ so that important error events occurred much more frequently. As seen in the figure, the estimated error probability has converged rather nicely after simulating only a few thousand packets. Hence, for this example, using the importance sampling method has sped up the simulation time by a few orders of magnitude.

Exercises

12.1 Consider the periodic sequence generated by the four-stage shift register in Figure 12.1. Suppose the sequence is converted to ± 1-valued sequence by mapping all 1s to -1s and all 0s to 1s. One period of the resulting sequence is

$$-1, -1, -1, -1, 1, 1, 1, -1, 1, 1, -1, -1, 1, -1, 1.$$

Calculate the autocorrelation function of this periodic sequence. *Hint:* Do not just treat this sequence as having finite length. The sequence is infinite in both directions. The finite sequence shown is just one period of the infinite periodic sequence.

12.2 Sketch the shift register described by the octal number 75. Find the sequence output by this shift register assuming that the shift register is initially loaded with all ones.

12.3 A certain N-stage shift register has its tap connections configured so that it produces an m-sequence. The shift register is initially loaded with the contents $c = (c_0, c_1, \ldots, c_{N-1})$, resulting in a periodic sequence of period $2^N - 1$. Prove that if the shift register is loaded with some other contents $c' \neq 0$, the new output sequence will be a cyclic shift of the original output sequence. *Hint:* Recall that for an m-sequence, the shift register must cycle through all nonzero states.

12.4 Suppose we create a binary ± 1 valued sequence of length N by drawing N independent realizations of a Bernoulli random variable to form one period of the sequence. Compute the autocorrelation function of this random Bernoulli sequence.

12.5 Suppose a pseudorandom sequence is constructed using the power residue method as described by

$$x_k = a x_{k-1} \bmod q, \quad k = 1, 2, 3, \ldots.$$

Find the period and the sequence that results for the following values of (a, q). For each case, assume the seed is $x_0 = 1$.

(a) $a = 4, \quad q = 9,$

(b) $a = 5, \quad q = 9,$

(c) $a = 2, \quad q = 5,$

(d) $a = 5, \quad q = 11.$

12.6 Suppose a pseudorandom sequence is constructed using the power residue method as discussed in Exercise 12.5. If $q = 11$, find a value of a that leads to a sequence with maximum possible period.

12.7 Find a transformation that will change a uniform random variable into each of the following distributions (see Appendix D, Summary of

Common Random Variables, for the definitions of these distributions if necessary):

(a) arcsine,

(b) Cauchy,

(c) Rayleigh,

(d) geometric.

12.8 Suppose we wish to generate a 10-msec realization of a zero-mean Gaussian random process with a PSD of

$$S(f) = \frac{1}{1 + (f/f_3)^2}.$$

(a) Find the bandwidth that contains 99 percent of the total power in the random process.

(b) Determine how many frequency samples are needed for the frequency domain method described in Section 12.2A.

12.9 Suppose we wanted to generate the random process whose PSD is given in Exercise 12.8 using the time domain method discussed in Section 12.2B.

(a) Find the transfer function of the analog filter that will produce the desired output PSD when the input is a zero-mean, white Gaussian random process.

(b) Use a bilinear transformation to convert the analog filter to a digital filter.

(c) What is the approximate duration of the impulse response of the digital filter if the 3-dB frequency of the random process is $f_3 = 1$ kHz?

12.10 Suppose we use a Monte Carlo simulation approach to simulate the probability of some rare event A. It is decided that we will repeat the simulation until the event A occurs 35 times. With what accuracy will we estimate p_A to within a 90 percent confidence level?

12.11 Prove that the importance sampling (IS) estimator of Equation 12.26 is unbiased. That is, show that

$$E[\hat{p}_{A, IS}] = p_A.$$

MATLAB Exercises

12.12 We wish to generate a periodic sequence of numbers that cycles through the integers from 1 to 100 in a pseudorandom fashion. Choose a pair of integers (a, q) that can be used in the power residue method to produce a sequence of the desired form. Write a MATLAB program to verify that the sequence produced does in fact cycle through each of the integers 1 to 100 exactly once each before the sequence repeats.

12.13 Let $X \sim N(2, 1)$, $Y \sim N(0, 1)$, $Z \sim N(0, 1)$, and $W = \sqrt{X^2 + Y^2 + Z^2}$. We desire to find $\Pr(W > 3)$. Write a MATLAB program to estimate this probability through Monte Carlo simulation techniques. If we want to be 90 percent sure that our estimate is within 5 percent of the true value, about how many times should we observe the event $\{W > 3\}$? Run your program and provide the estimate of the desired probability. Can you find the probability analytically?

12.14 Write a MATLAB program to generate a realization of the random process from Exercise 12.8 (using the frequency domain method). Use a periodogram to estimate the PSD of the process using the realization of the process you generated. Does your PSD estimate agree with the PSD that the process is designed to possess?

12.15 Write a MATLAB program to generate a realization of the random process from Exercise 12.9 (using the time domain method). Use a periodogram to estimate the PSD of the process using the realization of the process you generated. Does your PSD estimate agree with the PSD that the process is designed to possess?

Review of Set Theory

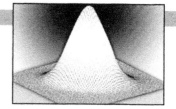

The purpose for reviewing set theory is to provide a mathematical structure for organizing methods of counting and grouping objects. Set theory may be used to define the probabilities of possible outcomes of experiments. There are two common methods for defining a set. The first method, known as the *roster method*, is to list the elements of a set. Synonyms for sets include class, aggregate, and collection. We will denote sets by capital letters, A, B, C, and so forth. The elements of a set will be indicated by lowercase letters, such as a, b, c, and so forth. If a is an element (or object, or member, or point) of A, then we denote this as $a \in A$. If a is not an element of A, the notation is $a \in A$. A second way of defining a set is called the *property method*, which describes some property held by all elements of the set but not held by objects that do not belong to the set.

DEFINITION A.1: A set A is said to be a *subset* of another set, B, if all elements of A are also in B, in which case we write $A \subseteq B$. With this definition, it is possible that the two sets are equal (i.e., they have all the same elements), in which case $A \subseteq B$ and at the same time $B \subseteq A$. If on the other hand, A is a subset of B and there are some elements of B that are not in A, then we say that A is a *proper subset* of B and we write $A \subset B$.

DEFINITION A.2: The *universal set*, S, is the set of all objects under consideration in a given problem, while the *empty set*, $\emptyset$, is the set that contains no elements.

DEFINITION A.3: The *complement* of a set A, written $\overline{A}$, is the set of all elements in S that are not in A. For two sets A and B that satisfy $A \subset B$, the *difference set*, written $B - A$, is the set of elements in B that are not in A.

Note that for any set A, $\emptyset \subseteq A \subseteq S$ and $A \subseteq A$. Also, if $A \subseteq B$ and $B \subseteq C$, then $A \subseteq C$. Finally, we also note the relationship $\overline{S} = \emptyset$.

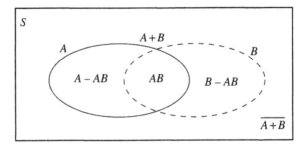

Figure A.1   A Venn diagram illustrating some of the concepts of sets.

DEFINITION A.4: For any two sets A and B the *union* of the two sets, $A \cup B$, is the set of all elements that are contained in either A or B, and the *intersection* of the two sets, $A \cap B$, is the set of all elements that are contained in both A and B. In the algebra of sets, the union operation plays the role of addition and so sometimes the notation $A + B$ is used, while the intersection operation plays the role of multiplication and hence the alternative notations $A \bullet B$ or AB are common.

Some of the concepts just presented are illustrated using a Venn Diagram in Figure A.1. The set A is contained within the thick solid line; the set B is within the dashed line; the set $A + B$ is the set of points inside either line, and the set AB is the set of points inside both. The set $A - AB$ is the set of points inside the solid line but not inside the dashed line, while the set $B - AB$ is the set of points inside the dashed line but not inside the solid line. The set $\overline{A + B}$ is the set of all points outside of both lines.

DEFINITION A.5: Two sets A and B are said to be *mutually exclusive*, or *disjoint*, if and only if they have no common elements, in which case $A \cap B = \emptyset$. A collection of sets $A_1, A_2, \ldots, A_n$ are said to be *exhaustive* if each element in the universal set is contained in at least one of the sets in the collection. For exhaustive sets, $A_1 \cup A_2 \cup \cdots \cup A_n = S$.

The following laws are consequences of the definitions we have just introduced. The reader should verify these laws to gain familiarity with the algebra of sets.

- Idempotent: $A \cup A = A$, $A \cap A = A$ for all sets A.
- Commutative: $A \cup B = B \cup A$, $A \cap B = B \cap A$ for all sets A and B.
- Associative: $A \cup (B \cup C) = (A \cup B) \cup C = A \cup B \cup C$, $A \cap (B \cap C) = (A \cap B) \cap C = A \cap B \cap C$ for all sets A, B, and C.
- Distributive: $A \cap (B \cup C) = (A \cap B) \cup (A \cap C)$, $A \cup (B \cap C) = (A \cup B) \cap (A \cup C)$ for all sets A, B, and C.

- Consistency: The three conditions $A \subseteq B$, $A \cap B = A$, and $A \cup B = B$ are all consistent or mutually equivalent.
- Universal bounds: $\varnothing \subseteq A \subseteq S$ for all sets A.
- Product: $\varnothing \cap A = \varnothing$, $S \cap A = A$ for all sets A.
- Sum: $\varnothing \cup A = A$, $S \cup A = S$ for all sets A.
- Involution: $\overline{(\overline{A})} = A$ for all sets A.
- Complementarity: $A \cup \overline{A} = S$, $A \cap \overline{A} = \varnothing$ for all sets A.
- DeMorgan's first law: $\overline{A \cup B} = \overline{A} \cap \overline{B}$ for all sets A and B.
- DeMorgan's second law: $\overline{A \cap B} = \overline{A} \cup \overline{B}$ for all sets A and B.

DeMorgan's laws can be stated as follows: To find the complement of an expression, replace each set by its complement and interchange additions with multiplications and multiplications with additions.

Review of Linear Algebra

The study of probability and random processes draws heavily upon concepts and results from elementary linear algebra. In the main text, we assume that the reader has a working knowledge of undergraduate-level linear algebra. The aim of this appendix is to provide a review for those who need to brush up on these concepts. This review is not intended to be an exhaustive treatment of the topic, but rather is a summary of selected concepts that are used within the text.

DEFINITION B.1: A matrix is a two dimensional array of numbers. We say that a matrix has size $m \times n$ if the array has m rows and n columns. If the matrix has only one row, then we refer to it as a *row vector*, while if there is only one column it is a *column vector*. A matrix with one row and one column is called a *scalar*. The elements of a matrix, B, are written as $b_{i,j}$, where the first subscript represents the row number and the second subscript gives the column number.

DEFINITION B.2: The *transpose* of a matrix is obtained by exchanging the row and column number of each element. That is, if matrix B has elements $b_{i,j}$, then the element in the ith row and jth column of B^T is $b_{j,i}$. Hence, if a matrix has size $m \times n$, then its transpose has size $n \times m$. Also, the transpose of a column vector is a row vector and the transpose of a row vector is a column vector. A matrix B is said to be *symmetric* if $B^T = B$.

DEFINITION B.3: The Hermitian transpose (or just Hermitian) of a matrix B is written B^H and is formed by taking the complex conjugate of the transpose. That is, if matrix B has elements $b_{i,j}$, then the element in the ith row and jth column of B^H is $b_{j,i}^*$. A matrix B is said to be *Hermitian symmetric* if $B^H = B$. Sometimes such a matrix is simply called a *Hermitian* matrix.

DEFINITION B.4: Addition and multiplication of matrices is defined as follows. If two matrices A and B have the same size (i.e., the same number of rows and columns), then their sum is defined as $C = A + B$ where $c_{i,j} = a_{i,j} + b_{i,j}$. That is, matrix addition is done on an element by element basis. If the two matrices do not have the same dimensions, they cannot be added. If A has size $m \times k$ and B has size $k \times n$, then their product, $C = AB$, will be an $m \times n$ matrix whose elements are given by

$$c_{i,j} = \sum_{l=1}^{k} a_{i,l} b_{l,j}. \tag{B.1}$$

In order for the matrix product AB to be defined, the number of columns in A must equal the number of rows in B.

It is also common to define two different products involving vectors, the so-called scalar (or dot) product and the matrix (or cross) product. We have no occasion to use the cross product in this text and so we do not consider it here. For two column vectors a and b (both with the same number of elements), the dot product is defined as $a \bullet b = a^H b$, where the standard definition of matrix multiplication as it applies to vectors is used. Two vectors are said to be *orthogonal* if their dot product is zero. Finally, the norm of a vector is $||b|| = \sqrt{b^H b}$.

With these definitions in place, the reader should be able to verify the following properties of matrix arithmetic.

- Commutative: $A + B = B + A$ for matrices for which the addition is defined. However, the same property does not usually hold for multiplication. That is, AB does not necessarily equal BA. In fact, BA may not even be defined.
- Associative: $A + (B + C) = (A + B) + C$ and $A(BC) = (AB)C$.
- Distributive: $A(B + C) = AB + AC$.
- Transposes of sums: $(A + B)^T = A^T + B^T$ and $(A + B)^H = A^H + B^H$.
- Transposes of products: $(AB)^T = B^T A^T$ and $(AB)^H = B^H A^H$.

In much of this text, many of the matrices we deal with are square. The following definition identifies some characteristics that can be applied to square matrices.

DEFINITION B.5: A matrix B is *diagonal* if its elements satisfy $b_{i,j} = 0$ for all $i \neq j$. A matrix is *upper triangular* if $b_{i,j} = 0$ for all $i > j$ and *lower triangular* if $b_{i,j} = 0$ for all $i < j$. Note that a diagonal matrix is simultaneously upper and lower triangular. Finally, a matrix is an $m \times m$ *identity* matrix if it is a diagonal matrix whose diagonal entries are all equal to 1. The letter I is reserved to represent an identity matrix.

An identity matrix has the form

$$I = \begin{bmatrix} 1 & 0 & 0 & \cdots & 0 \\ 0 & 1 & 0 & \cdots & 0 \\ 0 & 0 & 1 & \cdots & 0 \\ \cdots & \cdots & \cdots & & \cdots \\ 0 & 0 & 0 & \cdots & 1 \end{bmatrix}. \tag{B.2}$$

Sometimes we use a subscript to indicate the dimensions of the identity matrix. For example, $I_{m \times m}$ would represent an identity matrix with m rows and columns. Note that the identity matrix is the identity with respect to matrix multiplication. That is, for any $m \times n$ matrix B, $I_{m \times m}B = BI_{n \times n} = B$. The identity with respect to matrix addition would be a matrix of all zeros.

DEFINITION B.6: The *inverse* of a square matrix B, written B^{-1} (if it exists), is a matrix that satisfies $BB^{-1} = B^{-1}B = I$. If the matrix B is not square, then it may have a left inverse that satisfies $B^{-1}B = I$ and a different right inverse that satisfies $BB^{-1} = I$. In fact, for nonsquare matrices, the left inverse and right inverse will not have the same dimensions.

The inverse of a square matrix need not exist, but if it does, it is unique and the left and right inverses are identical. If the inverse of a matrix does not exist, then the matrix is said to be *singular*, while if the inverse exists, the matrix is *nonsingular*, or *invertible*.

The inverse of a matrix plays an important role in the solutions of simultaneous linear equations. Consider the following system of n linear equations in n unknowns, $x_1, x_2, \ldots, x_n$:

$$a_{1,1}x_1 + a_{1,2}x_2 + \cdots + a_{1,n}x_n = b_1,$$

$$a_{2,1}x_1 + a_{2,2}x_2 + \cdots + a_{2,n}x_n = b_2,$$

$$\cdots$$

$$a_{n,1}x_1 + a_{n,2}x_2 + \cdots + a_{n,n}x_n = b_n. \tag{B.3}$$

By defining A as the $n \times n$ matrix of coefficients whose elements are $a_{i,j}$ and the column vectors $x = (x_1, x_2, \ldots, x_n)^T$ and $b = (b_1, b_2, \ldots, b_n)^T$, this system of equations can be written in matrix forms as

$$Ax = b. \tag{B.4}$$

Multiplying both sides by the inverse of A leads us to the solution

$$x = A^{-1}b. \tag{B.5}$$

Hence, if the coefficient matrix is invertible, the system has a unique solution. On the other hand, if the coefficient matrix is singular, then A^{-1} does not exist and the set of equations does not have a unique solution. This would be the case if some of the equations in Equation B.3 were linearly dependent (redundant), in which case the system would have more than one solution, or if some of the equations were inconsistent, which would lead to no solution. In either case of redundant or inconsistent equations, the rows of the A matrix will be linearly dependent and the inverse will not exist. Conversely, if the rows of A are linearly independent, the matrix will be invertible.

A few properties of matrices that are related to the inverse are

- Inverse of transposes: $(A^T)^{-1} = (A^{-1})^T$.
- Inverse of Hermitian transposes: $(A^H)^{-1} = (A^{-1})^H$.
- Inverse of products: $(AB)^{-1} = B^{-1}A^{-1}$.
- Inverse of identities: $I^{-1} = I$.

A single quantity that is extremely useful in characterizing the invertibility of a matrix is its determinant. The determinant can be rather difficult to define in a simple manner, but it has many useful properties. We use a recursive definition, which may seem rather cryptic, but is probably the simplest definition and is also consistent with how determinants are often calculated.

DEFINITION B.7: Let B be an $n \times n$ matrix with elements $b_{i,j}$. Define $B_{(i,j)}$ to be the $(n-1) \times (n-1)$ matrix obtained by removing the ith row and jth column from B. Then the determinant of B is defined recursively according to

$$\det(B) = \sum_{j=1}^{n}(-1)^{i+j}\det(B_{(i,j)}), \quad \text{for any } i = 1, 2, \ldots, n. \tag{B.6}$$

This recursion, together with the definition that for a 1×1 matrix $B = [b]$, $\det(B) = b$, is sufficient to calculate the determinant of any $n \times n$ matrix.

To see how this works, consider a 2×2 matrix:

$$\det\left(\begin{bmatrix} a & b \\ c & d \end{bmatrix}\right) = a\det([d]) - b\det([c]) = ad - bc. \tag{B.7}$$

This result was obtained using $i = 1$ in Equation B.6. We could also have used $i = 2$ and achieved the same result:

$$\det\left(\begin{bmatrix} a & b \\ c & d \end{bmatrix}\right) = -c\det([b]) + d\det([a]) = -cb + da. \tag{B.8}$$

For a general 3×3 matrix, the determinant works out to be

$$\det\left(\begin{bmatrix} a & b & c \\ d & e & f \\ g & h & i \end{bmatrix}\right) = a\det\left(\begin{bmatrix} e & f \\ h & i \end{bmatrix}\right) - b\det\left(\begin{bmatrix} d & f \\ g & i \end{bmatrix}\right) + c\det\left(\begin{bmatrix} d & e \\ g & h \end{bmatrix}\right)$$

$$= a(ei - fh) - b(di - fg) + c(dh - eg). \tag{B.9}$$

Probably the most important property of determinants is that if $\det(B) = 0$, then the matrix B is singular, and conversely if $\det(B) \neq 0$, then B is invertible. This, along with some other important properties, is listed next. We will not prove any of these properties in this review.

- Invertibility: $\{\det(B) = 0\} \Longleftrightarrow \{B \text{ is singular}\}$.
- Row exchange: If the matrix A is formed by exchanging any two rows in the matrix B, then $\det(A) = -\det(B)$.
- Identity matrices: For any identity matrix, $\det(I) = 1$.
- Triangular matrices: If B is a triangular matrix,

$$\det(B) = \prod_{i=1}^{n} b_{i,i}.$$

That is, the determinant is the product of the diagonal elements. Note that diagonal matrices are a special case and this property applies to them as well.
- Products of matrices: $\det(AB) = \det(A)\det(B)$.
- Inverses: $\det(B^{-1}) = 1/\det(B)$.
- Transposes: $\det(B^T) = \det(B)$.
- Hermitian transposes: $\det(B^H) = (\det(B))^*$.

In addition to computing determinants of matrices, we will also find need throughout the text to compute eigenvalues and eigenvectors of square matrices.

DEFINITION B.8: For a square matrix B, the scalar λ is an eigenvalue and the vector x is a corresponding eigenvector if

$$Bx = \lambda x. \tag{B.10}$$

Note that the previous equation can be written in the slightly different form $(B - \lambda I)x = 0$. The eigenvector x will be nonzero only if the matrix $B - \lambda I$ is singular. If it were nonsingular, we could multiply both sides by its inverse to obtain the trivial solution $x = 0$. Hence, the eigenvalues of the matrix B must be solutions to the equation

$$\det(B - \lambda I) = 0. \tag{B.11}$$

This is the so-called characteristic equation for the matrix B. For an $n \times n$ matrix, B, the characteristic equation will be an nth order polynomial equation in λ, and hence an $n \times n$ matrix will have n eigenvalues (although some of them may be repeated). Corresponding to each eigenvalue, λ_k, is an eigenvector, x_k. Note that the eigenvector is not unique since if x_k satisfies Equation B.10, then any multiple of x_k will also satisfy the same equation and hence will also be an eigenvector. In order to resolve this ambiguity, it is common to normalize the eigenvectors so that $||x||^2 = 1$, but the vector is still an eigenvector even if it is not normalized. In the case of repeated eigenvalues, there may also be corresponding eigenvectors that differ by more than just a scale constant.

Before listing the important properties of eigenvalues and eigenvectors, it is necessary to include one more definition.

DEFINITION B.9: A matrix B is *positive definite* if $z^H B z > 0$ for any vector z and it is *negative definite* if $z^H B z < 0$ for all z. If $z^H B z \geq 0$ then the matrix is referred to as *positive semi-definite*, and if $z^H B z \leq 0$ the matrix is *negative semi-definite*.

With this definition in place, we now list the following properties of eigenvalues and eigenvectors.

- Trace of a matrix: $\text{trace}(B) = \sum_{k=1}^{n} b_{k,k} = \sum_{k=1}^{n} \lambda_k$. That is, the sum of the eigenvalues is equal to the sum of the diagonal elements of a matrix, also known as its *trace*.
- Determinant of a matrix:

$$\det(B) = \prod_{k=1}^{n} \lambda_k$$

That is, the product of the eigenvalues is the determinant of a matrix. As a result, any singular matrix must have at least one eigenvalue that is zero.
- Triangular matrices: If a matrix B is triangular (or diagonal), the eigenvalues are just the diagonal entries, $\lambda_k = b_{k,k}$.
- Positive and negative definite matrices: If B is positive definite, then its eigenvalues are all real and positive; whereas, if B is positive semi-definite, all its eigenvalues are nonnegative. Similarly, if B is negative definite, then the eigenvalues are negative; whereas, if B is negative semi-definite, the eigenvalues are nonpositive.
- Linear independence: If the eigenvectors $x_1, x_2, \ldots, x_n$ are nonzero and correspond to distinct (not repeated) eigenvalues $\lambda_1, \lambda_2, \ldots, \lambda_n$, then the eigenvectors are linearly independent.
- Diagonal form: Suppose B is an $n \times n$ matrix and has n linearly independent eigenvectors. Construct a matrix S whose columns are the n eigenvectors and a

diagonal matrix Λ whose diagonal entries are the eigenvalues. Then the matrix B can be factored as $B = S\Lambda S^{-1}$.

- Powers of a matrix: A direct result of the previous property is that for matrices with linearly independent eigenvectors, $B^k = S\Lambda^k S^{-1}$. Furthermore, since Λ is diagonal, Λ^k is computed by raising each diagonal entry to the kth power.

In many of the applications encountered in this text, the matrices we are dealing with are Hermitian. These matrices posess additional properties that are not necessarily shared by all matrices, making Hermitian matrices particularly convenient to work with.

- Positive semi-definite: Any Hermitian matrix has all real eigenvalues and is at least positive semi-definite. Furthermore, if the matrix is also nonsingular, then it will be positive definite.
- Orthogonal eigenvectors: Eigenvectors of a Hermitian matrix that correspond to different eigenvalues are orthogonal.
- Spectral decomposition: Any Hermitian matrix can be decomposed into the form

$$B = U\Lambda U^H = \sum_{k=1}^{n} \lambda_k x_k x_k^H, \tag{B.12}$$

where U is the matrix whose columns are the eigenvectors (normalized).

Review of Signals and Systems

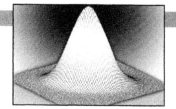

This appendix provides a summary of some important results in the area of signal representation and linear time–invariant systems. Any engineering student embarking on a serious study of probability and random processes should be familiar with these concepts and hence a rigorous development is not attempted here. Rather this review is intended as a brief refresher for those who need it and also as a quick reference for some important results that are used throughout the text. In this appendix, attention is focused on deterministic signals and systems in both continuous and discrete time.

DEFINITION C.1: Consider a periodic signal $x(t)$ whose period is T_o. That is $x(t) = x(t + T_o)$ for all t. The inverse of the period $f_o = 1/T_o$ is called the *fundamental frequency* of $x(t)$ and any frequency, $f_n = nf_o$, that is a multiple of the fundamental frequency is called a *harmonic*.

Any periodic signal (subject to some mild constraints known as the *Dirichlet conditions*) can be represented as a linear combination of complex exponential signals, $\exp(j2\pi f_n t)$, whose frequencies are at the harmonics of the signal. That is, if $x(t)$ is periodic with period T_o, then

$$x(t) = \sum_{k=-\infty}^{\infty} x_n e^{j2\pi nf_o t}. \tag{C.1}$$

This is known as the *Fourier series* expansion and the series coefficients can be computed according to

$$x_n = \frac{1}{T_o} \int_{T_o} x(t) e^{-j2\pi nf_o t} \, dt. \tag{C.2}$$

Since the signal is periodic, the integral in the previous expression can be taken over any convenient interval of length T_0. In general, the series coefficients are complex numbers and it is common to express them in terms of their magnitude and phase, $x_n = |x_n| \exp(j\angle x_n)$. The Fourier series coefficients display the frequency content of periodic signals.

For signals that are not periodic, the Fourier transform can be used to display the frequency content of a signal. The Fourier transform of a signal is given by

$$X(f) = \int_{-\infty}^{\infty} x(t)e^{-j2\pi ft}\, dt, \tag{C.3}$$

and the inverse Fourier transform is

$$x(t) = \int_{-\infty}^{\infty} X(f)e^{j2\pi ft}\, df. \tag{C.4}$$

Sometimes we use the notation $x(t) \leftrightarrow X(f)$ to indicate that $x(t)$ and $X(f)$ are a Fourier transform pair. A table of some common Fourier transform pairs is provided in Appendix E, Mathematical Tables, Table E.1. Some of the more important properties of Fourier transforms are listed next.

- Linearity: If $x(t) \leftrightarrow X(f)$ and $y(t) \leftrightarrow Y(f)$, then $ax(t) + by(t) \leftrightarrow aX(f) + bY(f)$ for any constants a and b.
- Symmetry: If $x(t)$ is real-valued, then $X(-f) = X^*(f)$. As a result, $|X(f)|$ must then be an even function of f and $\angle X(f)$ must be an odd function of f. In addition, if $x(t)$ is both real and even, then $X(f)$ will be real and even.
- Time shifting: If $x(t) \leftrightarrow X(f)$, then $x(t - t_0) \leftrightarrow e^{-j2\pi f t_0} X(f)$. As a consequence, shifting a signal in time does not alter the magnitude of its Fourier transform.
- Differentiation: If $x(t) \leftrightarrow X(f)$, then $\dfrac{dx(t)}{dt} \leftrightarrow j2\pi f X(f)$.
- Integration: If $x(t) \leftrightarrow X(f)$, then $\int_{-\infty}^{t} x(u)du \leftrightarrow \dfrac{X(f)}{j2\pi f} + \dfrac{1}{2}X(0)\delta(f)$. The term $X(0)$ that appears in this expression is the *direct current* (DC) value of the signal.
- Time and frequency scaling: If $x(t) \leftrightarrow X(f)$, then $x(at) = \dfrac{1}{|a|}X\left(\dfrac{f}{a}\right)$ for any constant $a \neq 0$.
- Parseval's relation: If $x(t) \leftrightarrow X(f)$, then $\int_{-\infty}^{\infty} |x(t)|^2\, dt = \int_{-\infty}^{\infty} |X(f)|^2\, df$. This is a statement of conservation of energy. That is, the energy in the time domain is equal to the energy in the frequency domain.
- Convolution: If $x(t) \leftrightarrow X(f)$ and $y(t) \leftrightarrow Y(f)$, then

$$x(t) * y(t) = \int_{-\infty}^{\infty} x(u)y(t - u)\, dt \leftrightarrow X(f)Y(f).$$

For signals in discrete time, $x[n]$, a *discrete-time Fourier transform* (DFT) is defined according to

$$X(f) = \sum_{n=-\infty}^{\infty} x[n]e^{-j2\pi fn}, \tag{C.5}$$

and the inverse DFT is

$$x[n] = \int_{-1/2}^{1/2} X(f)e^{j2\pi fn}\, df. \tag{C.6}$$

Since $X(f)$ is periodic with a period of 1, the integral in Equation C.5 can be taken over any interval of length 1. It is common to view the DFT using discrete frequency samples as well. In that case, the definition of the DFT and its inverse is modified to give the N-point DFT:

$$X[m] = \sum_{n=0}^{N-1} x[n]e^{-j2\pi mn/N}, \tag{C.7}$$

$$x[n] = \frac{1}{N} \sum_{m=0}^{N-1} X[m]e^{j2\pi mn/N}. \tag{C.8}$$

Alternatively, by replacing $\exp(j2\pi f)$ with z in the definition of the DFT, we get the z-transform:

$$X(z) = \sum_{n=-\infty}^{\infty} x[n]z^{-n}. \tag{C.9}$$

The inverse z-transform is given by a complex contour integral,

$$x[n] = \frac{1}{2\pi j} \oint X(z)z^{n-1}\, dz, \tag{C.10}$$

where the contour of integration is any closed contour that encircles the origin in the counterclockwise direction and is within the region of convergence of $X(z)$. Because of the complicated nature of the inverse transform, it is common to compute these inverse transforms via tables. A table of some common z-transform pairs is provided in Appendix E, Mathematical Tables, Table E.2.

These various transform representations of signals are particularly useful when studying the passage of signals through linear time–invariant (LTI) systems.

DEFINITION C.2: Suppose when $x(t)$ is input to a system, the output is $y(t)$. The system is said to be *time-invariant* if the input $x(t - t_0)$ produces an output of $y(t - t_0)$.

That is, a time delay in the input produces the same time delay on the output but no other changes to the output. Furthermore, suppose the two inputs $x_1(t)$ and $x_2(t)$ produce the two outputs $y_1(t)$ and $y_2(t)$, respectively. Then, the system is *linear* if the input $ax_1(t) + bx_2(t)$ produces the output $ay_1(t) + by_2(t)$ for any constants a and b. Identical definitions apply to discrete time systems as well.

A direct consequence of the linearity of a system is the concept of superposition, which states that if the input can be written as a linear combination of several terms $x(t) = a_1x_1(t) + a_2x_2(t) + \cdots + a_nx_n(t)$, then the corresponding output can be written as the same linear combination of the corresponding outputs $y(t) = a_1y_1(t) + a_2y_2(t) + \cdots + a_ny_n(t)$.

Any LTI system can be described in terms of its impulse response, $h(t)$. If the input is a delta (impulse) function, $\delta(t)$, the output is then the impulse response $y(t) = h(t)$. For any LTI system, the input/output relationship is given in terms of the impulse response according to the convolution integral

$$y(t) = x(t) * h(t) = \int_{-\infty}^{\infty} x(u)h(t-u)\,du = \int_{-\infty}^{\infty} x(t-u)h(u)\,du. \tag{C.11}$$

If the input is a complex exponential at some frequency f (i.e., $x(t) = \exp(j2\pi ft)$), then the corresponding output is then

$$y(t) = e^{j2\pi ft} \int_{-\infty}^{\infty} h(u)e^{-j2\pi fu}\,du = e^{j2\pi ft}H(f). \tag{C.12}$$

That is, the output will also be a complex exponential whose magnitude and phase have been adjusted according to the complex number $H(f)$. This function of frequency is called the *transfer function* of the system and is the Fourier transform of the impulse response. Since complex exponentials form eigenfunctions of any LTI system, when studying LTI systems, it is convenient to decompose signals into linear combinations of complex exponentials. If $x(t)$ is a periodic signal it can be written as a linear combination of complex exponentials through its Fourier series representation,

$$x(t) = \sum_k x_k e^{j2\pi kf_0 t}. \tag{C.13}$$

Using the concept of superposition together with the previous result, we find that the output of an LTI system when $x(t)$ is input is

$$y(t) = \sum_k x_k H(kf_0)e^{j2\pi kf_0 t} = \sum_k y_k e^{j2\pi kf_0 t}. \tag{C.14}$$

Hence, the Fourier series coefficients of the input and output of an LTI system are related by the simple form

$$y_k = x_k H(kf_0). \tag{C.15}$$

A similar relationship holds for the Fourier transforms of nonperiodic signals. Taking Fourier transforms of both sides of Equation C.11 and using the convolution property of Fourier transforms results in

$$Y(f) = X(f)H(f).$$ (C.16)

Identical relationships hold for the DFT and z-transforms as well.

Summary of Common Random Variables

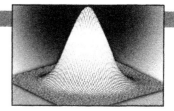

This appendix provides a quick reference of some of the most common random variables. Special functions that are used in this appendix are defined in the following list.

- Gamma function: $\Gamma(\alpha) = \int_0^\infty u^{\alpha-1} e^{-u}\, du,\ Re[\alpha] > 0$.
- Incomplete gamma function: $\gamma(\alpha, \beta) = \int_0^\beta u^{\alpha-1} e^{-u}\, du,\ Re[\alpha] > 0$.
- Beta function: $B(a, b) = \int_0^1 u^{a-1}(1-u)^{b-1}\, du = \dfrac{\Gamma(a)\Gamma(b)}{\Gamma(a+b)}$.
- Incomplete beta function: $\beta(a, b, x) = \int_0^x u^{a-1}(1-u)^{b-1}\, du,\ 0 < x < 1$.
- Modified Bessel function of order m: $I_m(x) = \dfrac{1}{2\pi} \int_0^{2\pi} e^{x\cos(\theta)} \cos(m\theta)\, d\theta$.
- Q-function: $Q(x) = \int_x^\infty \frac{1}{2\pi} \exp\left(-\frac{u^2}{2}\right) du$.
- Marcum's Q-function: $Q(\alpha, \beta) = \int_\beta^\infty u \exp\left(-\dfrac{\alpha^2 + u^2}{2}\right) I_0(\alpha u)\, du$.

D.1 Continuous Random Variables

D.1.1 Arcsine

For any $b > 0$,

$$f_X(x) = \frac{1}{\pi\sqrt{b^2 - x^2}} \quad -b < x < b. \tag{D.1}$$

$$F_X(x) = \begin{cases} 0 & x < -b \\ \dfrac{1}{2} + \dfrac{1}{\pi} \sin^{-1}\left(\dfrac{x}{b}\right) & -b \le x \le b \\ 1 & x > b \end{cases} \tag{D.2}$$

$$\mu_X = 0, \quad \sigma_X^2 = \frac{b^2}{2}. \tag{D.3}$$

Note:

(1) Formed by a transformation $X = b\cos(2\pi U + \theta)$, where b and θ are constants and U is a uniform random variable over $[0, 1)$.

D.I.2 Beta

For any $a > 0$ and $b > 0$,

$$f_X(x) = \frac{1}{B(a, b)} x^{a-1}(1 - x)^{b-1}, \quad 0 < x < 1. \tag{D.4}$$

$$F_X(x) = \begin{cases} 0 & x < 0 \\ \dfrac{\beta(a, b, x)}{B(a, b)} & 0 \le x \le 1 \\ 1 & x > 1 \end{cases} \tag{D.5}$$

$$\mu_X = \frac{a}{a + b}, \quad \sigma_X^2 = \frac{ab}{(a + b)^2(a + b + 1)}. \tag{D.6}$$

D.I.3 Cauchy

For any $b > 0$,

$$f_X(x) = \frac{b/\pi}{b^2 + x^2}. \tag{D.7}$$

$$F_X(x) = \frac{1}{2} + \frac{1}{\pi} \tan^{-1}\left(\frac{x}{b}\right). \tag{D.8}$$

$$\Phi_X(\omega) = e^{-b|\omega|}. \tag{D.9}$$

Notes:

(1) Both the mean and variance are undefined.
(2) Formed by a transformation of the form $X = b\tan(2\pi U)$, where U is uniform over $[0, 1)$.

D.1.4 Chi-Square

For integer $n > 0$,

$$f_X(x) = \frac{x^{n/2-1}}{2^{n/2}\Gamma(n/2)}e^{-x/2}, \quad x \geq 0. \tag{D.10}$$

$$F_X(x) = \begin{cases} 0 & x < 0 \\ \dfrac{\gamma(n/2, x/2)}{\Gamma(n/2)} & x \geq 0 \end{cases}. \tag{D.11}$$

$$\Phi_X(\omega) = \frac{1}{(1 - 2j\omega)^{n/2}}. \tag{D.12}$$

$$\mu_X = n, \quad \sigma_X^2 = 2n. \tag{D.13}$$

Notes:

(1) The chi-square random variable is a special case of the gamma random variable.
(2) The parameter n is referred to as the number of degrees of freedom of the chi-square random variable.
(3) The chi-square random variable is formed by a transformation of the form $X = \sum_{k=1}^{n} Z_k^2$, where the Z_k are independent and identically distributed (IID), zero-mean, unit variance Gaussian random variables.

D.1.5 Erlang

For any integer $n > 0$ and any $b > 0$,

$$f_X(x) = \frac{b^n x^{n-1} e^{-bx}}{(n-1)!}, \quad x \geq 0. \tag{D.14}$$

$$F_X(x) = \begin{cases} 0 & x < 0 \\ \dfrac{\gamma(n, bx)}{(n-1)!} & x \geq 0 \end{cases}. \tag{D.15}$$

$$\Phi_X(\omega) = \frac{1}{(1 - j\omega/b)^n}. \tag{D.16}$$

$$\mu_X = n/b, \quad \sigma_X^2 = n/b^2. \tag{D.17}$$

Notes:

(1) The Erlang random variable is a special case of the gamma random variable.
(2) The Erlang random variable is formed by summing n IID exponential random variables.
(3) The CDF can also be written as a finite series

$$\frac{\gamma(n, bx)}{(n-1)!} = 1 - e^{bx} \sum_{k=0}^{n-1} \frac{(bx)^k}{k!}, \quad x \geq 0. \tag{D.18}$$

D.1.6 Exponential

For any $b > 0$,

$$f_X(x) = b e^{-bx}, \quad x \geq 0. \tag{D.19}$$

$$F_X(x) = \begin{cases} 0 & x < 0 \\ 1 - e^{-bx} & x \geq 0 \end{cases}. \tag{D.20}$$

$$\Phi_X(\omega) = \frac{1}{1 - j\omega/b}. \tag{D.21}$$

$$\mu_X = 1/b, \quad \sigma_X^2 = 1/b^2. \tag{D.22}$$

Notes:

(1) The exponential random variable is a special case of the Erlang and gamma random variables.
(2) The exponential random variable possesses the memoryless property,

$$f_X(x|X > a) = f_X(x - a). \tag{D.23}$$

D.1.7 F

For any integers $n > 0$ and $m > 0$,

$$f_X(x) = \frac{\left(\frac{n}{m}\right)^{n/2}}{B\left(\frac{n}{2}, \frac{m}{2}\right)} x^{\frac{n}{2}-1} \left(1 + \frac{n}{m}x\right)^{-\frac{n+m}{2}}, \quad x > 0. \tag{D.24}$$

$$\mu_X = \frac{m}{m-2} \quad \text{for} \quad m > 2, \qquad \sigma_X^2 = \frac{m^2(2n + 2m - 4)}{n(m-2)^2(m-4)} \quad \text{for} \quad m > 4. \tag{D.25}$$

Notes:

(1) If U and V are independent chi-square random variables with n and m degrees of freedom, respectively, then $F = (U/n)/(V/m)$ will be an F random variable with n and m degrees of freedom.

D.1.8 Gamma

For any $a > 0$ and $b > 0$,

$$f_X(x) = \frac{b^a x^{a-1} e^{-bx}}{\Gamma(a)}, \quad x \geq 0. \tag{D.26}$$

$$F_X(x) = \frac{\gamma(a, bx)}{\Gamma(a)}. \tag{D.27}$$

$$\Phi_X(\omega) = \frac{1}{(1 - j\omega/b)^a}. \tag{D.28}$$

$$\mu_X = a/b, \quad \sigma_X^2 = a/b^2. \tag{D.29}$$

Note:

(1) The gamma random variable contains the chi-square, Erlang, and exponential random variables as special cases.

D.1.9 Gaussian

For any μ and any $\sigma > 0$,

$$f_X(x) = \frac{1}{\sqrt{2\pi\sigma^2}} \exp\left(-\frac{(x-\mu)^2}{2\sigma^2}\right). \tag{D.30}$$

$$F_X(x) = 1 - Q\left(\frac{x-\mu}{\sigma}\right). \tag{D.31}$$

$$\Phi_X(\omega) = \exp\left(j\omega\mu - \frac{1}{2}\omega^2\sigma^2\right). \tag{D.32}$$

$$\mu_X = \mu, \quad \sigma_X^2 = \sigma^2. \tag{D.33}$$

D.1.10 Gaussian-Multivariate

For any n element column vector μ and any valid $n \times n$ covariance matrix C,

$$f_X(x) = \frac{1}{(2\pi)^{n/2}\det(C)} \exp\left(-\frac{1}{2}(X-\mu)^T C^{-1}(X-\mu)\right). \tag{D.34}$$

$$\Phi_X(\omega) = \exp\left(j\mu^T\omega - \frac{1}{2}\omega^T C\omega\right). \tag{D.35}$$

$$E[X] = \mu, \quad E[(X-\mu)(X-\mu)^T] = C. \tag{D.36}$$

D.1.11 Laplace

For any $b > 0$,

$$f_X(x) = \frac{b}{2}\exp(-b|x|). \tag{D.37}$$

$$F_X(x) = \begin{cases} \frac{1}{2}e^{bx} & x < 0 \\ 1 - \frac{1}{2}e^{-bx} & x \geq 0 \end{cases}. \tag{D.38}$$

$$\Phi_X(\omega) = \frac{1}{1+(\omega/b)^2}. \tag{D.39}$$

$$\mu_X = 0, \quad \sigma_X^2 = 2/b^2. \tag{D.40}$$

D.1.12 Log-Normal

For any μ and any $\sigma > 0$,

$$f_X(x) = \frac{1}{x\sqrt{2\pi\sigma^2}}\exp\left(-\frac{(\ln(x)-\mu)^2}{2\sigma^2}\right), \quad x > 0. \tag{D.41}$$

$$F_X(x) = \begin{cases} 0 & x < 0 \\ 1 - Q\left(\frac{\ln(x)-\mu}{\sigma}\right) & x \geq 0 \end{cases}. \tag{D.42}$$

$$\mu_X = \exp\left(\mu + \frac{\sigma^2}{2}\right), \quad \sigma_X^2 = [\exp(\sigma^2) - 1]\exp(2\mu + \sigma^2). \tag{D.43}$$

Notes:

(1) The log-normal random variable is formed by a transformation of the form $X = \exp(Z)$, where Z is a Gaussian random variable with mean μ and variance σ^2.

(2) It is common to find instances in the literature where σ is referred to as the standard deviation of the log-normal random variable. This is a misnomer. The quantity σ is not the standard deviation of the log-normal random variable, but rather is the standard deviation of the underlying Gaussian random variable.

D.1.13 Nakagami

For any $b > 0$ and $m > 0$,

$$f_X(x) = \frac{2m^m}{\Gamma(m)b^m} x^{2m-1} \exp\left(-\frac{m}{b}x^2\right), \quad x \geq 0. \tag{D.44}$$

$$F_X(x) = \begin{cases} 0 & x < 0 \\ \dfrac{\gamma\left(m, \dfrac{m}{b}x^2\right)}{\Gamma(m)} & x \geq 0 \end{cases}. \tag{D.45}$$

$$\mu_X = \frac{\Gamma(m+1/2)}{\Gamma(m)}\sqrt{\frac{b}{m}}, \quad \sigma_X^2 = b - \mu_X^2. \tag{D.46}$$

D.1.14 Rayleigh

For any $\sigma > 0$,

$$f_X(x) = \frac{x}{\sigma^2} \exp\left(-\frac{x^2}{2\sigma^2}\right), \quad x \geq 0. \tag{D.47}$$

$$F_X(x) = \begin{cases} 0 & x < 0 \\ 1 - \exp\left(-\dfrac{x^2}{2\sigma^2}\right) & x \geq 0 \end{cases}. \tag{D.48}$$

$$\mu_X = \sqrt{\frac{\pi\sigma^2}{2}}, \quad \sigma_X^2 = \frac{(4-\pi)\sigma^2}{2}. \tag{D.49}$$

Notes:

(1) The Rayleigh random variable arises when performing a Cartesian to polar transformation of two independent, zero-mean Gaussian random variables.

That is, if Y_1 and Y_2 are independent zero mean Gaussian random variables with variances of σ^2, then $X = \sqrt{Y_1^2 + Y_2^2}$ follows a Rayleigh distribution.

(2) The Rayleigh random variable is a special case of the Rician random variable.

D.1.15 Rician

For any $a \geq 0$ and any $\sigma > 0$,

$$f_X(x) = \frac{x}{\sigma^2} \exp\left(-\frac{x^2 + a^2}{2\sigma^2}\right) I_0\left(\frac{ax}{\sigma^2}\right), \quad x \geq 0. \tag{D.50}$$

$$F_X(x) = \begin{cases} 0 & x < 0 \\ 1 - Q\left(\frac{a}{\sigma}, \frac{x}{\sigma}\right) & x \geq 0 \end{cases}. \tag{D.51}$$

$$\mu_X = \sqrt{\frac{\pi\sigma^2}{2}} \exp\left(-\frac{a^2}{4\sigma^2}\right) \left[\left(1 + \frac{a^2}{2\sigma^2}\right) I_0\left(\frac{a^2}{4\sigma^2}\right) + \frac{a^2}{2\sigma^2} I_1\left(\frac{a^2}{4\sigma^2}\right)\right]. \tag{D.52}$$

$$\sigma_X^2 = 2\sigma^2 + a^2 - \mu_X^2. \tag{D.53}$$

Notes:

(1) The Rician random variable arises when performing a Cartesian to polar transformation of two independent Gaussian random variables. That is, if Y_1 and Y_2 are independent Gaussian random variables with means of μ_1 and μ_2, respectively, and equal variances of σ^2, then $X = \sqrt{Y_1^2 + Y_2^2}$ follows a Rician distribution, with $a = \sqrt{\mu_1^2 + \mu_2^2}$.

(2) The ratio a^2/σ^2 is often referred to as the Rician parameter or the Rice factor. As the Rice factor goes to zero, the Rician random variable becomes a Rayleigh random variable.

D.1.16 Student t

For any integer $n > 0$,

$$f_X(x) = \frac{1}{B(n/2, 1/2)\sqrt{n}} \left(1 + \frac{x^2}{n}\right)^{-\frac{n+1}{2}}. \tag{D.54}$$

$$\mu_X = 0, \quad \sigma_X^2 = \frac{n}{n-2} \quad \text{for} \quad n > 2. \tag{D.55}$$

Notes:

(1) This distribution was first published by W. S. Gosset in 1908 under the pseudonym "A. Student." Hence, this distribution has come to be known as the student's *t*-distribution.

(2) The parameter n is referred to as the number of degrees of freedom.

(3) If X_i $i = 1, 2, \ldots, n$ is a sequence of IID Gaussian random variables and $\hat{\mu}$ and $\widehat{s^2}$ are the sample mean and sample variance, respectively, then the ratio $T = (\hat{\mu} - \mu)/\sqrt{\widehat{s^2}/n}$ will have a *t*-distribution with $n - 1$ degrees of freedom.

D.1.17 Uniform

For any $a < b$,

$$f_X(x) = \frac{1}{b - a}, \quad a \leq x < b. \tag{D.56}$$

$$F_X(x) = \begin{cases} 0 & x < a \\ \dfrac{x - a}{b - a} & a \leq x \leq b \\ 1 & x > b \end{cases}. \tag{D.57}$$

$$\Phi_X(\omega) = \frac{e^{jb\omega} - e^{ja\omega}}{j\omega(b - a)}. \tag{D.58}$$

$$\mu_x = \frac{a + b}{2}, \quad \sigma_X^2 = \frac{(b - a)^2}{12}. \tag{D.59}$$

D.1.18 Weibull

For any $a > 0$ and any $b > 0$,

$$f_X(x) = abx^{b-1} \exp(-ax^b), \quad x \geq 0. \tag{D.60}$$

$$F_X(x) = \begin{cases} 0 & x < 0 \\ 1 - \exp(-ax^b) & x \geq 0 \end{cases}. \tag{D.61}$$

$$\mu_X = \frac{\Gamma\left(1 + \dfrac{1}{b}\right)}{a^{1/b}}, \quad \sigma_X^2 = \frac{\Gamma\left(1 + \dfrac{2}{b}\right) - \left[\Gamma\left(1 + \dfrac{1}{b}\right)\right]^2}{a^{2/b}}. \tag{D.62}$$

Note:

(1) The Weibull random variable is a generalization of the Rayleigh random variable and reduces to a Rayleigh random variable when $b = 2$.

D.2 Discrete Random Variables

D.2.1 Bernoulli

For $0 < p < 1$,

$$P_X(k) = \begin{cases} 1 - p & k = 0 \\ p & k = 1 \\ 0 & \text{otherwise} \end{cases}.$$

$$\quad (D.63)$$

$$H_X(z) = 1 - p(1 - z) \quad \text{for all } z. \quad (D.64)$$

$$\mu_X = p, \quad \sigma_X^2 = p(1 - p). \quad (D.65)$$

D.2.2 Binomial

For $0 < p < 1$ and any integer $n > 0$,

$$P_X(k) = \begin{cases} \binom{n}{k} p^k (1 - p)^{n-k} & k = 0, 1, 2, \ldots, n \\ 0 & \text{otherwise} \end{cases}. \quad (D.66)$$

$$H_X(z) = (1 - p(1 - z))^n \quad \text{for any } z. \quad (D.67)$$

$$\mu_x = np, \quad \sigma_X^2 = np(1 - p). \quad (D.68)$$

Note:

(1) The binomial random variable is formed as the sum of n independent Bernoulli random variables.

D.2.3 Geometric

For $0 < p < 1$,

$$P_X(k) = \begin{cases} (1-p)p^k & k \geq 0 \\ 0 & k < 0 \end{cases}.$$ (D.69)

$$H_X(z) = \frac{1-p}{1-pz} \quad \text{for} \quad |z| < 1/p.$$ (D.70)

$$\mu_X = \frac{p}{1-p}, \quad \sigma_X^2 = \frac{p}{(1-p)^2}.$$ (D.71)

D.2.4 Pascal (or Negative Binomial)

For $0 < q < 1$ and any integer $n > 0$,

$$P_X(k) = \begin{cases} 0 & k < n \\ \binom{k-1}{n-1}(1-q)^n q^{k-n} & k = n, n+1, n+2, \ldots \end{cases}.$$ (D.72)

$$H_X(z) = \left(\frac{(1-q)z}{1-qz} \right)^n, \quad \text{for} \quad |z| < 1/q.$$ (D.73)

$$\mu_X = \frac{n}{1-q}, \quad \sigma_X^2 = \frac{nq}{(1-q)^2}.$$ (D.74)

D.2.5 Poisson

For any $b > 0$,

$$P_X(k) = \begin{cases} \dfrac{b^k}{k!}e^{-b} & k \geq 0 \\ 0 & k < 0 \end{cases}.$$ (D.75)

$$H_X(z) = \exp(b(z-1)), \quad \text{for all } z.$$ (D.76)

$$\mu_X = b, \quad \sigma_X^2 = b.$$ (D.77)

Mathematical Tables

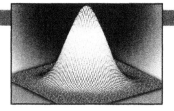

E.1 Trigonometric Identities

$$\sin^2(x) + \cos^2(x) = 1. \tag{E.1}$$

$$\cos(x \pm y) = \cos(x)\cos(y) \mp \sin(x)\sin(y). \tag{E.2}$$

$$\sin(x \pm y) = \sin(x)\cos(y) \pm \cos(x)\sin(y). \tag{E.3}$$

$$\cos(x)\cos(y) = \frac{1}{2}\cos(x+y) + \frac{1}{2}\cos(x-y). \tag{E.4}$$

$$\sin(x)\sin(y) = \frac{1}{2}\cos(x-y) - \frac{1}{2}\cos(x+y). \tag{E.5}$$

$$\sin(x)\cos(y) = \frac{1}{2}\sin(x+y) + \frac{1}{2}\sin(x-y). \tag{E.6}$$

$$\exp(jx) = \cos(x) + j\sin(x). \tag{E.7}$$

$$\cos(x) = \frac{e^{jx} + e^{-jx}}{2}. \tag{E.8}$$

$$\sin(x) = \frac{e^{jx} - e^{-jx}}{2j}. \tag{E.9}$$

E.2 Series Expansions

$$\frac{1}{1-x} = \sum_{k=0}^{\infty} x^k, \quad \text{for } |x| < 1. \tag{E.10}$$

$$\frac{1 - x^{n+1}}{1 - x} = \sum_{k=0}^{n} x^k, \quad \text{for all } x. \tag{E.11}$$

$$\frac{1}{(1 - x)^{n+1}} = \sum_{k=n}^{\infty} \binom{k}{n} x^{k-n} = \sum_{k=0}^{\infty} \binom{k+n}{n} x^k, \quad \text{for } |x| < 1. \tag{E.12}$$

$$(x + y)^n = \sum_{k=0}^{n} \binom{n}{k} x^k y^{n-k}, \quad \text{for all } x, y. \tag{E.13}$$

$$\exp(x) = \sum_{k=0}^{\infty} \frac{1}{k!} x^k, \quad \text{for all } x. \tag{E.14}$$

$$\cos(x) = \sum_{k=0}^{\infty} \frac{(-1)^k}{(2k)!} x^{2k}, \quad \text{for all } x. \tag{E.15}$$

$$\sin(x) = \sum_{k=0}^{\infty} \frac{(-1)^k}{(2k + 1)!} x^{2k+1}, \quad \text{for all } x. \tag{E.16}$$

$$\ln(1 - x) = -\sum_{k=1}^{\infty} \frac{1}{k} x^k, \quad \text{for } |x| < 1. \tag{E.17}$$

$$Q(x) = \frac{1}{2} + \frac{1}{\sqrt{2\pi}} \sum_{k=0}^{\infty} \frac{(-1)^{k+1}}{k! 2^k (2k + 1)} x^{2k+1}, \quad \text{for all } x. \tag{E.18}$$

$$I_m(x) = \sum_{k=0}^{\infty} \frac{1}{k!(k + m)!} \left(\frac{x}{2}\right)^{2k+m}, \quad \text{for all } x. \tag{E.19}$$

E.3 Some Common Indefinite Integrals

Note: For each of the indefinite integrals, an arbitrary constant may be added to the result.

$$\int x^n dx = \begin{cases} \dfrac{x^{n+1}}{n + 1} & n \neq -1 \\ \ln(x) & n = -1 \end{cases}. \tag{E.20}$$

$$\int b^x dx = \frac{b^x}{\ln(b)} \quad b \neq 1. \tag{E.21}$$

$$\int \ln(x)dx = x\ln(x) - x. \tag{E.22}$$

$$\int \sin(x)dx = -\cos(x). \tag{E.23}$$

$$\int \cos(x)dx = \sin(x). \tag{E.24}$$

$$\int \tan(x)dx = -\ln(|\cos(x)|). \tag{E.25}$$

$$\int \sinh(x)dx = \cosh(x). \tag{E.26}$$

$$\int \cosh(x)dx = \sinh(x). \tag{E.27}$$

$$\int \tanh(x)dx = \ln(|\cosh(x)|). \tag{E.28}$$

$$\int e^{ax}\sin(bx)dx = e^{ax}\left(\frac{a\sin(bx) - b\cos(bx)}{a^2 + b^2}\right). \tag{E.29}$$

$$\int e^{ax}\cos(bx)dx = e^{ax}\left(\frac{b\sin(bx) + a\cos(bx)}{a^2 + b^2}\right). \tag{E.30}$$

$$\int x^n e^{bx}dx = e^{bx}\sum_{k=0}^{n}\frac{(-1)^k}{b^{k+1}}\frac{n!}{(n-k)!}x^{n-k} \quad (n \geq 0). \tag{E.31}$$

$$\int x^n \ln(bx)dx = x^{n+1}\left(\frac{\ln(bx)}{n+1} - \frac{1}{(n+1)^2}\right) \quad (n \neq -1). \tag{E.32}$$

$$\int \frac{1}{x^2 + b^2}dx = \frac{1}{b}\tan^{-1}\left(\frac{x}{b}\right) \quad (b > 0). \tag{E.33}$$

$$\int \frac{1}{\sqrt{b^2 - x^2}}dx = \sin^{-1}\left(\frac{x}{b}\right) \quad (b > 0). \tag{E.34}$$

$$\int \frac{1}{\sqrt{x^2 + b^2}}dx = \log(x + \sqrt{x^2 + b^2}) = \sinh^{-1}\left(\frac{x}{b}\right) \quad (b > 0). \tag{E.35}$$

$$\int \frac{1}{\sqrt{x^2 - b^2}}dx = \log\left|x + \sqrt{x^2 - b^2}\right| = \cosh^{-1}\left(\frac{x}{b}\right) \quad (b > 0). \tag{E.36}$$

$$\int \frac{1}{ax^2 + bx + c}dx = \begin{cases} \frac{1}{\sqrt{b^2 - 4ac}}\ln\left|\frac{2ax + b - \sqrt{b^2 - 4ac}}{2ax + b + \sqrt{b^2 - 4ac}}\right| & b^2 > 4ac \\ \frac{2}{\sqrt{4ac - b^2}}\tan^{-1}\left(\frac{2ax + b}{\sqrt{4ac - b^2}}\right) & b^2 < 4ac \end{cases}. \tag{E.37}$$

$$\int \frac{1}{\sqrt{ax^2 + bx + c}} dx = \begin{cases} \dfrac{1}{\sqrt{a}} \ln \left| 2ax + b + 2\sqrt{a(ax^2 + bx + c)} \right| & a > 0 \\[3mm] \dfrac{1}{\sqrt{-a}} \sin^{-1} \left(\dfrac{-2ax - b}{\sqrt{b^2 - 4ac}} \right) & a < 0 \end{cases} . \qquad (E.38)$$

E.4 Some Common Definite Integrals

$$\int_0^\infty x^n e^{-x} dx = \Gamma(n+1) = n! \quad \text{for integer } n \geq 0. \qquad (E.39)$$

$$\int_{-\infty}^\infty e^{-x^2} dx = \int_0^\infty x^{-1/2} e^{-x} dx = \Gamma(1/2) = \sqrt{\pi}. \qquad (E.40)$$

$$\int_0^\infty x^{n-1/2} e^{-x} dx = \Gamma(n+1/2) = \frac{(2n)!}{2^{2n} n!} \sqrt{\pi}, \quad \text{for integer } n \geq 1. \qquad (E.41)$$

$$\int_{-\infty}^\infty \text{sinc}(x) dx = \int_{-\infty}^\infty \text{sinc}^2(x) dx = 1. \qquad (E.42)$$

$$\frac{1}{2\pi} \int_0^{2\pi} \cos^n(x) dx = \frac{1}{2\pi} \int_0^{2\pi} \sin^n(x) dx = \begin{cases} 0 & n \text{ odd} \\[2mm] \dbinom{n}{n/2} \dfrac{1}{2^n} & n \text{ even} \end{cases} . \qquad (E.43)$$

$$\int_{-\infty}^\infty \frac{1}{x^2 + b^2} dx = 2 \int_0^\infty \frac{1}{x^2 + b^2} dx = \frac{\pi}{b}, \quad b > 0. \qquad (E.44)$$

$$\int_{-b}^b \frac{1}{\sqrt{b^2 - x^2}} dx = 2 \int_0^b \frac{1}{\sqrt{b^2 - x^2}} dx = \pi, \quad b > 0. \qquad (E.45)$$

E.5 Definitions of Some Common Continuous Time Signals

$$\text{Step function: } u(x) = \begin{cases} 1 & x > 0 \\ 0 & x < 0 \end{cases} . \qquad (E.46)$$

$$\text{Rectangle function: rect}(x) = \begin{cases} 1 & |x| < 1/2 \\ 0 & |x| > 1/2 \end{cases}. \tag{E.47}$$

$$\text{Triangle function: tri}(x) = \begin{cases} 1 - |x| & |x| \le 1 \\ 0 & |x| > 1 \end{cases}. \tag{E.48}$$

$$\text{Sinc function: sinc}(x) = \frac{\sin(\pi x)}{\pi x}. \tag{E.49}$$

E.6 Fourier Transforms

Table E.1   Common Fourier Transform Pairs

Signal (time domain)	Transform (frequency domain)
$\text{rect}(t/t_0)$	$t_0 \text{sinc}(ft_0)$
$\text{tri}(t/t_0)$	$t_0 \text{sinc}^2(ft_0)$
$\exp\left(-\dfrac{t}{t_0}\right) u(t)$	$\dfrac{t_0}{1 + j2\pi ft_0}$
$\exp\left(-\dfrac{|t|}{t_0}\right)$	$\dfrac{2t_0}{1 + (2\pi ft_0)^2}$
$\text{sinc}(t/t_0)$	$t_0 \text{rect}(ft_0)$
$\text{sinc}^2(t/t_0)$	$t_0 \text{tri}(ft_0)$
$\exp(j2\pi f_0 t)$	$\delta(f - f_0)$
$\cos(2\pi f_0 t + \theta)$	$\dfrac{1}{2}\delta(f - f_0)e^{j\theta} + \dfrac{1}{2}\delta(f + f_0)e^{-j\theta}$
$\delta(t - t_0)$	$\exp(-j2\pi ft_0)$
$\text{sgn}(t)$	$\dfrac{1}{j\pi f}$
$u(t)$	$\dfrac{1}{2}\delta(f) + \dfrac{1}{j2\pi f}$
$\exp(-(t/t_0)^2)$	$\sqrt{\pi t_0^2}\exp(-(\pi ft_0)^2)$

E.7 z-Transforms

Table E.2   Common z-Transform Pairs

Signal	Transform	Region of convergence
$\delta[n]$	1	all z
$u[n]$	$\dfrac{1}{1-z^{-1}}$	$|z| > 1$
$nu[n]$	$\dfrac{z^{-1}}{\left(1-z^{-1}\right)^2}$	$|z| > 1$
$n^2 u[n]$	$\dfrac{z^{-1}(1+z^{-1})}{\left(1-z^{-1}\right)^3}$	$|z| > 1$
$n^3 u[n]$	$\dfrac{z^{-1}(1+4z^{-1}+z^{-2})}{\left(1-z^{-1}\right)^4}$	$|z| > 1$
$b^n u[n]$	$\dfrac{1}{1-bz^{-1}}$	$|z| > |b|$
$nb^n u[n]$	$\dfrac{bz^{-1}}{\left(1-bz^{-1}\right)^2}$	$|z| > |b|$
$n^2 b^n u[n]$	$\dfrac{bz^{-1}(1+bz^{-1})}{\left(1-bz^{-1}\right)^3}$	$|z| > |b|$
$b^n \cos[\Omega_o n]u[n]$	$\dfrac{1-b\cos(\Omega_o)z^{-1}}{1-2b\cos(\Omega_o)z^{-1}+bz^{-2}}$	$|z| > |b|$
$b^n \sin[\Omega_o n]u[n]$	$\dfrac{b\sin(\Omega_o)z^{-1}}{1-2b\cos(\Omega_o)z^{-1}+bz^{-2}}$	$|z| > |b|$
$\dfrac{u[n-1]}{n}$	$\ln\left(\dfrac{1}{1-z^{-1}}\right)$	$|z| > 1$
$\dbinom{n+m}{m} b^n u[n]$	$\dfrac{1}{\left(1-bz^{-1}\right)^{m+1}}$	$|z| > |b|$
$\dfrac{b^n}{n!}u[n]$	$\exp(bz^{-1})$	all z

E.8 Laplace Transforms

Table E.3 Common Laplace Transform Pairs

Function	Transform	Region of convergence
$u(t)$	$1/s$	$\text{Re}[s] > 0$
$\exp(-bt)u(t)$	$\dfrac{1}{s+b}$	$\text{Re}[s] > -b$
$\sin(bt)u(t)$	$\dfrac{b}{s^2+b^2}$	$\text{Re}[s] > 0$
$\cos(bt)u(t)$	$\dfrac{s}{s^2+b^2}$	$\text{Re}[s] > 0$
$e^{-at}\sin(bt)u(t)$	$\dfrac{b}{(s+a)^2+b^2}$	$\text{Re}[s] > -a$
$e^{-at}\cos(bt)u(t)$	$\dfrac{s+a}{(s+a)^2+b^2}$	$\text{Re}[s] > -a$
$\delta(t)$	1	all s
$\dfrac{d}{dt}\delta(t)$	s	all s
$t^n u(t), \quad n \geq 0$	$\dfrac{n!}{s^{n+1}}$	$\text{Re}[s] > 0$
$t^n e^{-bt} u(t), \quad n \geq 0$	$\dfrac{n!}{(s+b)^{n+1}}$	$\text{Re}[s] > -b$

E.9 Q-Function

Table E.4 lists values of the function $Q(x)$ for $0 \leq x < 4$ in increments of 0.05. To find the appropriate value of x, add the value at the beginning of the row to the value at the top of the column. For example, to find $Q(1.75)$, find the entry from the column headed by 1.00 and the row headed by 0.75 to get $Q(1.75) = 0.04005916$.

Table E.4   Values of $Q(x)$ for $0 \leq x < 4$ (in increments of 0.05)

$Q(x)$	0.00	1.00	2.00	3.00
0.00	0.50000000	0.15865525	0.02275013	0.00134990
0.05	0.48006119	0.14685906	0.02018222	0.00114421
0.10	0.46017216	0.13566606	0.01786442	0.00096760
0.15	0.44038231	0.12507194	0.01577761	0.00081635
0.20	0.42074029	0.11506967	0.01390345	0.00068714
0.25	0.40129367	0.10564977	0.01222447	0.00057703
0.30	0.38208858	0.09680048	0.01072411	0.00048342
0.35	0.36316935	0.08850799	0.00938671	0.00040406
0.40	0.34457826	0.08075666	0.00819754	0.00033693
0.45	0.32635522	0.07352926	0.00714281	0.00028029
0.50	0.30853754	0.06680720	0.00620967	0.00023263
0.55	0.29115969	0.06057076	0.00538615	0.00019262
0.60	0.27425312	0.05479929	0.00466119	0.00015911
0.65	0.25784611	0.04947147	0.00402459	0.00013112
0.70	0.24196365	0.04456546	0.00346697	0.00010780
0.75	0.22662735	0.04005916	0.00297976	0.00008842
0.80	0.21185540	0.03593032	0.00255513	0.00007235
0.85	0.19766254	0.03215677	0.00218596	0.00005906
0.90	0.18406013	0.02871656	0.00186581	0.00004810
0.95	0.17105613	0.02558806	0.00158887	0.00003908

Numerical Methods for Evaluating the Q-Function

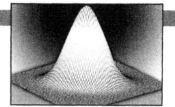

In this appendix, we give an overview of several methods available for numerically evaluating the CDF of a Gaussian random variable and related integrals. Recall that for a zero-mean unit variance Gaussian random variable, the CDF is given by the integral

$$F_X(x) = \int_{-\infty}^{x} \frac{1}{\sqrt{2\pi}} \exp\left(-\frac{t^2}{2}\right) dt. \tag{F.1}$$

The Q-function is the complement of this integral:

$$Q(x) = \int_{x}^{\infty} \frac{1}{\sqrt{2\pi}} \exp\left(-\frac{t^2}{2}\right) dt. \tag{F.2}$$

Many math packages, such as MATLAB, have internal routines for evaluating related integrals, usually the error function or complementary error function. Given the most common definitions of these functions,

$$\text{erf}(x) = \frac{2}{\sqrt{\pi}} \int_{0}^{x} \exp(-t^2) dt, \tag{F.3}$$

$$\text{erfc}(x) = \frac{2}{\sqrt{\pi}} \int_{x}^{\infty} \exp(-t^2) dt, \tag{F.4}$$

the Q-function can then be written in terms of these functions as

$$Q(x) = \frac{1}{2}\text{erfc}\left(\frac{x}{\sqrt{2}}\right) = \frac{1}{2} - \frac{1}{2}\text{erf}\left(\frac{x}{\sqrt{2}}\right). \tag{F.5}$$

For situations where internally defined functions are not available, several numerical techniques are available for efficiently evaluating the Q-function. Recall the symmetry relationship

$$Q(x) = 1 - Q(-x). \qquad (F.6)$$

Hence, any routine for evaluating the Q-function needs to work only on positive values of x. To start with, we consider the Taylor series expansion of the Q-function about the point $x = 0$,

$$Q(x) = \frac{1}{2} + \frac{1}{\sqrt{2\pi}} \sum_{k=0}^{\infty} \frac{(-1)^{k+1}}{k!2^k(2k+1)} x^{2k+1}. \qquad (F.7)$$

This series is convergent for all $x \geq 0$ but will converge faster for smaller values of x. A good approximation can be obtained by truncating the series to a sufficient number of terms. Since the series is alternating, the truncation error is bounded by the first term neglected. Figure F.1 shows the Q-function along with its approximations using the Taylor series truncated to various numbers of terms. It is seen from this figure that for $x > 2$, a large number of terms may be needed for the Taylor series to converge.

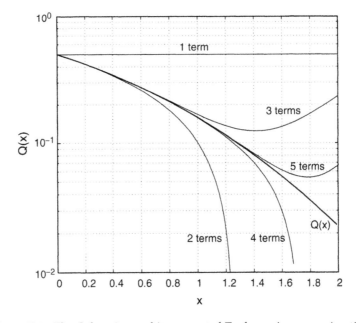

Figure F.1   The Q-function and its truncated Taylor series approximation.

For larger values of x, it is common to use the following asymptotic series expansion:

$$Q(x) = \frac{1}{\sqrt{2\pi}x} \exp\left(-\frac{x^2}{2}\right) \left\{1 + \sum_{n=1}^{\infty} \frac{(-1)^n(2n)!}{2^n n!} x^{-2n}\right\}, \quad \text{for } x > 0. \qquad (F.8)$$

Since the series has negative powers of x, the larger x is, the faster the series will converge. Also, as with the Taylor series expansion, since this is a convergent alternating series, the truncation error will be bounded by the first neglected term. Further, the sign of the error will be the same as that of the first neglected term. As a result, for large values of x, the Q-function can be upper- and lower-bounded by

$$\frac{1}{\sqrt{2\pi}x}\left(1 - \frac{1}{x^2}\right)\exp\left(-\frac{x^2}{2}\right) \leq Q(x) \leq \frac{1}{\sqrt{2\pi}x}\exp\left(-\frac{x^2}{2}\right). \qquad (F.9)$$

These two bounds are shown in Figure F.2. From the figure as well as from the expressions in the previous equation, it is clear that as $x \to \infty$, both bounds are asymptotically tight. Therefore,

$$Q(x) \sim \frac{1}{\sqrt{2\pi}x}\exp\left(-\frac{x^2}{2}\right). \qquad (F.10)$$

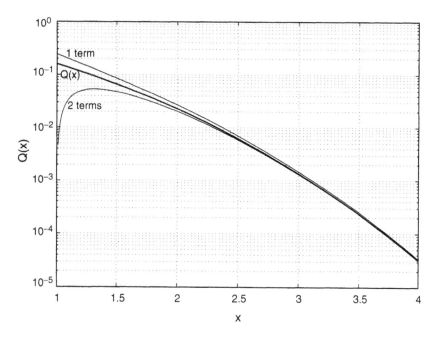

Figure F.2   Upper and lower bounds on the Q-function.

From the figure, it appears that this approximation is fairly accurate for $x > 4$. If more accuracy is desired, more terms can be included in the asymptotic expansion.

In addition to the Taylor series and asymptotic series expansions, there are also a few continued fraction expansions for the Q-function as follows.

$$Q(x) = \frac{1}{\sqrt{2\pi}} \exp\left(-\frac{x^2}{2}\right) \left\{ \cfrac{1}{x + \cfrac{1}{x + \cfrac{2}{x + \cfrac{3}{x + \cfrac{4}{x + \cdots}}}}} \right\}, \quad \text{for} \quad x > 0, \quad \text{(F.11)}$$

$$Q(x) = \frac{1}{2} - \frac{1}{\sqrt{2\pi}} \exp\left(-\frac{x^2}{2}\right) \left\{ \cfrac{x}{1 - \cfrac{x^2}{3 + \cfrac{2x^2}{5 - \cfrac{3x^2}{7 + \cfrac{4x^2}{9 - \cdots}}}}} \right\}, \quad \text{for} \quad x \geq 0, \quad \text{(F.12)}$$

A number of polynomial and rational approximations of the Q-function are available. Among them, the following seems to offer the best accuracy:

$$Q(x) = \frac{1}{\sqrt{2\pi}} \exp\left(-\frac{x^2}{2}\right) (b_1 t + b_2 t^2 + b_3 t^3 + b_4 t^4 + b_5 t^5), \quad t = \frac{1}{1 + px}, \quad \text{(F.13)}$$

where,

$p = 0.2316419,$
$b_1 = 0.319381530,$
$b_2 = -0.35656378,$
$b_3 = 1.7814779,$
$b_4 = -1.821256,$
$b_5 = 1.3302744.$

The error in this approximation is less than 7.5×10^{-8} for all $0 \leq x < \infty$.

Note that any desired accuracy can be obtained by computing the Q-function via numerical integration. Using the definition directly, the Q-function has an infinite limit which is inconvenient for performing numerical integration. For small to moderate values of x, this problem can be circumvented by rewriting the Q-function as

$$Q(x) = \frac{1}{2} - \frac{1}{\sqrt{2\pi}} \int_0^x \exp\left(-\frac{t^2}{2}\right) dt. \tag{F.14}$$

For large values of x, it may be more efficient to work with the standard definition and truncate the upper limit to form the approximation

$$Q(x) \approx \frac{1}{\sqrt{2\pi}} \int_x^{x+c} \exp\left(-\frac{t^2}{2}\right) dt. \tag{F.15}$$

where the constant c is chosen to insure the desired accuracy. For $c > 2$ and $x > 1.5$, the relative error in this approximation can be shown to be bounded by

$$\frac{\varepsilon(x)}{Q(x)} = \frac{Q(x+c)}{Q(x)} \leq \exp\left(-\left(\frac{c^2}{2} + 3\right)\right). \tag{F.16}$$

For example, choosing $c = 3.5$ will guarantee a relative accuracy of less than 10^{-4} (i.e., four digits of accuracy). Finally, we note an alternative form of the Q-function that has finite limits,

$$Q(x) = \frac{1}{\pi} \int_0^{\pi/2} \exp\left(-\frac{x^2}{2\sin^2\theta}\right) d\theta. \tag{F.17}$$

Since the integrand is fairly well-behaved and the limits of integration are finite, numerical integration on this form is particularly convenient and can be performed to any desired accuracy.

Index

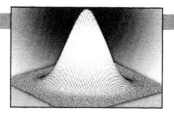

Printed and bound by CPI Group (UK) Ltd, Croydon, CR0 4YY

03/10/2024

01040314-0003